P9-DMY-547

Cryochemistry

Cryochemistry

Edited by

Martin Moskovits
Department of Chemistry
University of Toronto
Toronto, Canada

Geoffrey A. Ozin
Department of Chemistry
University of Toronto
Toronto, Canada

A Wiley-Interscience Publication
John Wiley & Sons
New York / London / Sydney / Toronto

Library of Congress Cataloging in Publication Data:
Main entry under title:

Cryochemistry.

"A Wiley-Interscience publication."
Bibliography: p.
Includes index.
1. Cryochemistry—Addresses, essays, lectures.
I. Moskovits, Martin. II. Ozin, Geoffrey A.
III. Title.

QD515.C79 541'.3686 76-11841
ISBN 0-471-61870-5

Printed in the United States of America

10 9 8 7 6 5 4 3 2 1

TWO LINDA

Contributors

L. ANDREWS, Associate Professor of Chemistry, University of Virginia, Charlottesville, Virginia.

J. BURDETT, Lecturer in Chemistry, The University, Newcastle-upon-Tyne, England.

D. M. GRUEN, Senior Scientific Officer, Argonne National Laboratory, Argonne, Illinois.

M. J. MCGLINCHEY, Assistant Professor of Chemistry, McMaster University, Hamilton, Ontario, Canada.

M. MOSKOVITS, Assistant Professor of Chemistry, University of Toronto, Toronto, Canada.

S. J. OGDEN, Lecturer in Chemistry, Southampton University, Southampton, Hampshire, England.

G. A. OZIN, Associate Professor of Chemistry, University of Toronto, Toronto, Canada.

P. S. SKELL, Professor of Chemistry, Pennsylvania State University, University Park, Pennsylvania.

P. TIMMS, Lecturer in Chemistry, University of Bristol, Cantocks Close, Bristol, England.

J. J. TURNER, Professor of Chemistry, The University, Newcastle-upon-Tyne, England.

Preface

Cryochemistry is a new field of research with an exciting future. Although most of the chemistry that has been achieved so far has been at a fundamental level, the industrial applications, particularly in the areas of organometallic synthesis and catalysis, are already quite apparent.

As the field develops it is becoming evident that the seemingly unrelated methods of the preparative and matrix cryochemists are intimately related. In areas such as reaction intermediates and products, reaction pathways, thermodynamics and kinetics, reaction feasibility and product yields, matrix and preparative cryochemistry cover much common ground, and an intelligent combination of the two techniques can complement and supplement each other.

It was, therefore, thought appropriate to develop a working knowledge of both topics in the same book, with the aim of fitting them into a coherent and logical scheme.

Soon, cryochemical techniques will enter the category of standard laboratory procedures, and it is only a matter of time before industrial evaporation plants are modified for large-scale chemical synthesis.

We believe that the material presented in this book can provide chemists with a comprehensive and timely introduction to the fundamentals of cryochemistry. The book may also serve as a first introduction to graduate students interested in the uses of reactive and transient species in chemical synthesis.

In conclusion, we express our gratitude to the eight contributors to this text for providing highly informative descriptions of the various aspects of cryochemistry as we know it today. We also thank our coworkers, Dr. D. Boal, Dr. A. Vander Voet, Dr. E. P. Kündig, Dr. G. Briggs, Dr. T. A. Ford, Mr. D. Friesen, Mr. W. Klotzbücher, Mr. D. McIntosh, Miss L. Hanlan, Mr. J. Hulse, and Mr. H. Huber, for their dedicated technical assistance, questions and criticisms that have helped us through the early stages of matrix Raman spectroscopy and matrix synthesis using transition metal vapors, and that enabled us to prepare this first text on cryochemistry. In addition, we are grateful to the excellent machine shop services of Mr. A. Campbell, Mr. M. Mitteldorf, Mr. K. Molnar, and Mr. Bob Torbet, the electronics genius of Mr. W. Panning, and the superb photographic skills of Mr. John Glover.

The generous financial assistance of the National Research Council of Canada, the Research Corporation, the Atkinson Foundation, the Connaught Fund, Liquid Carbonic, Erindale College, and the University of Toronto is also acknowledged.

Finally, we sincerely thank Mrs. Elinor Foden, whose excellent typing, proofreading, and good cheer under stress did much to lighten the burden of preparing the manuscript and Dr. A. Vander Voet, whose careful and intelligent reading of the final proofs eliminated much nonsense from the pages that follow.

GEOFFREY A. OZIN
MARTIN MOSKOVITS

Toronto, Canada
June 1976

Contents

Introduction 1

G. A. Ozin and M. Moskovits

In the past five years there has been a renaissance in matrix-isolation research that has been brought about mainly by inorganic and organometallic chemists. Matrix isolation was originally devised to trap chemical species, reactive or otherwise, for spectroscopic observations[1] and was very much the domain of the physical chemist or physicist. Recently, the emphasis of matrix-isolation research has shifted away from the more physical aspects of chemistry towards those of synthesis, structure, and reactivity.[2] This exciting renewal of interest was one of the incentives for writing this text.

A number of chemists who have been among the pioneers in these areas of cryogenic chemistry have agreed to combine their efforts in writing this book in an attempt to demonstrate the synthetic versatility of these techniques, which provide new routes to important and novel compounds.

The range of interests of the contributors is wide, which accounts for the wide application of cryogenic techniques to fields as diverse as organic and organometallic synthesis, homogeneous and heterogeneous catalysis, inorganic and organometallic photochemistry, polymerization reactions, metal aggregation phenomena, reactive intermediates, chemisorption, intermolecular interactions, and others.

What do we mean by the term, "cryogenic chemistry"? The preparative chemist would say that he has been employing cryogenic methods by using slush baths or liquid nitrogen. The cooling of samples for the purposes of sharpening spectral lines, to improve resolution, for slowing down exchange processes, and for investigating phase changes are just a few examples that come to mind. However, we do not include experiments of this kind in our definition of cryogenic chemistry. We restrict the term to include the study of chemical species, reactions, or processes that would not take place or could not be studied at leisure except by using low-temperature methods. The scale is not restricted by this definition and may involve microscopic quantities of material usually for spectroscopic purposes or macroscopic quantities for use on a preparative scale.

By using cryogenic techniques on a routine basis, chemists are no longer restricted to conventional methods of solving a synthetic problem. The chemist can now devise experiments taking advantage of starting materials that might be regarded as esoteric or even unattainable from a synthetic point of view.

For clarity of presentation the chemistry to be described has been divided into chapters covering the macroscopic and microscopic approaches, although this division is somewhat artificial in that either technique can often be employed to obtain the same end product. An obvious connection between the two methods exists and can be exploited.

To place the field in perspective and to do justice to the pioneering work that helped make this technique a reality, it is appropriate to begin with a brief résumé of the historical developments as we think they occurred. G. N. Lewis's[3] phosphorescence studies of aromatic molecules in low-temperature, glassy media in 1941 probably mark the birth of the matrix-isolation technique. The utilization of inert-gas matrix supports for the spectroscopic study of reactive species, particularly free radicals, was proposed simultaneously by G. Porter and G. Pimentel in 1954 and subsequently pioneered by G. Pimentel and his associates.[1] The first infrared spectroscopic observations of a free radical (HCO) represent a landmark in matrix photochemistry research.[4]

Once the exciting possibilities of the matrix isolation technique had been demonstrated, a number of important extensions of the method began to appear. Following Pimentel's suggestion that the matrix experiment might be suitable for studying the constituents of high-temperature vapors, Linevsky[5] in 1961 demonstrated that $(LiF)_n$ ($n = 1,2,3$) species could be trapped by condensing a molecular beam of lithium fluoride, effusing from a high-temperature Knudsen cell, with an inert-gas matrix. Since then, this kind of high-temperature experiment has found many synthetic applications. Probably the first successful preparative scale application of the high-temperature technique is the work of P. Skell and coworkers[6] in 1963, who employed the cocondensation procedure to study the reactions of carbon vapor species (C_1, C_2, and C_3 in the ground and excited electronic states) with a large variety of organic substrates. Skell's carbon vapor work has since proved to be an exceptionally fruitful area of chemical research and has served as a prototype for preparative scale cryogenic syntheses.

The work of Margrave[7] with gaseous SiF_2, Timms[8] with the gaseous boron and silicon subhalides, and Timms[9] and Skell[10] with gaseous elements marked the beginning of a new phase in the exploitation of low-temperature condensation of high-temperature species as a synthetic method.

Shortly after Linevsky's pioneering experiments with high-temperature molecular species, Andrews and Pimentel in 1966 published their alkali metal atom technique for the matrix production and stabilization of free radicals and other reactive species. Some of their early experiments are represented by the following equations:

$$Li + CH_3I \rightarrow CH_3 \cdots\cdot LiI$$

$$Li + NO \rightarrow LiON$$

$$Li + O_2 \rightarrow Li^+O_2^-$$

These reactions can be considered to be the prototypes for the study of transition metal atom matrix chemistry that was to begin in 1970–1971 with a flurry of activity in the laboratories of DeKock,[12] Weltner,[13] Turner,[14] Ogden,[15] and Ozin.[16]

At about the same time, matrix-isolation laser Raman spectroscopy was developed in the laboratories of Claassen,[17] Nibler,[18] Ozin,[19] and Andrews;[20] and more recently matrix Mössbauer spectroscopy by Barrett[21] and Bos[22] and matrix magnetic circular dichroism (MCD) spectroscopy by Thompson[23] have since proven to be a useful addition to matrix-infrared, uv-visible, and esr spectroscopic techniques.

In 1962, Sheline and coworkers demonstrated the viability of studying the photochemical reactions of matrix-isolated carbonyl complexes in organic glasses at 77°K.[24] The adaptation of this technique to study photochemically generated organometallic complexes and their reactive intermediates in inert gas matrices originated with the work of Rest and Turner[25] in 1969. Since then, the general area of organometallic matrix photochemistry has received considerable attention.

On first glancing through the text, a reader new to this field may find some of the chemistry somewhat unorthodox. This is understandable in view of the fact that much of the chemistry is new and many of the concepts and principles are fairly recent additions to the chemical literature. Therefore, a brief preview of the cryogenic chemistry presented in the remainder of this text is probably in order. Some reactions that have recently been studied are shown below. These should give the reader a taste of what is to come.

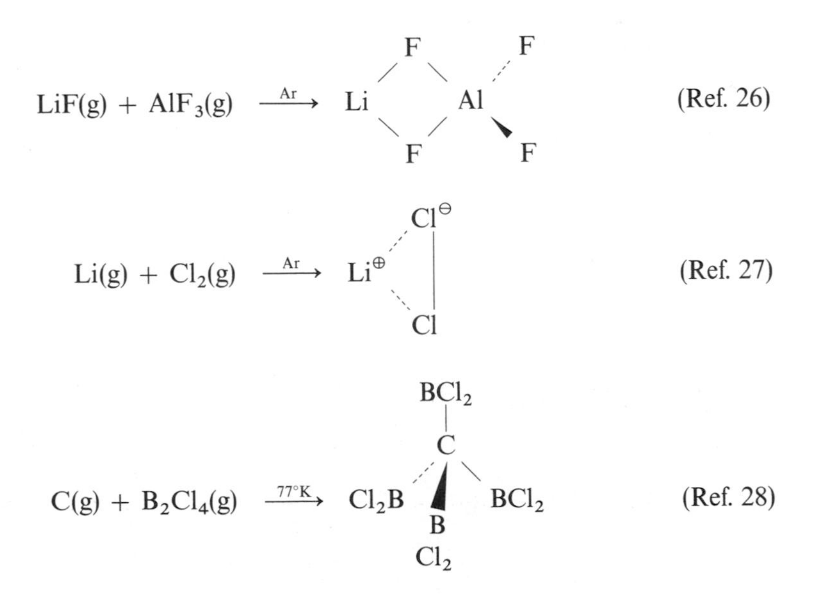

$$Li(g) + C_3(g) \xrightarrow{77°K} Li_2C{=}C{=}CLi_2 \qquad \text{(Ref. 29)}$$

$$Ge(g) + O_2(g) \xrightarrow{Kr} O{=}Ge{=}O \qquad \text{(Ref. 30)}$$

$$SiF_2(g) + HC{\equiv}CH(g) \xrightarrow{77°K} \begin{matrix} CH_2\text{—}SiF_2 \\ | \quad\;\; | \\ CH_2\text{—}SiF_2 \end{matrix} \qquad \text{(Ref. 31)}$$

$$BF(g) + HC{\equiv}CH(g) \xrightarrow{77°K} (FB)_2(CH{=}CH)_2 \qquad \text{(Ref. 32)}$$

$$BF(f) + B_2F_4(g) \xrightarrow{77°K} (F_2B)_2B(BF_2)_2B(BF_2)_2 \qquad \text{(Ref. 33)}$$

$$Cr(CO)_6 \xrightarrow[Xe]{h\nu} (CO)_5Cr \cdots\cdot Xe \qquad \text{(Ref. 34)}$$

$$Ni(CO)_4 \xrightarrow[Ar/N_2]{h\nu} Ni(CO)_3(N_2) \qquad \text{(Ref. 35)}$$

$$Ni(g) + CO(g) + N_2 \xrightarrow{Ar} Ni(CO)_n(N_2)_{4-n}\ (n = 0\text{–}4) \qquad \text{(Ref. 36)}$$

$$Ni(g) + C_2H_4(g) \rightarrow Ni(C_2H_4)_3 \qquad \text{(Ref. 37)}$$

$$Pd(g) + O_2(g) + N_2(g) \xrightarrow{Ar} (O_2)Pd\text{—}N{\equiv}N \qquad \text{(Ref. 38)}$$

$$Mo(g) + C_4H_6 \xrightarrow{77°K} (C_4H_6)_3Mo \qquad \text{(Ref. 39)}$$

$$Ni(g) + CH_2{=}CHCH_2Cl \xrightarrow{77°K} [(C_3H_5)Ni(\mu\text{-}Cl)]_2 \qquad \text{(Ref. 40)}$$

$$Fe(g) + C_8H_{12}(g) \xrightarrow{77°K} Fe(C_8H_{12})_2 \qquad \text{(Ref. 41)}$$

$$Cu(g) + Cu(g) \xrightarrow{Ar} Cu_2 \qquad \text{(Ref. 42)}$$

$$Cu(g) + CO(g) \rightarrow Cu(CO)_3 \qquad \text{(Ref. 42)}$$

$$Ir(g) + CO(g) \rightarrow Ir(CO)_4 + Ir_2(CO)_8 \qquad \text{(Ref. 43)}$$

$$2Ag(CO)_3 \xrightarrow{CO} Ag_2(CO)_6 \qquad \text{(Ref. 44)}$$

Most new techniques result from the availability of a new technology. An introduction to the technology of cryochemistry at the matrix and preparative levels, therefore, is given in Chapters 2 and 3. A number of selected examples are employed to illustrate each technique. Some of the areas covered include methods for achieving, measuring, and controlling low temperatures; the kinetic and thermodynamic factors controlling low-temperature reactions and product yields; the methods of generating and handling reactive and transient species; and finally the chemical and physical properties of inert and reactive matrix supports relevant to cryochemistry.

Detailed designs of state-of-the-art cryochemical apparatus; associated vacùum equipment; and high-temperature, photochemical, discharge accessories as well as monitoring devices are presented. These are intended to encompass the majority of applications that one is likely to encounter and if necessary are easily modified to suit a new set of circumstances.

Preparative cryochemistry employing main group and transition metal elemental vapors form the theme of Chapters 4 and 5. The main thrust is towards the synthesis and reactivity.of new organometallic and organic materials. The reaction chemistry of Li, Na, K, Mg, Ca, B, Al, C, Si, Ge, Cu, Ag, and Au vapors with a selection of organic and inorganic substrates is considered in detail in Chapter 4, the Group IB metals being included because their cryochemistry so far seems to resemble that of the main group elements.

Preparative cryochemistry involving transition metal vapors is considered in Chapter 5. Research in this new area has led to a variety of fascinating sandwich compounds and olefin complexes, many of which have proved to be useful catalysts as well as important starting materials for organic and organometallic synthesis. Besides being catalyst precursors, transition metal vapors themselves have been found to behave as efficient catalysts in a number of isomerization, hydrogenation, disproportionation, and polymerization reactions.

Matrix scale chemistry utilizing alkali metal vapors as starting materials forms the subject of Chapter 6. One might have envisaged that the chemistry of this group of metals would be limited to a few predictable reaction: this is not, however, found to be the case in practice. For example, alkali metal vapors have been shown to be versatile halogen and oxygen abstraction reagents leading to interesting and often unique inorganic and organic free radicals. They can also serve as electron-transfer agents and photoelectron sources, thereby providing new pathways to novel anionic species. In addition to synthesis, other interesting aspects of Chapter 6 are the utilization of ligand and metal isotopes and mixed metals for product characterization, the chemical applications of the recently developed matrix Raman technique, laser photochemistry, and a preview of matrix resonance Raman spectroscopy.

Matrix isolation studies involving main group metals constitute Chapter 7. In addition to some cocondensation reactions of the elemental vapors of Groups II, III, and IV, cryochemical reactions of their respective halide, oxide, and oxyhalide vapors are also described.

A variety of topics associated with the matrix chemistry of transition metal atoms is surveyed in Chapters 8 and 9. Product identification and characterization are generally achieved by infrared, Raman, uv-visible, or esr spectroscopy. Some of the areas covered in Chapter 8 include methods of determining product stoichiometry; the mode of bonding and nature of metal-ligand interactions; vibrational frequency, force field, and intensity calculations for isotopically labeled species; matrix effects, molecular distortions, and the elucidation of electronic and molecular properties. The methodology of matrix-isolation reaction kinetics is also described with reference to specific cryochemical reactions. The application of matrix esr spectroscopy to metal-atom cocondensation is also covered in this chapter, as little attention is given to this aspect of the field in other parts of the book. Finally, the thermodynamic stabilities of the reaction products are considered as well as the relationships between metal-vapor reaction chemistry and surface chemistry.

Chapter 9 forms a sequel to Chapter 8, the aim being to demonstrate how transition metal diatomic molecules, binuclear complexes, and higher cluster systems can be conveniently synthesized using metal-atom techniques. This is an important area of research as the aggregation properties of monatomic or monomeric species on condensation are fundamental to the understanding of the majority of cryochemical reactions. The topic is introduced by considering the factors that influence the isolation of metal atoms at low temperatures and the methods by which the surface and bulk diffusion of metal atoms can be exploited for the synthesis and characterization of M_n and M_nL_m species. Many of these cluster systems have proven to be impossible or cumbersome to prepare by more conventional routes.

From the preceding résumé it should be clear that matrix isolation studies of metal cocondensation reactions are important for understanding the processes that occur at the preparative level. Chapter 10 recognizes the need to understand these processes at the atomic level, through our knowledge of the absorption, fluorescence, esr, and Mössbauer spectra of matrix-isolated metal atoms. The spectroscopic identification and characterization of matrix-isolated atoms form the central theme of this chapter, emphasis being placed on the correlation between gaseous atomic spectra and the spectra of atoms isolated in noble gas matrices as well as matrix-perturbation effects on the spectra. The optical spectra of matrix-isolated atoms are considered in detail, particularly matrix perturbations of energy levels, spin-orbit coupling, and oscillator strengths.

The final chapter is concerned with matrix photochemistry and the intimate relationship between photochemical and cocondensation matrix syntheses as well as between matrix and solution photochemistry. The different categories of photochemical synthesis are outlined; the important phenomenon of the cage effect and its effects on matrix photochemistry as well as the chemical physics of the photolysis process are examined in detail. The production of charged species in matrices, the photochemistry of transition metal carbonyls, and the new technique of polarized matrix photochemistry constitute just some of the other interesting areas described in this chapter.

In this text we wish to present, in as comprehensive a manner as possible, some of the important developments in cryogenic chemistry of the past five years. These have provided exciting new resources for the chemist to exploit.

REFERENCES

1. I. Norman and G. Porter, *Nature*, **174**, 508 (1954); E. Whittle, D. A. Dows, and G. C. Pimentel, *J. Chem. Phys.*, **22**, 1943 (1954).
2. See for example, P. S. Skell, J. J. Havel, and M. J. McGlinchey, *Account Chem. Res.*, **6**, 97 (1973); P. L. Timms, *Account Chem. Res.*, **6**, 118 (1973); G. A. Ozin and A. Vander Voet, *Account Chem. Res.*, **6**, 313 (1973); K. Klabunde, *Account Chem. Res.*, (1975) (in press); J. S. Ogden and J. J. Turner, *Chem. Brit.*, **7**, 1861 (1971); P. L. Timms, *Advan. Inorg. Radiochem.*, **14**, 121 (1972); and *Angew. Chem. Int.* **14**, 5 (1975). (entire number).
3. G. N. Lewis and D. Lipkin, *J. Amer. Chem. Soc.*, **64**, 2801 (1942).
4. G. C. Pimentel, G. E. Ewing, and W. E. Thompson, *J. Chem. Phys.*, **32**, 927 (1960).
5. M. J. Linevsky, *J. Chem. Phys.*, **34**, 587 (1961).
6. P. S. Skell and L. D. Wescott, Jr., *J. Amer. Chem. Soc.*, **85**, 1023 (1963).
7. J. M. Basler, P. L. Timms, and J. L. Margrave, *Inorg. Chem.*, **5**, 729 (1966).
8. P. L. Timms, *J. Amer. Chem. Soc.*, **89**, 1629 (1967); P. L. Timms, *Inorg. Chem.*, **7**, 387 (1968).
9. P. L. Timms, *Chem. Commun.*, 1033 (1969); P. L. Timms, *Endeavour*, **27**, 133 (1968); P. L. Timms, *Chem. Commun.*, 1525 (1968); P. L. Timms, *J. Chem. Educ.*, **49**, 782 (1972).
10. P. S. Skell and J. J. Havel, *J. Amer. Chem. Soc.*, **93**, 6687 (1971).
11. L. Andrews and G. C. Pimentel, *J. Chem. Phys.*, **44**, 2361 (1961).
12. R. L. DeKock, *Inorg. Chem.*, **10**, 1205 (1971).
13. J. L. Slater, R. K. Sheline, K. C. Lin, and W. Weltner, Jr., *J. Chem. Phys.*, **55**, 5129 (1971).

14. J. K. Burdett and J. J. Turner, *Chem. Commun.*, 885 (1971).

15. J. H. Darling and J. S. Ogden, *Inorg. Chem.*, **11**, 666 (1972).

16. H. Huber, E. P. Kündig, M. Moskovits, and G. A. Ozin, *Nature Phys. Sci.*, **235**, 98 (1972).

17. J. S. Shirk and H. H. Claassen, *J. Chem. Phys.*, **54**, 3237 (1971).

18. J. W. Nibler and D. A. Coe, *J. Chem. Phys.*, **55**, 5133 (1971).

19. H. Huber, G. A. Ozin, and A. Vander Voet, *Nature*, **232**, 166 (1971); G. A. Ozin and A. Vander Voet, *J. Chem. Phys.*, **56**, 4768 (1972).

20. R. R. Smardzewski and L. Andrews, *J. Chem. Phys.*, **57**, 1327 (1972); L. Andrews and A. Hatzenbuhler, ibid., **56**, 3398 (1972); L. Andrews, ibid., **57**, 51 (1972).

21. T. K. McNab and P. H. Barrett, in I. J. Graverman (Ed.), Mössbauer Effect Methodology, Vol. 7, Plenum, New York, 1971.

22. A. Bos and A. T. Howe, *J. Chem. Soc., Faraday II*, **70**, 440, 451 (1974), A. Bos, A. T. Howe, L. W. Becker, and B. W. Dale, *Cryogenics*, **14**, 47, (1974).

23. A. Thompson, paper presented at the "Meldola Lecture Symposium," University College, London, 1974.

24. I. W. Stolz, G. R. Dobson, and R. K. Sheline, *J. Amer. Chem. Soc.*, **84**, 3589, (1962); ibid., **85**, 1013 (1963).

25. M. A. Graham, M. Poliakoff, and J. J. Turner, *J. Chem. Soc., A*, 2939 (1971).

26. S. J. Cyvin, B. N. Cyvin, and A. Snelson, *J. Phys. Chem.*, **75**, 2609 (1971).

27. L. Andrews and W. F. Havard, Jr., *Inorg. Chem.*, **14**, 767 (1975).

28. J. E. Dobson, P. M. Tucker, F. G. A. Stone, and R. Schaeffer, *J. Chem. Soc., A*, 1882 (1969).

29. R. J. Lagow and L. A. Shimp, *J. Amer. Chem. Soc.*, **95**, 1343 (1973).

30. A. Bos, J. S. Ogden, and L. Orgee, *J. Phys. Chem.*, **78**, 1763 (1974).

31. C. S. Liu, J. L. Margrave, J. C. Thompson, and P. L. Timms, *Can. J. Chem.*, **50**, 459 (1972).

32. P. L. Timms, *J. Amer. Chem. Soc.*, **90**, 4585 (1968).

33. R. W. Kirk, D. L. Smith, W. Airey, and P. L. Timms, *J. Chem. Soc. Dalton*, 1392 (1972).

34. R. N. Perutz and J. J. Turner, *J. Amer. Chem. Soc.*, **97**, 4791 (1975).

35. A. J. Rest, *J. Organometal. Chem.*, **40**, C76 (1972).

36. E. P. Kündig, M. Moskovits, and G. A. Ozin, *Can. J. Chem.*, **51**, 2737 (1973).

37. G. A. Ozin and E. P. Kündig, unpublished work.

38. W. Klotzbücher and G. A. Ozin, *J. Amer. Chem. Soc.*, **97**, 3965 (1975).

39. P. S. Skell, E. M. Van Dam, and M. P. Silvon, *J. Amer. Chem. Soc.*, **96**, 627 (1974).

40. M. J. Piper and P. L. Timms, *Chem. Commun.*, 50 (1972).

41. R. Mackenzie and P. L. Timms, *J. Chem. Soc., Chem. Commun.*, 650 (1974).

42. E. P. Kündig, M. Moskovits, and G. A. Ozin, *J. Amer. Chem. Soc.*, **97**, 2097 (1975).

43. L. Hanlan and G. A. Ozin, *J. Amer. Chem. Soc.*, **96**, 6324 (1974).

44. D. McIntosh, M. Moskovits, and G. A. Ozin, *Inorg. Chem.*, 1975 (in press).

Techniques of Matrix Cryochemistry

2

Martin Moskovits and Geoffrey A. Ozin

1. INTRODUCTION

Matrix isolation is the technique whereby one can immobilize an unstable or reactive chemical species by cocondensing it at very low temperatures with copious quantities of an inert substance such as, for example, a noble gas. In this manner the species of interest may be isolated long enough to allow its observation by one or more spectroscopic techniques. Since the most plentiful noble gas, argon, melts at 83°K and since a much revered rule of thumb, due originally to Tammann, states that isolation is impossible above a temperature approximately one-half the melting point, one can immediately see that matrix isolation implies cryogenic chemistry at temperatures below 77°K, which is commonly regarded as the lowest temperature used in conventional chemistry.

To do matrix isolation, then, one must do several things. One must create a cold surface and measure its temperature, and one must have ways of producing and depositing materials of interest onto this cold surface with accurate knowledge of the quantities involved. Finally, one must be able to adapt conventional spectroscopic techniques to matrix isolation. Details of how to do these things form the substance of this chapter, together with a few theoretical considerations that must be borne in mind while doing so.

2. ACHIEVING LOW TEMPERATURES

There are four methods in common use for cooling to cryogenic temperatures a substrate on which a matrix is to be deposited. These are (a) the double Dewar cryostat, (b) the continuous flow cryostat, (c) the Joule–Thomson open-cycle cryostat, and (d) the closed-cycle refrigerator. The first two use liquid helium or hydrogen as refrigerants, while the last two use pressurized gas from either cylinders or an internal reservoir. Traditionally one achieves liquid helium temperature (4.2°K) by means of double-Dewar cryostats. These vessels are commonly made of either stainless steel or Pyrex glass. The former is sturdier, more permanent, and more expensive. In this cryostat the specimen to be cooled, for example, a cesium iodide ir window, is mounted within an annular copper block using indium gaskets for proper thermal contact; the copper block is either mounted, soldered, or glued to the bottom surface of the inner chamber of the cryostat that contains liquid helium. A narrow vacuum gap separates the inner chamber from an intermediate chamber containing liquid nitrogen, while this second chamber is separated from the outer jacket by another evacuated space. The liquid nitrogen heat shield is mandatory in order to prevent rapid boil-off of helium via radiation heating. One finds, in fact, that in some cryostats no liquid helium can be collected at all without the liquid-nitrogen-filled inner chamber. Despite this design precaution, one normally finds boil-off rates of several tenths of liters of He per hour due to heat radiation falling on the sample mount and other unshielded portions of the cryostat. (Dewars designed especially for refrigerant storage have evaporation rates that are substantially lower than those discussed above.)

Several designs for double cryostats are reported in the literature.[1] An example is shown in Figure 1*a*.

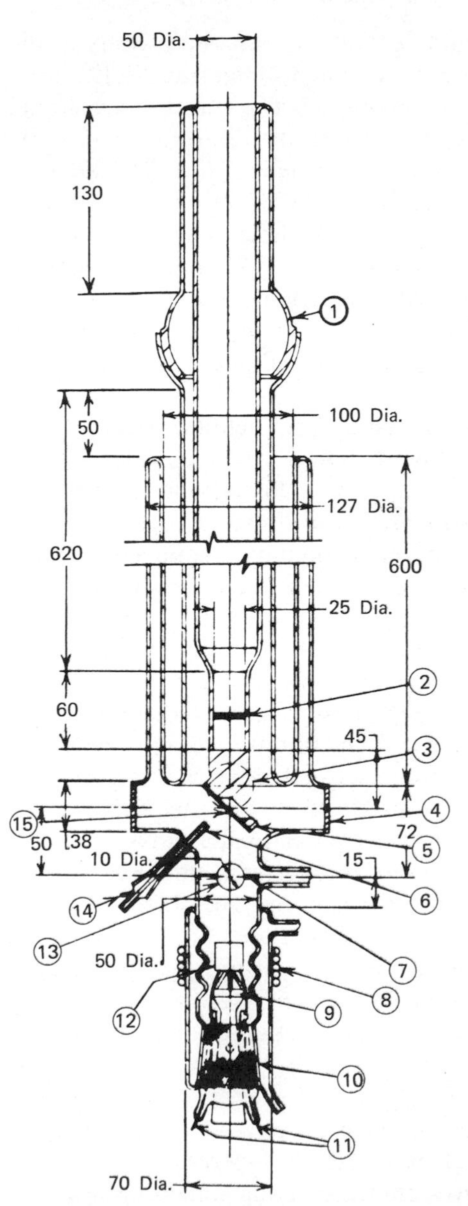

Figure 1a. Double cryostat after Linevsky [*J. Chem. Phys.* **34**, 587 (1961)], reproduced by permission. Labeled parts are as follows: (1) 102/75 ball and socket joint, (2) kovar to glass seal, (3) pumping line, (4) AgCl window, (5) doughnut, (6) matrix jet, (7) nickel disk, (8) induction coil, (9) quartz stage, (10) 45/50 tapered joint, (11) thermocouple leads, (12) Knudsen cell, (13) shutter, (14) 14/35 tapered joint, (15) silver chloride cold window.

Transferring liquid helium from storage Dewars to the cryostat requires special precautions. A vacuum-jacketed transfer line must be used in order to minimize helium evaporation. The transfer line should ideally also have an outer portion through which the returning helium gas can flow so as to precool and shield the central helium line. The helium gas that boils off in the cryostat should be recovered in order to reduce operating costs.[2] Several transfer lines are available commercially, as, for example, the Air Products' Helitran systems[3] and the Oxford Instruments' TTL series.[4] Transfer lines must be purged with helium before use in order to prevent solid air or water plugs from forming in them; and the cryostat must be precooled with liquid N_2 followed by removal of excess liquid nitrogen before helium is introduced into it.

Dangers do exist in using liquid helium. Helium Dewars can form solid air plugs if air is allowed to flow into their inlets or vent outlets. Dewars have been known to rupture explosively under such circumstances. Users should familiarize themselves with the technology required for using liquid helium. This information is commonly supplied by Dewar manufacturers. Danger of explosion is, of course, increased if one uses liquid hydrogen: the extra precautions necessary with that refrigerant are discussed in Ref. 5.

Liquid helium is transferred into the cryostat by pressurizing the storage Dewar with helium, thereby forcing refrigerant through the transfer line and into the interior of the cryostat. Naturally there must be an evacuated area separating the cooled specimen from the atmosphere. The vacuum must be low enough so that thermal conduction is at a minimum. This criterion is usually met at approximately 10^{-4} torr. One strives to achieve, however, as good a vacuum as one's budget and apparatus constraints allow in order to minimize residual gases from condensing on the cold surface. Our experience indicates that with a little care one can obtain a vacuum of 10^{-7} torr at the same cost as the more conventional $10^{-4} - 10^{-5}$ range.

One should also scrupulously ensure that leaks, even minor ones, are absent. The results of more than one matrix isolation study have been shown to arise from matrix reaction with atmospheric impurities slowly leaking into the system.

In continuous-flow cryostats one essentially removes the storage portion of the double cryostat, placing a relatively small sample holder at the end of a transfer line. This makes the operating portion of the cryostat very small and very convenient for handling and fitting into a spectrometer sample area. These units consume a little more helium than double cryostats on a per hour basis, but because with continuous-flow units one transfers out of the storage Dewar only as much helium as one needs, one generally saves helium through their use. Both Air Products and Oxford Instruments make continuous-flow cryostats (Figure 1*b*). Both companies supply a heater

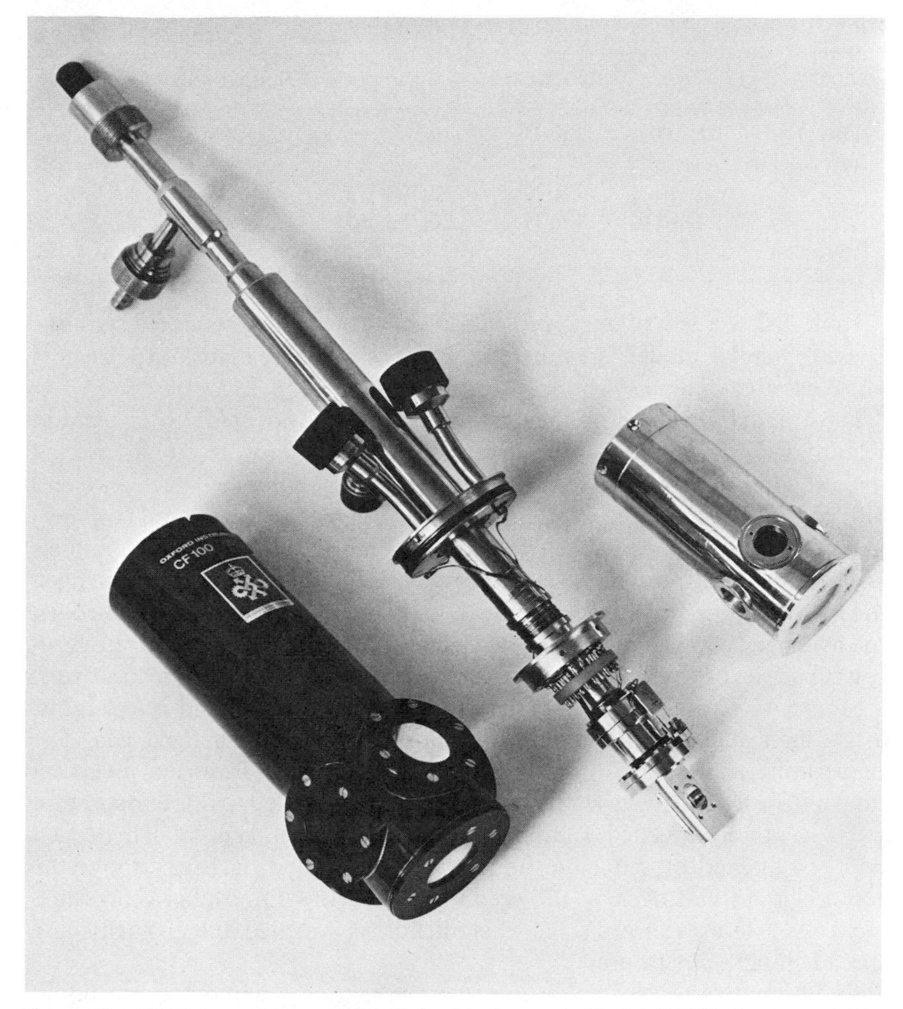

Figure 1b. Continuous-flow cryostat (Oxford Instruments Model CF100) showing cold tip, radiation shield, and vacuum shroud with optical windows.

and thermocouple with their products for controlling and measuring the temperature of the sample-mounting block. One usually buys these items complete with transfer-line and refrigerant flowmeters. Moreover, both companies supply vacuum shrouds that fit around the cryostat with ports conveniently located so that appropriate windows can be mounted on them. These continuous-flow cryostats can be used with liquid nitrogen as well as liquid helium.

Table 1. Specifications of Continuous-Flow Cryostats

Company	Type	Refrigerant	Consumption	Temp. Range	Capacity
Oxford Instruments	CF 100 series	Liq He	0.9 1/hr (4.2°K)	3–300°K	360 mW at 4.2°K
		Liq N_2	—	65–300°K	—
Air Products	LT-3-110 series	Liq He	0.75 1/hr (4.2°K)	—	1.5 W at 4.4°K

Table 1 lists a few of the specifications of two continuous-flow cryostats. These or similar models are adequate and desirable for matrix experiments when a source of liquid helium is readily available.

Those individuals who either cannot get or do not wish to use liquid helium or hydrogen can obtain temperatures as low as 3.6°K from a Joule–Thomson cryostat. Geist and Lashmet[6] have described such a system with which they have achieved temperature as low as 15°K. Several commercial models are also available. Air Products[3] makes several models: models AC–2 and AC–2L use nitrogen gas and hydrogen gas, and liquid nitrogen and hydrogen gas, respectively, while their model AC–3L uses liquid nitrogen and hydrogen and helium gas. The AC–2, AC–2L, and AC–3L have refrigeration capacities of 4 W at 23°K, 6 W at 22°K and 500 mW at 4.4°K, respectively.

Model AC–2 is a two-stage refrigerator that uses no liquid refrigerants. In the first stage, Joule–Thomson expansion liquefies nitrogen gas. The temperature achieved in this stage is below the Joule–Thomson inversion temperature for hydrogen (193°K). Hydrogen is precooled to this temperature and is eventually liquefied. Temperatures below the boiling point of hydrogen may be obtained by reducing the pressure of the hydrogen gas above the pool of liquid hydrogen in the second stage below atmospheric pressure. Model AC–3L uses a third stage where helium is liquefied, thereby achieving liquid helium temperatures.

An alternative way of achieving low temperatures without using liquid helium is to use a closed-cycle refrigerator. Although these systems are somewhat costly, their operating cost is very small. These systems also have further drawbacks in that they will not achieve temperatures much below 10°K (i.e., their refrigeration capacity is low at 10°K) and they are heavier, bulkier, and noisier than the cryostats described above. These systems are, however, more than adequate for doing most matrix work and they have the unique convenience of yielding cryogenic temperatures at the flick of a switch. Several closed-cycle refrigerators are available, among them the Model 21 Cryocooler made by Cryogenics Technology Inc.,[7] the Displex Model CS 202 made by Air Products,[3] and the Model GB 02 Cryogenic Refrigerator made by Cryomech Inc.[8] A Displex is shown in Figure 1*c*.

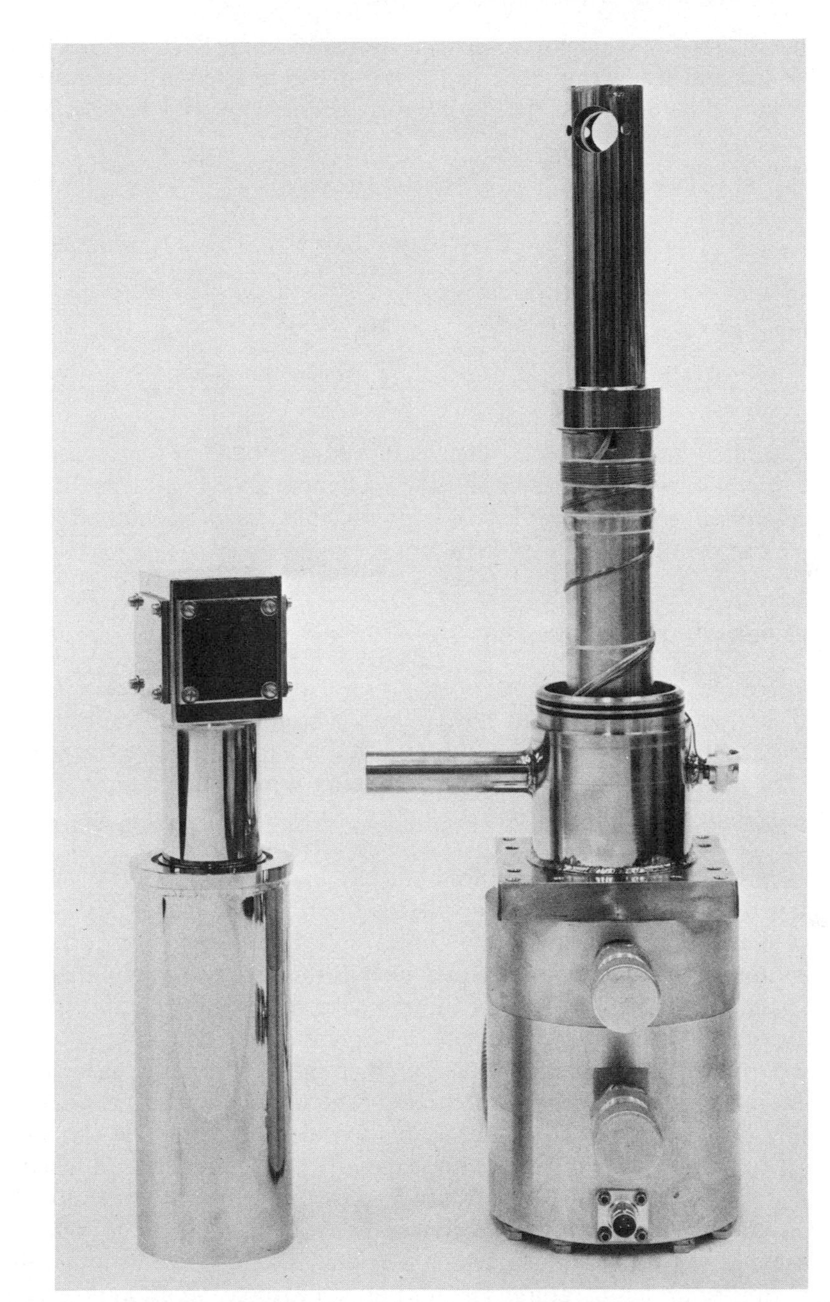

Figure 1c. Closed-cycle refrigerator (Displex) showing cold head with radiation shield and vacuum shroud (courtesy of Air Products and Chemicals Inc.).

Table 2 lists some operating specifications for these systems. In these systems the cold head weighing from 12 to 25 lb is connected to a compressor via flexible, prepressurized tubing. In this manner the cold head can be

Table 2. Specifications of Closed-Cycle Refrigerators

Manufacturer	System	Lower Temp. Limit (°K)	Cold Head Weight (lb)
CTI	Model 21	10	12
Air Products	Displex CS 202	10	16
Cryomech	GB 02	7.5	25

easily handled and positioned. The manufacturers of these items usually supply them with sample heater and thermocouple already attached to the sample-mounting block. They also supply several types of vacuum shrouds suitable for various types of spectroscopy.

3. TEMPERATURE MEASUREMENT

Four thermometric methods are commonly used in the 1–50°K region. These are (a) thermocouples, (b) solid state (diode) thermometers, (c) capacitance cryogenic temperature sensors, and (d) hydrogen vapor pressure thermometers. The criteria on which a good thermometer is judged are ample signal, high sensitivity, and good reproducibility. Linearity is not essential so long as one can calibrate the thermometer against a known standard (such as the hydrogen vapor pressure thermometer), and in fact all of the sensors mentioned above are quite nonlinear. It is the lack of sufficient sensitivity at low temperatures that makes it impossible to use the more popular conventional thermometric techniques such as the chromel-alumel thermocouple and the platinum-resistance thermometer at cryogenic temperatures. Sensitivity is defined as dA/dT where A is the output signal characteristic of the particular technique, for example, thermal EMF for thermocouples and resistance for resistance thermometers. Several (usually gold) alloy compositions have been found that give usable sensitivity at liquid He temperatures. These are listed in Table 3 for comparison. One of these is usually supplied with commercial He cryostats. The thermocouples are operated in the conventional manner with one end attached with an appropriate adhesive (see below) to the cold surface, while the other is kept at ice-water or the water triple-point temperature. The output is measured on an appropriate milli- or microvolt meter or potentiometer. Digital dc voltmeters are very convenient for this purpose. In purchasing one, one should make sure that

Table 3. Sensitivity and Useful Range of Selected Temperature Sensors

Material	Useful Range (°K)	Sensitivity (μV per °K) At 4°K	At 20°K	At 77°K	At 300°K
Copper-gold (4% Co) AC	0.3–300	4.0	16.4	35.0	42.9
Chromel-gold (0.07% Fe) P–AF	0.3–300	12.2	16.3	17.6	22.3
Constantan-copper, T	30–700	1.3	5.5	16.1	39.3
Constantan-iron, E	20–1000	2.1	8.5	25.8	59.3
Alumel-chromel, K	30–1700	0.8	4.1	15.7	39.8
Pt–Rh (13–40%) vs Pt–Rh (1–20%)	200–2000	—	—	—	5.6–6.4
Ir vs Ir–Rh	—	—	—	—	1.26
Pt–Ir vs Pd	—	—	—	—	7.02
Ga–As	0.3–400	5000	2000	25000	—
Ge	0.1–300	240 Ω	5 Ω	0.24 Ω	—

the instrument has at least 10 μV sensitivity. Suppliers of these instruments are too numerous to list. In addition, most commercial cryogenics outlets supply digital voltmeters for the convenience of their customers.

Semiconductor temperature sensors take advantage of the temperature dependence of the forward-biased voltage at constant current of a solid state diode. Silicon and gallium-arsenide are the two most popular semiconductors used for the purpose. The temperature sensitivities of these items vary from 20 to 50 mV/°K for Si and 2 to 10 mV/°K for GaAs in the 1–10°K range with 10 μA current. These temperature sensors are available in various configurations and encapsulations ranging in size from 5.8 to 1.27 mm in their largest dimensions from LakeShore Cryotronics Inc., Eden, New York, which also supplies constant-current sources. To use them one simply passes a known constant current through the diode and measures the forward voltage on a millivoltmeter as an indication of the temperature, which is read off a temperature-voltage table.

The glass-ceramic capacitance thermometer[9] exhibits a temperature-dependent capacitance. Measuring their capacitance on an appropriate bridge therefore provides a measure of the sensors' temperature. These sensors are designed to operate below 80°K and are particularly insensitive to magnetic fields, making them useful in matrix esr experiments. Sensitivities of 250 pF/degree at 4.2°K are available in commercial devices. LakeShore Cryotronics supplies these items also.

The hydrogen manometer measures hydrogen vapor pressure from which the temperature of the manometer's sensing end can be deduced. Its range is limited to temperatures below the boiling point of hydrogen (21°K). Air Products and other suppliers of cryogenic equipment sell these items. They can also be "home-built."

4. TEMPERATURE CONTROL

Obtaining known and constant temperatures above the bottom temperature of one's cryostat is mandatory in carrying out matrix annealing and warm-up experiments during which the matrix-isolated species can diffuse and react. To do this most cryostats are fitted with small heaters that are operated simultaneously with the refrigerant in order to provide a steady state at a desired temperature. Resistive heating is most often used, and its output is regulated simply with a variable autotransformer (Variac) or potentiometer. The desired temperature is obtained by manually raising the heating current carefully. Alternatively, one can use a commercial or home-made temperature controller. These instruments use either a thermocouple or a diode sensor to provide them with input generating heater current proportional to the difference between the desired temperature, which is set on a dial, and the measured temperature of the cold surface. Temperature regulation to 0.01°K is easily obtained. LakeShore Cryotronics, Air Products, and Cryogenics Technology all offer these instruments for sale to be used each with their own sensors. A circuit used successfully in our laboratories is given in Figure 2. It used a gold 0.07 atomic percent Fe versus chromel thermocouple input and is intended for a 40 Ω heater, but can be adapted to other heaters with a minor adjustment of the output power resistor.

5. LOW-TEMPERATURE ADHESIVES AND SOLDERS

Thermocouples or other temperature sensors as well as sample-holder and matrix-isolation substrates must be anchored in place using a suitable bonding agent that has high thermal conductivity, a sufficiently high tolerance to temperature cycling, low thermal expansion, and no phase transitions or other change in physical properties that would make it fracture at low temperatures. Moreover, thermocouple wires should preferably be anchored with a dielectric material.

The most popular adhesive having these properties is GE 7031 varnish (General Electric Co., Insulating Materials Dept., Schenectady, N.Y.). After properly degreasing both the thermocouple and the surface to which it is bonded, a thin film of the varnish is brushed onto the surface. The thermocouple wires are then laid on the surface, and another coat of the varnish is

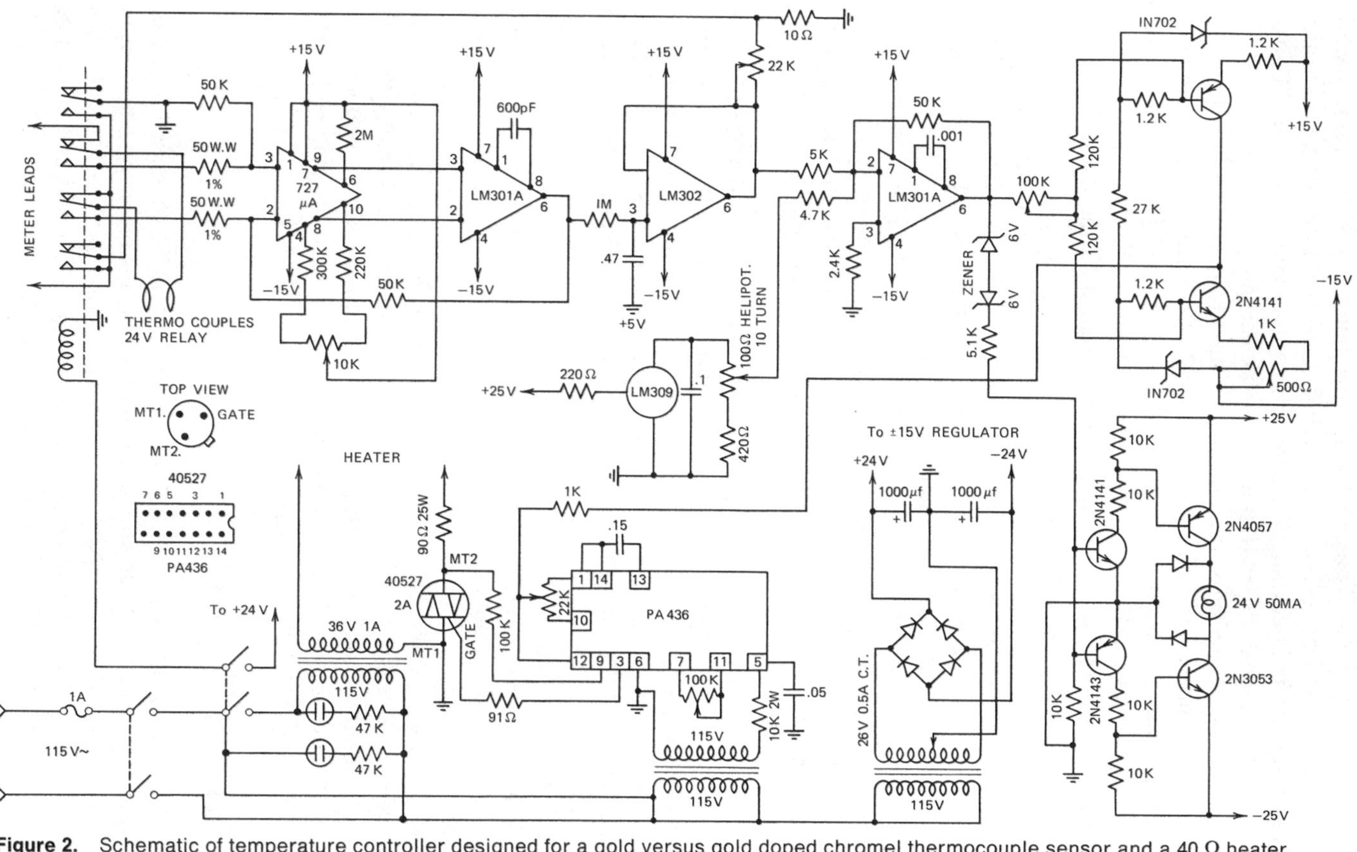

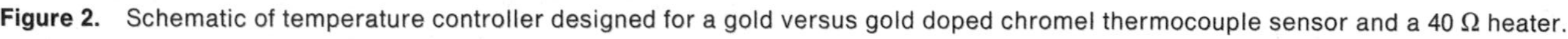

Figure 2. Schematic of temperature controller designed for a gold versus gold doped chromel thermocouple sensor and a 40 Ω heater.

applied. The two parts must be held together for several minutes while the Xylene carrier evaporates, leaving behind the hardened varnish. D. L. Swartz[10] pointed out that Xylene dissolves epoxy, so that GE 7031 varnish should not be used to bond to epoxy surfaces.

Several epoxy adhesives have also been found to have good low-temperature bonding properties, for example, Stycast 2850FT with catalyst, obtainable from Emerson and Cumings (U.K.) Ltd., London, and Hysol Epoxi-patch, obtainable from the Dexter Corporation, Olean, N.Y. These are found to have rather poorer thermal conductivity[11] than does the General Electric varnish. They also result in permanent bonds.

The temperature errors resulting from the bonding of a thermocouple to a cold surface have been analyzed by Kopp and Stack.[12] Those authors find that errors result from both bulk thermal resistance through the bonding agent and, more importantly, interface resistance.[13]

The temperature error is also reduced by increasing the length of the thermocouple wires as well as the length of wire bonded to the body. Twisting the thermocouple wires together near the junction is also frowned upon by Kopp and Stack, who point out that two electrical junctions are thereby formed.

Permanent metal-to-metal bonds can be made using silver solders of various kinds. Occasionally one wishes to use a low-melting solder in order to allow the parts to be taken apart easily. The two most commonly used solders for that purpose are Wood's metal (Bi 50, Pb 25, Sn 12.5, Cd 12.5), which melts at 70°C, and Indium, which melts at 156°C. Several other alloys melting between 57 and 330°C are listed by Warren and Bader,[14] as well as the fluxes commonly used with them.

A method for bonding a nonmetal to a metal in such a way as to form a strong but easily demountable joint is described by M. A. Brown.[11] He uses a conductive silver composition (duPont 4666, E. I. duPont de Nemours and Co., Wilmington, Del.), which is applied to the dielectric material (in his case an alumina rod) and baked at 500°C for 15 min. Finally, Wood's metal is used to bond the coated end of the alumina rod to a copper base. This method may be used to bond esr sapphire rods to the cold end of the cryostat in doing matrix esr experiments.

6. MATRIX SUBSTRATES

Because the matrix-isolated products are generally examined spectroscopically, the substrate on which the matrix is formed must be transparent in the appropriate wavelength region and a good thermal conductor. Table 4 gives a partial list of materials commonly used as infrared and uv-visible matrix substrates and their spectral ranges.

Table 4. Commonly Used Matrix Substrates and Their Transparent Spectral Regions

Material	Range (μ)	Type of Spectroscopy
NaCl	0.25–16	ir
KBr	0.25–25	ir
CsI	1–50	ir
CsBr	1–37	ir
BaF_2	0.13–13.5	ir, uv
KRS 5	0.5–40	ir
CaF_2	0.15–8	uv, ir
Sapphire (uv grade)	0.14–10	uv, ir
LiF	0.12–7	uv, ir
MgF_2	0.13–9	uv, ir

Sapphire rod is used for matrix esr work and a polished aluminum surface for matrix Raman. Although uv-quality silica or quartz is often used as uv-visible windows, its low thermal conductivity makes it a poor matrix substrate. A note of warning should be sounded regarding KRS–5, which is a mixed thallium bromide chloride salt. Thallium compounds are very toxic, and great care should be used in polishing these substrates.

Thermal contact between the substrates and the cryostat is usually ensured by sandwiching the desired material between two annular copper pieces, using indium wire gaskets to enhance thermal contact, then screwing this "plate-holder" into the cold end of the cryostat using indium foil gaskets. Figure 3 shows such an assembly in detail.

For matrix esr, a high purity sapphire rod may be used. This is bonded to a copper cup with a threaded end using epoxy cement or using Brown's method as shown in Figure 4.

7. MATRIX MATERIALS

Only two universal characteristics are shared by all matrix materials. These are their rigidity at the desired deposition temperature and their transparency in the desired spectral region. All other qualities vary from experiment to experiment. Thus, in some (perhaps most) studies, one may wish as little interaction as possible between the guest and the host species as, for example, in spectroscopic studies of isolated metal atoms, free radicals, or reactive intermediates, while in other cases one may wish the cocondensate to react with the matrix material, thereby isolating the product in a matrix of one of the reagents, for example, in the formation of $Pt(CO)_4$, $Pd(CO)_4$, or $Ni(O_2)_2$ in pure CO or O_2.

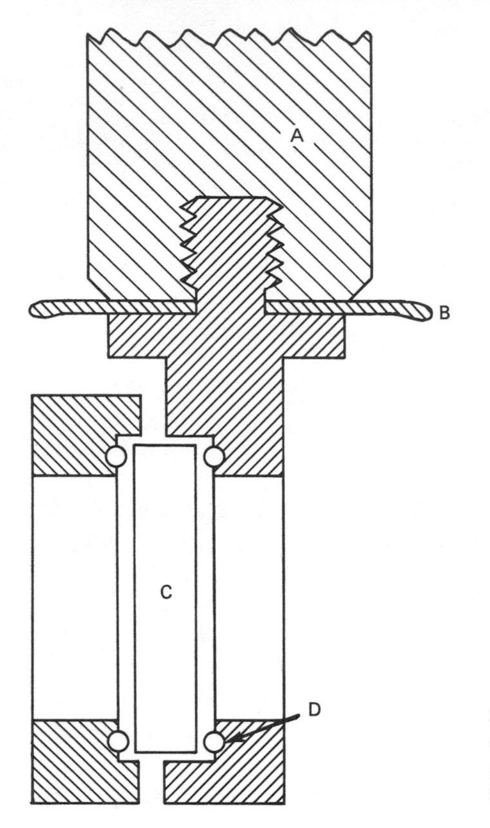

Figure 3. Detail of mount used to connect optical windows to cold tip. (*a*) Cold tip, (*b*) indium gasket, (*c*) optical plate, (*d*) indium wire gasket.

7.1 Matrix Rigidity

The melting point of the matrix material is generally a good guide to its rigidity, and by rigidity one usually means the lack of appreciable diffusion of trapped species. One should, however, look with some skepticism upon Tamman's rule, which states that diffusion is negligible (presumably over a time period of several hours) at temperatures below one-half the melting point. We, for instance, have found that Cu atoms and even species as large as $Ag(CO)_3$ diffuse to a measurable extent at 30°K in solid argon and CO, respectively. Others have reported similar observations.[15]

Table 5 lists the melting points of several commonly used matrix materials. In choosing a matrix appropriate to one's needs, one should also keep in mind the volatility of the matrix. The matrix material should, on one hand, possess a sufficiently high vapor pressure at room temperature to allow it to be sprayed on the cold tip, while on the other its vapor pressure at cryogenic

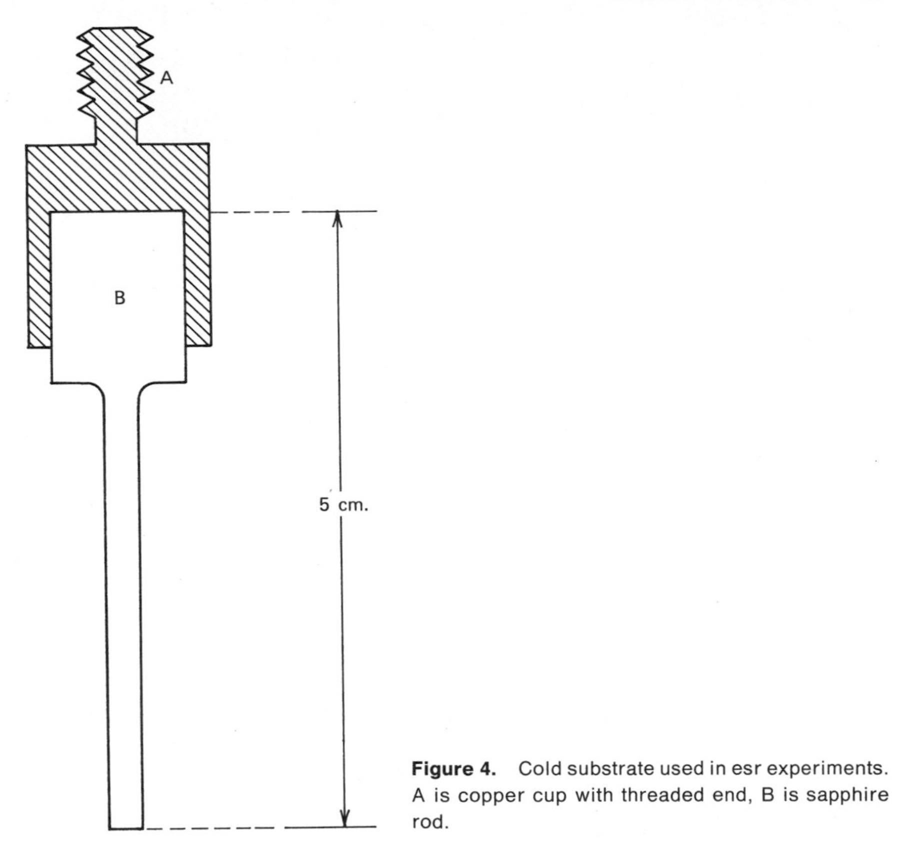

Figure 4. Cold substrate used in esr experiments. A is copper cup with threaded end, B is sapphire rod.

temperatures is sufficiently low to prevent its vaporizing while under ambient vacuum conditions. Referring to Table 5, then, one can see that Ar has a vapor pressure of 10^{-3} torr at 39°K. Its upper temperature limit, therefore, is restricted to about that temperature under normal vacuum conditions.

7.2 Thermal Properties of Matrices

When a matrix is condensed from the gas phase, the latent heat of fusion of the matrix material is released and must be absorbed by the cryostat, which acts as an isothermal heat sink. This energy must, moreover, be conducted to the cryostat through previously deposited layers of the matrix. Thus both the latent heat of fusion (L_f) and the thermal conductivity (λ) of the matrix material are, in principle, important quantities since they may determine how quickly the matrix solidifies, hence the amount of time available to

Table 5. Properties of Matrix Materials

	T_d (°K)	m.p. (°K)	b.p. (°K)	Vapor Pressure = 10^{-3} mm at (°K)	L_f (Jmole^{-1})[c]	$-U_0$ (Jmole^{-1})[d]	λ(20°K) (Wm^{-1} °K^{-1})	λ(5°K)	$\mu \times 10^{18}$ (esu cm)	$Q \times 10^{26}$ (esu cm^2)	α (Å^3)	ε' (20°K)[c]	n	IP (eV)
Ne	10	24.6	27.1	11	335	1874	0.4	—	0	0	0.395	—	1.23 (4°K)	21.56
Ar	35	83.3	87.3	39	1190	7724	1.3	2.0	0	0	1.641	1.63	1.29 (20°K)	15.76
Kr	50	115.8	119.8	54	1640	11155	1.2	0.7	0	0	2.480	1.88	1.28 (4°K)	14.00
Xe	65	161.4	165.0	74	2295	16075	2	—	0	0	4.044	2.19	1.49 (40°K)	12.13
CH_4	45	90.7	111.7	48	971	—	0.1	—	0	0	2.60	1.79	1.33 (90°K)	12.65
SF_6	—	222.7	209.4(sub)	—	—	—	—	—	0	0	4.52	—	—	—
N_2	30	63.2	77.4	34	721	6904	0.4	5.0	0	−1.4	1.767	—	1.22 (4°K)	15.5
O_2	26	54.4	90.2	40	444	—	—		0	−0.4	1.598	—	1.25 (4°K)	12.07
Cl_2	—	172.2	239.1	118	6820	32400	—		0	+6.1	4.61	—	—	11.48
CO	35	68.1	81.6	38	836	7950	—		0.112	−2.5	1.977	—	—	14.01
NO	—	109.6	121.4	66	2310	—	—		0.153	−1.8	1.74	—	—	9.26
CO_2	63	216.6	194.6(sub)	106	8339	26987	—		0	−4.3	2.63	—	1.22 (4°K)	13.78
N_2O	—	182.4	184.7	99	6535	24267	—		0.167	−3.5	3.00	—	—	12.91
SO_2	—	197.6	263.1	138	—	—	—		1.63	4.4	3.89	—	—	12.34
H_2S	—	187.7	212.5	104	—	—	—		0.97	—	3.78	—	—	10.42
C_2H_6	—	89.9	184.5(sub)	—	—	—	—		0	−0.8	4.47	—	—	11.55
C_2H_4	—	104.0	169.5	—	—	—	—		0	+2.0	4.22	—	1.44 (63°K)	10.46
C_2H_2	—	191.4	189.6(sub)	—	—	—	—		0	3.0	3.49	—	—	11.40
CF_4(14)[b]	—	89.2	145.2	—	—	—	—		0	0	3.86	—	—	14.6
CHF_3(23)	—	118.0	191.2	—	—	—	—		1.65	—	3.09	—	—	13.86
$CClF_3$(13)	—	92.2	191.8	—	—	—	—		0.50	—	—	—	—	12.91
$CBrF_3$(13Bl)	—	105.2	215.4	—	—	—	—		0.65	—	—	—	—	11.89
$CHClF_2$(22)	—	113.2	232.4	—	—	—	—		1.42	—	—	—	—	12.45
CCl_2F_2(12)	—	115.2	243.4	—	—	—	—		0.51	—	—	—	—	12.31
$CHCl_2F$(21)	—	138.2	282.1	—	2736	—	—		—	—	—	—	—	—
C_2F_6(116)	—	172.6	195	—	—	—	—		—	—	—	—	—	—
$CClF_2CF_3$(115)	—	167.2	234.5	—	—	—	—		—	—	—	—	—	—
$CClF_2CClF_2$(114)	—	179.2	2770.0	—	—	—	—		—	—	—	—	—	—
C_4F_8(C318)	—	231.8	267.4	—	—	—	—		—	—	—	—	—	—

[a] Reproduced by permission from H. E. Hallam (Ed.), *Vibrational Spectroscopy of Trapped Species*, Wiley, New York, 1973.
[b] Number given to Arcton, Freon, etc.
[c] Latent heat of fusion.
[d] Lattice energy.

guest species to polymerize or otherwise combine and react. Values of these quantities for some matrix materials are listed in Table 5. One should note, however, that for the noble gases, λ is a rapidly varying function of temperature at low temperatures. For Ar,[16] for instance, λ increases from 6 to 38 mW cm^{-1} $°K^{-1}$ as one proceeds from 2 to 8°K and decreases with temperature thereafter approximately as $1/T$. The temperature difference of the surface of the matrix above that of its interior can be calculated assuming rapid establishment of a thermal steady state as follows:

The thickness, l, of a matrix formed when a constant deposition rate of n moles/sec is deposited for a length of time, t, on an initially bare, cold surface of area, A, is given by

$$l = \frac{nt}{\rho A} \tag{1}$$

where ρ is the molar density of the matrix material. The power liberated in this process is nL_f cal sec^{-1}, and, at steady state, this quantity equals the power conducted through the matrix, that is,

$$L_f n = \frac{A\lambda(T - T_0)}{l} \tag{2}$$

where λ is the thermal conductivity in cal cm^{-1} sec^{-1} $°K^{-1}$ and T and T_0 are the surface and interior temperatures, respectively. Substituting for l from (1) into (2) we get

$$T = T_0 + \frac{L_f n^2 t}{\lambda \rho A^2} \tag{3}$$

That is, the temperature rises as the square of the deposition rate and linearly with time. Substituting typical values for the parameters in equation (3), one finds, however, that with normal deposition rates the surface temperature does not rise more than a degree even after several hours' deposition.

Experience indicates, however, that the heat of fusion and the thermal conductivity are often not the quantities that determine guest diffusion upon condensation. In metal-vapor cocondensation experiments, the radiation reaching the matrix from the deposition source (whose temperature often exceeds 1000°C) is a far more important source of heat than that of fusion. Moreover, association with host molecules to form Van der Waals complexes reduces the apparent diffusion of guest species, at least for the more massive matrix gases, below what one would expect on the basis of a simple diffusing system. Thus, although L_f increases markedly as one proceeds from Ne to Xe, while the average values of λ for these noble gas solids at 5°K decrease,[17] the concentration of metal dimers is found to *decrease* as one, in turn, isolates metal atoms in Ne, Ar, Kr, and Xe.

7.3 Matrix Morphology

Little formal attention has been given to the morphology of solid matrices. It is not known, for instance, if most matrices are glassy, polycrystalline, or single crystal. Obviously, this entire range of structures may be had under properly selected conditions. Because the matrix substrate is often a single crystal (alkali halide or sapphire), it is conceivable that a certain degree of eptaxy is possible, and, in fact, Weltner[18] obtained an anisotropic matrix esr spectrum of BO in Ne and Ar, indicating orientation of the product with respect to the sapphire rod axis. The abnormally high rates of diffusion of species in Ar and CO matrices, on the other hand, tend to suggest that the matrix is either glassy or polycrystalline, that is, that diffusion proceeds either via large interstitial holes or grain boundaries.[19] Despite this lack of specific information regarding the structure of the matrix, it is important to consider certain basic data.

7.4 The Noble Gas Solids

The crystal structures at various temperatures of the noble gas solids have been thoroughly studied and reviewed by Pollack.[16] Except for helium, all the noble gases crystallize in a face-centered, cubic structure. It has been shown, however, that argon will sometimes assume a hexagonal, close-packed structure, especially in the presence of certain impurities such as nitrogen, oxygen, and carbon monoxide.[20]

In a perfect fcc crystal there are three sites that can accommodate a guest species. These are, in decreasing order of size, the substitutional, the octahedral, and the tetrahedral. The substitutional site is formed when a host atom is removed and a guest species substituted for it. Assuming the host crystal has a lattice parameter (fcc unit cell edge length) equal to a_0, the substitutional site can accommodate without appreciable strain a guest atom or molecule whose "equivalent spherical" diameter is $a_0/2^{1/2}$. The octahedral and tetrahedral sites are interstitial sites, and, as their names imply, are enclosed by six and four host atoms forming an octahedron and tetrahedron, respectively. They are situated in the center and in one corner of the fcc unit cell as shown in Figure 5. The largest spheres that can be accommodated in the two sites without significant lattice distortion have diameters equal to $0.293a_0$ and $0.159a_0$, the tetrahedral site being the smaller. Table 6 lists lattice parameters and the diameters of the three matrix sites for the noble gases.

Molecules with larger dimensions than those of the three trapping sites listed previously can, of course, be accommodated by introducing strain in the host crystal, by forming multiple vicinal vacancies or by forcing the matrix to adopt a glassy structure. No systematic study of these processes in the context of matrix isolation has yet been undertaken. The phenomenon of matrix splitting due to multiple trapping sites or local site symmetry (see

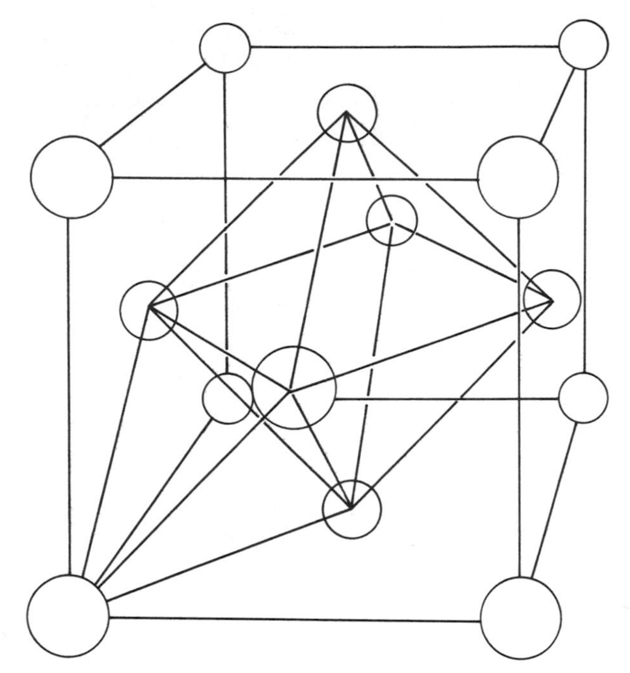

Figure 5. A face-centered cubic cell showing the octahedral and tetrahedral cavities. The size of the atoms occupying the lattice positions has been reduced for clarity. They should, in fact, be contiguous along face diagonals.

Table 6. Lattice Parameters and Site Diameters for the Noble Gas Solids

Solid (4°K)	a_0	Substitutional Hole (Å)	Octahedral Hole (Å)	Tetrahedral Hole (Å)
Ne	4.46	3.16	1.31	0.71
Ar	5.31	3.75	1.56	0.85
Kr	5.64	3.99	1.65	0.90
Xe	6.13	4.34	1.80	0.97

Chapter 8), however, tends to indicate that there exist definite trapping sites and order within the matrix, signifying at least some degree of crystallinity.

7.5 Spectra of Matrix Materials

A matrix material must be absorption-free in the spectral region of interest. The noble gas solids fulfil this condition more adequately than any other substance, since all are essentially transparent from about 100 cm^{-1} to the vacuum uv. These solids do have phonon spectra below 100 cm^{-1}, however, which may occasionally be observed in matrix Raman experiments. In solid Ar, for instance, these occur at 64 cm^{-1} and 42 cm^{-1}.[21] Spectra of various other matrix materials are reproduced and discussed by Hallam and Scrimshaw.[22]

8. PRODUCING MATRIX-ISOLATED SPECIES

In the preceding pages we have laid the technical foundation needed in trapping reactive species for subsequent spectroscopic examination. We now survey briefly some of the methods used to produce these entities. These techniques range from the very simple and straightforward, such as producing metal atoms by directly heating a metal ribbon, to elegant chemical means for producing free radicals.

8.1 Direct Vaporization

Metal atoms are most easily produced by directly heating a filament, ribbon, or rod of the metal. A low-voltage, high-current transformer is usually required for this purpose. In our laboratory, for example, we use 5 V, 300 A transformers and either water-cooled copper tubing or heavy-gauge multistrand copper cable to lead the current to the water-cooled electrodes between which the filament is suspended. Several metals cannot be deposited in this way. Manganese and chromium, for example, cannot be worked into wire or sheet and thus cannot be made into filaments, while the high conductivities of copper, silver, and gold make these metals difficult to deposit from an unsupported filament. These metals can be deposited from Knudsen cells or, in the case of the latter group, by supporting them from refractory metal (Ta, Mo, W) filaments.

A Knudsen cell is a small chamber constructed from a refractory material (Table 7) with an opening perforated in it to allow outflow of the enclosed substance. It can either be directly heated by passing current through the walls of the cell or indirectly heated by means of heater windings wrapped around it. Knowledge of the temperature, vapor pressure of the enclosed material, and orifice size allows one to calculate the rate of efflux of evaporand by using the equation

$$N = \frac{p}{(2\pi MRT)^{1/2}} \tag{4}$$

where N is the number of moles of evaporand per cm^2 of orifice per second leaving the Knudsen cell and M is the gram molecular weight.

Several Knudsen cell designs are shown in Figure 6. Figure 7 shows several "double" Knudsen cells. In a double cell the two chambers are operated at different temperatures. The temperatures of the two chambers are either regulated independently as shown in Figure7*a* or together as in Figure 7*c*,

Table 7. Properties of Refractory Materials

Material	Electrical Resistivity at 0°C (Ω-cm × 10^6)	Temperature Coefficient (°C^{-1} × 10^5)	Thermal Conductivity (cal cm^{-1} °C^{-1} sec^{-1})	Emissivity	Melting Point (°C)
Nickel	6.14	692	0.190 (1000°C)	~0.4 (1000°C)	1453
Niobium	16	343	0.163 (600°C)	0.37 (2000°C)	2468
Tantalum	12.4	382	0.172 (2500°C)	~0.4	2996
Molybdenum	5.03	473	0.289 (1000°C)	~0.35 (1000°C)	2610
Tungsten	4.89	510	0.263 (1000°C)	~0.4	3410
Stainless steel (304)	72.0	—	0.052 (500°C)	—	1400–1450
Graphite	464.0[a]	49.6[a]	0.012	—	3652[b]
Monel (regular)	48.7	248	0.090 (538°C)	—	1300–1350
Platinum	9.81	396	0.213 (1000°C)	0.3 (1300°C)	1769
BN (polycrystalline)	—	—	0.05 (1000°C)	—	1650[c]
SiO_2 (fused)	—	—	0.0033	—	2230
BeO	—	—	0.15 (500°C)	—	2530
Al_2O_3	—	—	0.10 (870°C)	—	2015
ZrO_2	—	—	0.004 (1760°C)	—	~2700

[a] The electrical conductivity of graphite depends on its thermal history. See text.
[b] Sublimes.
[c] Decomposes.

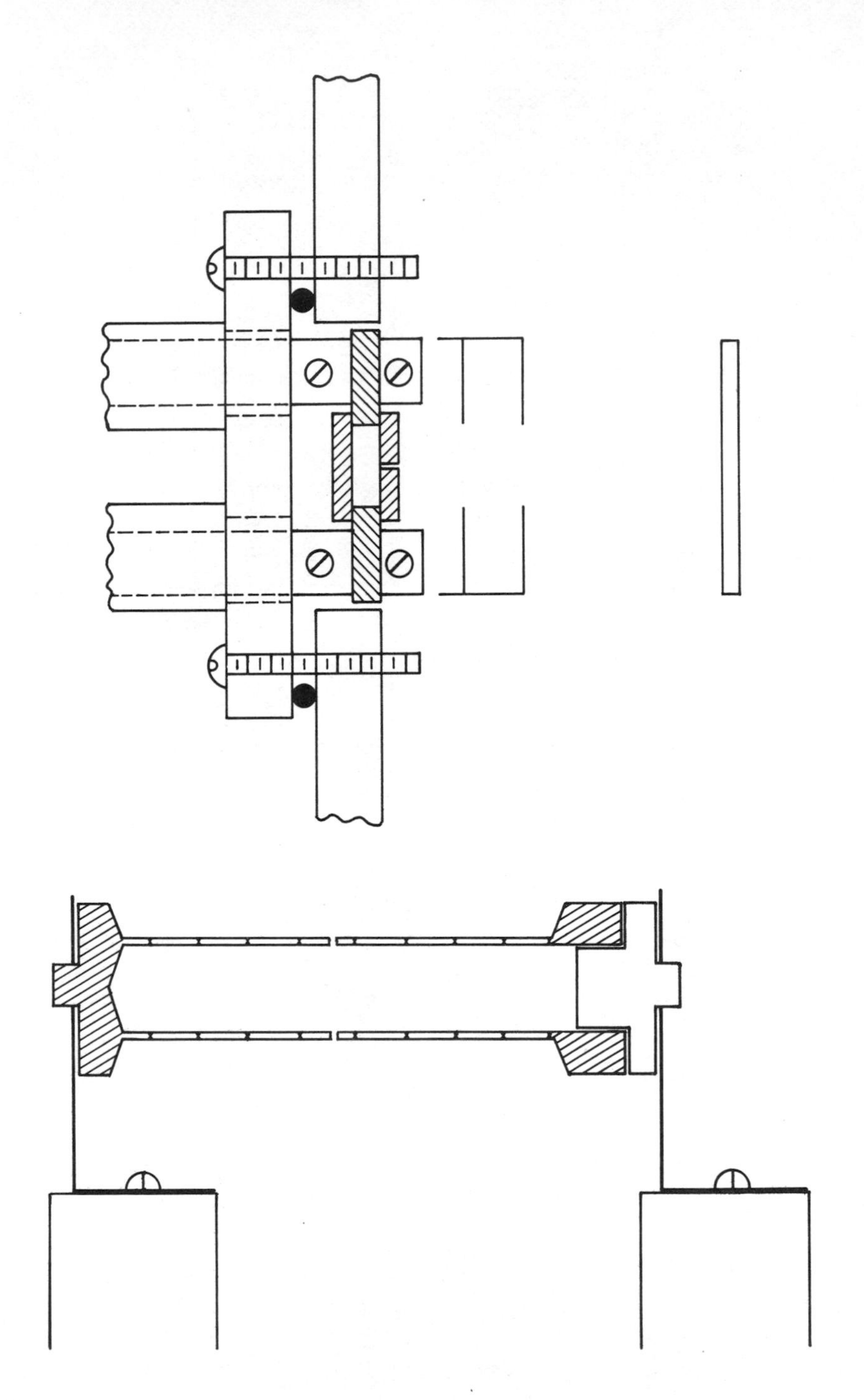

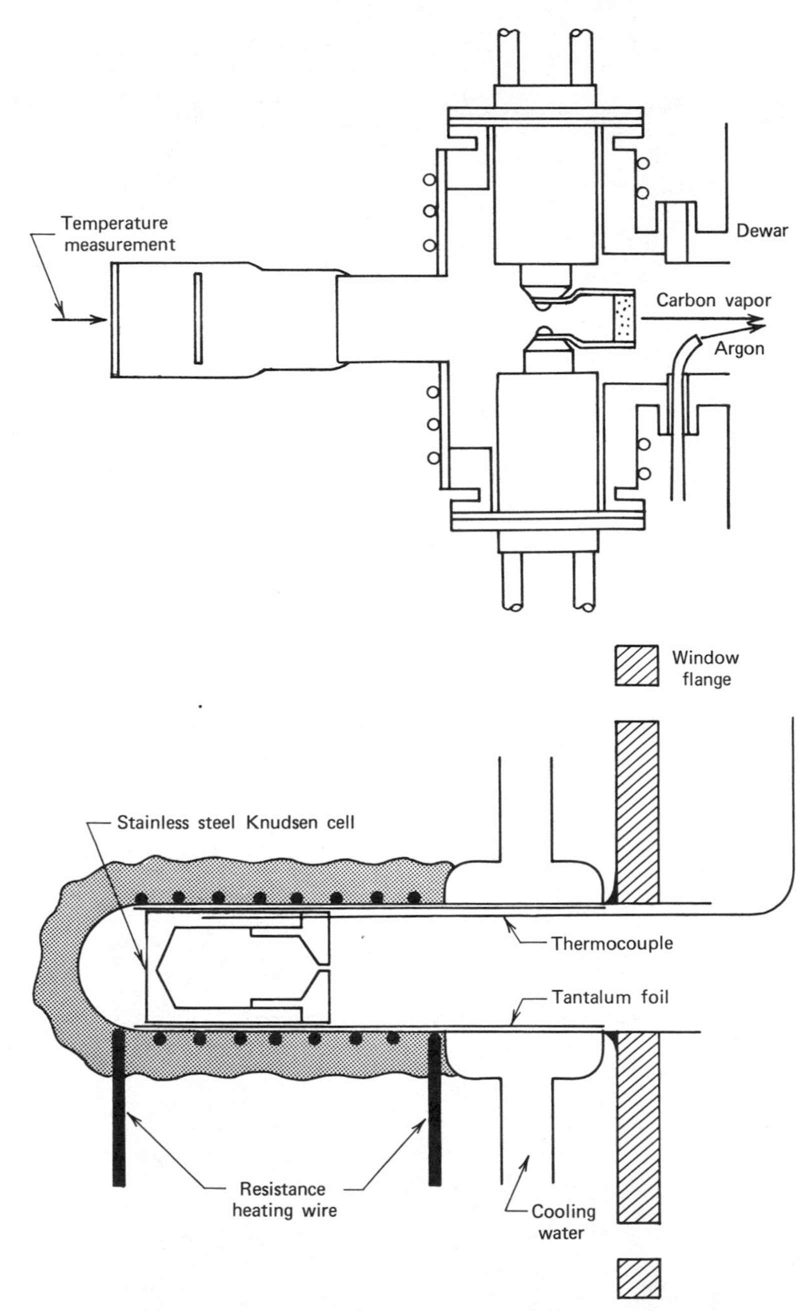

Figure 6. Knudsen cells. (*a*) Directly heated cell used by Brewer et al. to vaporize carbon C (courtesy of *The Journal of Chemical Physics*). (*b*) Directly heated tantalum cell supported on tantalum brackets. (*c*) Directly heated tantalum cell used by Weltner for vaporizing carbon showing water-cooled copper electrodes holding cell in place (courtesy of *The Journal of Chemical Physics*). (*d*) An indirectly heated Knudsen cell (courtesy of B. Meyer).

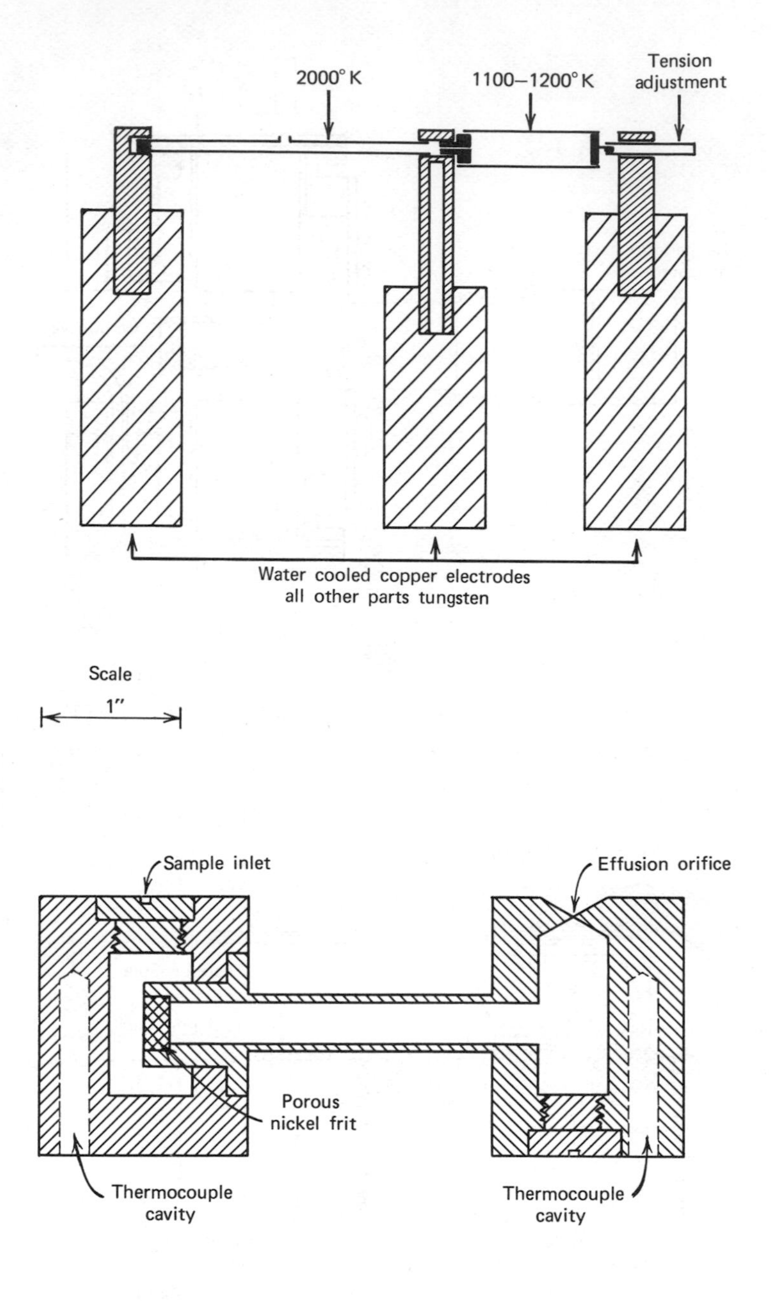
2000° K
1100–1200° K
Tension adjustment
Water cooled copper electrodes
all other parts tungsten
Scale
1″
Sample inlet
Effusion orifice
Porous nickel frit
Thermocouple cavity
Thermocouple cavity

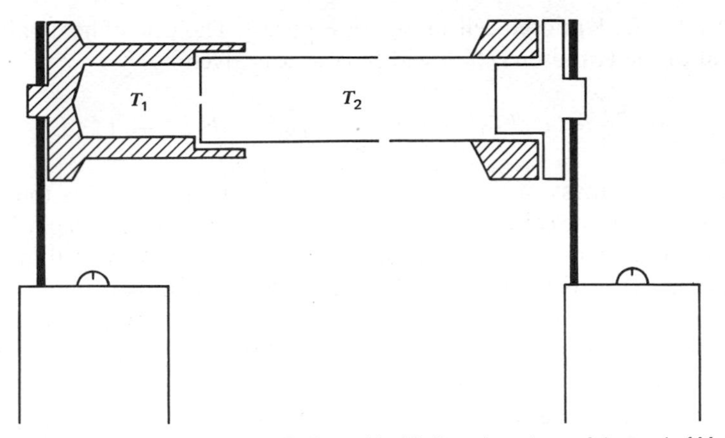

Figure 7. Double Knudsen cells. (*a*) Cell used by Weltner (courtesy of *Journal of Molecular Spectroscopy*). (*b*) Cell used by Berkowitz et al. (courtesy of *Journal of Physical Chemistry*). (*c*) Directly heated cell requiring only two electrodes used by Ozin et al.

where passage of current through the assembly causes the two chambers to heat up to different temperatures, the wall of the second chamber being thinner than that of the first. Vapor effusing from the first is superheated in the second before effusing into the vacuum chamber. In this way polymeric vapors may be broken down into smaller fragments. For example, phosphorus, which vaporizes as P_4 at low temperatures, may be converted to P_2 in the double Knudsen cell. Berkowitz et al. employed the "double oven" shown in Figure 7*b* to break up ethyllithium polymer[23] and LiF polymer[24] into fragments, and Weltner et al.[25] used the double Knudsen cell shown in Figure 7*a* to enrich a BS_2, B_2S_2, and B_2S_3 mixture in the BS_2 component.

Occasionally a metallic species one wishes to vaporize from a metal Knudsen cell alloys with it and thereby destroys it (e.g., Cu in a Ta Knudsen cell, or Si in stainless steel). A liner made of a nonmetallic refractory such as boron nitride or aluminum oxide usually remedies the situation.

8.2 The Temperature of Self-Heated Knudsen Cells and Filaments

When a Knudsen cell or filament is self-heated by passing a current, *i*, through it, a steady-state temperature profile arises as a result of the energy gained by resistive heating and the energy lost through radiation and conduction. This temperature profile, as well as the maximum temperature obtainable for a cell or filament of given geometry, can be calculated theoretically. Such a calculation can be used as a starting point in deciding upon design parameters for these devices, and a brief outline of such an analysis follows.

Consider the Knudsen cell shown in Figure 8. The rate of heat gain by an element of the Knudsen cell of width dx at x is given by

$$mc\frac{dT}{dt} = i^2\rho(T)\frac{dx}{a} + ka\frac{d^2T}{dx^2}\,dx - \varepsilon\sigma b(T^4 - T_R^4)\,dx$$

where m is the mass of the element; c is its heat capacity; i is the current passing through the cell; $\rho(T)$ is the resistivity of the cell material, which is a function of temperature; k is the thermal conductivity of the cell metal; ε is the emissivity of the metal; σ is the Stefan–Boltzmann constant, T_R is the temperature of the surroundings (room temperature); a is the cross-sectional area that equals $2\pi rt$ for a cylindrical Knudsen cell, πr^2 for a solid cylindrical wire filament and wt for a ribbon filament of width w and thickness t; and b is the perimeter of the cell, that is, $2\pi r$ for a cylindrical Knudsen cell or wire filament and $2w$ for a thin ribbon filament.

At steady state, $dT/dt = 0$ and

$$\frac{d^2T}{dx^2} = \frac{\varepsilon\sigma b}{ka}(T^4 - T_R^4) - \frac{i^2\rho(T)}{ka^2} \tag{5}$$

The resistivity of many metals shows a linear temperature dependence over a large temperature range. Accordingly, $\rho(T)$ may be written in the form

$$\rho(T) = \rho_0 + \alpha_T$$

With this expression, equation (5) becomes

$$\frac{d^2T}{dx^2} = \frac{\varepsilon\sigma b}{ka}(T^4 - T_R^4) - \frac{i^2}{ka}(\rho_0 + \alpha T) \tag{6}$$

Equation (6) may be solved numerically by assuming the two boundary conditions $T = T_R$ (or some other known temperature) at $x = \pm l/2$ and $dT/dx = 0$ at $x = 0$. As an example, the temperature profile for a tantalum

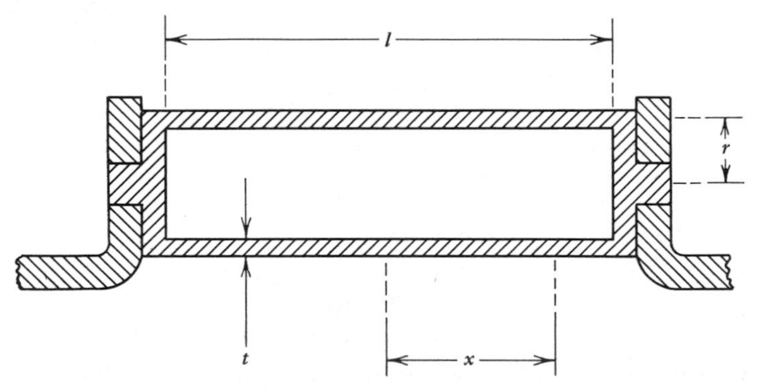

Figure 8. Details of a Knudsen cell used in calculating its temperature profile.

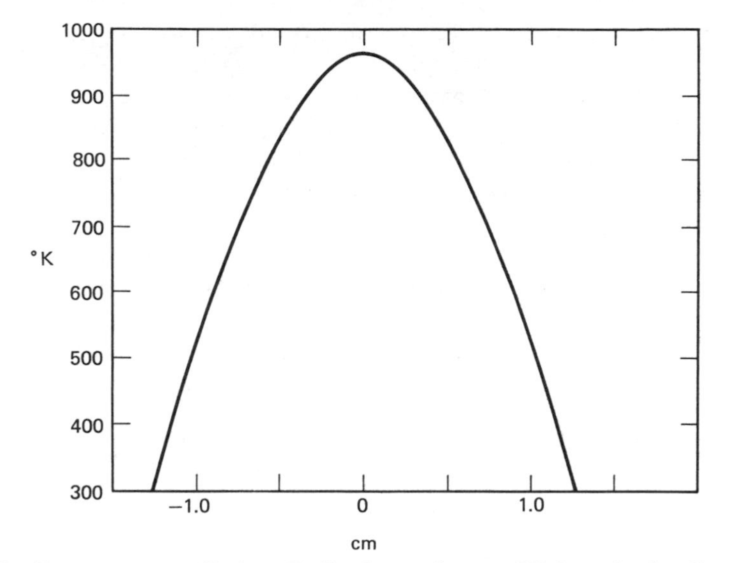

Figure 9. Temperature profile for a Ta Knudsen cell run at 200 Amps having dimensions given in the text.

Knudsen cell 2.54 cm long, 0.635 cm in diameter, and 0.0254 cm wall thickness is plotted in Figure 9. The data given in Table 7 were used in the calculation. Heat conduction by the vapor contained within the Knudsen cell was omitted from the calculation. This could have an important temperature equalizing effect that would "flatten" the temperature profile of the Knudsen cell. Calculations also indicate that the maximum temperature achieved by the Knudsen cell or filament is approximately proportional to the square of the current passed through it. A word of caution is required with regard to graphite Knudsen cells or filaments. Mrozowski[26] and others have shown that the electrical resistivity of polycrystalline graphite depends in a complex way on both temperature and thermal history. For example, a graphite specimen not previously heated to high temperatures was found to have a resistivity of approximately 3.5 mΩ-cm at 1000°C. That value dropped to approximately 0.7 mΩ-cm if the rod was previously heated to 3000°C for 10 min or so. The resistivity of specimens heat-treated in that fashion behaves more acceptably, showing a decrease in resistivity with temperature between 0 and 1000°C and a gradual increase in that quantity thereafter.

8.3 Induction Heating

Knudsen cells can also be heated indirectly by means of an external heating coil or by induction heating. The latter method involves producing a high-power (~15 kW) radio frequency (~300 kHz) field in the region of the

Knudsen cell using a water-cooled coil around the cell that induces eddy currents in the metal cell material, which heats up as a result of dissipative processes within it. Using this procedure, one gains the advantage of not having to pass heater wires into the vacuum chamber at the cost of being restricted to a glass vacuum system at least in the immediate vicinity of the RF source. Snelson, for instance, has described an induction-heated Knudsen cell with which he has produced and studied monomeric LiF,[27] alkaline earth halides,[28] OBCl and OBBr,[29] and other systems. His effusion cell is shown in Figure 10. The induction coils are wrapped around an elongated tube in which the orifice is situated. An effusion barrier can be inserted into the tube to ensure temperature homogeneity. The sample is not directly heated but rather heated by conduction. In this manner the vapors at the point of effusion are at a higher temperature than the material being vaporized.

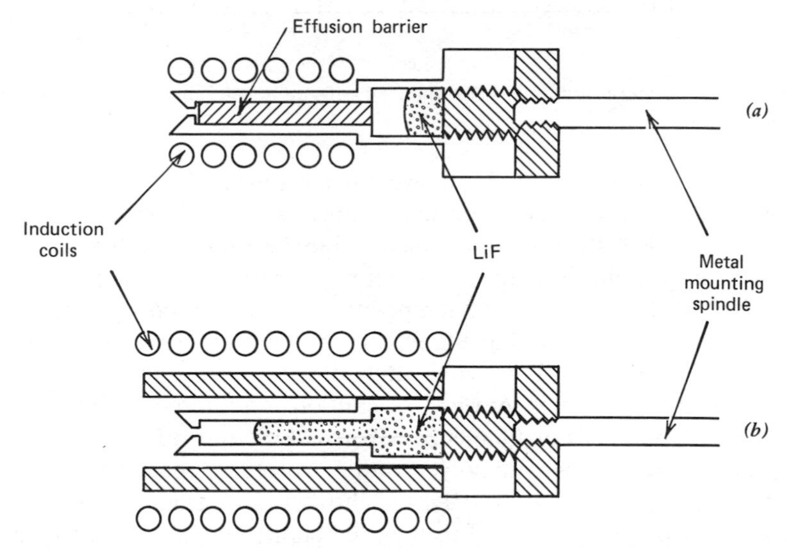

Figure 10. Inductively heated graphite effusion cells as used by Snelson (courtesy of *Journal of Physical Chemistry*).

Weltner and his coworkers have used induction heating to vaporize carbon[30] and pyrolyze B_2O_3 to form B_2O_2.[31] His carbon source is shown in Figure 11. Margrave[32] has described in several articles an extension of induction heating called levitation induction heating whereby metal atoms can be generated from a induction-heated molten metallic sphere and simultaneously levitated above a water-cooled RF coil. Levitation occurs as a result of the magnetization induced within the sample. In this way one produces

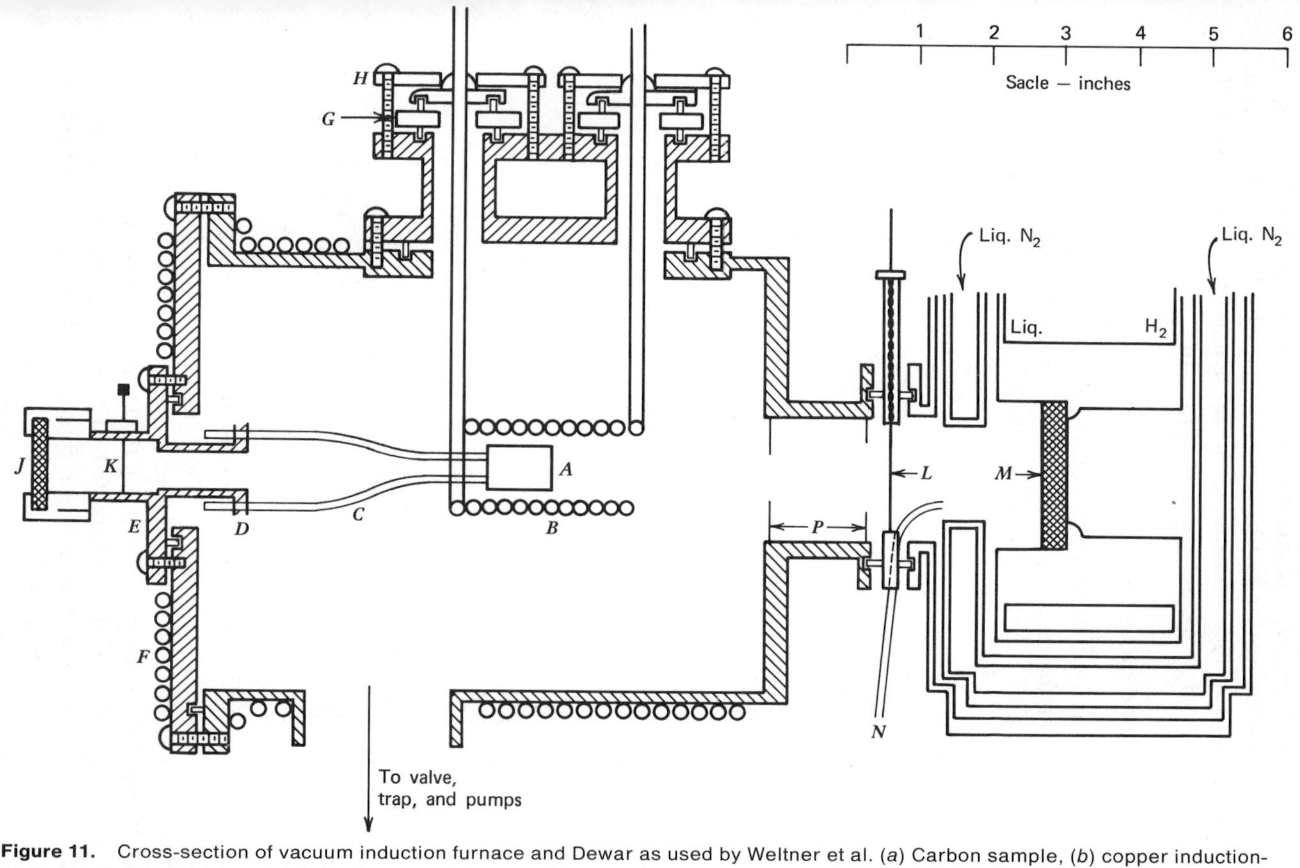

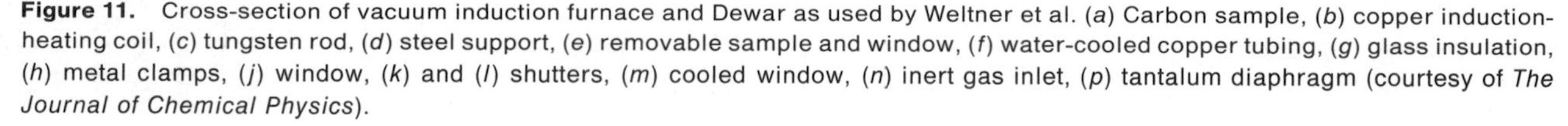
Figure 11. Cross-section of vacuum induction furnace and Dewar as used by Weltner et al. (*a*) Carbon sample, (*b*) copper induction-heating coil, (*c*) tungsten rod, (*d*) steel support, (*e*) removable sample and window, (*f*) water-cooled copper tubing, (*g*) glass insulation, (*h*) metal clamps, (*j*) window, (*k*) and (*l*) shutters, (*m*) cooled window, (*n*) inert gas inlet, (*p*) tantalum diaphragm (courtesy of *The Journal of Chemical Physics*).

metal atoms free from contamination by the support material. It is also a useful technique when working with high-melting metals such as W, Ta, Mo, and Pt. Naturally, this method is limited to those materials that absorb RF radiation, that is, mainly metals and semiconductors.

8.4 Sputtering and Electron Bombardment

Sputtering and electron bombardment also allow one to vaporize metals as well as many other materials without contamination by support material. In the former, atoms or molecules are "sputtered" when a suitable target of the desired material is ion-bombarded. Shirk and Bass,[33] for example, used a device shown in Figure 12 with which they produced and studied Cu, Ag, Cd,

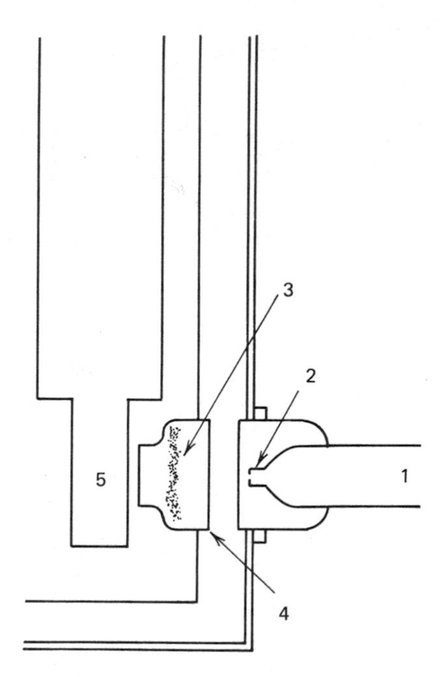

Figure 12. Sputtering source of Shirk and Bass. (1) Discharge tube, (2) nozzle, (3) metal sample, (4) quartz sample-holder, (5) cold window (courtesy of *The Journal of Chemical Physics*).

and Fe atoms as well as SiO monomer. In their deposition system, low-pressure (0.05–1 torr) Xe gas was microwave (2450 MHz) discharged in a quartz discharge tube, and the discharge products (presumably Xe^{+}) were allowed to impinge upon a metal target in the form of a cup or wire grid. The target was either grounded or connected to a high voltage supply. Alternatively, one can use a hollow cathode sputtering device as described by Gruen et al.[34] and shown in Figure 13. In this device a low-pressure inert-gas discharge is maintained between a hollow metal cylindrical cathode and a coaxial metal anode. Metal atoms sputtered from the cathode are swept

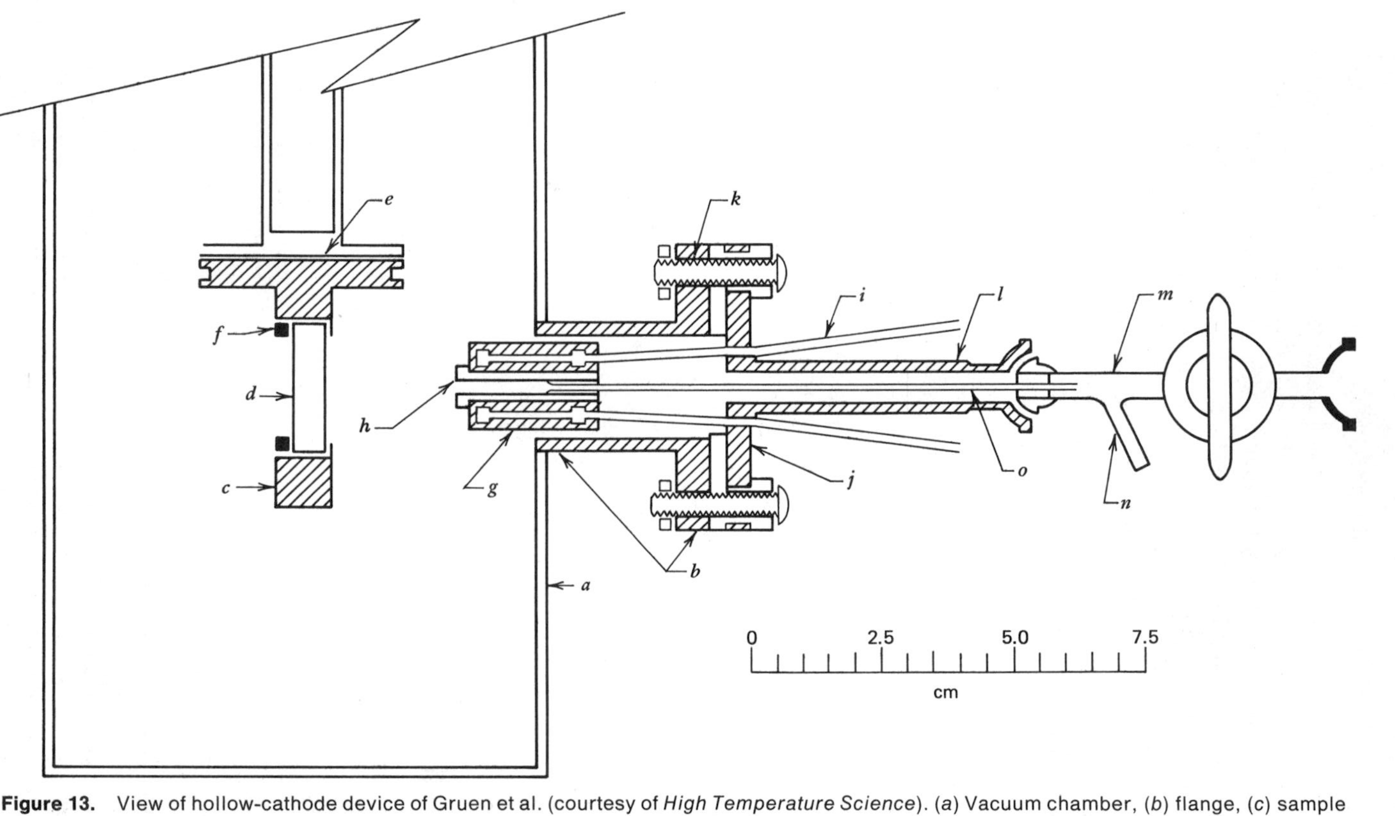

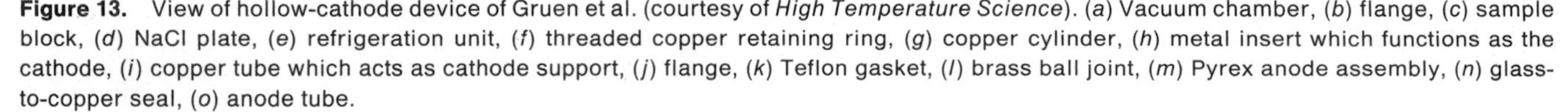

Figure 13. View of hollow-cathode device of Gruen et al. (courtesy of *High Temperature Science*). (*a*) Vacuum chamber, (*b*) flange, (*c*) sample block, (*d*) NaCl plate, (*e*) refrigeration unit, (*f*) threaded copper retaining ring, (*g*) copper cylinder, (*h*) metal insert which functions as the cathode, (*i*) copper tube which acts as cathode support, (*j*) flange, (*k*) Teflon gasket, (*l*) brass ball joint, (*m*) Pyrex anode assembly, (*n*) glass-to-copper seal, (*o*) anode tube.

towards the matrix isolation apparatus by a continuous stream of inert gas or inert gas/reactive gas mixtures, thereby allowing one to observe either pure metal spectra or the spectra of gas-metal cocondensation products.

In electron bombardment vaporization, high-energy (~3 keV) electrons are thermionically produced from a hot tungsten filament and then electrostatically accelerated toward the target material held in a water-cooled crucible. The electron beam is magnetically deflected and focused onto the target by means of a permanent magnet. Commercial electron beam systems are available. Varian Associates,[35] for instance, sells several types of "*e*-gun" sources that they claim can be used to deposit various metallic and nonmetallic substances, including refractories such as Al_2O_3, SiO_2, Ta, Pt, Mo, C, and others. Carstens and Gruen[36] describe an electron-beam sublimator in which the electron beam, emitted by a resistively heated W filament, is accelerated by a 3kV potential, bent through 270°, and focused by a permanent magnet onto a tip of wire made of the desired metal. The wire is stored on a spool and advanced by a bellows drive at the desired rate of vaporization. This device was used to matrix-deposit Ti, Cu, and Pt (Figure 14).

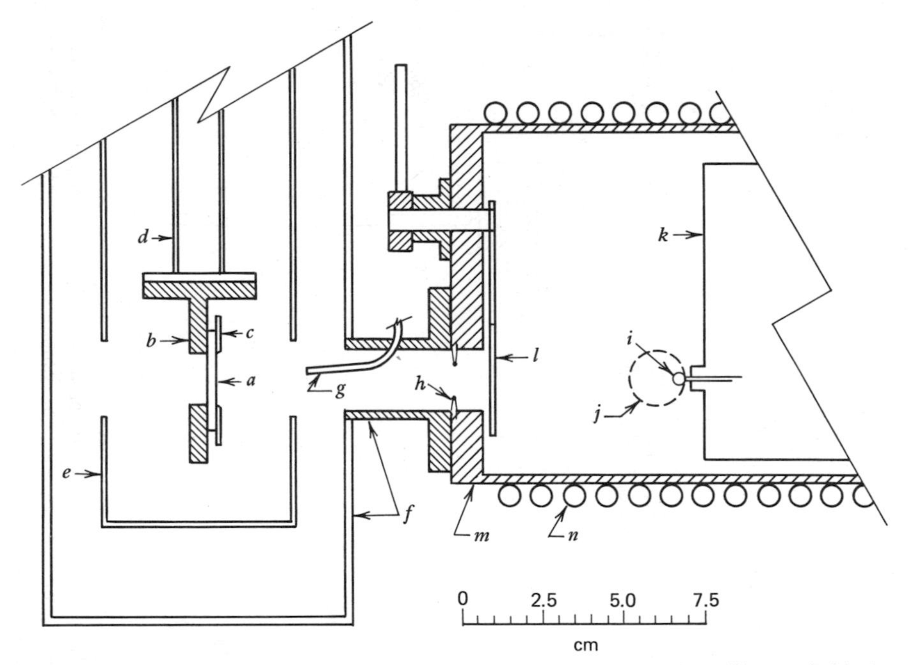

Figure 14. Electron beam sublimator (after Gruen). (*a*) Deposition plate, (*b*) sample block, (*c*) cover plate, (*d*) refrigerator, (*e*) heat shield, (*f*) deposition chamber, (*g*) gas inlet tube, (*h*) monitor plate, (*i*) sample wire and molten bead, (*j*) window (perpendicular to plane of drawing), (*k*) electron beam sublimator cover, (*l*) shutter, (*m*) chamber, (*n*) cooling coils (courtesy of *Reviews in Scientific Instruments*).

It has been pointed out by Skell, Timms,[37] and others that electron-beam evaporation and presumably sputtering give rise to electronically excited atoms. Depending on the dimensions of the apparatus and the lifetime of the excited species, these may or may not decay to their ground states before being condensed upon the cold surface of the matrix-isolation apparatus. The presence of stray high-energy electrons within the apparatus may also bring about chemical processes that would otherwise not have occurred. In an attempt to obviate these problems while at the same time obtaining deposition that is essentially free from contamination by support material, von Gustorf et al.[38] and others have used focused laser beams to vaporize various metals. The method suffers several drawbacks, among which are the inefficiency of vaporization due to the fact that most metals are good reflectors as well as good absorbers of visible radiation, and the geometrical constraints placed on the system by the fact that the window through which the laser beam enters the vacuum chamber must remain uncoated by the material being vaporized.

8.5 Pyrolysis

Aside from vaporizing pure substances, Knudsen cells and hot tubes can be used to pyrolyze substances in order to trap the pyrolysis products. Using a resistively heated Pt tube, Snelson has prepared the radicals, CF_2 and CF_3,[39] by pyrolyzing C_2F_4 and CF_3I, respectively, and CH_3[40] by pyrolyzing CH_3I and $(CH_3)_2Hg$. His apparatus is shown in Figure 15.

Pyrolysis can be used to produce H atoms. Weltner et al.[41] have shown that passing H_2 at 10^{-4} torr through a tungsten Knudsen cell heated to 2800°K produces large quantities of atomic hydrogen. These were used to make alkaline earth and transition metal hydrides by simultaneously vaporizing metal from a second Knudsen cell. The apparatus is shown in Figure 16. Weltner and his coworkers have also produced ZrO, TiO, and HfO[42] by decomposing the dioxides of the same metals at elevated temperatures.

In another study, a series of rare-earth monoxides and dioxides were produced by Weltner and DeKock[43] by pyrolyzing the commonly found oxides in a tungsten cell.

8.6 Knudsen Cell Reactions

A much richer source of interesting species is obtained by reacting two or more substances within a Knudsen cell either by mixing solid substances together or by flowing a gas through a solid heated in a Knudsen cell and allowing the products to effuse out. In this manner, Weltner[44] and others have succeeded in producing the species AlO, AlO_2, and Al_2O_2 by reducing Al_2O_3 with Al in a resistively heated tungsten cell.

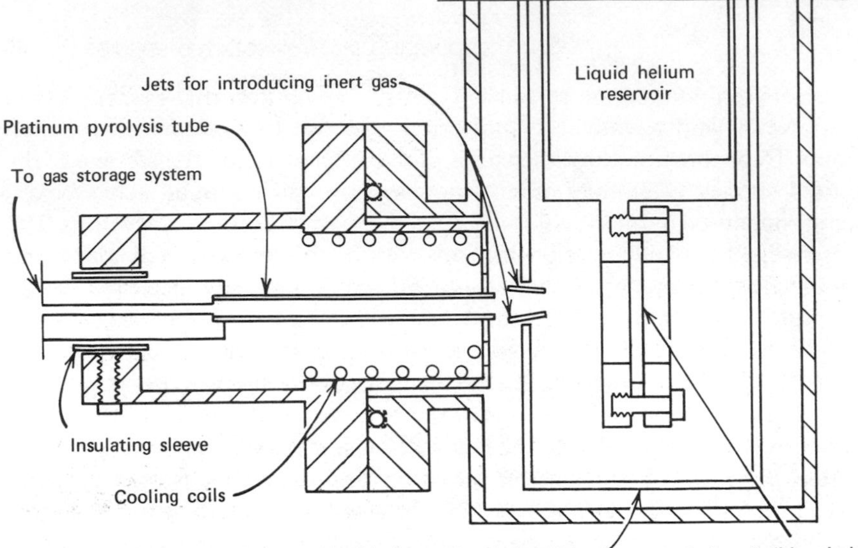

Figure 15. Pyrolysis attachment to matrix cryostat (after Snelson, courtesy of *High Temperature Science*).

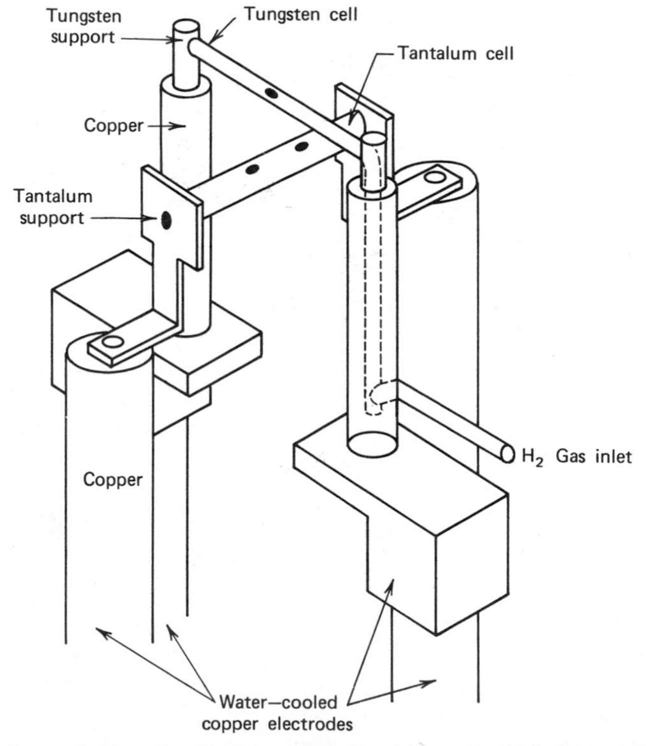

Figure 16. Apparatus for simultaneous production of beams of alkaline-earth metal and hydrogen atoms (after Weltner, courtesy of *The Journal of Chemical Physics*).

By reacting boron and graphite in a graphite cell at about 2650°K, Weltner[45] succeeded in making BC_2, B_2C, and BC. BS_2[25] was made by reacting ZnS with boron in a double oven as mentioned previously.

Snelson[46] has produced AlF by reacting Al with AlF_3, OAlF by reacting AlF_3 with Al_2O_3,[47] and $LiAlF_4$ and $NaAlF_4$ via the reaction of AlF_3 with LiF and NaF,[48] respectively. Solid-solid and solid-liquid reactions have been used to produce atomic species that are normally difficult to obtain. Andrews,[49] for instance, reacted Li metal with CsCl in a Knudsen cell to obtain a stream of Cs atoms, while Scheer and Fine[50] were able to produce beams of various alkali metal atoms, M, by heating a 1:2 molar ratio M_2CrO_4/Si mixture in a stainless steel cell to 650°C.

An ingenious method for vaporizing carbon has been reported by Weltner et al.[51] In order to attain temperatures in excess of 2900°K needed to vaporize carbon, they first converted a tantalum cell to tantalum carbide by heating the cell filled with graphite to 2400°K for 1 hr. The temperature could then be raised as high as 3300°K.

Several other Knudsen cell reactions are listed below:

$$TiF_3 + Ti \rightarrow TiF_2(g) \qquad \text{(Ref. 52)}$$

$$SiO_2 + Si \rightarrow SiO(g) \qquad \text{(Ref. 53)}$$

$$MgF_2 + B_2O_3 \rightarrow BOF \qquad \text{(Ref. 54)}$$

$$P_4N_3 \rightarrow PN(g) \qquad \text{(Ref. 55)}$$

Gas flow through a hot Knudsen cell packed with a reagent has been used by several workers to produce species such as OBCl,[29] obtained by Snelson by flowing gaseous BCl_3 through hot boric oxide. Several other reactions are listed below:

$$GeF_4(g) + Ge \rightarrow GeF_2 \qquad \text{(Ref. 56)}$$

$$SiF_4(g) + Si \rightarrow SiF_2 \qquad \text{(Ref. 57)}$$

$$BF_3(g) + B \rightarrow BF \qquad \text{(Ref. 58)}$$

$$Ni + Cl_2(g) \rightarrow NiCl_2 \qquad \text{(Ref. 59)}$$

$$W(C) + CS_2 \rightarrow CS \qquad \text{(Ref. 60)}$$

where W(C) stands for a carbon-packed tungsten Knudsen cell.

9. PRODUCTION OF SPECIES IN ARCS AND PLASMAS

Whereas hydrogen atoms can be readily produced by thermally dissociating H_2, other atomic species such as O and N are not produced in great quantity under easily attainable pyrolytic conditions, and other means must be resorted to. These include dissociation in discharges and arcs and by photolysis.

A glow discharge can be struck between two metal electrodes sealed in a glass tube, provided that the tube contains a low-pressure gas, and that sufficiently high (ac or dc) voltage is applied between the electrodes. In a self-sustaining dc glow discharge[61] that can be set up in a gas whose pressure is approximately 1 torr, electrons are emitted from the cathode primarily as a result of cation bombardment. These electrons are accelerated to high velocities over a relatively short distance called the cathode dark space, over which most of the voltage applied to the tube is dropped. Electron-cation collisions in the cathode dark space cause an electron avalanche, multiplying greatly the number of "free" electrons present in the tube. Beyond this region the current is carried almost exclusively by electrons, and the major portion of the discharge is taken up by the so-called positive column where the field strength is relatively low (of the order of 1 V/cm), that is, the ionizing and bond-dissociating conditions are relatively mild. Electron bombardment within the discharge can be used to rupture bonds, thereby forming reactive species. For example, CS can be made in this way[62,63] by flowing an Ar/CS_2 mixture through a glow discharge tube similar to that shown in Figure 17.

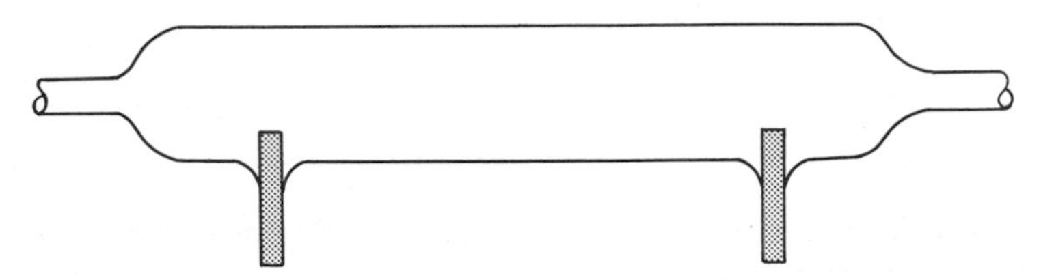

Figure 17. Glow-discharge tube showing gas inlet and outlet and metal electrodes.

In an ac discharge the cathode and anode regions and electron direction are interchanged rapidly; otherwise the discharge remains essentially unchanged. A popular extension of the ac discharge is the so-called electrodeless discharge, which is generated in a low-pressure gas within a glass tube by placing the tube within a metal waveguide that often takes the form of a cylindrical cavity connected to a moderate power microwave source, usually operating at 2450 MHz. The electromagnetic field in the waveguide ionizes the gas within the tube and accelerates the electrons sufficiently to cause an avalanche and form a plasma. By eliminating the electrodes, the microwave discharge bypasses some of the problems encountered in conventional electrically generated plasmas, such as electrode sputtering. It is also generally more convenient than the latter. Electrodeless discharges have also been used to energize vacuum-uv resonance lamps, which are discussed later.

Electrodeless discharges have been used to make Cl atoms[64] by passing a Cl_2/Ar mixture through a discharge tube. Cocondensing the resultant Cl in a Cl_2/Ar matrix yielded Cl_3. Hydrogen and deuterium atoms[65] and N atoms[66] have also been made in this fashion and trapped in low-temperature matrices.

The arc discharge has also been used to generate reactive species. The arc is a self-sustaining discharge of *low* voltage and high current. It differs from a glow discharge by its low cathode voltage drop (about 10 V) and high current density. Arcs are classified as low and high intensity depending on the current conditions. As the current is increased, the arc crater spreads over the entire area of the anode tip and eventually gives rise to a jet of gaseous material known as the tail flame, emanating from the anode surface. It is this jet that characterizes a high-intensity arc. Figure 18 shows this transition. The high current density at the anode leads to very rapid evaporation from its surface and a plasma temperature in the range 9000–10,000°K is obtained. Clearly the possibility of vaporizing a species by doping the carbon anode with the desired substance exists. Kana'an and Margrave[67] have thoroughly

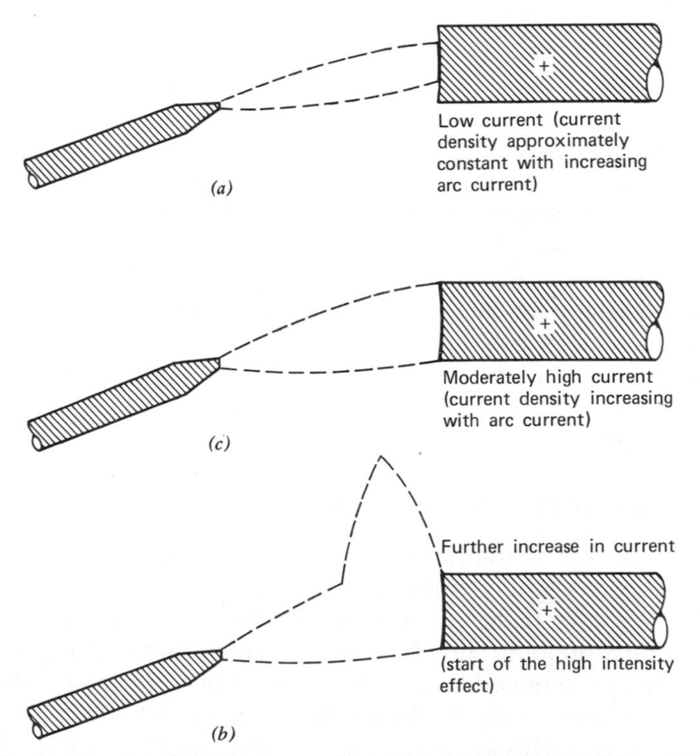

Figure 18. Appearance of a carbon arc as the total current is increased through transition to the high-intensity arc. (*a*) Low-intensity region, (*b*) just below transition point, (*c*) just above transition point (after Margrave and Kana'an, courtesy of *Advances in Inorganic and Radiochemistry*).

reviewed the subject of arc discharges, including various methods of forming arcs and the chemistry possible therewith.

The carbon arc has been used by Skell[68] and his coworkers to produce and study the chemistry of C_1, C_2, and C_3, the three predominant species produced by the carbon arc. Skell found that arced carbon had a different distribution of monatomic, diatomic, and triatomic carbon than did thermally vaporized carbon. A more important difference, and one that characterizes discharge or photolytically produced species from thermally produced ones, is the fact that a large fraction of the carbon atoms produced by the arc was in an electronically excited ^1S or ^1D state as compared to the almost exclusively ^3P ground state of thermally produced carbon. The ^1S and ^1D excited states of atomic carbon are sufficiently long-lived to undergo collisionless flight to the cold surface so that their chemistry as well as that of the ^3P ground state may be investigated. The chemistry of the excited state and ground state can, of course, be radically different. The ^1S state of C_1, for instance, can react with alkanes[69] to yield unsaturated hydrocarbons, for example,

$$C(^1S) + C_3H_8 \longrightarrow \triangle + (CH_3)_3CH + \rangle\!\!= + \;=\!\!\diagdown\!\!=$$

while $C(^3P)$ can be isolated in solid alkane matrices without reacting. Atoms in excited states can also be produced photochemically. ^1D sulfur atoms were produced[70] by photolyzing COS with the Cd resonance line (2288 Å). In high-pressure Ar, the excited sulfur atoms were collisionally deactivated to the ^3P ground state. Again each of the two states displays its own unique chemistry: $S(^1D)$ inserts into paraffinic CH bonds while $S(^3P)$ does not. As a further example we note that excited oxygen (^1D or ^1S) can be produced from N_2O by irradiation with uv from a medium-pressure Hg source.[71] Further discussion of photochemical production of species is carried out in the next section.

9.1 Photochemical Production of Species

Perhaps half of the matrix-isolation spectroscopy literature to date is concerned with photolytically produced species. Briefly, the technique has been applied in two ways. In the first, a suitable starting material is deposited with large quantities of argon, and the matrix is irradiated with an appropriate source of ultraviolet or visible radiation. In this manner Jacox and Milligan[72] produced the radical CF_2 via the reaction, $CF_2N_2 \xrightarrow{h\nu} N_2 + CF_2$, and the ions[73] CCl_3^+ and $HCCl_2^+$ from the reaction, $HCCl_3 \xrightarrow{h\nu} CCl_3^+, HCCl_2^+$, while Rest and Turner[74] produced $Ni(CO)_3$ by photolyzing $Ni(CO)_4$. In the second, two substances are simultaneously deposited in an inert matrix and the matrix photolyzed. So, for instance, Breeze and Turner[75] made $Cr(CO)_5^-$ and $Co(CO)_4^-$ by photoionizing an alkali metal codeposited with $Cr(CO)_6$ and $Co(CO)_3NO$, respectively, in a CO/Ar matrix with medium-pressure Hg lamp radiation.

The product that one gets depends, of course, on the wavelength of the photolyzing radiation. For efficient photolysis to take place, the source must emit photons whose energies are sufficiently large to cause the desired result. So, for instance, the uv component ($\lambda = 184.9$ nm) in a medium-pressure mercury lamp is energetic enough to photodetach a CO from $Ni(CO)_4$ to form $Ni(CO)_3 + CO$, but not energetic enough to ionize CO. The 121.6 nm line of a hydrogen resonance lamp, however, can photoionize CO to CO^+, thereby forming $Ni(CO)_3^-$ as the anion in the above system.[76] Clearly, the photolysis source is of great importance and warrants the following discussion.

9.2 Photolysis Sources

Continuous radiation from the near uv to the near ir can be obtained from a number of commercially available lamps, some of which are listed in Table 8. Lamps such as high-pressure mercury, high-pressure Xe, and deuterium are called continuum sources because they emit radiation at all wavelengths in the range indicated in Table 8. They all, however, have intense lines superimposed over the continuous background. Should only a specific portion of the spectrum be desired, these lamps have to be used with suitable filters or monochromators. Several inexpensive monochromators are sold commercially for that purpose. Low- and medium-pressure mercury lamps, on the other hand, have much of their power concentrated in the 184.9 and 253.7 nm lines of Hg.

Table 8. The Spectral Range in the Output of Several Lamps

Type	Range (nm)
High-pressure Hg	245–near ir
High-pressure Xe	150–900
High-pressure Xe and Hg	150–1400
Deuterium	170–400
Xe	147–210
Kr	125–180
Ar	107–160
Ne	75–100
C arc	300–ir

For vacuum-uv photolysis, one usually uses a resonance lamp. These are available in two types, closed and flow. In the closed resonance lamp, a suitable gas such as H_2 is sealed in a pyrex tube with a getter such as Ba. A LiF window is glued to one end of the tube, and a discharge is struck in the tube by placing it within a concentric microwave cavity to which a microwave source (usually of the order of 100 W) is attached. A schematic of such a source is shown in Figure 19*a*.

The flow resonance lamp is essentially the same except that a continuous flow of gas is used. This obviates the need for a getter and allows optimization of the output conditions. It also allows one to change gases easily, thereby changing the wavelength of the radiation output.

Several resonance lamp designs have been published. Warneck[77] describes a microwave-powered H_2 lamp with which he obtained up to 10^{16} photons/

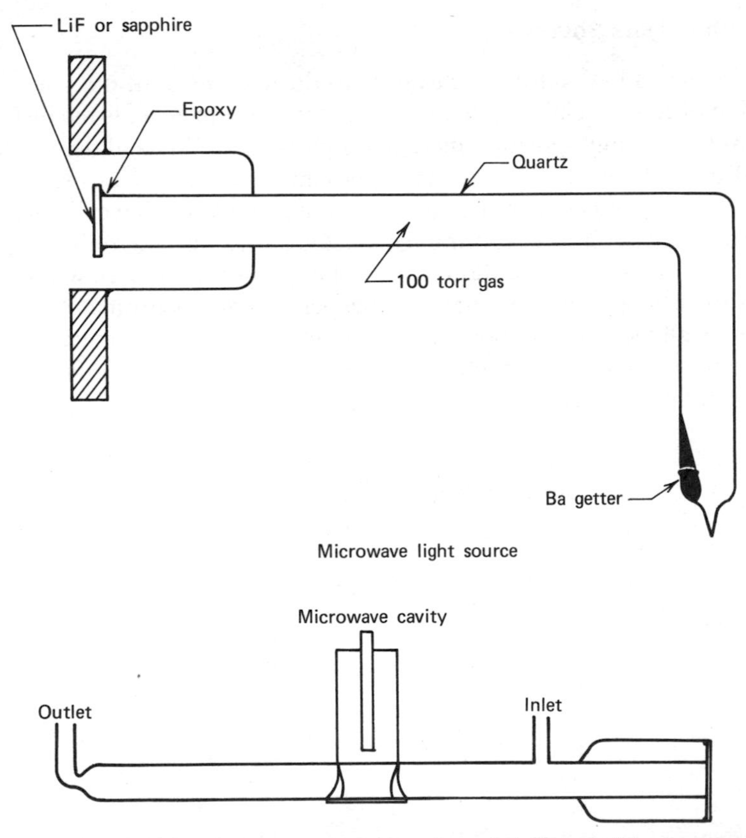

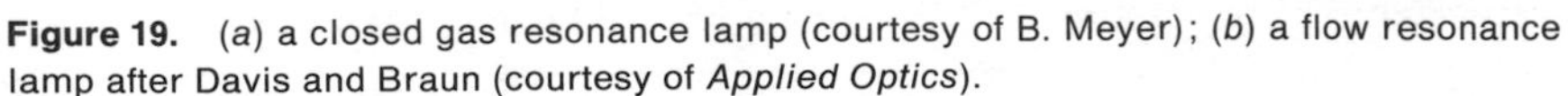

Figure 19. (*a*) a closed gas resonance lamp (courtesy of B. Meyer); (*b*) a flow resonance lamp after Davis and Braun (courtesy of *Applied Optics*).

sec in the 1100–1850 Å spectral region. He found that the design of the waveguide placed around the discharge tube as well as the hydrogen pressure greatly influence the spectral content of the output. Davis and Braun[78] describe a flow-type resonance lamp that they have used to obtain resonance lines from H, Xe, Kr, O, N, Cl, Br, C, and S. A schematic of their lamp is shown in Figure 19*b*, while Table 9 lists the wavelengths and intensities of the lines obtained from various gas mixtures. A microwave source often used is a 2450 MHz, 125 W commercial instrument such as the Microtherm Unit (Model CMD–10) manufactured by Raytheon Co.

Proper microwave cavity design is essential in obtaining efficient vacuum-uv output from these lamps, and several good designs have been published,[79] many of which are commercially available.

Table 10 lists some species generated by photolysis.

In closing, it should be pointed out that occasionally photolysis takes place unexpectedly as a result of radiation from the source of the spectrometer when the spectrum of the matrix-isolated species is recorded. So, for instance, Turner et al.[89] noticed that matrix-isolated $Fe(CO)_4$ formed in CO/Ar matrices by photolyzing $Fe(CO)_5$ would “reverse” photolyze to $Fe(CO)_5$ in the ir spectrometer beam. Placing a germanium filter between the source and sample was found to halt the process, which was presumably induced by the uv or visible portion of the source radiation.

10. MEASURING RATES OF DEPOSITION

Matrix isolation usually involves the cocondensation of a matrix gas with some other species that is often not gaseous at room temperature if it exists at all. In performing these experiments, then, one requires a method for metering the matrix gas flow and guest species deposition rate. These are discussed in turn.

10.1 Gas Flow

Gas flow can be adequately controlled by means of a fine needle valve. The valve material should be selected in accordance with the corrosive properties of the gases that will flow through it. These valves can be bought with a micrometer handle so that reproducible settings can be achieved. The valve can be calibrated by measuring the rate of pressure drop in a bulb of known volume being pumped through the valve for various valve settings. The gas flow is a function of the pressure in the bulb. Fortunately, so little gas is required during any one matrix experiment that the bulb pressure remains essentially constant during the run. And, since the flow of gas through the valve roughly follows Poiseuille’s law, the flow rate is approximately proportional to the pressure difference across the valve.

Table 9. Atomic Lines Produced by Microwave Excitation in a Flow Lamp

Atomic Species	Gas Mixture	Emission Line (Å)	Transition	Relative Intensity	Absolute Intensity × 10^{-15}
H	2% H_2 in He	1215.7	$(2p)^2P_{1/2} \rightarrow (1s)^2S_{1/2}$	—	0.76[b]
Xe	3% Xe in He	1469.6	$(5p^56s)^2P_{3/2} \rightarrow (5p^6)^1S_0$	—	1.00[b]
Kr	3% Kr in He	1235.8	$(4p^55s)^2P_{3/2} \rightarrow (4p^6)^1S_0$	1.00	1.20[b]
		1164.9	$P_{1/2} \rightarrow S_0$	0.20	
O	10% O_2 in He	1302.2	$(2p^33s)^3S_1^0 \rightarrow (2p^4)^3P_2$	0.61	0.20[b]
		1304.9	$^3S_1^0 \rightarrow {}^3P_1$	0.97	
		1306.0	$^3S_1^0 \rightarrow {}^3P_0$	1.00	
N	1% N_2 in He	1745.2	$(2p^23s)^2P \rightarrow (2p^3)^2P^0$	0.78	1.40[c]
		1742.7		1.00	
		1494.7	$^2P \rightarrow {}^2D^0$	0.15	0.35[b]
		1492.6		0.30	
		1411.9	$^2D \rightarrow {}^2P^0$	0.06	—
Cl	0.1% Cl_2 in He	1396.5	$(3p^44s)^4P_{3/2} \rightarrow (3p^5)^2P_{1/2}^0$	0.35	1.04[b]
		1389.9	$^4P_{1/2} \rightarrow {}^2P_{1/2}^0$	1.00	
		1379.6	$^4P_{3/2} \rightarrow {}^2P_{3/2}^0$	0.13	
		1363.5	$^2P_{3/2} \rightarrow {}^2P_{1/2}^0$	0.17	
		1351.7	$^2P_{1/2} \rightarrow {}^2P_{1/2}^0$	0.05	
		1347.3	$^2P_{3/2} \rightarrow {}^2P_{3/2}^0$	0.03	

Br	0.1% Br_2 in He	1633.6	$(4p^45s)^4P_{3/2} \rightarrow (4p^5)^2P^0_{1/2}$	1.00	1.20[c]
		1582.4	$^4P_{1/2} \rightarrow {}^2P^0_{1/2}$	0.50	—
		1576.5	$^4P_{5/2} \rightarrow {}^2P^0_{3/2}$	0.92	—
		1540.8	$^4P_{3/2} \rightarrow {}^2P^0_{3/2}$	0.25	—
		1575.0	$^2P_{3/2} \rightarrow {}^2P^0_{3/2}$	0.46	—
		1531.9	$^2P_{1/2} \rightarrow {}^2P^0_{1/2}$	0.17	—
		1488.6	$^2P_{3/2} \rightarrow {}^2P^0_{1/2}$	0.15	—
		1449.9	$^2P_{1/2} \rightarrow {}^2P^0_{3/2}$	0.03	—
		1384.6	$^4P_{1/2} \rightarrow {}^2P^0_{1/2}$	0.01	—
C	1% CH_4 in He	1930.9	$(2s^22p^3s)^1P^0_1 \rightarrow (2s^22p^2)^1D_2$	1.00	0.98[c]
		1658.1			
		1657.0	$^3P^0 \rightarrow {}^3P$	0.52	—
		1656.3			
		1560.3			
		1560.7	$^3D_0 \rightarrow {}^3P$	0.58	—
		1561.4			
S	0.2% H_2S + He	1914.9	$(3p^34s)^5S^0_2 \rightarrow (3p^4)^3P_1$	0.16	—
		1900.3	$^5S^0_2 \rightarrow {}^3P_2$	0.52	—
		1826.3	$^3S^0_1 \rightarrow {}^3P_0$	1.00	0.50[c]
		1820.4	$^3S^0_1 \rightarrow {}^3P_1$	0.64	—
		1807.3	$^3S^0_1 \rightarrow {}^3P_2$	0.18	—
		1667	$^1D^0_2 \rightarrow {}^1D_2$	0.13	—

[a] D. Davis and W. Braun, *Appl. Opt.*, **7**, 2071 (1968). Reproduced by permission.
[b] Line intensity determined by CO_2 actinometry.
[c] Line intensity determined from monochromator traces.

Table 10. A Selection of Matrix Species Generated by Photolysis

Species	Reaction	Ref.
A. Medium pressure Hg		
$Ni(CO)_3$	$Ni(CO)_4 \xrightarrow{h\nu} Ni(CO)_3 + CO$	74
CF_2	$CF_2N_2 \xrightarrow{h\nu} CF_2 + N_2$	72
GeH_2Br, $GeHBr$	$GeH_3Br \xrightarrow{h\nu} GeH_2Br, GeHBr$	80
NCO	$HNCO \xrightarrow{h\nu} NCO + H$	81
PF_2	$PF_2H \xrightarrow{h\nu} PF_2 + H$	82
$Cr(CO)_5^-$	$Cr(CO)_6 + \text{alkali metal} \xrightarrow{h\nu} Cr(CO)_5^-$	75
$Co(CO)_4^-$	$Co(CO)_3NO + \text{alkali metal} \xrightarrow{h\nu} Co(CO)_4^-$	75
Cyclobutadiene	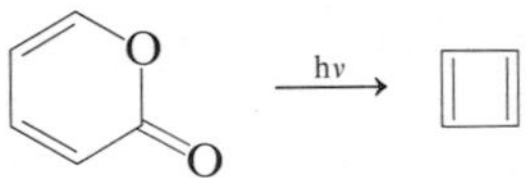 (2-pyrone structure, O, O) $\xrightarrow{h\nu}$ (cyclobutadiene structure) $+ CO_2$	83
B. Resonance lamps		
CCl_3^+	$HCCl_3 \xrightarrow[\text{(H resonance)}]{1216\ \text{Å}} CCl_3^+$	84
$HCCl_2$, $HCCl_2^+$, $HCCl_2^-$	$HCCl_3 \xrightarrow[\text{(Ar resonance)}]{1067\ \text{Å}} \left\{ HCCl_2, HCCl_2^+, HCCl_2^- \right\}$	84
GeH_3	$GeH_4 \xrightarrow[\text{(H resonance)}]{1216\ \text{Å}} GeH_3$	85
CuO	$Cu + O_2/Xe \xrightarrow[\text{(Xe resonance)}]{1470\ \text{Å}} CuO + O$	86
NCO NCO	$\left\{ HNCO/Ar, HN_3/CO, N_2O/HCN/Ar \right\} \xrightarrow[\text{(H resonance)}]{1216\ \text{Å}} NCO$	87
CCl_2	$CH_2Cl_2 \xrightarrow[\text{(H resonance)}]{1216\ \text{Å}} CCl_2 + H_2$	88
$Ni(CO)_3^-$ $Cr(CO)_5^-$	$\left\{ Ni(CO)_4/Ar, Cr(CO)_6/Ar \right\} \xrightarrow[\text{(H resonance)}]{1216\ \text{Å}} \left\{ Ni(CO)_3^- + CO^+, Cr(CO)_5^- + CO^+ \right\}$	76

More direct ways of measuring gas flows exist that do not require the assumption that the valve handle settings do not change from run to run (a poor assumption, especially for spring-loaded needle valves). The simplest flow gauge is a vertically mounted glass tube containing a small light bead that is raised by the flowing gas to a level determined by the flow rate. These devices can be purchased from compressed-gas supply companies. They are, however, too coarse for most matrix experiments unless unusually large matrix flow rates are required for prolonged periods of time. Electronic flow gauges, usually referred to as mass flow meters, may be used, although they too seem to be a bit on the coarse side. In a mass flow meter, the moving gas moves a light metallic core whose position is sensed inductively and converted to an electrical signal displayed on a meter. These instruments can also be bought from most compressed gas companies. At the time of this writing, the most sensitive meter has a 0–5 sccm, that is, approximately 13 mmole/hr, full-scale range, which would make it adequate for average matrix runs but too coarse for runs involving slow flows. Bernoulli gauges can also be used: in these the pressure drop across a capillary through which the gas is flowing is recorded. The pressure drop is measured manometrically and is a direct function of the flow. Microflow meters of this sort are available commercially, or one can use a differential pressure transducer similar to that described immediately below to accomplish the same result.

A method that we have used successfully to measure gas flows requires measuring accurately the small decrease in pressure due to gas outflow from a bulb of known volume during deposition. In order to do so accurately, we use a pressure transducer manufactured by Validyne Corp.[90] This device measures differential (or absolute) pressure by detecting the small movement of an elastic diaphragm due to the pressure gradient across it. The transducer comes in several selectable pressure ranges, and the meter range can be selected to 1% of the transducer's full scale. This feature together with a 10,000/1 zero suppression allows small pressure changes to be measured with high accuracy. So, for instance, with a 0–500 torr transducer, a 0–5 torr full scale can be displayed and flow rates as low as 0.01 mmole/hr can be monitored.

10.2 Deposition of Guest Species

As stated previously, the rate of effusion of high-temperature species from a Knudsen cell can be determined directly from the cell orifice size and the vapor pressure of the species within the cell. Clearly the vapor-pressure–temperature relation must be known for the substance of interest. Data for a wide range of substances including the elements have been tabulated.[91] Because the vapor pressure of most substances depends so critically on temperature, the Knudsen cell temperature must be accurately known and the cell should be uniform in temperature. The latter criterion is better met with an indirectly heated rather than a directly heated cell, while the temperature can be measured with a thermocouple spot-welded to the cell or with an optical pyrometer if the cell is incandescent and a viewing port available.

Should vapor pressure data not be available, one can determine the rate of efflux by weighing the cell beforehand and again after effusion has been allowed to proceed for a length of time. By doing so at several temperatures, one is essentially measuring the vapor-pressure curve.

The rate of deposition of opaque substances such as metals from a filament can be measured by the rate of darkening of a viewing port situated so as to receive some of the metal flux. By keeping the filament temperature constant, the filament acts simultaneously as a light source and heater. A photodetector placed outside the deposition vacuum cell measures the time rate of decrease of light transmission, which is related to deposit thickness and extinction coefficient. Refractive indices and extinction coefficients are available for many metals, while the expressions relating transmittances to deposit thickness, refractive index, and extinction coefficient are given by Heavens[92] and others. Since the sampling location and the matrix are not one and the same, one must determine the geometrical factor relating the deposit flux at the two locations. Deposition rates can also be deduced spectroscopically from the absorbance of bands associated with the substance after it has been deposited at a constant rate for a known length of time, provided that extinction data are available for it, or the species to be deposited may be reacted with a matrix material giving rise to known products whose extinction coefficients are known. For example, the deposition rate of Ni vaporized from a filament may be deduced from the absorbance of the resulting $Ni(CO)_4$ band when the metal atoms are cocondensed with CO. As a further example, Skell and his coworkers[68] determined the relative abundance of C_1, C_2, C_3, and C_4 in carbon vapor by cocondensing the vapor with Cl_2 and determining the relative concentration of the chlorocarbons produced.

One of the most direct methods for measuring the deposition rate of substances with high melting points is to weigh the deposit with a quartz crystal microbalance.[93] This technique makes use of the fact that the resonance frequency of a quartz crystal wafer (the so-called AT cut is the one most often used) vibrating in a thickness shear mode diminishes in a predictable way when mass adheres to one or both of its faces. The mass frequency relationship is given by[94]

$$\frac{\delta f}{f} = \frac{k\,\delta m}{A} \tag{7}$$

where δf is the magnitude of the frequency decrease, f the resonance frequency, δm the mass increment causing the frequency decrease, A the crystal face area, and k is a constant that depends on, among other things, the thickness of the crystal and the density of quartz. For a 5 MHz fundamental AT-cut crystal, k is approximately 12 cm^2/g. Thus, assuming a crystal area of 1.3 cm^2 (as in those commonly available commercially), equation (7) predicts a sensitivity of 20 ng/Hz. A frequency stability of approximately 1 Hz over several hours is possible routinely, while greater stability is available with a little care.

Commercially available crystals are approximately 1.75 cm in diameter and 0.025 cm in thickness. They are provided with either aluminum or gold electrodes deposited on the two faces and mounted with two wire clips onto a two-prong electrical mount (Figure 20). Only a simple oscillator circuit

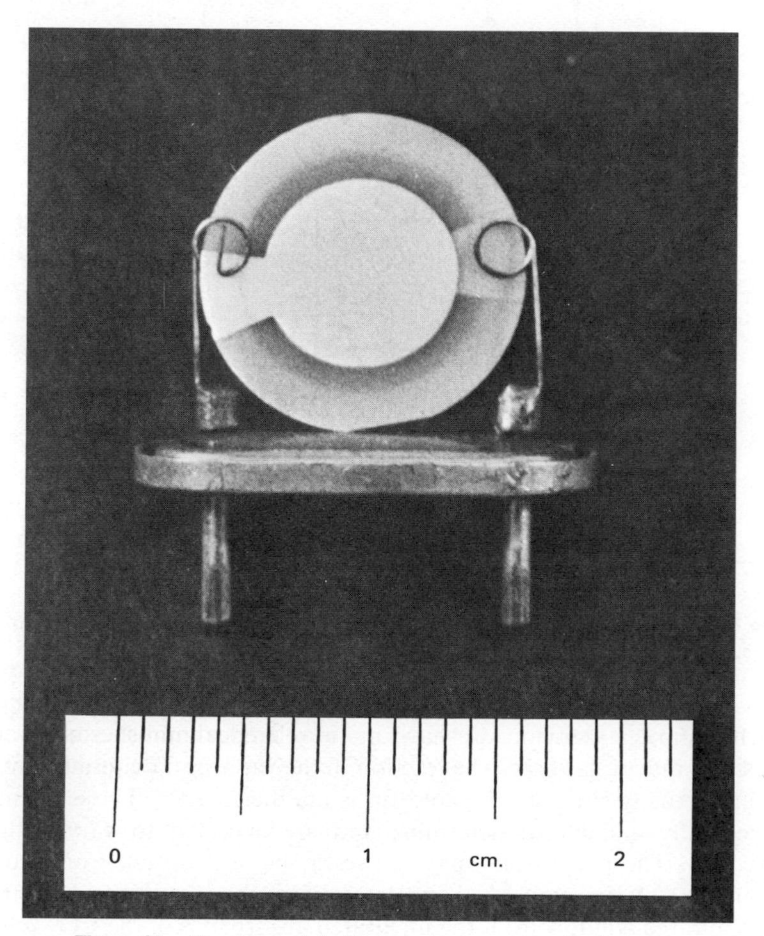

Figure 20. Quartz crystal with mount as used in microbalance.

as shown in Figure 21 is required to cause the crystal to oscillate. The resonance frequency is measured on a commercial frequency counter. Because of its small size the crystal can be incorporated directly into the deposition cell so that it receives the backward-going flux of material being deposited from either a filament or Knudsen cell while simultaneously depositing on the cold tip in the forward direction. Here too a geometric factor must be determined relating the material flux received by the cold tip and crystal. This is most easily done by placing a second crystal in place of the cold tip in a calibration run.

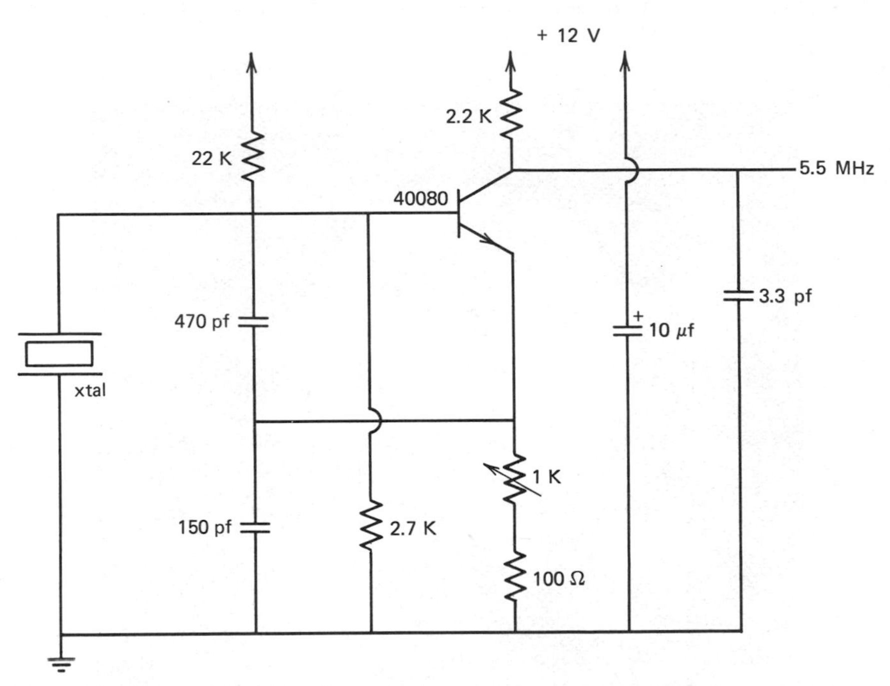

Figure 21. Schematic for 5.5 MHz oscillator used to drive quartz crystal in microbalance applications.

In the above discussion, we have made reference almost exclusively to deposition rate or gas flow. The fraction of the vaporized material or matrix gas that actually sticks to the cold tip is another matter. These quantities are generally difficult to determine and are expected to vary with the matrix gas. The corrections may, however, be substantial. For example, condensing CO_2 on a cold plate and determining the quantity actually frozen on the infrared window from the integrated absorbance of the CO_2 infrared band, it was determined[95] that only approximately 15% of the CO_2 passed into the vacuum shroud condensed on the ir window. Clearly the sticking fraction depends on the experimental design as well as on the sticking coefficient (i.e., the fraction of molecules that stick upon the first collision) of the gas.

11. CONCLUSION

In this chapter we have limited ourselves to species produced outside the matrix and the technology needed for their production in the matrix. The chemistry and character of these species after isolation form the substance of the remainder of the book.

Needless to say, we have discussed only a small fraction of the methods that researchers in matrix isolation have used to generate molecules and follow their chemistry. New methods are published at regular intervals. Andrews,[96] for example, recently described a technique for generating ions such as CCl_3^+ and Cl_3^+ in matrices by proton bombardment of Ar/CCl_4 matrices and simultaneous charge neutralization by thermionic electrons. Creative ideas like these seem to be the rule in the matrix-isolation methods of the past. The future does not seem to hold less promise.

REFERENCES

1. F. A. Mauer, in A. M. Bass and H. P. Broida (Eds.), *Formation and Trapping of Free Radicals*, Chap. 5, Academic, New York, 1960; L. J. Schoen, L. E. Kuentzel, and H. P. Broida, *Rev. Sci. Instrum.*, **29**, 663 (1958); W. J. Durig and I. L. Mador, *Rev. Sci. Instrum.*, **23**, 421 (1952); P. Rowland Davies, *Discuss. Faraday Soc.*, **48**, 181 (1969).
2. R. F. Brown and A. C. Rose-Innes, *J. Sci. Instrum.*, **33**, 420 (1956).
3. Air Products and Chemicals Inc., P.O. Box 538, Allentown, Pa. 18105.
4. Oxford Instrument Corp., 100 Cathedral St., Annapolis, Md. 21401.
5. H. P. Hernandez, J. W. Mark, and R. D. Watt, *Rev. Sci. Instrum.*, **28**, 528 (1957).
6. J. M. Geist and P. K. Lashmet, *Advan. Cryog. Eng.*, **5**, 324 (1960); ibid., **6**, 73 (1961).
7. Cryogenic Technology Inc., Kelvin Park, 266 Second Ave., Waltham, Mass. 02154.
8. Cryomech Inc., 314 Ainsley Dr., Syracuse, N. Y. 13210.
9. W. N. Lawless, *Rev. Sci. Instrum.*, **42**, 561 (1971); L. G. Rubin and W. N. Lawless, ibid., **42**, 571 (1971).
10. D. L. Swartz, *Phys. Today*, **75**, 27 (1974).
11. M. A. Brown, *Cryogenics*, **10**, 439 (1970).
12. J. Kopp and G. A. Stack, *Cryogenics*, **11**, 22 (1971).
13. W. Little, *Can. J. Phys.*, **37**, 337 (1959).
14. W. H. Warren, Jr., and W. G. Bader, *Rev. Sci. Instrum.*, **40**, 180 (1969).
15. B. Miles, *Angew. Chem.*, **7**, 507 (1968).
16. G. L. Pollack, *Rev. Mod. Phys.*, **36**, 748 (1964).

17. G. K. White and S. B. Woods, *Phil. Mag.*, **3**, 785 (1958).

18. L. B. Knight, Jr., W. C. Easley, and W. Weltner, Jr., *J. Chem. Phys.*, **54**, 1610 (1971).

19. D. F. McIntosh, M. Moskovits, and G. A. Ozin, *J. Amer. Chem. Soc.*, (1976) in press.

20. C. S. Barrett and L. Meyer, *J. Chem. Phys.*, **42**, 107 (1965).

21. G. O. Jones and J. M. Woodfine, *Proc. Phys. Soc.*, **86**, 101 (1965).

22. H. E. Hallam and G. F. Scrimshaw, in H. E. Hallam (Ed.), *Vibrational Spectroscopy of Trapped Species*, Wiley, New York, 1973.

23. J. Berkowitz, D. A. Bafus, and T. L. Brown, *J. Phys. Chem.*, **65**, 1380 (1961).

24. J. Berkowitz, H. A. Tasman, and W. A. Chupka, *J. Chem. Phys.*, **36**, 2170 (1962).

25. J. M. Brom, Jr., and W. Weltner, Jr., *J. Mol. Spectrosc.*, **45**, 82 (1973).

26. S. Mrozowski, *Phys. Rev.*, **85**, 609 (1952).

27. A. Snelson, *J. Phys. Chem.*, **73**, 1919 (1969).

28. A. Snelson, *J. Phys. Chem.*, **70**, 3208 (1966).

29. A. Snelson, *High Temp. Sci.*, **4**, 318 (1972).

30. W. Weltner, Jr., P. N. Walsh, and C. L. Angell, *J. Chem. Phys.*, **40**, 1299 (1964).

31. W. Weltner, Jr., and J. R. W. Warn, *J. Chem. Phys.*, **37**, 292 (1962).

32. A. K. Chaudhuri, D. W. Connell, L. A. Ford, and J. L. Margrave, *High Temp. Sci.*, **2**, 203 (1970).

33. J. S. Shirk and A. M. Bass, *J. Chem. Phys.*, **49**, 5156 (1963).

34. D. H. W. Carstens, J. F. Kozlowski, and D. M. Gruen, *High Temp. Sci.*, **4**, 301 (1972).

35. Varian Vacuum Division, 611 Hansen Way, Palo Alto, Calif. 94303.

36. D. H. Carstens and D. M. Gruen, *Rev. Sci. Instrum.*, **42**, 1194 (1971).

37. P. Timms, private communication.

38. E. A. K. von Gustorf, O. Jaenicke, O. Wolfbeis, and C. R. Eady, *Angew. Chem.* (*Int. Edit.*), **14**, 278 (1975).

39. A. Snelson, *High Temp. Sci.*, **2**, 70 (1970).

40. A. Snelson, *J. Phys. Chem.*, **74**, 537 (1970).

41. L. B. Knight, Jr., and W. Weltner, Jr., *J. Chem. Phys.*, **54**, 3875 (1971).

42. W. Weltner, Jr., and D. McLeod, Jr., *J. Phys. Chem.*, **69**, 3488 (1965).

43. R. L. DeKock and W. Weltner, Jr., *J. Phys. Chem.*, **75**, 514 (1971).

44. L. B. Knight and W. Weltner, Jr., *J. Chem. Phys.*, **55**, 5066 (1971).

45. W. C. Easley and W. Weltner, Jr., *J. Chem. Phys.*, **52**, 1489 (1970).

46. A. Snelson, *J. Phys. Chem.*, **71**, 3203 (1967).

47. A. Snelson, *High Temp. Sci.*, **5**, 77 (1973).

48. S. J. Cyvin, B. N. Cybin, and A. Snelson, *J. Phys. Chem.*, **75**, 2609 (1971).

49. E. C. Spiker and L. Andrews, *J. Chem. Phys.*, **58**, 713 (1973).

50. M. D. Scheer and J. Fine, *J. Chem. Phys.*, **37**, 107 (1962).

51. K. R. Thompson, R. L. DeKock, and W. Weltner Jr., *J. Amer. Chem. Soc.*, **93**, 4688 (1971).

52. J. L. Margrave, J. W. Hastie, and R. H. Hauge, *J. Chem. Phys.*, **51**, 2648 (1969).

53. J. S. Anderson and J. S. Ogden, *J. Chem. Phys.*, **51**, 4189 (1969).

54. A. Snelson, *High Temp. Sci.*, **4**, 141 (1972).

55. P. L. Timms, *Angew. Chem.* (*Int. Edit.*), **14**, 273 (1975).

56. H. Huber, E. P. Kündig, G. A. Ozin and A. Vander Voet, *Can. J. Chem.*, **52**, 95 (1974).

57. J. M. Bassler, P. L. Timms, and J. L. Margrave, *Inorg. Chem.*, **5**, 729 (1966).

58. P. L. Timms, *J. Amer. Chem. Soc.*, **89**, 1629 (1967).

59. D. E. Milligan and M. E. Jacox, *J. Chem. Phys.*, **51**, 4143 (1969).

60. L. P. Blanchard and P. LeGoff, *Can. J. Chem.*, **35**, 89 (1957).

61. A. M. Howatson, *An Introduction to Gas Discharges*, Pergammon, New York, 1965.

62. H. Huber, M. Moskovits, and G. A. Ozin, unpublished work.

63. K. J. Klabunde, C. M. White, and H. F. Efner, *Inorg. Chem.*, **13**, 1778 (1974).

64. L. Y. Nelson and G. C. Pimentel, *J. Chem. Phys.*, **47**, 3671 (1967).

65. V. E. Bondybey and G. C. Pimentel, *J. Chem. Phys.*, **56**, 3832 (1972).

66. H. P. Broida and M. Peyron, *J. Chem. Phys.*, **28**, 725 (1958).

67. A. S. Kana'an and J. L. Margrave, *Advan. Inorg. Radiochem.*, **6**, 143 (1964).

68. P. S. Skell, J. J. Havel, and M. J. McGlinchey, *Account Chem. Res.*, **6**, 97 (1973).

69. P. S. Skell and R. R. Engel, *J. Amer. Chem. Soc.*, **88**, 4883 (1966).

70. A. R. Knight, O. P. Strausz, S. M. Malm, and H. E. Gunning, *J. Amer. Chem. Soc.*, **86**, 4243 (1964).

71. H. Yamazaki and R. J. Cvetanovic, *J. Chem. Phys.*, **39**, 1902 (1963).

72. D. E. Milligan, D. E. Mann, and M. E. Jacox, *J. Chem. Phys.*, **41**, 1199 (1964).

73. M. E. Jacox and D. E. Milligan, *J. Chem. Phys.*, **55**, 3403 (1971).

74. A. J. Rest and J. J. Turner, *Chem. Commun.*, 1026 (1969).

75. P. A. Breeze and J. J. Turner, *J. Organometal. Chem.*, **44**, C7 (1972).

76. J. K. Burdett, *Chem. Commun.*, 763 (1973).

77. P. Warneck, *Appl. Opt.*, **1**, 721 (1962).

78. D. Davis and W. Braun, *Appl. Opt.*, **7**, 2071 (1968).

79. F. C. Fehsenfeld, K. M. Evenson, and H. P. Broida, *Rev. Sci. Instrum.*, **36**, 294 (1965).

80. R. J. Isabel and W. A. Guillroy, *J. Chem. Phys.*, **57**, 1116 (1972).

81. M. E. Jacox and D. E. Milligan, *J. Chem. Phys.*, **40**, 2457 (1964).

82. J. K. Burdett, L. Hodges, V. Dunning, and J. H. Current, *J. Chem. Phys.*, **74**, 4053 (1970).

83. C. Y. Lin and A. Krantz, *Chem. Commun.*, 1111 (1972).

84. M E. Jacox and D. E. Milligan, *J. Chem. Phys.*, **54**, 3935 (1971).

85. G. R. Smith and W. A. Guillroy, *J. Chem. Phys.*, **56**, 1423 (1972).

86. A. M. Bass and J. Shirk, *J. Chem. Phys.*, **52**, 1892 (1970).

87. M. E. Jacox and D. E. Milligan, *J. Chem. Phys.*, **47**, 5157 (1967).

88. J. A. Shirk, *J. Chem. Phys.*, **55**, 3608 (1971).

89. M. Poliakoff and J. J. Turner, *J. Chem. Soc.*, *Dalton*, 2276 (1974).

90. Validyne Engineering Corp., 19414 Londelius St., Northridge, Calif. 91324.

91. F. Rosebury, *Handbook of Electron Tube and Vacuum Techniques*, Addison-Wesley, Reading, Mass., 1965.

92. O. H. Heavens, *Optical Properties of Thin Solid Films*, Dover, New York, 1965.

93. M. Moskovits and G. A. Ozin, *J. Appl. Spectros.*, **26**, 481 (1972).

94. C. D. Stockbridge, *Vacuum Microbal. Tech.*, **5**, 147, 179, 199 (1966).

95. M. Moskovits and J. Hulse, unpublished data.

96. L. Andrews, R. O. Allen, and J. M. Grzybowski, *J. Chem. Phys.*, **61**, 2156 (1974).

Techniques of Preparative Cryochemistry

3

P. L. Timms

1. INTRODUCTION

This chapter is concerned with the practical details of using reactive gaseous species formed at high temperatures as preparative reagents by means of cryochemistry. The theoretical background is considered in as far as it helps in understanding the ways in which apparatus must be designed and experiments must be conducted. Similarly, only sufficient chemistry of individual species is reviewed to put the described experiments into context.

Short-lived atomic and molecular species of high reactivity can be formed in suitable systems through energetic chemical changes or when energy is supplied by the use of high temperatures, photolysis, or electrical discharges. If the species can be generated at temperatures below about 200°C in the gas phase or in solution, it is often possible to carry out reactions with them in situ. However, if the species can only be formed at high temperatures, in situ reactions are usually impossible because of thermal decomposition of the other reactants and the products. Then the preparation of the species and its reaction with other compounds have to be carried out separately. This is best achieved by forming the high-temperature species in an evacuated vessel so that it can pass to the walls without intermolecular collisions. If the walls are cold, they can be used to collect not only the high-temperature species but also any other condensible vapor sprayed into the vacuum chamber. By condensing the species and a compound at the same time on the cold surface, interaction can then occur without risk of thermal decomposition. The synthetic importance of this idea was first demonstrated by Skell and Westcott[1] who condensed carbon vapor, formed under vacuum in a carbon arc, with organic compounds on a liquid-nitrogen-cooled surface. The surface acted as an efficient cryo-pump so that the pressure due to vapors in the apparatus was kept very low and thermal decomposition of the organic compounds on the hot carbon was avoided. This cocondensation method has been used extensively since,[2–6] although a useful alternative procedure is to condense the high-temperature species into a cooled solution of a compound in an inert solvent of low vapor pressure.[7,8]

2. THE GASEOUS SPECIES OF INTEREST

A species that can be formed at a high temperature should meet two requirements if it is to be considered for use in low-temperature chemical synthesis. First, it must contain an element or group in a state that is likely to be very reactive for thermodynamic and kinetic reasons. Second, it must be possible to form the species at the rate of at least a few millimoles an hour, preferably in a fairly pure state and without accompanying generation of noncondensible gases. If gases that are not condensible at $-196°C$ are liberated, vacuum conditions can only be maintained by using fast vacuum pumps or by cryopumping on to a surface cooled to liquid hydrogen or helium temperature.

The largest class of species clearly meeting the above thermodynamic and kinetic requirements are the atoms of the elements. With the exception of the noble gases, atoms of all the elements are endothermic. As shown in Table 1, the heats of formation of atoms range from about 60 kJ/mole to over 800 kJ/mole. Most of the metals and metalloids vaporize predominantly as atoms, but a few, notably carbon, give high concentrations of molecular species in addition to the atoms.

Table 1. Heats of Formation ($H^{0}_{298°K}$) of Some Gaseous Atoms of Interest in Cryochemistry (values in kJ/mole of atoms)

H	—	—	—	—	—	—	—	—	—	—	—	—	—	—	—
213															
Li	Be	—	—	—	—	—	—	—	—	—	—	B	C	N	O
161	326											576	715	469	243
Na	Mg	—	—	—	—	—	—	—	—	—	—	Al	Si	P	S
108	149											326	446	314	221
K	Ca	Sc	Ti	V	Cr	Mn	Fe	Co	Ni	Cu	Zn	Ga	Ge	As	Se
90	177	326	473	515	397	281	416	425	430	339	125	272	372	288	210
Rb	Sr	Y	Zr	Nb	Mo	(Tc)	Ru	Rh	Pd	Ag	Cd	In	Sn	Sb	Te
82	164	410	611	774	659	649	669	577	381	286	111	244	301	259	200
Cs	Ba	La	Hf	Ta	W	Re	Os	Ir	Pt	Au	Hg	Tl	Pb	Bi	—
78	178	435	703	781	837	791	728	690	566	368	62	180	197	207	
Rare earths and lanthanides															
Ce	Pr	Nd	(Pm)	Sm	Eu	Gd	Tb	Dy	Ho	Er	Tm	Yb	Lu	—	—
405	356	321	293	209	182	341	364	297	314	315	245	167	427		
Th		U	—	—	—	—	—	—	—	—	—	—	—	—	—
566		488													

The second large class of very reactive high temperature species are the small molecules containing an element in an unusual low-valency state. Many such species are known to spectroscopists from studies of vapors at high temperatures. However, the species are often formed only as minor components in equilibrium or nonequilibrium reactions, and comparatively few can be made on a scale or with the purity required for use in low-temperature synthesis. Table 2a lists some of the more important species

Table 2. Some Molecular Species of Interest in Cryochemistry

a. Low-valency species		
BF, BCl	CS	
	SiO, SiS	PF_2
	SiF_2, $SiCl_2$, $SiBr_2$	
b. Coordinately unsaturated species: metal-salt vapors, for example		
$FeCl_2$, CuCl	B_2O_2	
ZnF_2, AlF_3	PN	
TiO, KOH		
KCN		
c. Strained or unsaturated organic compounds		
$CH_2 = SO_2$		
$CH_2 = SiMe_2$		

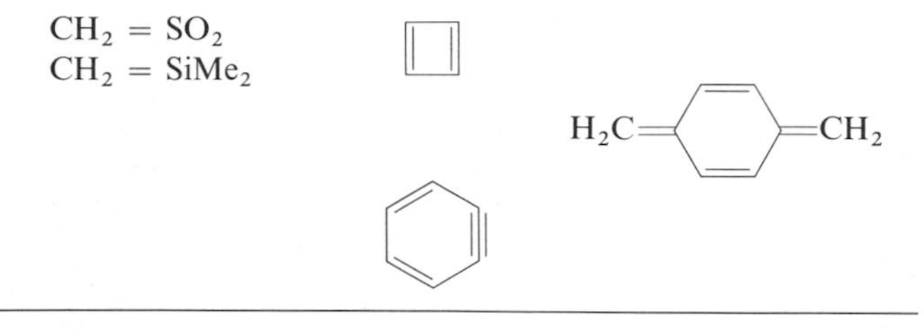

for synthetic work. The table does not include those reactive species that can be prepared at high temperatures but for which there are available excellent alternative methods of preparation at ordinary temperatures. For example, difluorocarbene, CF_2, can be prepared by high-temperature methods, but it is much more easily prepared under mild conditions from precursors such as CF_2N_2 and is often formed in situ.

The third class of species comprises molecules containing an element in an unusual low coordination state. Thus, vapors of metal salts contain the metal ion in a low coordination state compared to that in the normal solid state of the salt. Similarly, boron monoxide, O=B—B=O, contains trivalent boron in a low coordination state. Some of the species that could be used synthetically are shown in Table 2b. Comparatively little work has yet been done with this class of compounds.

The fourth class of species are highly unsaturated or strained organic molecules formed by rapid pyrolysis of organic or organometallic compounds at temperatures of 200–1000°C: Table 2c gives some examples. Most of the species result from nonequilibrium decomposition of more complex molecules. The species are usually thermodynamically unstable at the temperature at which they are prepared, but the rapidity with which the species are swept out of the hot zone prevents decomposition.

3. KINETIC AND THERMODYNAMIC CONSIDERATIONS

Preparative cryochemistry has developed on an empirical basis, and no detailed studies of the kinetic and thermodynamic factors controlling the low temperature reactions have yet been made. Thus, the following discussion is based on ideas that have been only partly tested by experimental observation and that may have to be modified in the future.

3.1 Kinetics

When a high-temperature species and an ordinary compound are condensed together on a cold surface, three types of processes can occur. First, the high-temperature species may polymerize without reacting with the compound. Second, the monomeric high-temperature species may react with the compound. Third and least commonly, the species may polymerize to form dimers, trimers, and so forth, which react more readily than the monomer with the compound.

For atoms and for those species listed in Table 2a and b, the rate of polymerization is generally high and often only diffusion controlled even at 20°K. However, even among closely related species such as the atoms of metals in the first transition series, there can be considerable differences in polymerization behavior. For example, it has been shown by spectroscopic studies of metal atoms condensed in noble gas matrices (see Chapter 9) that manganese atoms appear to form clusters readily. The manganese atoms may be slightly less polarizable than atoms of the other two metals, allowing more rapid diffusion through the matrix or across the surface of the matrix during condensation. Although some of the organic species shown in Table 2c do not polymerize at −196°C, they polymerize below 0°C.

A high-temperature species reacts with a compound on a cold surface only if the energy barriers are very low; otherwise polymerization of the species is usually the dominant process. The desired reaction can be made to compete with the polymerization in two ways, either by condensing the species with a very large excess of the compound or by carrying out the condensation at a higher temperature.

The use of excess compound increases the number of compound-species interactions and inhibits the polymerization of the species. Customarily in the cocondensation technique, a fivefold to hundredfold mole excess of the compound is condensed with the high-temperature species to limit unwanted polymerization. The greater the excess the better the yield of product based on the amount of the species used. However, two practical problems arise. First, the compound condensed with the species may not be available in large amounts; and second, the separation of the product from a huge excess of unreacted compound may be difficult. If the compound is in limited supply, it is usually very difficult to convert it efficiently into a desired product by reaction with a high-temperature species. Condensation of the species with a stoichiometric or less than stoichiometric amount of a compound generally results in substantial loss of the compound as intractable products, and there is a low yield of simple molecular products. The compound can be diluted with an inert volatile compound, for example, an alkane, and the mixture condensed with the high-temperature species. The fraction of the compound that is converted to desired products is usually higher under these conditions than if the compound is used neat. The diluent may hinder polymerization of the species and help to prevent strong chemisorption of reaction products on the surface of the polymerized species. In some reactions the role of the diluent is poorly understood. For example, in the reaction between iron atoms and trimethylphosphine (Section 7.1), dilution of the trimethylphosphine with *n*-hexane gives a much lower yield of the desired product than dilution with *n*-pentane. Both hexane and pentane appear chemically inert in the reaction, and the only obvious difference between them is their freezing point, which may influence the mobility of the atoms and molecules on condensation.

In the solution technique when the species is condensed into a neat liquid compound of low vapor pressure or into a solution of the compound in a solvent of low vapor pressure, rapid stirring is necessary to disperse the species and to prevent it polymerizing on the surface of the liquid. In the absence of stirring, it is possible to metal plate the surface of a cold liquid by condensing metal atoms on it. The yield of products in a reaction between a high-temperature species and a compound in solution seems to decline sharply if the concentration of the compound is reduced below 0.5 *M*.

There are two reasons why low temperatures have been commonly used to carry out reactions of high-temperature species. First, a very cold surface is an excellent cryopump that enables the pressure in the apparatus to be kept low. Second, if unstable products are formed, low temperatures hinder their decomposition. The latter effect is particularly marked when matrix isolation conditions are used at temperatures below 30°K. The polymerization of high-temperature species is usually a reaction involving little or no activation energy, whereas the reactions of species with compounds may have higher activation energies. Thus, low temperatures tend to favor the polymerization of the species rather than productive reactions of the species. This leads to the conclusion that reactions of high-temperature species and compounds should be conducted at as high a temperature as is compatible with maintenace of vacuum and the stability of the products.

As discussed more fully in Section 5.3, in cocondensation experiments heat radiation may raise the surface temperature of the condensate, and this may be beneficial to the efficiency of the reaction. Higher temperatures of reaction are most easily achieved by condensing the species into a solution of a compound. Provided that the species condense easily, the pressure of vapor above the solution can be kept below the vapor pressure of the solution by vacuum pumping. This is in contrast to the situation during cocodensation when the pressure of the vapor is above the equilibrium vapor pressure at the surface temperature. Thus, an adequate vacuum can be maintained at a higher surface temperature for an evaporating solution than for a condensing vapor.

3.2 Thermodynamics

Most high-temperature species are thermodynamically unstable at low temperatures with respect to polymerization. In the case of metal atoms, the difference in energy between the atoms and the solid metal is given by the values in Table 1. When metal atoms are isolated at low temperatures in a matrix of a noble gas or other inert substance, the energy of the atoms relative to the bulk metal is only slightly different from the values in Table 1, that is, the kinetic energy lost on condensation amounts to only a small part of the total energy. If the isolated metal atoms are then allowed to diffuse to a limited extent so that they can interact with molecules of a compound also isolated in the matrix, any reaction that occurs yields a matrix isolated product. Under these conditions, the energy change in the reaction can be calculated simply in terms of the bond energy changes. Thus, the heat evolved in the reaction,

$$\mathrm{M(i) + X(i) \rightarrow MX(i)}$$

(i = matrix-isolated form, close in energy to the gaseous form) is approximately equal to the M—X bond dissociation energy. This means that most simple addition reactions of metal atoms are thermodynamically favored under these conditions. Under preparative conditions, the products cannot remain matrix isolated. Then the above reaction can be followed by dissociation if MX(s) is endothermic,

$$\mathrm{MX(i) \rightarrow MX(s) \rightarrow M(s) + X(s)}$$

(s = solid, liquid, or gaseous state close in energy to the standard state of the substance): MX(s) is observed only as a product if its decomposition is kinetically hindered. Thus, many more compounds can be made by addition reactions under matrix-isolation conditions than can be obtained under preparative conditions.

For example, Ozin[9] has shown that nickel atoms react with N_2 in noble gas matrices to give matrix-isolated $Ni(N_2)_4$. However, if nickel is condensed on a liquid-nitrogen-cooled surface in the presence of gaseous nitrogen, there is no evidence for the formation of $Ni(N_2)_4$. This compound is presumably thermodynamically unstable with respect to solid nickel and N_2, and is not kinetically stable at $-196°C$. In other cases, endothermic products of atom reactions do not decompose at $-196°C$ and can be handled in preparative quantities by suitable low temperature techniques (see Section 5.5). The product of the reaction between iron vapor and benzene at $-196°C$ is so endothermic that it may explode on warming.[10] A number of addition reactions of carbon atoms reported by Skell and his coworkers[2] have produced endothermic products that are stable at ordinary temperatures. The decomposition of the organic compounds formed is much more kinetically hindered than is the decomposition of metal compounds.

Metal atoms are useful halogen abstractors,[4,11] but their superiority over bulk metals in this role seems to depend more on kinetics than thermodynamics. Gold atoms dehalogenate ethyl iodide or ethyl bromide on condensation at $-196°C$ to give mainly butane and a gold halide, but there is no reaction with ethyl chloride. Although solid gold does not react at an appreciable rate with ethyl halides, the reaction

$$2Au(s) + 2C_2H_5X \rightarrow 2AuX + C_4H_{10}$$

is just exothermic for ethyl iodide and ethyl bromide but endothermic for ethyl chloride. Thus, the gold atoms appear able to bring about only the reactions that are thermodynamically possible with solid gold: the advantage of using the metal atoms in this case is the great rapidity of the reaction.

The thermodynamic arguments that have been put forward for metal atoms apply equally to low-temperature reactions of other high-temperature species, although the differences in energies between the species and their polymerized forms may be less well defined than for metal atoms. There is some experimental evidence that high-temperature species, that polymerize on condensation to form ionic or semi-ionic solids, are less reactive towards nonpolar compounds than are most species. This may be due to very rapid polymerization aided by relatively long-range electrostatic forces. For example, AlF appears to undergo none of the reactions at low temperatures with organic compounds that are characteristic of the less polar BF.

The uncertainties in the conclusions in this section highlight the need for fundamental research on low-temperature reactions to determine the relative roles of thermodynamic, kinetic, and diffusion effects. A detailed study of a few reactions under a variety of conditions is required.

4. THE FORMATION OF HIGH-TEMPERATURE SPECIES

Gaseous high-temperature species can be prepared by evaporation of pure liquid or solid substances, by reactions between liquid or solid substances, by reactions between a gas and a solid substance, or by reactions involving one or more gases. Each of these types of preparation requires different experimental techniques that are discussed below.

There are basically two ways of supplying heat energy to the substances that are precursors of a high-temperature species. Heat can be transferred to the substances from the walls of a container, for example, a crucible that is heated externally. This method requires that there is no chemical reaction between the container and the substances being heated, which can be a severe limitation because of the enhanced reactivity of substances at high temperatures. The problem can be overcome by using a heating method that supplies energy only to the substances forming the high-temperature species. Such "containerless" methods include heating by electron bombardment, by laser beams, in electric discharges and arcs, and by passing current through a wire of the evaporant.

4.1 Vaporization of the Metals and the Less Volatile Nonmetals

The vapor pressures of the elements are now known fairly accurately from a variety of independent sources. Data on the elements that might be used in cryogenic synthesis are shown in Table 3. Under a vacuum of 10^{-3} torr or better, practical rates of vaporization are achieved when the vapor pressure of the element exceeds 5×10^{-3} torr. Tables of rates of evaporation per unit area of metal surface versus temperature are given in Dushman.[12]

Both the amount of heat radiated and the amount of metal evaporated from unit area of surface increase with temperature, but the latter increases more rapidly than the former. For example, the rate of evaporation of nickel increases a hundredfold between 1500 and 1900°C, but the rate of heat loss by radiation increases only twofold over this temperature range. The rate of radiative heat loss is proportional to T^4, but the rate of evaporation is proportional to ($\log T - 1/T$). Thus, a metal should be evaporated at as high a temperature as possible to minimize the amount of heat radiated for a given weight of metal evaporated. This is an important principle that influences the choice of evaporation techniques for preparative cryochemistry. There are numerous articles and books on the subject of metal-evaporation techniques.[12–15] The methods described here are those which are most

Table 3. Melting Points (m.p.) and Vapor Pressures (v.p.) of the Metals (temperatures °C)

		Li	Be											(B)	(C)	
m.p.		181	1285											2150	3860	
v.p. (torr)	10^{-3}	465	1095											1865	2290	
	10^{-2}	535	1225											2025	2460	
	10^{-1}	625	1375											2250	2660	
	1	745	1555											2510	2900	
		Na	Mg											Al	(Si)	
m.p.		98	650											659	413	
v.p. (torr)	10^{-3}	235	377											1082	1470	
	10^{-2}	289	439											1215	1630	
	10^{-1}	357	509											1365	1815	
	1	641	605											1555	2055	
		K	Ca	Sc	Ti	V	Cr	Mn	Fe	Co	Ni	Cu	Zn	Ga	Ge	
m.p.		62	850	1538	1667	1917	1903	1244	1538	1495	1452	1084	419	30	937	
v.p. (torr)	10^{-3}	161	522	1230	1575	1685	1265	835	1340	1382	1380	1132	292	1005	1055	
	10^{-2}	208	597	1375	1735	1840	1400	935	1475	1520	1525	1255	344	1130	1395	
	10^{-1}	267	691	1560	1935	2050	1550	1060	1645	1685	1695	1415	408	1260	1555	
	1	345	802	1795	2175	2290	1735	1215	1855	1905	1905	1615	485	1470	1775	
		Rb	Sr	Y	Zr	Nb	Mo	Tc	Ru	Rh	Pd	Ag	Cd	In	Sn	Sb
m.p.		39	770	1500	1855	2497	2610	—	2310	1960	1552	961	221	156	232	630
v.p. (torr)	10^{-3}	129	465	1465	2170	2445	2305	—	2145	1855	1315	932	217	835	1105	475
	10^{-2}	173	537	1630	2390	2655	2525	—	2345	2035	1470	1027	265	955	1245	533
	10^{-1}	227	625	1830	2650	2895	2785	—	2585	2245	1645	1162	320	1080	1410	612
	1	295	730	2080	2970	3175	3115	—	2850	2505	1875	1332	392	1245	1610	757

		Cs	Ba	La*	Hf	Ta	W	Re	Os	Ir	Pt	Au	Hg	Tl	Pb	Bi
m.p.		194	710	920	2222	2997	3410	3180	3045	2454	1769	1063	−39	304	327	271
v.p. (torr)	10^{-3}	114	527	1560	2175	2800	2975	2805	2685	2285	1905	1252	16	530	625	587
	10^{-2}	145	610	1725	2395	3055	3230	3165	2915	2495	2095	1395	46	609	715	672
	10^{-1}	209	711	1925	2655	3350	3540	3405	3185	2765	2315	1565	80	706	832	777
	1	280	852	2175	2965	3700	3800	3800	3525	3085	2585	1755	125	825	975	897
*Lanthanides			Ce	Pr	Nd	Pm	Sm	Eu	Gd	Tb	Dy	Ho	Er	Tm	Yb	Lu
m.p.			705	935	1024	~ 1080	1072	826	1315	1365	1407	1461	1495	1600	842	1652
v.p.	10^{-3}		1522	1275	1165	—	653	530	1190	1275	1000	1050	1050	755	485	1410
	10^{-2}		1700	1425	1300	—	742	610	1325	1425	1115	1165	1175	845	555	1570
	10^{-1}		1905	1615	1495	—	847	708	1485	1615	1260	1330	1330	960	645	1705
	1		2165	1845	1725	—	985	827	1680	1845	1435	1525	1525	1095	780	1995
Actinides			Th		U											
m.p.			1695		1132											
v.p.	10^{-3}		2165		1735											
	10^{-2}		2400		1825											
	10^{-1}		2685		2150											
	1		3030		2450											

Table 4. Useful Methods of Forming Vapors of Some Elements for Preparative Cryochemistry

Element	Evaporation Method
Aluminum	For small quantities wrap Al wire around heavy tungsten filament and resistance heat; evaporate large amounts from TiB_2 crucible or off a resistance heated TiB_2–BN strip.
Boron	Electron bombardment; molten boron is very reactive.
Calcium	Sublimes easily from Mo-wire basket or alumina crucible.
Carbon	Carbon arc quite efficient; evaporate small amounts from resistance-heated filaments.
Chromium	Sublimes from Mo- or W-wire baskets or from pure alumina crucible.
Cobalt	Evaporates smoothly from alumina crucible.
Copper	Can be evaporated in very large amounts from alumina crucibles as the metal refluxes well.
Germanium	Use a graphite crucible or a resistance heated strip of graphite tape.
Gold Indium	Very easy evaporation from an alumina crucible.
Iridium	Small quantities from a resistance-heated W wire; electron bombardment probably better.
Iron	Alumina crucibles satsifactory, but they do not last many times; use mild steel as the source of iron for the most rapid evaporation.
Lanthanum (and lanthanides)	Sm, Eu, Dy, Er, and Tm can be sublimed from W-wire baskets; electron bombardment or laser heating is best for those metals that melt before evaporating.
Lithium	Easily evaporated from stainless steel crucible.
Magnesium	Easily sublimed from Mo-wire basket or from a stainless steel or alumina crucible.
Manganese	Use alumina crucible with the metal just molten.
Molybdenum	Sublime from resistance heated wire or strip at 2500°C; electron bombardment.
Nickel	Evaporates well from pure alumina crucibles.
Niobium	Simplest by electron bombardment.
Palladium	Clean and rapid evaporation from alumina crucible.
Platinum Rhodium	Alloys with tungsten, but small amounts evaporate well off tungsten wire.
Ruthenium	Resistance-heated wire can be sublimed slowly; electron bombardment.

Table 4 (*Continued*)

Element	Evaporation Method
Silicon	Use beryllia crucible with external W-wire heater; slow evaporation from heated silicon or silicon carbide rod; electron bombardment evaporation rapid.
Silver	Very easy evaporation from alumina crucible.
Tantalum	Simplest by electron bombardment.
Tin	Vaporizes well from alumina crucible.
Titanium	Slow evaporation from Ti–Mo alloy wire resistance heated; quite fast by electron bombardment.
Uranium	Small quantities evaporate off a W-wire or boat.
Vanadium	Resistance heated W or Mo boat; slow sublimation from a W-wire basket.
Zinc	Very fast evaporation from alumina crucible.
Zirconium	Simplest by electron bombardment.

compatible with the requirements of chemical synthesis. Table 4 shows methods of evaporation that work well for particular metals: the table summarizes the information discussed in detail in the following two subsections.

Evaporation from a Heated Container

Some metals can be vaporized directly from wire coils or boats made of W, Ta, or Mo that are heated resistively. Various shapes of coils and boats used for thin-film deposition are available commercially, but coils are easily made in the laboratory. Metals like chromium that sublime can be evaporated efficiently from the simple coil shown in Figure 1. The metal vapor sprays out of the bottom of the coil. The alumino-silicate wool (composition ca. $Al_2O_3 \cdot 2SiO_2$, trade name Kao-Wool in Great Britain) over an inner radiation shield of 0.05 mm Mo foil, minimizes radiative heat losses. Boats of the refractory metals can be used to contain metals that melt before vaporization (Figure 2). Heat losses can be reduced by a combination of radiation shields and insulating wool as in the apparatus of Figure 1. The amount of metal that can be vaporized from a boat is limited in two ways. First, the electrical resistance of that portion of the boat containing the molten metal is lower than other portions of the boat, so that most heat is generated where it is least effective in vaporizing metal. Second, many molten metals alloy with the refractory metals, and this may cause melting or mechanical failure unless the weight of the refractory exceeds that of the alloying metal. These problems can be overcome by continuously feeding the evaporant metal on to a heated

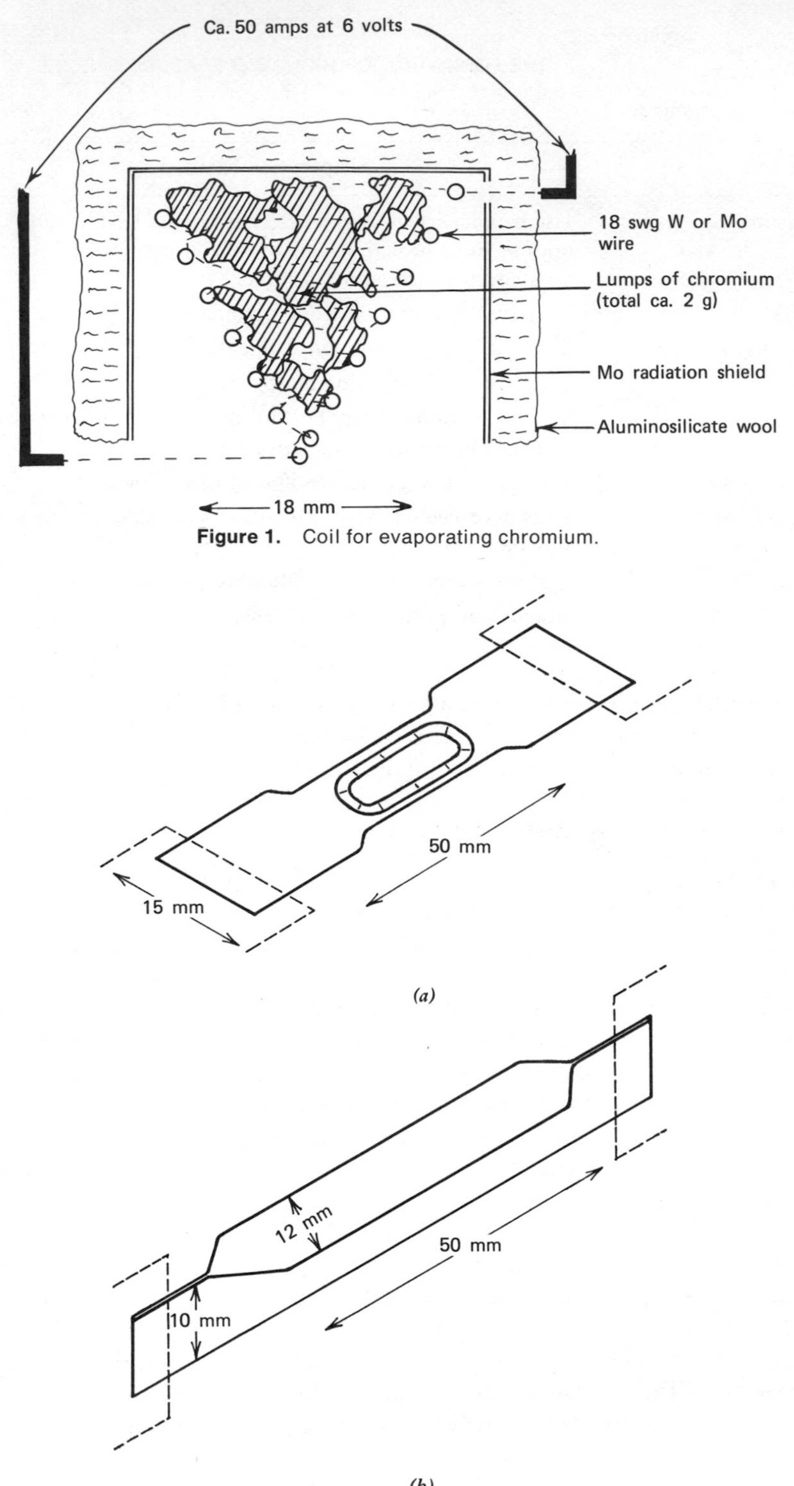

Figure 1. Coil for evaporating chromium.

Figure 2.

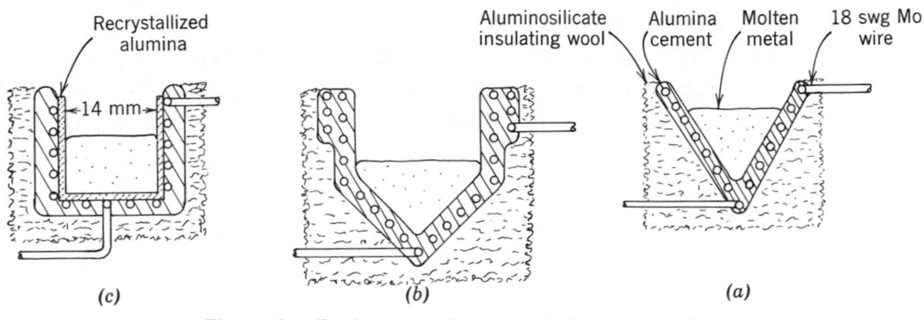

Figure 3. Resistance-wire-wound alumina crucibles.

strip or shallow boat of a refractory metal so that the rate of evaporation matches the rate of addition. The technique is used commonly in thin-film technology[13] but has not yet been applied to preparative cryochemistry, although the technique offers the best way of obtaining a uniform rate of formation of metal vapor over a long period.

Resistance-heated refractory metal wires can be used to heat crucibles made of the refractory oxides. Typical arrangements are shown in Figure 3. Alumina is the most convenient oxide to use: it is commercially available in the form of a cement that can be painted on to molybdenum or tungsten coils, or as ready-made crucibles that can be heated to 1800°C and are fairly thermally shock resistant. Alumina is attacked by the most electropositive metals at high temperatures, but it can be used to vaporize Cr, Mn, Fe, Co, Ni, Pd, Cu, Ag, Au, In, and Sn. A crucible of only 5 ml capacity (Figure 3*c*) permits vaporization of these metals at the rate of 10–50 g/hr. Other oxides can replace alumina for special operations. Thus, small beryllia crucibles can be used to vaporize silicon, and thoria crucibles to vaporize titanium; but such crucibles are much more costly and less rugged than alumina crucibles.

Graphite can be used to contain some metals, such as germanium and tin, that form no stable carbides. Other metals, such as titanium, zirconium, and aluminum, react with graphite to form a refractory carbide layer that limits further attack by the molten metal on the graphite. Aluminum is evaporated on a large scale from resistance-heated strips made of a hot pressed mixture of boron nitride and titanium diboride, a material that is completely unattacked by the molten metal.

Evaporation by Containerless Methods

Metals that have vapor pressures greater than about 10^{-2} torr at their melting points can be sublimed at a useful rate by heating them in the form of wires, strips, or rods to just below their melting points. Evaporation of titanium from wires of a Ti/Mo alloy is used in titanium sublimation pumps.[13] Figure 4*a* shows a simple arrangement used at Bristol for evaporating small amounts of titanium (100–200 mg/hr) from a short length of the commercially available 18 swg Ti/Mo wire by passing a current of about 50 amp. The evaporation of tungsten and molybdenum from resistance-heated wires has been used in chemical synthesis.[16,17] Figure 4*b* shows a way of mounting a hairpin of the wire to be evaporated. The attraction of the direct sublimation method is its

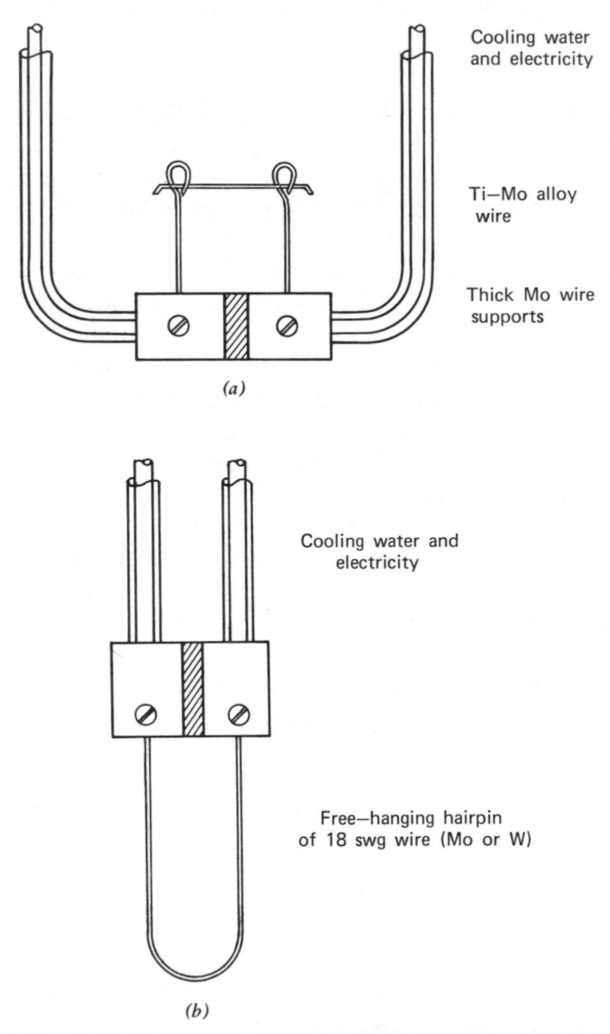

Figure 4. Methods of evaporating metal from wires by resistance heating.

simplicity, but it is not an ideal procedure for preparative cryochemistry. The rate of evaporation is restricted by the need to avoid melting the metal. A large area of metal must be used to obtain worthwhile amounts of vapor, and thus there is much accompanying heat radiation. Normally, about 10% of the weight of a wire or rod of a metal can be evaporated before failure occurs due to necking and local overheating.

In principle, electric arcs represent a good method of evaporation for chemical synthesis as they create a small area of intensely hot material that vaporizes rapidly. Arc evaporation is simplest for materials that sublime, for example, C, Cr, or Mg: a carbon arc suitable for preparative cryochemistry[2] is shown in Figure 5*a*. However, there is an extensive technology of melting metals under vacuum using dc arcs and water-cooled containers for the metal:[14] such methods may be applicable to vaporization of metals for synthetic chemistry, and Figure 5*b* shows a possible apparatus for this purpose. Some electronic excitation of the metals produced in the arc is to be expected, but this may have advantages for synthetic purposes.[2]

One of the most widely used methods for evaporation of metals in commercial practise is electron bombardment. Electrons are drawn off a hot

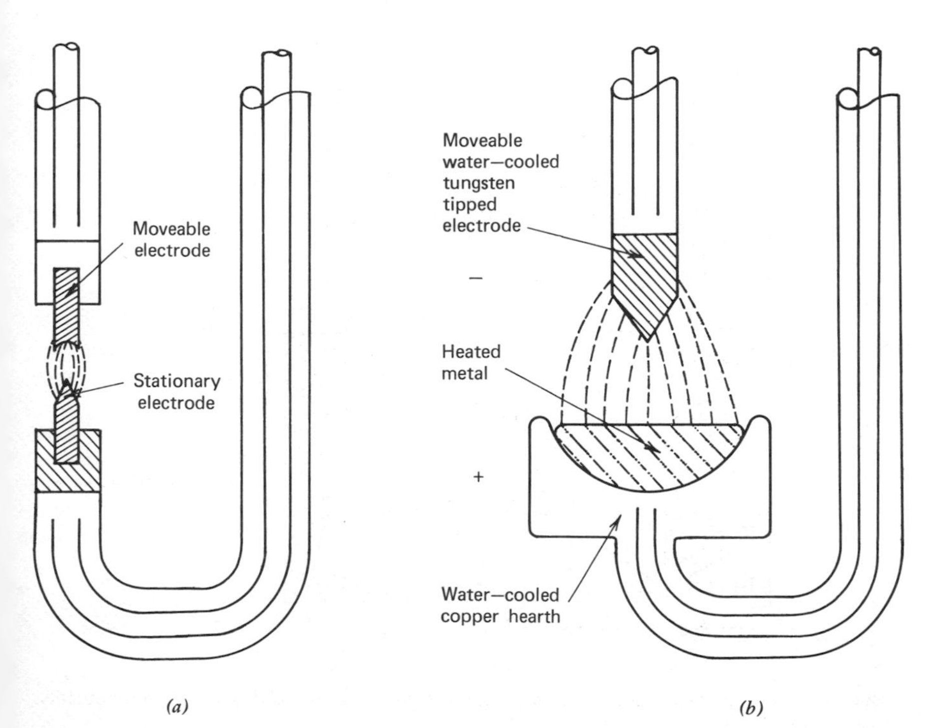

Figure 5. Evaporation from arcs. (*a*) An ac carbon arc, (*b*) a dc arc for vaporizing metals.

filament by an applied potential and are made to impinge on the metal to be heated, supplying heat energy proportional to the electrical power in the beam of electrons. With a beam of electrons focused electrostatically or magnetically, an energy density of 100 watts/mm^2 is easily obtained, and this is sufficient to vaporize any material very rapidly. There are two drawbacks to the method when it is applied to preparative cryochemistry. First, the hot metal reflects some of the electrons striking it. These stray electrons may interact chemically with vapors of compounds being condensed with the metal vapor. Second, electron bombardment is more sensitive to the pressure in the apparatus than the other evaporation methods that have been

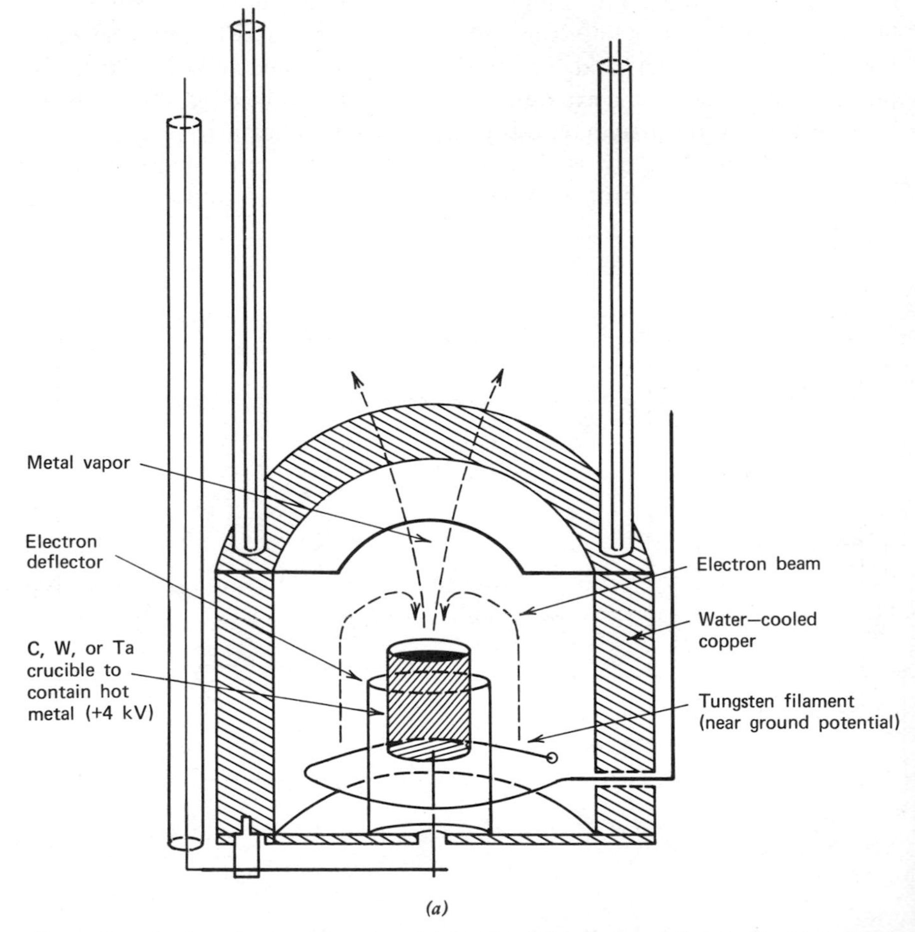

Figure 6. Electron beam evaporators (*a*) Electrostatically focused gun, (*b*) magnetically focused gun.

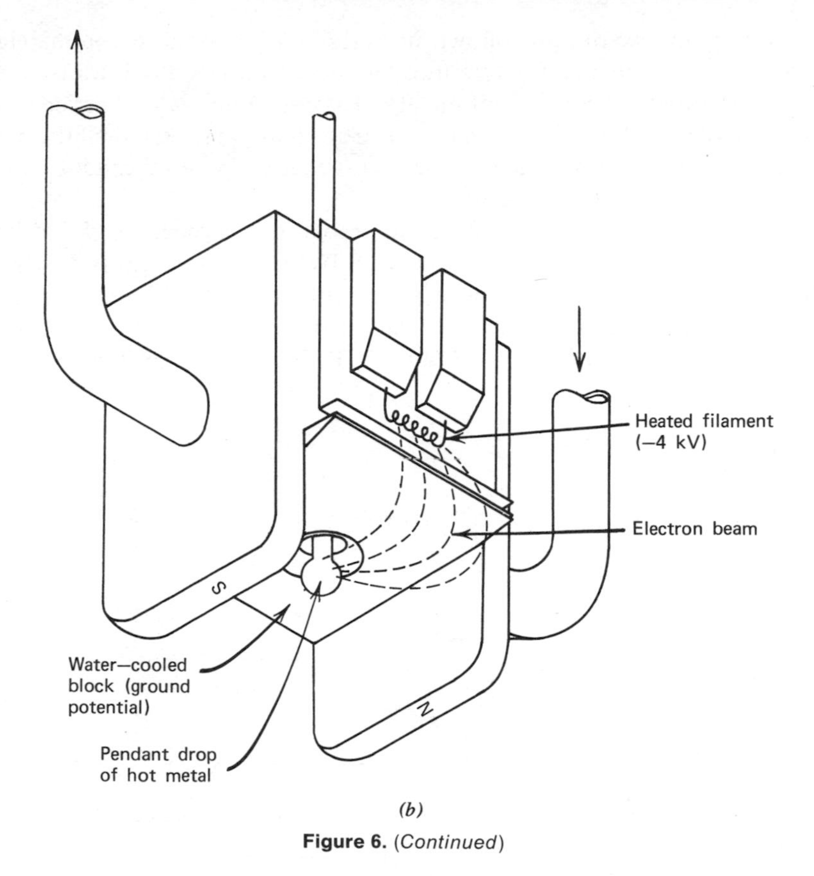

(b)

Figure 6. (*Continued*)

described. At pressures above 10^{-3} torr electrical discharges may short out the electron beam and interrupt the evaporation of metal.

The number of stray electrons is minimized by a design of electron bombardment heater like that shown in Figure 6*a*. Electrons, drawn off the resistance-heated tungsten filament by a high positive potential applied to the metal charge, are focused by the electrostatic field around the aperture above the metal. In Figure 6*a*, the hot metal is in contact with a pedestal of a refractory metal, but the pedestal must be water cooled when evaporating the most corrosive metals. The device will evaporate 0.1–1 g of most metals in an hour with 100–500 watts of electron beam power. Green[18] has used a similar device for the evaporation of titanium, although in his evaporator the relative potentials were different from those in Figure 6*a*, with the metal at earth potential and the filament at a high negative potential (see Figure 18).

The type of electron gun shown in Figure 6*b* is produced commercially with power outputs of a few hundred watts to hundreds of kilowatts, and it gives very rapid evaporation of metals. However, there is no barrier to the escape of the 15–25% of the incident electrons that are reflected off the metal surface. For this reason, such an electron-beam evaporator cannot be recommended for use in cryochemistry.

Laser beam evaporation of metals in preparative cryochemistry has been described by von Gustorf,[19] using a 200W neodymium doped yttrium aluminum garnet laser. The laser beam enters the apparatus through a window that is protected from deposition of metal vapor by a stream of condensible vapor as shown in Figure 7. If the laser beam is focused, the

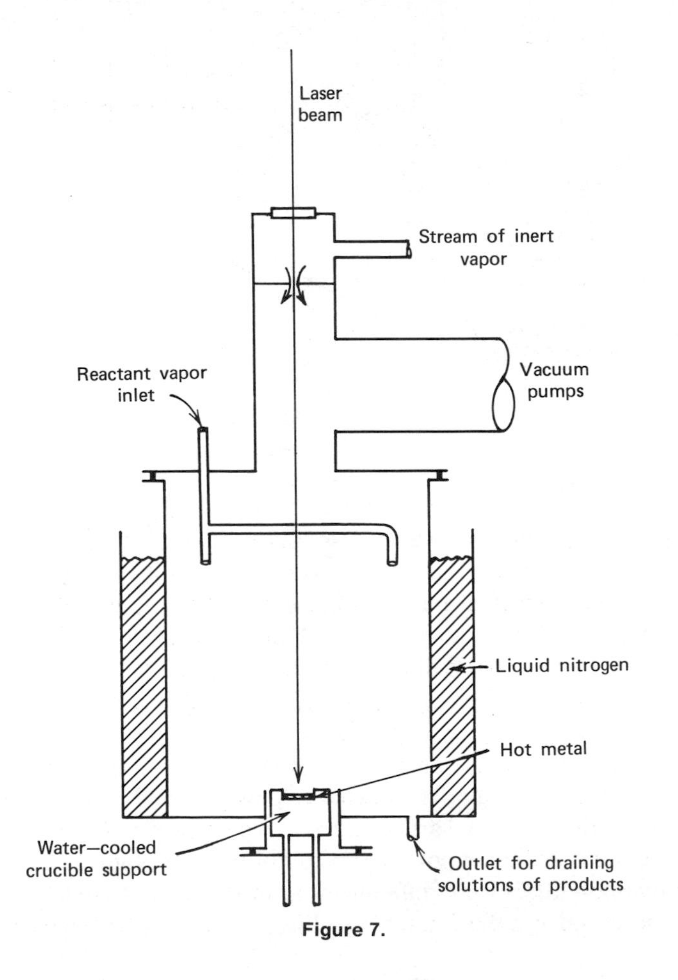

Figure 7.

possible energy density is very high. However, much of the incident radiation may be reflected from metal surfaces causing wastage of power. This difficulty, coupled with the high initial cost of a laser, limits the appeal of this method of evaporation.

Metals can be atomized by ion bombardment, the process being termed "sputtering." Usually, the metal is maintained at a negative potential and is bombarded by argon ions created in an electric discharge. The method has not been used in preparative scale work, but it has been used on a spectroscopic scale.[20] The argon is a nuisance in a vacuum system as it is not cryopumped by a liquid-nitrogen-cooled surface, but in the spectroscopic work condensation took place on a surface cooled below 40°K. The metal atoms formed by sputtering have quite high thermal energies, and they may be more chemically reactive than atoms formed by equilibrium vaporization from the surface of a heated metal. A major attraction of sputtering over other methods of forming metal atoms is that it is as easy to sputter many of the refractory metals as to sputter volatile metals. Thus, almost no variations in experimental procedure are required when changing from chromium to tungsten or from vanadium to tantalum.

The technique of levitation of molten metals by high-frequency currents is well established,[15] but it has not yet been applied to the evaporation of metals in cryochemistry. It may be found that a levitated drop of molten metal becomes unstable when it is evaporating rapidly under vacuum.

4.2 The Preparation of Molecular Species

The methods described above for evaporation of metals can often be applied to the preparation of those molecular species which are formed purely from condensed phases, that is, by evaporation of a solid or liquid or by a reaction in the condensed phase. For example, silicon monoxide can be formed as a gas by heating a mixture of silicon and silica to 1100–1200°C using alumina, molybdenum, or tantalum crucibles, or almost any of the containerless methods described in the preceding section. Table 5 lists some molecular species that are prepared using metal-evaporation techniques.

Gas-Solid Reactions

Some molecular species, notably the boron and silicon subhalides, are best formed by passing a gas over a solid at a high temperature. The reaction between a gas and a solid above 1000°C can be very fast so that equilibrium concentrations of species are approached with a contact time of a few milliseconds. This permits fast streaming rates of gas through a column of solid contained in a quartz, mullite, alumina, or graphite tube: the practical problem of how to avoid blowing away the solid is considered in Section 6.

Table 5. The Formation of Some Molecular Species

Vapor Species	Formation Temperature (°C under vacuum)	Method
BF	1800	Pass BF_3 over solid boron in graphite container at a pressure of 1 torr.
BCl	1100	Pass B_2Cl_4 through an alumina tube under flash-thermolysis conditions.
B_2O_2	1200	Heat powdered boron (99% pure) in a limited supply of low-pressure oxygen until visible combustion ceases, then heat solid to 1200°C in alumina crucible under vacuum.
SiF_2, $SiCl_2$, $SiBr_2$	1250	Pass SiX_4 over silicon at 1 torr pressure in a graphite, alumina, or mullite tube.
SiO	1200	Heat lumps of commercial high-purity "SiO" or a mixture of silicon and powdered quartz in alumina, Mo, or Ta crucibles; efficient SiO generators commercially available.
SiS	1100	Heat FeS and excess silicon in alumina crucible.
CS	—	Pass CS_2 vapor through discharge beween electrodes at 1 torr pressure; sulfur condenses out leaving CS as moderately long-lived species.
Metal chlorides (e.g., KCl, $MnCl_2$, $NiCl_2$)	600–800	Heat anhydrous chloride in alumina, vitreous carbon, or graphite crucibles.
Metal fluorides (e.g., LiF, ZnF_2, AlF_3)	700–1000	Heat anhydrous fluorides in vitreous carbon or graphite crucibles.
KCN	600	Heat compound in vitreous carbon crucible; liquid creeps.
TiO	1800	Electron bombardment heating of "electronic grade" commercial TiO.

Gas-solid reactions can only be used to form species that are not readily dissociated. For example, the equilibrium,

$$Si(s) + SiX_4(g) \rightarrow 2SiX_2(g)$$

is shifted to the right at high temperatures and low pressures when X is any of the halogens. However, SiI_2 can dissociate further to Si(s) and iodine atoms at temperatures above 1000°C, so that the result of passing SiI_4 over hot silicon is to form iodine vapor with very little SiI_2(g).

Decomposition of Gases: Flash Thermolysis

The successful formation of reactive species by thermal decomposition of vapors of compounds requires careful control of pressure and heating time. For example, BCl is best made by thermal decomposition of diboron tetrachloride at 1000°C. The decomposition can follow two pathways,

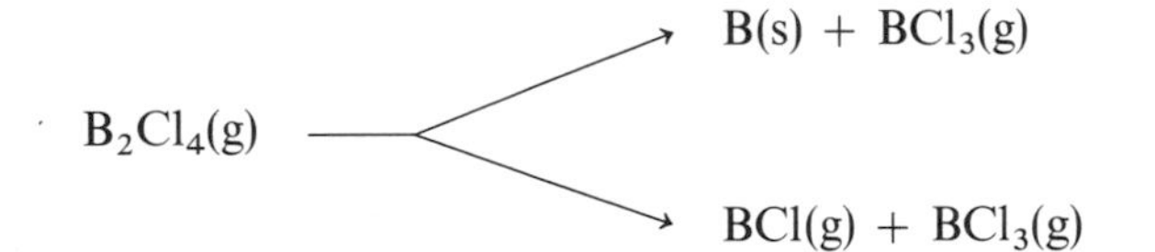

The thermodynamically favored process is the formation of solid boron and boron trichloride, and BCl can be decomposed heterogeneously to give those products. Solid boron does not form readily except in collisions of molecules with a solid surface, whereas BCl must result mainly from gas-gas collisions between thermally excited molecules. Gas-wall collisions predominate at very low pressures and gas-gas collisions at higher pressures. Thus, BCl is obtained in highest yield by passing B_2Cl_4 vapor rapidly through a narrow bore quartz tube heated to 1000°C and allowing the gases to emerge into an evacuated cryogenic reactor (see Figure 12). The flow of B_2Cl_4 is adjusted so that the highest pressure in the tube is a few torr. Lower rates of flow cause more boron to be deposited as the pressure is lower, and at higher flow rates much of the B_2Cl_4 passes through without decomposition.

The problems associated with the flash thermolysis of organic compounds are similar to those for diboron tetrachloride. Carbon deposition occurs in preference to the desired homogeneous gas phase decomposition if the pressure is too low, the residence time too long, or the temperature too high. The conditions required for the flash thermolysis of a compound are often given in the chemical literature as $T°$ at pressure P torr. The pressure is usually measured at the beginning of the cracking zone, although the pressure decreases along the tube in the direction of vacuum pumping. In order to reproduce the reported cracking it may be very important to establish the same pressure profile, giving the same residence time and number of intermolecular collisions.

Formation of Species in Electrical Discharges

Although electrical discharges can be used to generate numerous unstable, reactive gaseous species, only a few of the species of interest in cryochemistry are best made this way. Electrons, ions, and reactive neutral species will exist in a discharge. If the neutral species have a long lifetime in the gas phase at a low pressure, they can be carried far enough from the discharge to become free of accompanying electrons and ions. Condensation and reaction of the neutral species with other compounds on a cold surface is then not complicated by secondary reactions induced by the charged species. Carbon monosulphide (see Section 6.1) and nitrogen atoms (active nitrogen) are two species with suitably long lifetimes that are most conveniently made in discharges.

5. THE DESIGN OF APPARATUS

The preceding section is concerned with sources of reactive species. This section considers factors influencing the design of apparatus to carry out reactions of these species with other compounds under vacuum and to isolate the products.

5.1 Materials of Construction

The apparatus used for preparative cryochemistry should be capable of being pumped to the 10^{-6} torr range to be sure that there is no appreciable air leak and to be able to remove most of the adsorbed gases and vapors, particularly water (when in operation, the pressure in the apparatus due to inert gases or vapors may be much higher, up to 10^{-2} torr). Apparatus should also be designed to be cleaned easily by physical and chemical means.

The above considerations make stainless steel or borosilicate glass the best materials for the main vacuum vessels in which the high-temperature species are prepared and reacted with other compounds at low temperatures. For safety reasons, stainless steel is to be preferred for vacuum vessels of greater than about 25 cm diameter that are cooled in liquid nitrogen. Stainless steel apparatus is readily made compatible with commercially available fast-vacuum-pumping systems. However, glass apparatus is usually cheaper to construct, and it permits easier viewing of the reactions than stainless steel. The lighter weight of glass apparatus is also an advantage since it is more easily handled and supported by conventional laboratory clamps.

Viton O-ring seals work well for demountable joints in stainless steel apparatus. Even for glass apparatus, there is some advantage is using Viton O-ring joints and Viton and Teflon O-ring stopcocks rather than greased glass joints and stopcocks. Adsorption of vapors by the greaseless apparatus is relatively slight, so that it is much easier to keep the apparatus clean and to distinguish between real vacuum leaks and "virtual" leaks caused by desorption of vapors.

Strong, vacuum-tight joints between glass and metal can be made with high-quality epoxy resins (Araldite, made by Ciba-Geigy and sold as a domestic adhesive in Great Britain, is particularly effective). Such joints are inert to organic solvents, and they withstand moderate heating and cooling: the joints may fail below $-20°C$ and can be broken cleanly above 200°C. As shown in Figures 11 and 17, epoxy resin can also be used to cement copper tubes carrying water and electricity into the vacuum system through stainless steel or glass flanges. The joints are robust and can give years of service, although they can be broken as needed by strong heating.

Copper is superior to aluminum or brass for electrical leads, terminal blocks, and so forth, inside the vacuum system. It can be cleaned chemically, and it is unharmed by accidental heating to red heat under vacuum. Vacuum-tight joints between copper parts can be made with soft or silver solders: the latter are more tolerant of chemical cleaning in addition to withstanding higher temperatures.

Electrical or thermal insulating materials that outgas strongly should be excluded from the vacuum system, for example, plasticized polyvinylchloride, sodium-silicate-based refractory cements, and asbestos. A more detailed account of the vacuum properties of materials can be found in Ref. 13.

5.2 Vacuum-Pumping Systems

A mercury or oil diffusion pump backed by a rotary pump is normally an ideal pumping combination for preparative cryochemistry. However, the required pumping speed depends critically on whether or not permanent gases are evolved during the formation or reaction of the high-temperature species. The cold surface on which the reaction between the species and other compounds is taking place acts as a cryopump for condensible vapors and is particularly efficient if cooled by liquid nitrogen. In the absence of appreciable amounts of permanent gases, the speed of pumping need only be 0.5–5 liters/sec from vessels of 1 to 20 liter capacity. These speeds require only small diffusion pumps connected to the vessels by 10–25 mm bore tubing. A much more elaborate pumping system is needed to handle permanent gas evolution. For example, liberation of 4 mmole/hr of a permanent gas in the reaction vessel requires a pumping speed of 20 liters/sec to maintain the pressure at 10^{-3} torr. The diffusion pump used then has to have a speed of at least 40 liters/sec and to be connected to the vessel by not more than 300 mm of 50 mm bore tubing. A valve to isolate the vessel from the pumping system has to have a corresponding large bore: glass stopcocks of sufficient conductance may not be available, and metal butterfly or gate valves must be used (see Figures 11 and 17).

Fairly high pumping speeds for permanent gases are particularly important in apparatus that uses electron bombardment or induction heating to form the high-temperature species. Even brief surges of pressure above 10^{-3} torr may cause vigorous discharges with these methods of heating.

A liquid-nitrogen-cooled trap between the diffusion pump and the reaction vessel is essential when the pump fluid is mercury, and it is desirable when the pump fluid is an oil. The trap should have a high conductance: it should also be easy to clean as it is almost inevitable that small amounts of vapor from the reaction vessel will be collected in the trap.

It is useful to check the ultimate vacuum obtainable in an apparatus by using a hot-filament-ionization gauge. If this type of gauge is not available, a Penning gauge or a McLeod gauge can be used. When the apparatus is being used, the pressure is conveniently monitored with two Pirani or thermocouple gauges, one attached to the reaction vessel and the other mounted between the diffusion pump and the rotary forepump. These gauges only detect pressures above 10^{-3} torr. Thus, any movement of the gauge connected to the reaction vessel indicates that the pressure is becoming too high. Provided that there is a liquid-nitrogen trap between the diffusion pump and the reaction vessel, a pressure rise in the foreline gauge shows that permanent gas is being generated in the reaction vessel.

5.3 The Reaction Vessel

The principal forms of reaction vessel that have been used for preparative cryochemistry are shown schematically in Figure 8. Detailed diagrams are shown later (also in Figure 7).

The cold finger apparatus (Figure 8*a*) is generally less satisfactory than the other forms of apparatus for preparative scale work because the surface area of the cold finger is limited, and any condensate that peels off falls into an uncooled portion of the apparatus. The apparatus (Figure 8*b*) is only useful for reactions of long-lived species, for example, SiF_2, as the gas has to flow around a corner before condensing in the trap.

Since the original work by Skell on carbon atoms, the majority of reactions of short-lived species have been carried out on a preparative scale in apparatus of the type shown in Figure 8*c*. The furnace forming the species is mounted inside the cooled reaction vessel, and vapors of other compounds

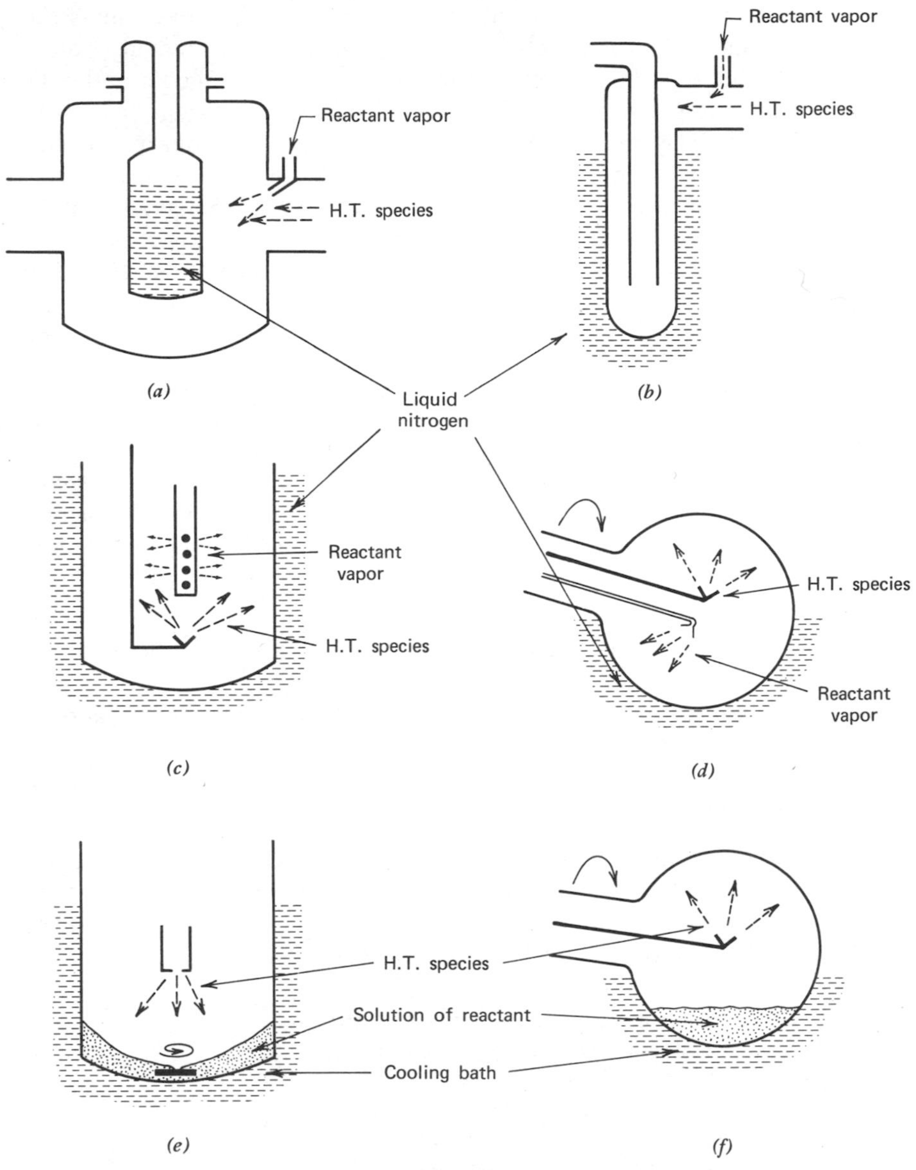
Reactant vapor
H.T. species
(a)
Reactant vapor
H.T. species
(b)
Liquid nitrogen
Reactant vapor
H.T. species
(c)
H.T. species
Reactant vapor
(d)
H.T. species
Solution of reactant
Cooling bath
(e)
(f)

Figure 8.

are sprayed into the vessel to condense with the species on the walls. Both cylindrical and spherical vessels have been used with radii ranging from 5 to 13 cm. Choosing the size of reactor depends partly on the desired scale of operation and partly on the complexity of the furnace system that has to be fitted inside. The scale of experiment that can be carried out in a reactor is limited by conduction of heat radiation from the furnace through the layer of condensed material on the cooled walls. When the temperature of the surface of the condensate rises too much, the surface ceases to act as a cryopump. The condensation efficiency declines sharply when the surface temperature corresponds to an equilibrium vapor pressure of about 10^{-4} torr. The more volatile the reactant compound the smaller the quantity that can be condensed.

The heat-transfer effect can be estimated by making the following assumptions:

1. The thermal conductivity of the cocondensate is about 0.025 watts/cm^2 for a 1°C temperature difference across 1 mm thickness. If the reaction vessel is glass, the thermal conductivity of the walls corresponds to roughly 0.5 mm thickness of cocondensate, although stainless steel vessels may have higher conductivities.
2. The furnace forming the high-temperature species acts as a point source radiating H watts.
3. Heat liberated on the cold surface through condensation of vapor is an insignificant fraction ($<10\%$) of the radiative heat flux: this approximation holds true for nearly all examples of preparative cryochemistry for which details have been published.

The surface temperature of the condensate is then $(196 - T)$ degrees, where $T = (d + 0.5)(320H/r^2)$, d being the thickness of the condensate in millimeters and r the distance in millimeters between the furnace and the cold walls. The figures given in Table 6 are calculated using this relationship.

Table 6. Relation of Distance from Heat Source to Surface-Temperature Rise

Distance (mm)	Surface-Temperature Rise (°C per mm thickness per watt radiated from heat source)
40	0.30
60	0.13
80	0.07
100	0.05

For example, assuming that a liter flask (radius, 62 mm) is used as the reaction vessel and it contains a furnace radiating 200 watts, then roughly 3.4 mm of benzene (100 g) or 0.6 mm of propene (20 g) could be condensed on the liquid-nitrogen-cooled walls before radiation would raise the vapor pressure of the condensate to 10^{-4} torr.

Any cracking of the layer of condensed material away from the cold walls drastically reduces the effective heat conduction from the surface of the cocondensate. Sometimes the addition of inert diluents along with the other condensing vapors can help to create a stable, adherent glass on the cold walls.

Many of the furnaces used to make high-temperature species do not act as point sources of either the species or heat radiation, but they give restricted beams that land on a relatively small part of the total area of a vessel. It is important to assess the maximum heat flux that will fall anywhere on the cold surface and to restrict the thickness of condensed layers so as not to exceed the limiting surface temperature locally. An over-heated area on the condensing surface generally still collects the high-temperature species, but not the vapor of the reacting compound—this condenses on cold portions of the surface. The overall effect is to waste the high-temperature species, but the pressure in the apparatus remains low because of efficient cryo-pumping by parts of the cold surface.

The rotating reaction vessel shown in Figure 8*d* is a type of apparatus first described by Green.[18] He used a 5-liter flask rotating at 50–80 rpm (see Figure 18). A single O-ring provided the rotary vacuum seal. Recently a similar apparatus has been marketed by G. V. Planer Ltd. that has a more robust and long-lived graphite- and Teflon-based rotary seal (see Figure 21). The rotary motion makes it possible to spray the high-temperature species and the reactant vapor on to the opposite sides of the flask. This has advantages for thermally unstable compounds. It is also possible to deposit material on the cold walls in a narrow band that is spread out by the rotary action (see Section 5.4). However, when species and reactants are condensed separately in the rotating vessel, the rate of condensation of the species has to be restricted so that less than a monolayer is deposited on the cold surface in each revolution. Otherwise, each layer of the high-temperature species tends to polymerize before it is covered by reactant molecules. A rotary apparatus can be used to cocondense species and reactants by spraying them on the same area of the flask, but then there is little advantage in using this type of apparatus compared with more easily constructed static apparatus. Nevertheless, the rotary apparatus does provide the best way of reacting species with solutions as described below.

The reaction of a high-temperature species with a solution containing a reactant was first carried out by Skell[6] using an apparatus of the type shown in Figure 8*e*. However, a rotary apparatus like that in Figure 8*f* is more versatile, since it permits the high-temperature species to be sprayed upwards or downwards and it gives good mixing.[7] Flask sizes of 1–5 liters have been used. The simplest form of the apparatus employs the motor unit of a conventional rotary evaporator with the addition of a rotary O-ring seal suitable for high-vacuum work (see Figure 19). The flask rotates in a cooling bath maintained at a temperature that corresponds to a vapor pressure above the solution of 10^{-4} torr or less (Section 5.4). The speed of rotation has to be sufficient to carry a film of solution over the top half of the flask, but not so fast that the solution is constrained in a narrow belt around the equator of the flask. For a 2-liter flask, speeds of rotation of 60–100 rpm are satisfactory. As considered in Section 3, the rate at which the high-temperature species can be condensed on the solution is limited by the efficiency of mixing. Greater rates of rotation of the flask do improve mixing and reduce the risk of saturating the surface of the liquid with the condensed species.

Heat radiation from a furnace in the flask causes some surface heating, but the amount the pressure rises depends on the speed of vacuum pumping. The operating pressure has to be kept below about 10^{-2} torr; otherwise vapors may be thermally cracked on the furnace and cause the pressure to rise to levels at which the high-temperature species does not travel to the cold surface.

5.4 Introduction of Vapors in Cocondensation: Solvents for Solution Reactions

The amount of the vapor of a compound or a mixture of compounds that is condensed with a high-temperature species in preparative cryochemistry typically ranges from about 50 mmoles up to several moles in an hour.

When a vapor stream is passed into a high-vacuum system, there is rapid expansion. For this reason, it is not very difficult to achieve a fairly uniform deposition of vapor onto a cold surface. As far as possible, the vapor inlet must be directed away from the furnace producing high-temperature species to avoid thermal cracking of the vapor. Some of the vapor distributors that have been used can be seen in Figures 11–18. As mentioned in Section 5.3, the rotary reaction vessel used by Green[18] permits the most complete separation between the condensing vapor and the condensing species.

If the compound to be condensed with the high-temperature species is a gas at room temperature, it is conveniently stored in a flask of between 1- and 5-liter capacity, and it is bled into the reaction vessel through a metal or Teflon-and-glass needle valve. The rate of addition is monitored by the fall in pressure in the flask measured with a barometer tube. When very large volumes of gas are to be used, as in the preparation of diboron tetrachloride from boron trichloride and a mole of copper atoms,[12] the bulk of the purified gas can be stored in a lecture bottle attached to a 5-liter flask. The flask is first filled with the gas, and the desired rate of flow into the reaction vessel is established by opening the needle valve and observing the fall in pressure in the flask. Then the needle valve from the lecture bottle is opened to give an input of gas to the flask that maintains the pressure at a constant value. Only slight adjustments of the two needle valves are required to keep the flow to the reaction vessel stable for a long period.

Compounds that are liquid at room temperature can often be bled from a calibrated reservoir through a needle valve into the vapor inlet system, where they vaporize. With liquids boiling below about 120°C it is important to have a constriction or a length of capillary tubing on the high vacuum side of the needle valve. This prevents the needle valve being cooled by evaporation. Teflon valves, in particular, are almost impossible to keep correctly adjusted if evaporation is occurring as Teflon has a very high coefficient of expansion. To facilitate the evaporation of liquids boiling above 150°C it may be necessary to heat the inlet tubes by wrapping them with resistance-heated nichrome wire or tape. Heat from the furnace forming the high-temperature species can be used to assist in vaporizing liquids in the inlet system shown in Figure 15.

It is much more difficult to monitor the rate of evaporation of a solid into a vacuum system than it is to monitor the rate of introduction of a liquid or gas. Whenever possible, the solid should be heated to its melting point and introduced as a liquid. Alternatively, the solid can be dissolved in an inert solvent of similar volatility and the solution allowed to evaporate in the inlet system.

Compounds that are not volatile or are too unstable to be evaporated, can be reacted with high-temperature species in two ways: either by spraying a solution of the compound on to the walls of the reaction vessel where the high-temperature species is condensing, or by condensing the species into a solution of the compound. The spraying method has been described by Green and Young,[22] and it has been used in their rotary apparatus.

A solution of the compound in a volatile or involatile solvent is injected onto a rapidly rotating Teflon cone as shown in Figure 9. The solution is thrown off the rim of the cone in a narrow band of droplets that land on the cold walls of the rotating flask and are conveyed as a film into the region where the high-temperature species is condensing. If a volatile solvent is used, it is thought that the solid may land on the walls in the form of a very fine powder entrained in a glassy deposit of solvent vapor. Droplets of solution that reach the walls probably freeze directly to a glass.

When the high temperature species is sprayed into a solution of a compound as shown in Figures 8 and 19, it is essential that the vapor pressure

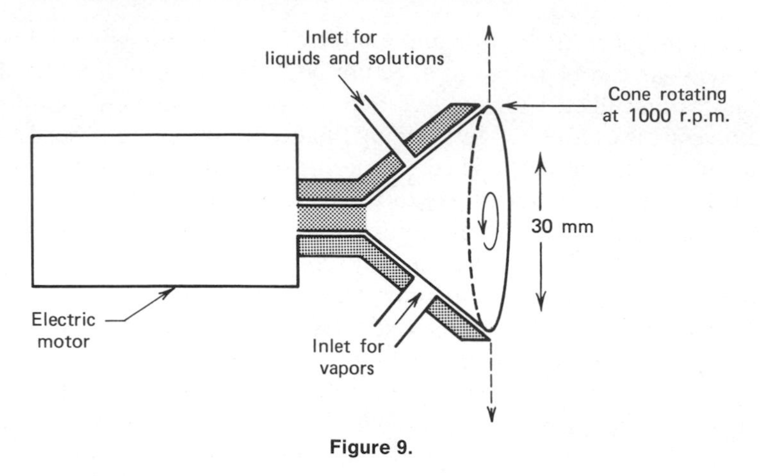

Figure 9.

Table 7. Solvents for Metal Atom Reactions

Solvent	Operating Temperature (°C)
Propane	−180 to −185
Isopentane	−145 to −160
Methylcyclohexane	−110 to −125
m-Diisopropylbenzene[a]	− 60 to −100
Bis(ethylhexyl)ether[a]	− 10 to − 60
Light liquid paraffin	+ 20 to 0
Silicone 704 (a phenylmethylsiloxane)[a]	+ 50 to − 20

[a] Some transition metal atoms react with these solvents.

of the solution is below 10^{-3} torr at the temperature of the cooling bath. Table 7 gives a list of liquids that are quite good solvents, are inert to many high-temperature species, and have low vapor pressures at their freezing points. The solutions must be at least 0.5 *M* in the compound to be reacted with the high-temperature species; otherwise the yields of products are very poor (see Section 3.1).

5.5 Recovery of Reaction Products

Many of the most interesting products that can be made by preparative cryochemistry are thermally unstable, and some decompose well below room temperature. Almost all of them are air sensitive. As a result, apparatus has to be designed carefully to permit recovery of products at low temperatures and in the total absence of air.

When a compound and a high-temperature species have been condensed together on a cold surface, the condensate contains (a) a large excess of the compound or a smaller excess of the compound plus a large excess of an inert diluent, (b) some of the polymerized form of the high-temperature species, and (c) the desired product probably mixed with reaction by-products.

If the product is sufficiently volatile, it can be removed from the reaction vessel by vacuum pumping. Usually the compound from which the product has been formed is more volatile and pumps off first. The successful recovery of unstable, volatile products may require a fast pumping speed between the reaction vessel and a liquid-nitrogen-cooled trap where the product can be collected. Then the reaction vessel can be warmed slowly from liquid-nitrogen temperature to a temperature at which the product has a vapor pressure of 10^{-2}–1 torr, and the product will be removed so rapidly that it will not have a chance to decompose. This technique has been important in studying low-temperature reactions of the boron and silicon subhalides (see Section 6.1). A Pirani or thermocouple gauge attached to the reaction vessel provides a useful indication of the pressure during the pump-out of products.

In a few cases it has proved advantageous to remove products by vacuum pumping even when they are not very volatile. For example, in the reaction of chromium with arenes, the bisarene chromium compounds can nearly always be evaporated from the condensate after the much more volatile excess arene has been pumped away. It may be necessary to warm the condensate and to have a very short path to a liquid-nitrogen-cooled surface. Nevertheless, this is simpler than using solution methods to recover these products as solvents also dissolve the involatile byproducts of the arene-chromium reactions.

The recovery of products, that are not very volatile but are soluble in organic solvents, is achieved by Schlenk techniques.[23] In order to get the product into solution at the end of a cocondensation reaction it may be necessary to wash the walls of the reaction vessel with a solvent. This is easy with small reaction vessels, for example, Figure 17, that can be removed from the vacuum pumping system and shaken after solvent has been added. However, with larger reaction vessels that are more permanently attached to the vacuum pumping system, it is often advantageous to have sprayed the vapor of a low-melting solvent on to the walls of the vessel before the cocondensation reaction is carried out. A uniform layer of condensed solvent, 0.5–1.0 mm thick, has only a slight effect on heat transfer to the cold walls; but at the end of the experiment, when the vessel is allowed to warm, the frozen solvent will melt and carry the condensate to the bottom of the reaction vessel. In the apparatuses shown in Figures 7, 17, and 18, solutions of products can be run out of the bottom of the reaction vessel by opening a stopcock or by removing a ground glass stopper with the vessel filled with an inert gas.

In vessels not equipped with outlets at the bottom, a stainless steel or glass tube must be inserted from the top when the vessel is filled with an inert gas, and the solution blown or sucked out. These various methods of recovering products in solution can all be used below room temperature, but it is generally difficult to isolate a product that decomposes rapidly below $-40°C$. The point in the recovery process at which the product is separated from excess reactant varies from case to case. If the reactant is an expensive or rare compound, then it is normally recovered at an early stage in the work-up to avoid diluting it with solvents. Some specific examples of procedures are given in Section 6.

Products that are made by reactions of high-temperature species with solutions are obviously in solution or in suspension at the end of a reaction, and this often facilitates their recovery. However, separation of products from oily, involatile solvents such as Silicone 704 can be difficult or impossible.

Any unreacted, finely divided metal present in the condensate at the end of a reaction may catalyze the decomposition of unstable products. This is of particular significance in the preparation of endothermic organometallic compounds from transition metal atoms and organic ligands. If the product is volatile it can be simply pumped away from the metal. When products are handled in solution, the finely divided metal must be removed by filtration as soon as possible. Filtration through fine glass frits or through filter paper is often successful, but the metal may be too finely divided and pass through or clog these filters. A short column of silica gel above a glass frit gives more effective filtration. Finely divided iron, cobalt, or nickel can be removed by "magnetic filtration". Nickel powder (200–300 mesh) is mixed with its own weight of silica gel: a short column of the mixture above a glass frit is put into a magnetic field of about 1000 G. The solution containing suspended ferromagnetic particles is passed through the filter bed, and the particles cling strongly to the magnetized nickel.

Some cryochemical reactions produce compounds that decompose at such low temperatures that they are almost impossible to isolate. Evidence that the products exist at all may come from a study of their decomposition products or from spectroscopic studies of the cocondensation reaction. However, the product can often be reacted with other compounds that are added to the reaction vessel at a low temperature before warm-up commences. Care must be taken to get good mixing between the unstable product and the added compound; otherwise decomposition of the product may still predominate. Rotary reaction vessels are well suited for achieving the required mixing.

The recovery of involatile, insoluble products from a reaction vessel usually requires that the whole vessel is transferred to an inert atmosphere glove box or glove bag and the product is scraped out. Adhesion of the condensate to the walls of the vessel can be minimized by precoating the cold walls of the reaction vessel with a thin layer of an inert, volatile compound before the cocondensation reaction takes place. Reaction vessels with parallel sides are obviously easier to scrape than spherical vessels.

5.6 Characterization of Products

Products obtained in preparative cryochemistry that are stable at room temperature can be characterized by the whole range of standard spectroscopic and analytical methods. However, difficulties may arise in the characterization of products that decompose well below room temperature, and specialized methods have to be used.

Mass spectrometry gives valuable information with compounds that exert a vapor pressure of at least 10^{-3} torr before they decompose. A glass-inlet system to a mass spectrometer of the type shown in Figure 10 can be used. A sample of 1–5 mg of the product is collected by vacuum-line techniques in a trap that can be attached to the inlet system. The sample is initially kept

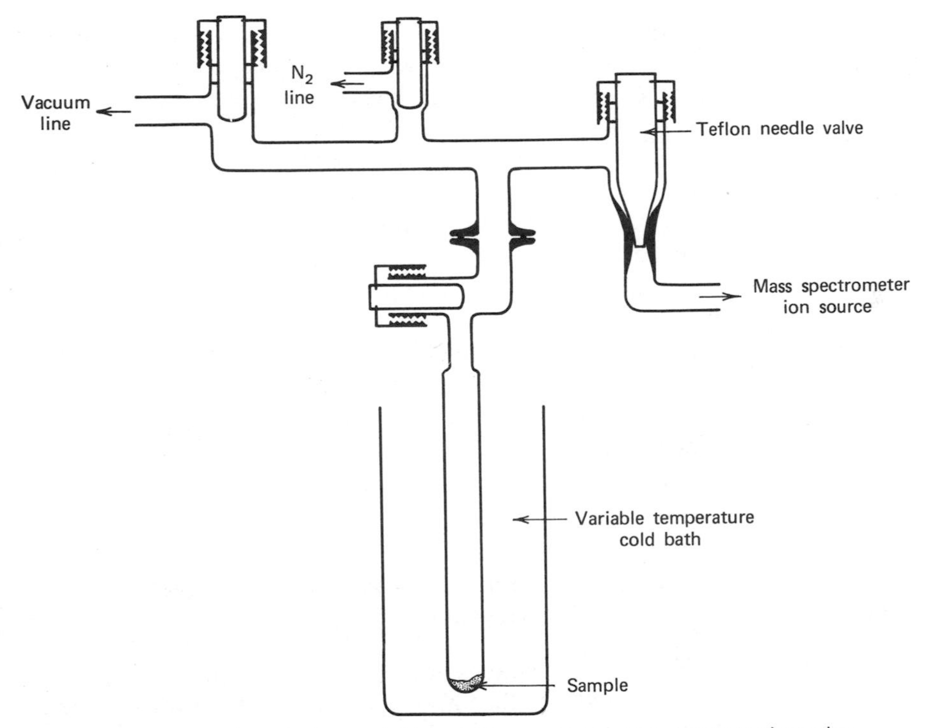

Figure 10. Introducing unstable volatile compounds to a mass spectrometer.

very cold. The Teflon-and-glass stopcock leading to the mass spectrometer is opened, and the sample gradually warmed until a spectrum of reasonable intensity is obtained. The required temperature corresponds to a vapor pressure of the sample of 10^{-2}–10^{-3} torr depending on the bore of the glass tubing leading to the mass spectrometer. After the spectrum has been run, part of the sample is pumped away into the pump trap, and another spectrum is run. The reproducibility of successive spectra is an important check on the homogeneity of the sample. The source of the mass spectrometer should be as near room temperature as possible to minimize thermal decomposition.

The infrared spectrum of some unstable products is best obtained by carrying out the cocondensation reaction on a cooled cesium iodide window. The technique is essentially that described in Chapter 8, except that the inert gas diluent is omitted and the high-temperature species and the reactant vapor are condensed in about the same ratio that would be used in preparative scale cryochemistry. The spectrum of the cocondensate at the deposition temperature (usually $-196°C$) is dominated by the bands from the excess reactant compound, but if the temperature of the window is raised with continuous vacuum pumping it is often possible to remove the reactant and to leave a film containing the product and some polymerized temperature species. In the reactions between transition metals and organic ligands, small amounts of condensed metal on the window do not interfere with the infrared spectrum of the organometallic products that may have been formed. The spectrum of the products can be obtained and, in some cases, the decomposition temperature of the complex observed as the window temperature is raised. This can be a useful guide to the practicability of making the compound on a preparative scale. The method is less satisfactory when the polymerized form of the high-temperature species has a very strong spectrum as this may dominate crucial parts of the spectral range.

Low-temperature nmr spectroscopy is a very powerful characterization method for products that are diamagnetic, contain appropriate elements, and are sufficiently soluble in an inert solvent to give a spectrum. Fourier-transform methods give enhanced sensitivity and permit more dilute solutions to be used. The transfer of a product to an nmr tube is carried out using vacuum line or Schlenk tube methods depending on the volatility of the product.

6. SOME EXPERIMENTAL PROCEDURES FOR HIGH-TEMPERATURE SPECIES DERIVED FROM THE NONMETALS

The detailed descriptions of apparatus in this section and in Section 7 are intended to give some insight into the important points of both construction and operation. Where possible, examples have been chosen from work done in the author's laboratory so that the descriptions can be based on first-hand experience. In a few cases, full experimental details are given. Summaries of the cryochemistry of some nonmetal species are shown as this information does not occur elsewhere in the book (see Chapter 7 on metals)

6.1 Experiments with High-Temperature Species Formed from Gases

Apparatus for Forming Short-Lived Species by Gas-Solid Reactions

Figure 11 shows an apparatus originally designed for studying reactions of boron monofluoride and silicon dichloride,[24] but which is suitable for work with a wide variety of high-temperature species formed from gases and vapors. The glass reaction vessel is made in two parts joined by an O-ring between ground-glass flanges. The upper part is rigidly fixed to a vacuum pumping system and to a vacuum pump-out line for volatile products. A short length of glass tubing and a 50 mm bore stainless steel butterfly valve connect the vessel to a liquid-nitrogen-cooled trap and a mercury diffusion pump. The pumping speed from the vessel is about 15 liters/sec. Volatile products can be pumped out through 50 mm bore tubing and a butterfly valve to a liquid-nitrogen-cooled collecting trap. Products can also be removed in solution by filling the vessel with nitrogen and spraying solvent on to the walls through a thin stainless steel tube inserted from the top of the vessel. The tube is then used to suck out the solution. The apparatus is not well designed for the recovery of involatile, insoluble products as it is not practicable to put the whole reaction vessel into a glove box or bag, and removal of the bottom half of the vessel in the total absence of air is quite difficult.

The furnace forming the high-temperature species is mounted to spray the species into the bottom half of the reaction vessel, which is cooled in liquid nitrogen. The induction heating device shown in Figure 11 is a very good method of heating boron or silicon in a graphite tube to prepare boron and silicon subhalides. A water-cooled copper work coil inside the vacuum vessel is connected to a 6 kW, 450 kHz generator. The copper tubes are sealed through the glass at the top of the reactor using epoxy-cement. The susceptor inside the coil is a 75 mm long, 25 mm diameter graphite tube of 12 mm bore, with a 3 mm bore nozzle outlet at the bottom. The susceptor contains 4–10 mesh pieces of crystalline boron or silicon with one or two larger lumps at the bottom to prevent small pieces dropping out of the nozzle exit. Thermal insulation of the susceptor and the outlet nozzle is effected by slipping on a stack of rings each cut with cork borers from 6 mm thick graphite felt: the felt is a relatively poor electrical conductor, so it is only slightly heated by the high-frequency current. To prevent heat radiation

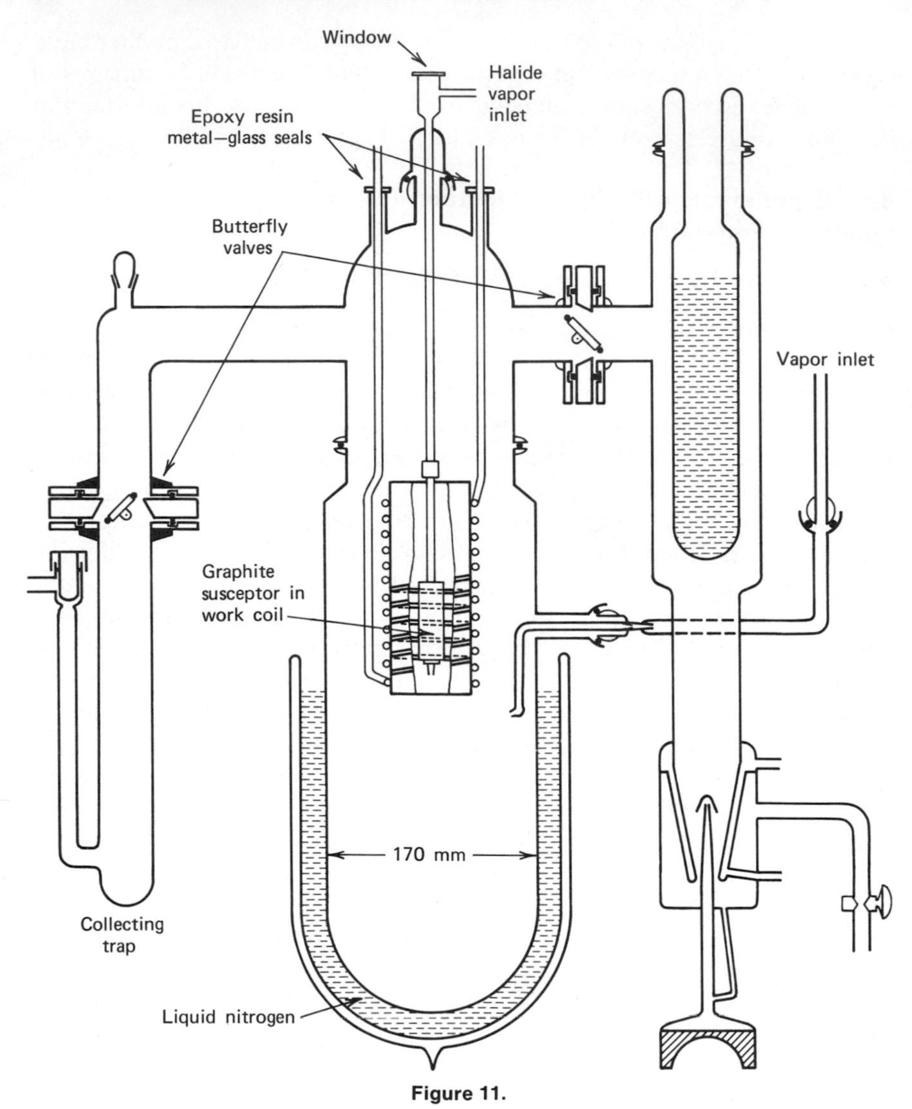

Figure 11.

escaping through the turns of the copper work coil, the outside of this is wrapped with an alumino-silicate wool. The susceptor is connected to a gas inlet system through a 6 mm diameter graphite tube, screwed into a copper tube that is sealed to glass: the relatively small diameter of the graphite tube prevents too much heat being conducted from the susceptor to the copper-glass seal. A great variety of other susceptors can be used in place of the graphite tube for special applications. For example, a quartz tube inserted through a stainless steel tube can be heated to 1200°C in the work coil, and an alumina tube surrounded by a molybdenum tube can be heated to 1700°C.

If an induction-heating unit is not available, resistance-heated, wire-wound ceramic-tube furnaces can be used to reach temperatures to 1700°C. When heating boron, it is almost essential to use graphite as the container material above 1600°C. Graphite cannot be heated satisfactorily in a metal-wire-wound furnace, and direct resistance heating of a graphite tube or spiral must be used. To be successful such graphite furnaces require careful engineering, and induction heating of graphite is much simpler.

Compounds to be condensed on the cold surface with the high-temperature species are sprayed in from a tube on one side of the reaction vessel. The end of the tube is ground at an angle that creates a satisfactory uniform deposition of vapor over the cold surface of the reaction vessel.

The following experimental procedure is typical of the use of the apparatus as shown in Figure 11 for studying reaction of boron monofluoride formed from boron and boron trifluoride.

The Preparation of $C_4Me_4B_2F_2$ *from 2-Butyne and BF.*[25,26] Boron monofluoride adds efficiently to 2-butyne on condensation at −196°C according to the equation,

$$2BF + CH_3C{\equiv}CCH_3 \xrightarrow[-196^\circ C]{\text{cocondense}} F{-}B \langle\ \rangle B{-}F$$

For the preparation, the apparatus is assembled as in Figure 11 with boron in the graphite susceptor. The reaction vessel is initially evacuated to below 5×10^{-5} torr, the pressure being measured by an ionization gauge. Then power is applied to the water-cooled work coil, and the temperature of the susceptor is gradually increased making certain that the pressure in the vessel does not rise above 10^{-3} torr so that a gas discharge is not formed. The temperature of the boron is measured by sighting with an optical pyrometer down the tube leading to the top of the susceptor. The time required to reach the operating temperature of 1800°C depends on the amount of outgassing from the susceptor, but is usually 20–30 min.

The reaction vessel below the susceptor is immersed in liquid nitrogen. The vapor of 2-butyne is bled into the vessel through the inlet at the rate of about 4 mmole/min. The flow of boron trifluoride through the column of hot boron in the susceptor is established at a rate of about 1 mmole/min. A slight pressure rise is observed during the cocondensation, but the pressure is easily kept below 5×10^{-4} torr. The run is continued for 30–40 min, continually topping up the liquid nitrogen level around the reaction vessel.

Sometimes the run has to be stopped prematurely because the nozzle outlet from the susceptor has become blocked with boron or boron carbide: this is less likely to occur if great care is taken to insulate the nozzle with graphite felt when the susceptor is assembled.

At the end of the run, the susceptor is allowed to cool for 20–30 min. The butterfly valve leading to the diffusion pump is closed, and the liquid nitrogen is taken away from the reaction vessel. Any volatiles released from the cold surface during warm-up are pumped into the collecting trap cooled in liquid nitrogen. While the excess 2-butyne is pumping away the pressure in the vessel may reach 0.2 torr, but when this has gone the pressure falls slowly as progressively less volatile materials are pumped out of the condensate. All the product can be assumed to have transferred to the collecting trap when the pressure in the reaction vessel, measured with a Pirani gauge, has fallen below 10^{-2} torr. The butterfly valve on the trap is closed and the materials in the trap are fractionated by trap-to-trap distillation on an attached vacuum line. The desired product, $C_4Me_4B_2F_2$, is among the least volatile substances in the trap, and it does not pump through a U-trap on the vacuum line cooled to $-30°C$. It is contaminated by traces of oily polymers of 2-butyne and by a compound, $C_4Me_4B_3F_5$. On standing at room temperature, the 2-butyne polymers tend to become involatile, and $C_4Me_4B_3F_5$ decomposes to $C_4Me_4B_2F_2$ and BF_3, and the crude product is easily freed from impurities.

In a 30-min run, about 50 mmoles (540 mg) of the 300 mmole charge of boron in the susceptor is vaporized as BF: the weight loss is measured by weighing the complete susceptor assembly before and after the experiment. The yield of $C_4Me_4B_2F_2$ is about 20 mmoles (3.4 g), 53% of theoretical. The involatile residue left on the walls of the reaction vessel is extremely air sensitive and may inflame, so that care has to be taken when dismantling the apparatus for cleaning. Only the bottom half of the reaction vessel which can be detached needs cleaning after each run. However, a thin film of condensed material builds up slowly on the glass of the upper half of the reaction vessel, and this has to be cleaned after a few runs otherwise it is difficult to out-gas the apparatus. The nozzle in the graphite susceptor has to be unscrewed or drilled out after each run and replaced by a fresh piece of graphite tube, as a nozzle used more than once invariably seems to block.

Scheme 1 summarizes some other reactions of boron monofluoride that have been carried out using the apparatus of Figure 11.

The apparatus of Figure 11 has also been used to study the chemistry of silicon dichloride.[29] This species is made by passing the vapor of silicon tetrachloride over silicon at 1200–1400°C: the higher the temperature the better the yield of silicon dichloride, but if a graphite container is used for the silicon, temperatures above 1400°C cause a vigorous reaction of silicon

Scheme 1. Some Cocondensation Reactions of Boron Monofluoride at −196°C

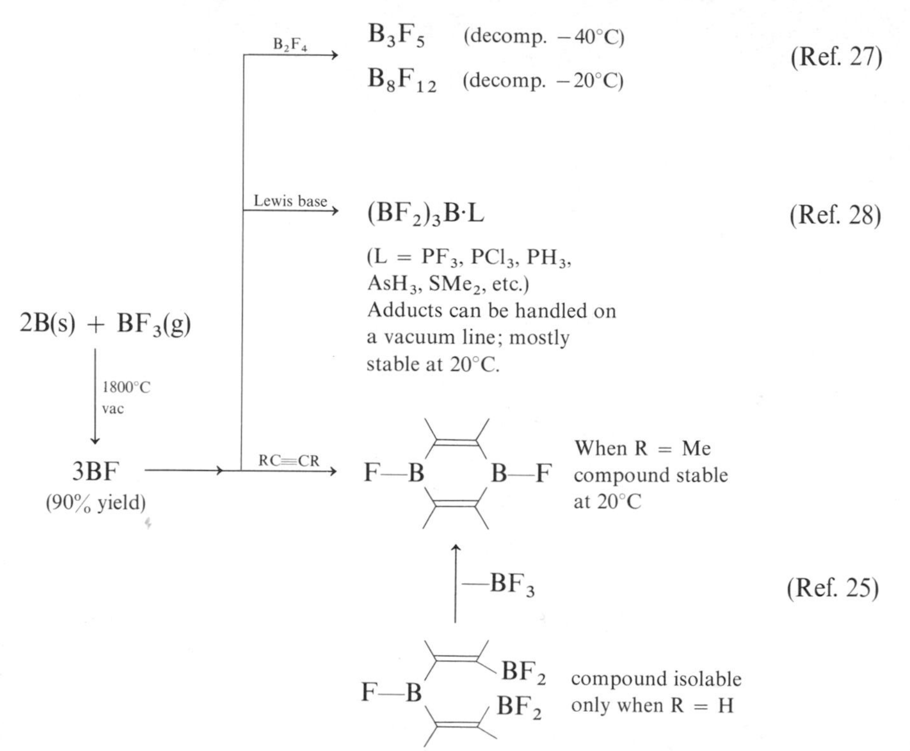

and carbon which breaks the container. Scheme 2 shows some of the chemistry of silicon dichloride. The following description of the reaction between silicon dichloride and diboron tetrachloride illustrates how an intermediate that cannot be isolated can be converted to a stable product.

The Preparation of $(SiCl_3)_2(BCl_2)BCO$.[28] Silicon dichloride reacts on condensation with diboron tetrachloride at −196°C, presumably according to the equation,

$$SiCl_2 + B_2Cl_4 \rightarrow SiCl_3BClBCl_2$$

Attempts to isolate this product have yielded only intractable materials. However, if the product is allowed to warm from −196°C in the presence of carbon monoxide, a carbonyl can be isolated whose formula suggests the initial presence of $SiCl_3BClBCl_2$, that is,

$$2SiCl_3BClBCl_2 \rightarrow B_2Cl_4 + (SiCl_3)_2(BCl_2)B \text{ decomposition on warming}$$

$$\downarrow CO$$

$$(SiCl_3)_2(BCl_2)BCO$$

Scheme 2. Some Cocondensation Reactions of Silicon Dichloride at −196°C

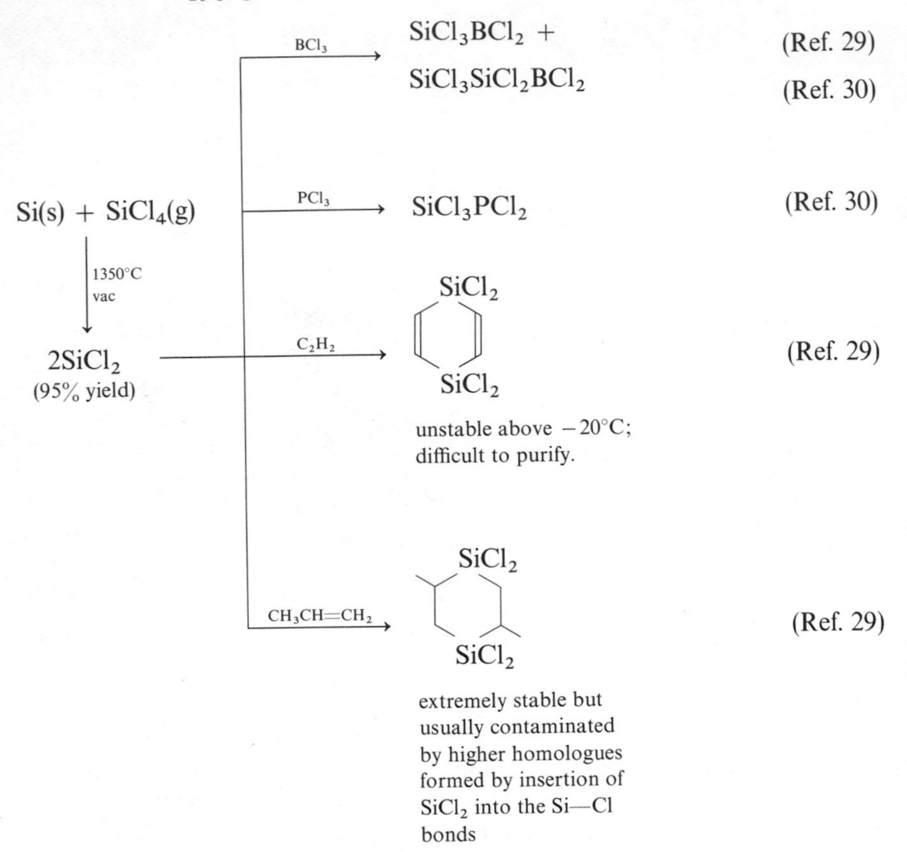

For the preparation, silicon granules are put into the graphite susceptor in the apparatus of Figure 11: the silicon should be at least 99.5% pure, less pure material may melt unexpectedly and run out of the bottom of the susceptor into the glass vessel. When the temperature of the silicon has been raised slowly to 1320–1350°C, the reaction vessel is cooled in liquid nitrogen. The vapor of silicon tetrachloride is passed through the susceptor at the rate of about 0.33 mmole/min, forming about 0.60 mmole/min of silicon dichloride. Simultaneously, diboron tetrachloride vapor is let into the vessel at the rate of 1.2 mmole/min. The run is continued for an hour: blocking of the nozzle of the susceptor is much less of a problem in the preparation of silicon dichloride than in the preparation of boron monofluoride. The susceptor is then allowed to cool, and the vessel is filled with dry carbon monoxide at a pressure of 300 torr. The liquid nitrogen cooling is removed, and the vessel allowed to warm to room temperature in the presence of the carbon monoxide. The condensate remains colorless during warm-up, and a colorless liquid collects at the bottom of the reaction vessel. After a few minutes at room temperature, the vessel is recooled in liquid nitrogen, and the carbon monoxide is pumped away slowly through the butterfly valve on the collecting trap. By this second cooling procedure, the difficulty of condensing vapors out of a noncondensible gas is largely overcome, as the vapors remain in the reaction vessel. Volatiles liberated when the vessel is finally warmed to room temperature are condensed in the collecting trap: pumping is continued for about an hour to be sure that all volatiles have been removed from the glassy residue in the reaction vessel. The desired product is obtained almost pure by allowing the contents of the collecting trap to pass through traps at -20 and -40°C: crystalline $(SiCl_3)_2(BCl_2)BCO$ condenses in the -40°C trap. The yield is 3 mmole (1.2 g), 15% of the theoretical.

Apparatus for Making Long-Lived Species by Gas-Solid Reactions

Simpler apparatus than that of Figure 11 can be used to prepare silicon difluoride and react it with compounds at a low temperature. This is because silicon difluoride is much longer lived in the gas phase than other silicon or boron subhalides, and it survives being passed along tubing and around corners at a pressure of 0.1–1 torr. In the apparatus of Figure 12, silicon difluoride is prepared by passing SiF_4 (99.9% pure) upwards through a column of pea-size silicon lumps heated to 1250°C in a 20 mm bore mullite tube. It is important that the top of the column of silicon is within the 1250°C hot zone of the furnace, otherwise disproportionation of SiF_2 to silicon and SiF_4 may occur on the surface of the cooler silicon lumps. A flow of SiF_4 of up to 100 mmole/hr can be used before silicon lumps are blown off the top of the column. The pressure at the top of the column at the above flow rate is about 0.5 torr, and the gas is a mixture of SiF_2 and SiF_4 in a mole ratio of about 5:3. Vapors to be condensed with the SiF_2 are introduced into the gas stream just before the cold trap.

The yields of products obtained in the reactions of SiF_2 are generally lower than those obtained with other carbenoids. Monomeric, ground-state SiF_2 does not react with many compounds on condensation at low temperatures, and much of the observed chemistry of the compound involves reactions of the polymers Si_2F_4, Si_3F_6 . . . , formed from SiF_2 in the condensed phase. High yields of products are obtained only with compounds that are powerful enough oxidizing agents to break the Si—Si bonds formed by polymerization of SiF_2, for example, I_2 or CF_3I. The preparation and

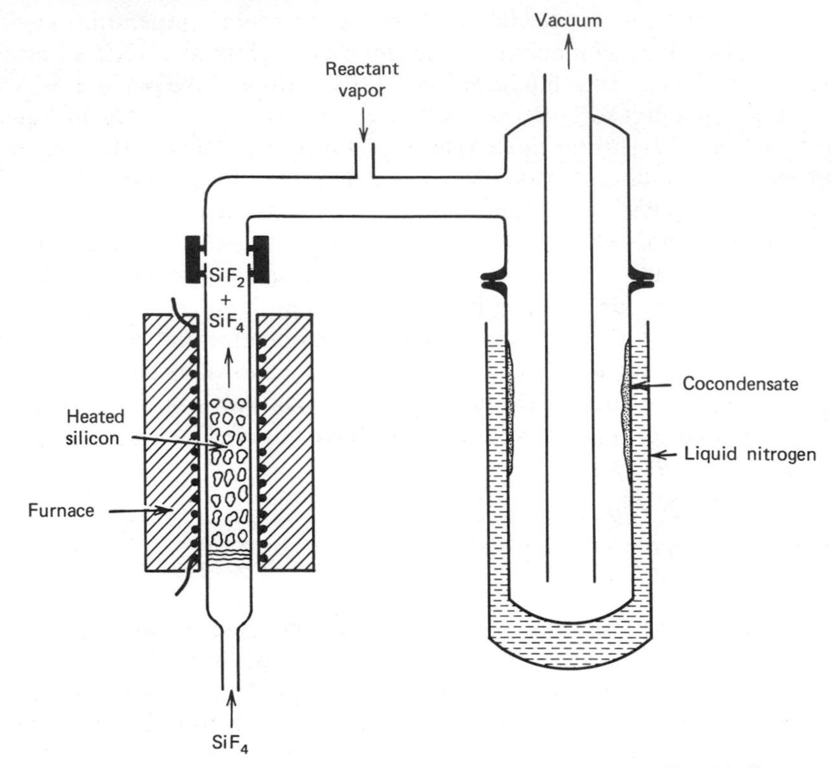

Figure 12. Apparatus for cryochemistry on silicon difluoride.[31]

chemistry of SiF_2 has been extensively reviewed.[24,31] Scheme 3 illustrates the classes of reactions that have been carried out in the apparatus of Figure 12.

Silicon difluoride can also be made in the apparatus of Figure 11 when it passes from the hot zone to the liquid-nitrogen-cooled walls without intermolecular collisions. In this apparatus, cocondensation of SiF_2 and BF_3 yields SiF_3BF_2 plus homologs[38] in the series $SiF_3(SiF_2)_nBF_2$ ($n = 1\text{–}4$), whereas no SiF_3BF_2 is formed when the apparatus of Figure 12 is used.[34] Clearly, a short-lived species is formed along with the long-lived ground state SiF_2, and this may be SiF_2 in an excited electronic state, although the experimental evidence does not rule out SiF_3 radicals as the active species.

Apparatus very similar to that in Figure 12 has been used to make HBS and $(FBS)_2$, which are also long-lived species,

$$H_2S + B(s) \xrightarrow{1250^\circ C} HBS + H_2 \qquad \text{(Ref. 39)}$$

$$SF_6 + B(s) \xrightarrow{1300^\circ C} (FBS)_2 + BF_3 \qquad \text{(Ref. 40)}$$

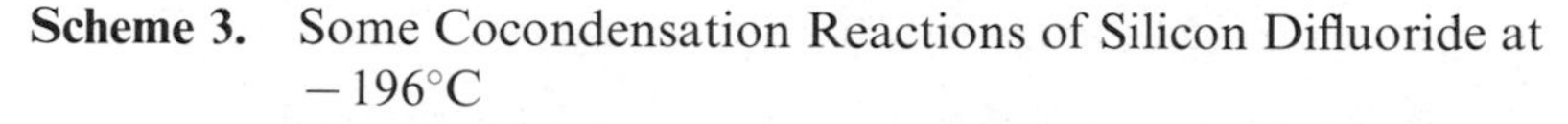

Scheme 3. Some Cocondensation Reactions of Silicon Difluoride at −196°C

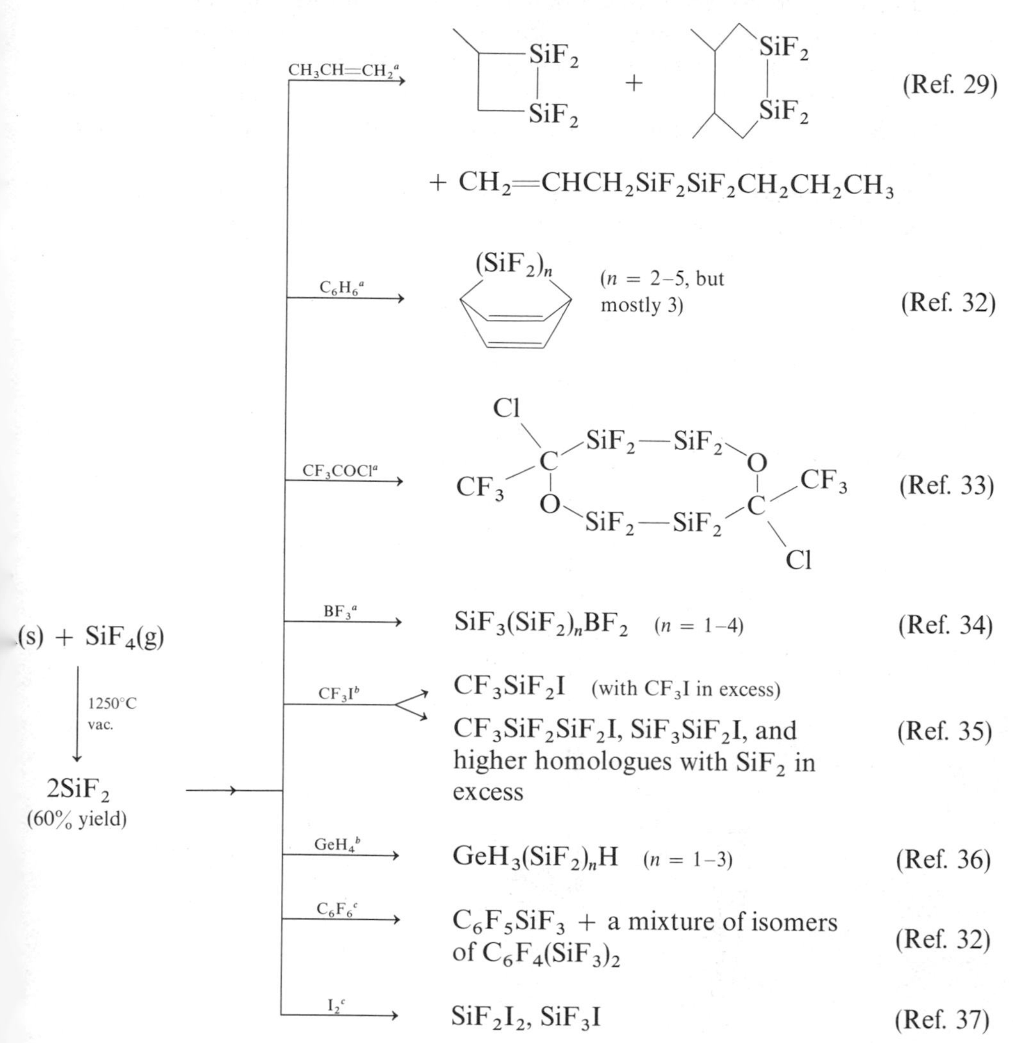

[a] Reactions in which only products containing two or more silicon atoms can be isolated. Yields, based on SiF_2, usually $<10\%$.
[b] Reactions in which both mono- and polysilicon products can be isolated.
[c] Reactions in which only monosilicon products can be isolated: the yield of iodosilanes can be made nearly quantitative.

Apparatus for Making Species by Flash Thermolysis of Vapors

Figure 13 shows apparatuses that have been used for thermally cracking gases. Diboron tetrachloride vapor has been cracked in the apparatus of Figure 13*a* to give boron monochloride.[26] As described in Section 4.2, cracking conditions have to be established that minimize the decomposition of B_2Cl_4 to solid boron and BCl_3. This is achieved by passing the vapor through a 20 mm long, 1 mm bore quartz tube heated to 1100°C by molybdenum-resistance windings, at a rate of 1–1.5 mmole/min. The pressure at the input end of the tube is then about 5 torr, while the gas expands

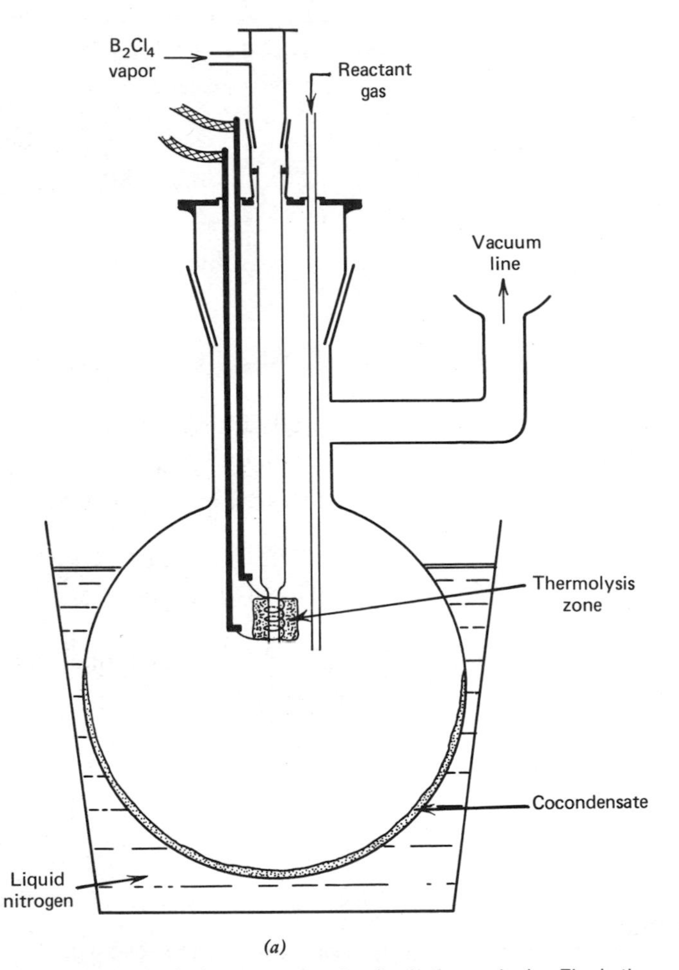

(*a*)

Figure 13. Apparatuses for making species by flash thermolysis, Flash thermolysis of (*a*) B_2Cl_4,[26] (*b*) P_2F_4,[41] (*c*) organic compounds.[42]

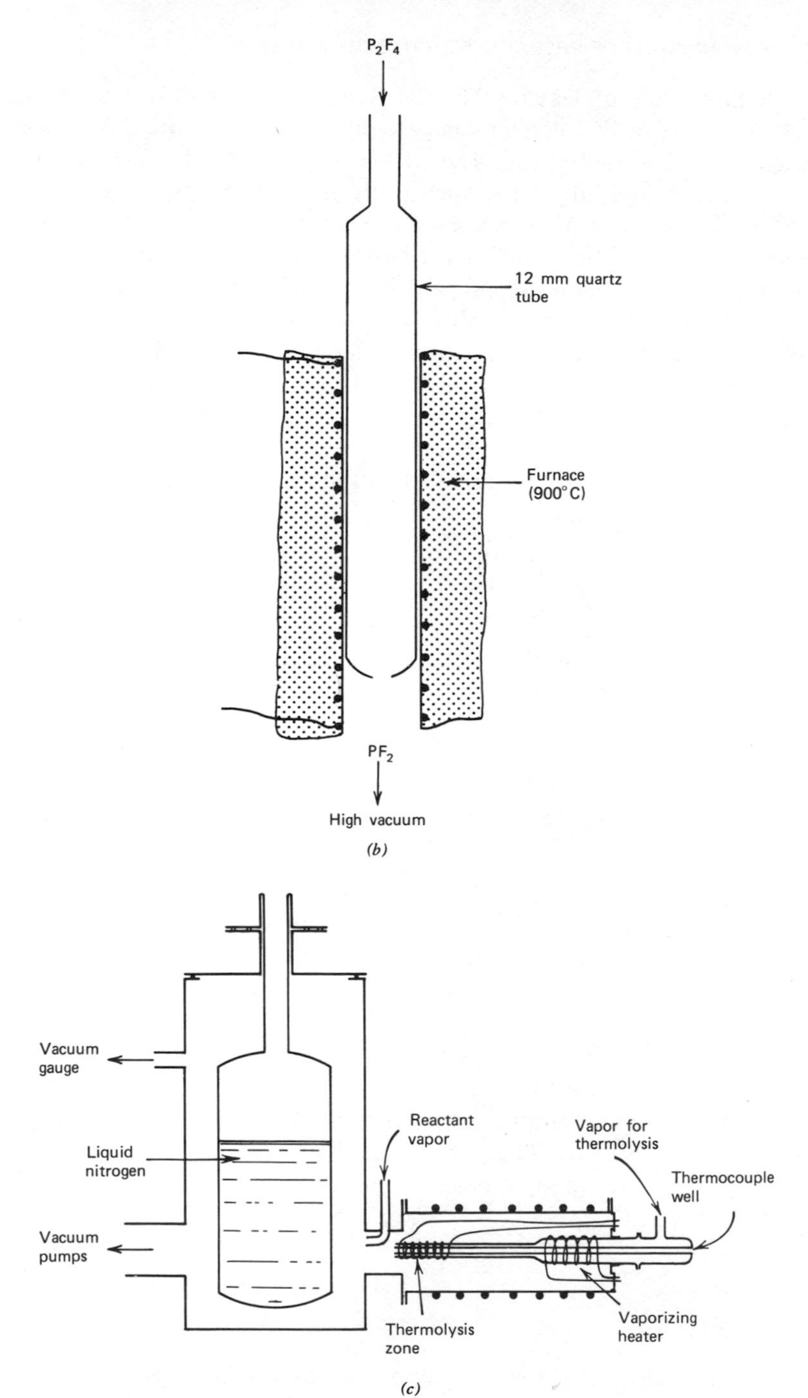

Figure 13. (*Continued*)

into a high vacuum on leaving the tube. A run of 20 min duration produces about 10 mmole of BCl: longer runs are difficult because the tube becomes blocked with deposited boron. The BCl emerging from the furnace is condensed on the liquid-nitrogen-cooled walls of the 1 liter reaction flask and vapors of reacting compounds are sprayed on to the walls simultaneously. The few reactions of BCl which have been studied are shown in Scheme 4. It should be noted that the preparation of BCl on the scale described depends on having an abundant supply of diboron tetrachloride. The preparation of this compound is described in Section 7.1. Boron monochloride has also been made from B_2Cl_4 by using an ac discharge (see Section 6.1). The apparatus of Figure 13*a* can be used to prepare SiF_2 from Si_2F_6 or $SiCl_2$ from Si_2Cl_6 using a tube temperature of 900–1000°C, although this method of preparation is less useful than the direct reduction of the tetrahalides by silicon (see above).

Scheme 4. Reactions of BCl (Ref. 26)

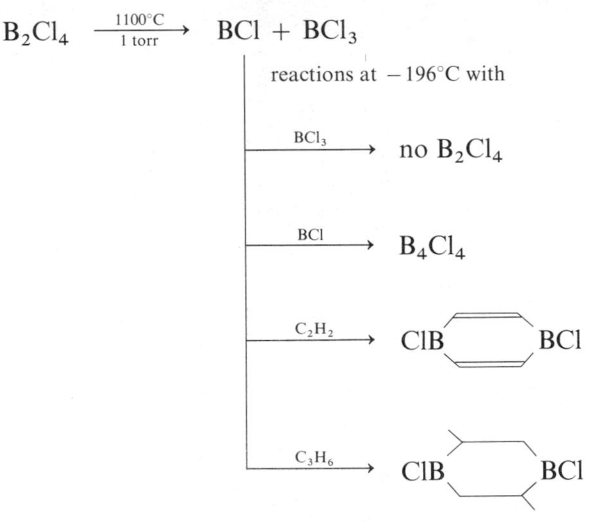

Figure 13*b* shows an alternative form of cracker tube to fit into the same evacuated flask assembly. This tube is suitable for cracking P_2F_4 to PF_2 at 800–900°C.[41] There are three possible modes of decomposition of P_2F_4,

$$P_2F_4 \rightarrow \begin{cases} 2PF_2(g) \\ P_4(g) + PF_3(g) \\ PF(g) + PF_3(g) \end{cases}$$

None of these reactions involves deposition of a solid in the hot zone so that the range of pressures that can be used for cracking P_2F_4 is much greater than for cracking B_2Cl_4. Provided that the residence time in the hot zone does not exceed a millisecond, PF_2 can be made in good yield from P_2F_4 at pressures of 0.01–2 torr. At higher pressures, cracking to P_4 becomes the dominant process. The only reaction of PF_2 that has been studied in detail is its reaction with itself on the cold surface to give a 10% yield of $P(PF_2)_3$ plus intractable polymers.

Unstable organic molecules have been made by flash thermolysis over a wide range of pressures and temperatures. The least stable molecules have usually been made by low-pressure thermolysis using an apparatus like that in Figure 13*c*.[42,43] This apparatus is very similar in basic design to that of Figure 13*a*, but the vacuum pumping speed is greater. This pumping capacity is necessary to remove hydrogen, which is a common byproduct of thermal decomposition of organic compounds. The hydrogen is, of course, not condensed by the liquid-nitrogen cold trap. The input pressure to the unconstricted thermolysis tube ranges from 1 to 0.01 torr for a throughput of organic vapor of 1–0.01 g per hour. Cracking products can be condensed with the vapors of other compounds on the liquid-nitrogen-cooled cold finger and reaction products pumped or washed off the finger at the end of the experiment.

Skell and Cholod[9] used an apparatus in which chloroform vapor passed through a short, resistance-heated graphite tube at 1500°C and emerged into an evacuated vessel containing a well-stirred solution of an alkene in an alkane cooled to −140 to −180°C. The dichlorocarbene formed reacted with the alkene.

Some longer-lived reactive species, for example, $Me_2Si{=}CH_2$,[44] have been formed by passing a mixture of the organic precursor and a large excess of an inert gas through a heated tube at atmospheric pressure. However, it is then difficult to quench the products because of the enthalpy of the gas stream. Commonly, other vapors are added to the gas stream after cracking, and the mixture is passed through traps in which products formed by gas phase reactions are collected.

The wide range of new chemistry that has resulted from flash thermolysis on organic compounds is surveyed in Ref. 6.

Apparatus for Preparing Species in Electrical Discharges

Figure 14*a* shows an apparatus used to prepare carbon monosulphide by passing an electrical discharge through carbon disulphide vapor,[21]

$$CS_2(g) \rightarrow CS(g) + S_8(s)$$

As CS is a fairly long-lived species in the gas phase, it can be passed at low pressure along glass tubing from the discharge zone to a cooled vessel almost without loss.

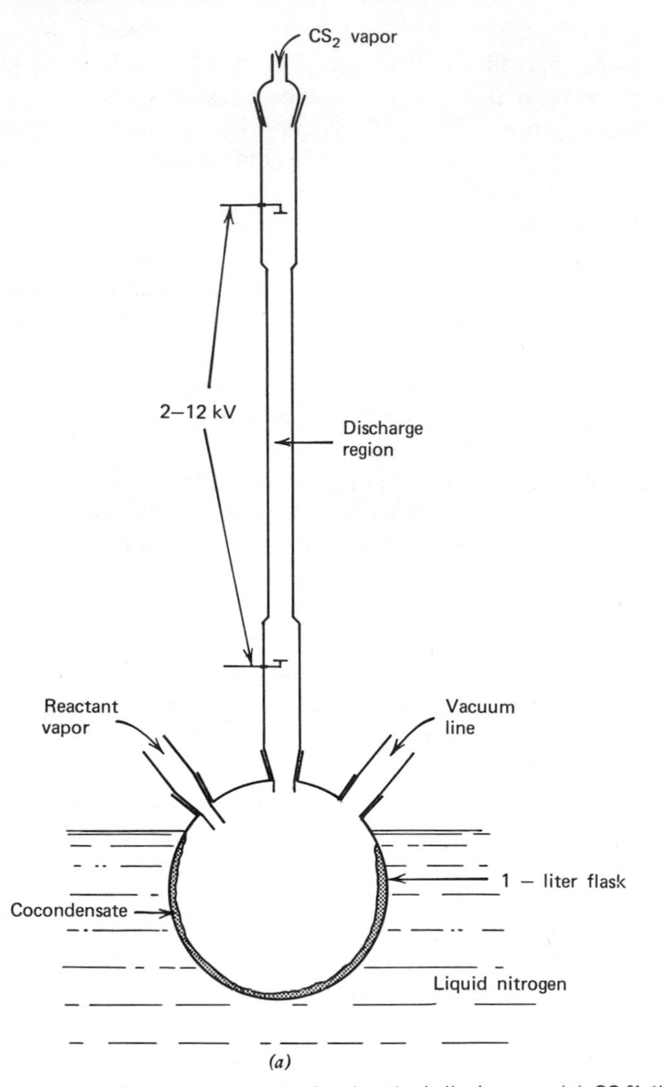

Figure 14. Apparatuses for making species in electrical discharges. (*a*) CS,[21] (*b*) BCl.[26]

The main part of the Pyrex discharge tube is 400 mm long and 18 mm diameter with electrodes sealed through the glass at either end. The vapor of CS_2 is passed through the tube at the rate of 40–150 mmole/hr giving a maximum pressure of a few torr at the upper end. The discharge is obtained by attaching a transformer rated at a maximum 12 kV and adjusting the power input to the transformer to give a discharge current of 10–30 mA.

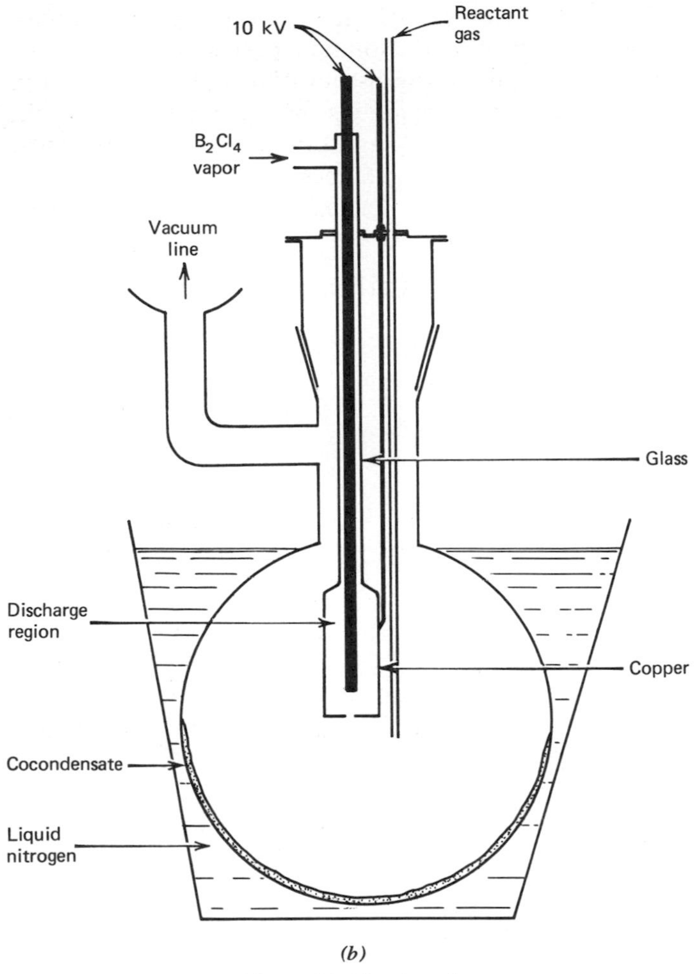

(b)

Figure 14. (*Continued*)

Under these conditions about 25% of the CS_2 is decomposed to CS: the sulfur plates out on the walls of the discharge tube and along the walls into the cooled flask. Reaction of CS with other compounds can be achieved in two ways. Using a 1-liter flask cooled in liquid nitrogen, CS and unchanged CS_2 can be condensed with the vapor of reactant compounds sprayed into the flask simultaneously. Alternatively, using a 250 ml flask the stream of CS and CS_2 vapors can be pumped over a well-cooled solution of a reactant compound in an inert solvent. Provided that the solution is thoroughly stirred, most of the CS is absorbed and reacts with the dissolved compound.

The solution must be cold enough that the vapor pressure above it is less than 1 torr but the viscosity of the solution must not be too high. Klabunde[21] recommends that a liquid-nitrogen-cooled metal U-trap is placed between the flask and the vacuum pumping system: any CS that passes through the cold flask is collected in the trap and can be allowed to polymerize safely, as the polymerization of CS is sometimes violent.

Some of the chemistry of CS that has been studied in the apparatus of Figure 14*a* is shown in Scheme 5.

Scheme 5. Some Reactions of Carbon Monosulphide (Ref. 21)

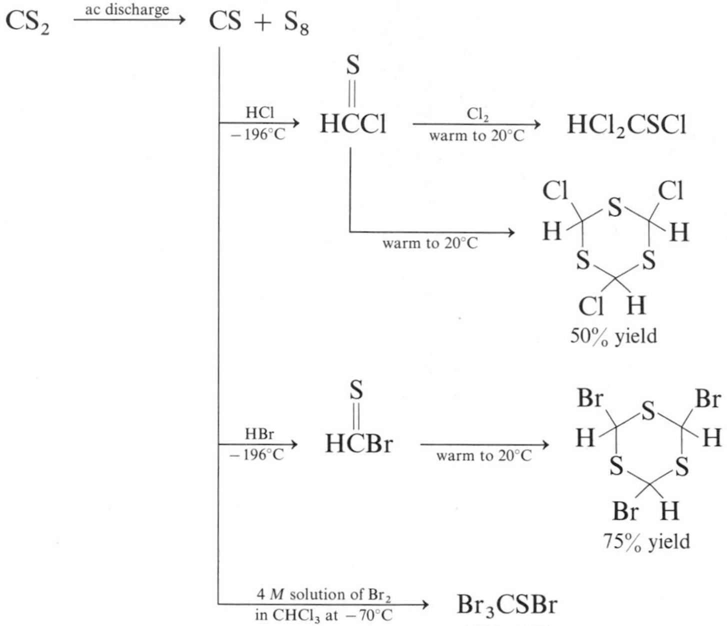

Figure 14*b* shows an apparatus that can be used to prepare BCl from an ac discharge in B_2Cl_4.[26] As BCl is much shorter lived than the isoelectronic CS, the species emerges directly from the discharge into a high-vacuum and thence on to a liquid-nitrogen-cooled surface. Some electrons escape from the discharge zone along with BCl, and these cause polymerization of the most sensitive organic compounds sprayed into the flask to condense with the BCl. However, it is possible to carry out each of the reactions shown in Scheme 4 with BCl made in the electrical discharge. With 10 kV applied between the inner and outer electrodes, B_2Cl_4 vapor is passed through the assembly at the rate of about 2 mmole/min, yielding about 0.7 mmole/min of BCl.

Comparatively little work has been done on low-temperature reactions of nitrogen atoms (active nitrogen). Havel and Skell[45] have bubbled nitrogen containing nitrogen atoms, formed in a microwave discharge at about 10 torr pressure, through cold liquid olefins. This seems to be an effective method for carrying out reactions that could be extended to studying other types of reactions of nitrogen atoms.

6.2 Experiments with High Temperature Species Formed from Condensed Phases

Generation of Species in a Heated Crucible

The apparatus shown in Figure 15 is suitable for reacting all types of high-temperature species that can be made in a crucible with vapors of reactant compounds on the liquid-nitrogen-cooled walls of the vacuum vessel. It can be used, for example, to study synthetic reactions of SiO or B_2O_2 among nonmetal species, and to study reactions of many transition metal atomic species as discussed in Section 7.

The glass reaction vessel is 300 mm long and 130 mm diameter, and it is evacuated through a 25 mm bore O-ring stopcock (J. Young and Co.). A demountable O-ring joint connects the vessel to a vacuum system comprising a rotary pump, mercury diffusion pump, and liquid-nitrogen-cooled cold trap. The pumping speed from the vessel is about 3 liters/sec. A pressure gauge mounted in the pump-out line gives a useful indication of pressure changes in the reaction vessel. A stainless steel ring, 60 mm i.d., 100 mm o.d., with an O-ring groove in it, is cemented to the top of the reaction vessel using epoxy cement. The O-ring mates with a flat stainless steel plate through which tubes bringing supplies into the vessel are cemented. A rim on the stainless steel ring locates the steel plate. This type of seal was designed originally for use by undergraduates in a teaching experiment on metal atom chemistry,[5] but it has proved so reliable that it has become a standard fitting on apparatus in the author's laboratory. Electrical power for resistively heating the crucible is brought into the vessel through two 6 mm diameter copper tubes that are cooled by water flowing through 3 mm copper tubes inside them. The crucible is attached to terminal blocks at the ends of the copper tubes, which are insulated from the stainless steel plate by ceramic washers. Currents of up to 100 A can be used with these electrical leads.

Volatile compounds to be condensed with the high-temperature species are pumped down the central glass tube and emerge through numerous holes to form a uniform layer on the cold walls of the reaction vessel. Compounds that are liquid at room temperature and boil below about 150°C can be simply run into the inlet system from a burette and evaporate down

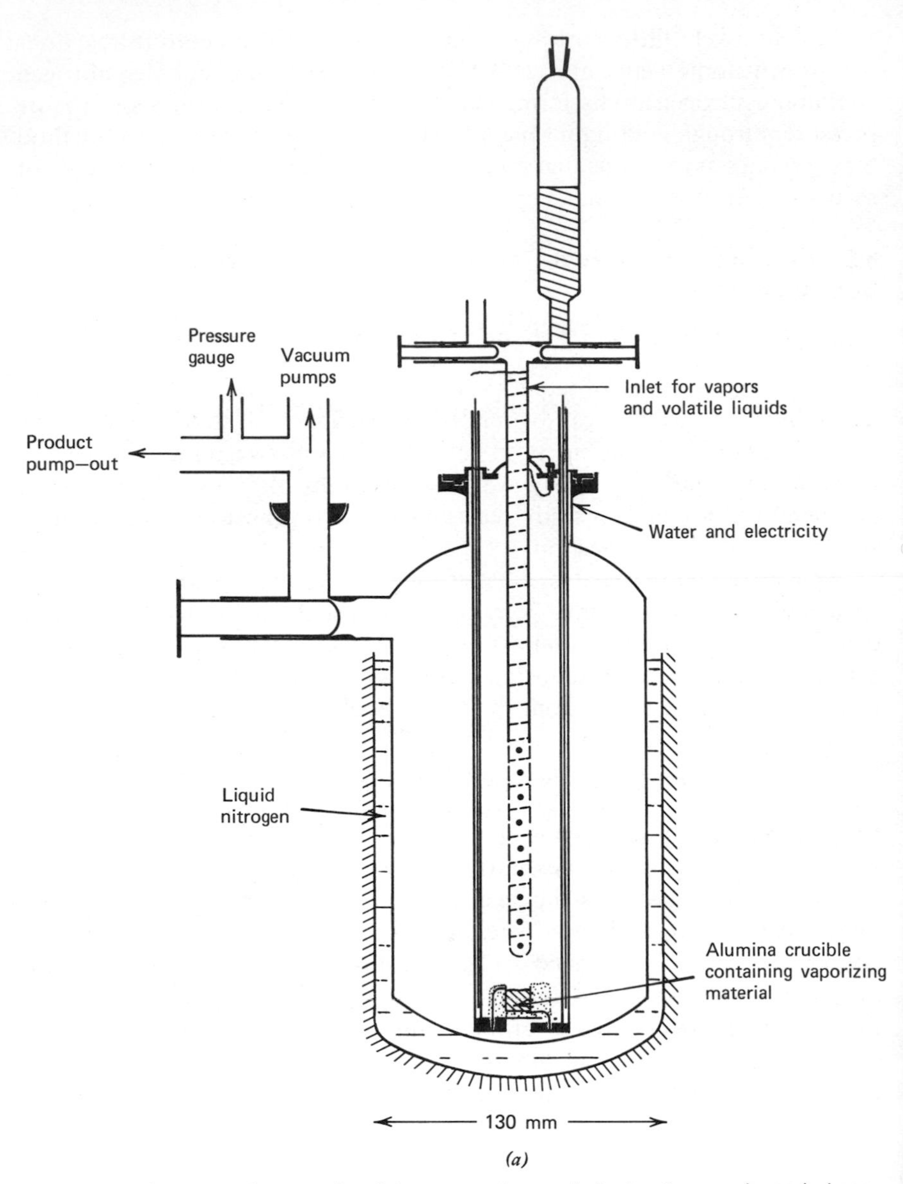

(a)

Figure 15. Apparatus for cryochemistry on species made by heating condensed phases. (*a*) Simple apparatus, (*b*) accessories to replace evaporator assembly from reaction vessel in (*a*).

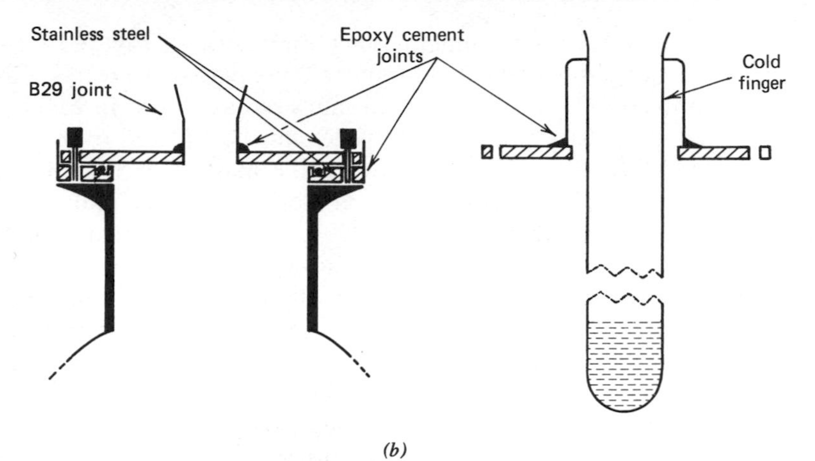

(b)

Figure 15. (*Continued*)

the distributor tube. With less volatile compounds (b.p. 150–300°C) the inlet system can be electrically heated by nichrome-resistance wire. Gaseous compounds are monitored into the reaction vessel from a gas burette system. With the walls of the vessel cooled in liquid nitrogen, compounds boiling over the temperature range −120°C to +300°C can added to the vessel at rates of up to 1 mole/hr while maintaining a high vacuum ($<10^{-3}$ torr) in the vessel. The bottom of the glass inlet tube is heated by radiation from the crucible just below it. If this is found to cause decomposition of a reactant compound, the bottom can be insulated with molybdenum or aluminum foil radiation shields. With extremely heat sensitive compounds, it is necessary to use an inlet tube that is water cooled.

When a cocondensation run has been completed, the reaction vessel can be filled with an inert gas and the evaporator assembly lifted out and replaced by either a blank stainless steel plate or a plate with a ground joint cemented to it as in Figure 15*b*. This procedure facilitates working up the product condensed on the walls of the vessel.

Preparation of SiO *and* B_2O_2. Silicon monoxide is made by heating a mixture of silicon and silica at 1250–1400°C under vacuum,

$$\mathrm{Si(s)} + \mathrm{SiO_2(s)} \rightarrow \mathrm{2SiO(g)}$$

In the laboratory it is most convenient to use "Silicon Monoxide," a material made commercially by condensing SiO vapor and that liberates SiO freely on heating under vacuum: it is used for condensing thin films of composition SiO on the surface of glass and ceramics.

To react gaseous SiO with compounds in the apparatus of Figure 15, 2–3 g of the commercial granular "Silicon Monoxide" are placed in a recrystallized alumina crucible of 15 mm bore and 25 mm deep as shown in Figure 3*c*. A cap of perforated 0.04 mm thick molybdenum sheet placed over the top of the granules reduces decrepitation when the material is heated under vacuum. The crucible is wrapped with a spiral of 1.2 mm thick molybdenum wire attached to the crucible with cement (99% alumina): thermal insulation is effected with a 15 mm thick layer of "Kao-wool" on the outside. Under high vacuum, a power input of about 300 W (35 A at 8 V) heats the crucible to 1400°C. At the start of an experiment, the apparatus is assembled as shown, and the reaction vessel is evacuated to about 10^{-5} torr. The crucible is heated slowly to about 1100°C: the temperature is estimated by sighting with an optical pyrometer through the side of the reaction vessel. After a few minutes outgassing of the "Silicon Monoxide" at this temperature, the reaction vessel is cooled in liquid nitrogen, the flow of reactant compound is begun, and the power to the crucible is raised to the level that will evaporate the contents at a rate of roughly 1 mmole/min (this power level has to be determined by a preliminary experiment). The flow of reactant compound is controlled at 5–10 mmole/min. After about an hour, the crucible is allowed to cool, and the reaction vessel is filled with dry nitrogen. The evaporator insert is removed, and a flange with a stoppered B29 glass joint put in its place. The vessel is reevacuated and allowed to warm to room temperature with continuous pumping. Excess of the more volatile reactant compounds can be pumped off at this stage, but if a less volatile reactant has been used it is easier to wash it away from the reactions products using organic solvents at room temperature. The products of reactions of SiO with organic compounds[46] are light colored, involatile, insoluble solids that rapidly absorb water from the atmosphere. They are best removed from the reaction vessel in a dry-nitrogen-filled glove box. Some of the reactions of SiO are summarized in Scheme 6.

A very similar procedure can be used to prepare gaseous B_2O_2 and react it with compounds at −196°C. The best precursor for B_2O_2 is partially oxidized solid boron. Finely divided boron is heated to 800°C under vacuum, and then oxygen is added cautiously: the boron glows but oxidation ceases at a composition of about B_3O. The solid so formed is ground in a mortar in a dry atmosphere and is then compressed into pellets that can be put into an alumina crucible to be heated to 1350°C under vacuum. The solid acts like an intimate mixture of anhydrous B_2O_3 and boron, and it liberates B_2O_2 according to the equation,

$$2B(s) + 2B_2O_3(l) \rightarrow 3B_2O_2(g)$$

The liquid-nitrogen cocondensation reactions of B_2O_2 are included in Scheme 6.

Scheme 6. Some Reaction of Silicon and Boron Suboxides at $-196°C$

$Si(s) + SiO_2(s) \rightarrow 2SiO(g)$ (Ref. 46)

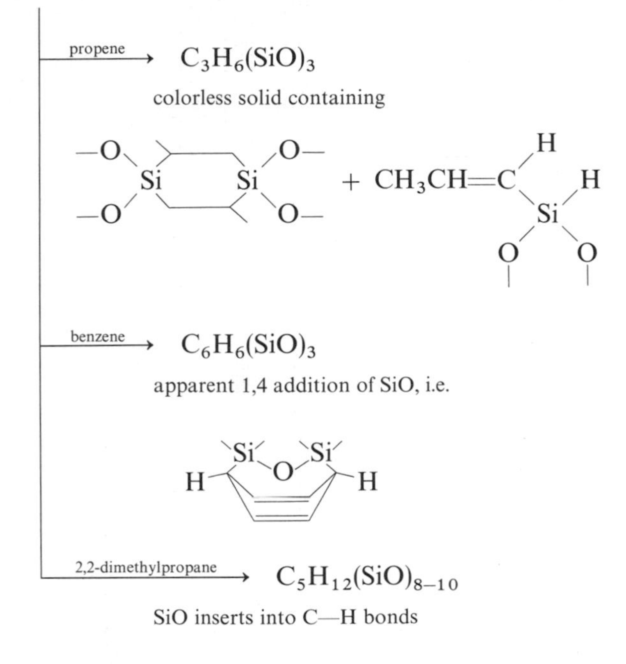

$2B(s) + 2B_2O_3(l) \rightarrow 3B_2O_2(g)$ (Ref. 26)

$\xrightarrow{BCl_3}$ B_2Cl_4 (0.6 mole from 1 mole B_2O_2)

$\rightarrow$ B_2F_4 (0.7 mole from 1 mole B_2O_2)

Evaporation of Carbon from an Arc

Figure 16 shows an apparatus used by Dobson et al.[47] for studying reactions of carbon vaporized in an arc: the apparatus is basically similar to that used by Skell and Westcott[1] in their pioneering research in cryochemistry.

The arc is struck between a lower stationary 6 mm diameter graphite rod and an upper moveable 12 mm diameter graphite rod. Both electrodes are clamped in water-cooled copper holders. A rack-and-pinion device on top of the reaction vessel allows the upper electrode to be moved by sliding its copper holder through an O-ring shaft seal. Vapors of compounds to be condensed with the evaporating carbon are sprayed into the vessel just below the arc. The electrical circuit used to supply power to the arc is shown in Figure 16*b*.

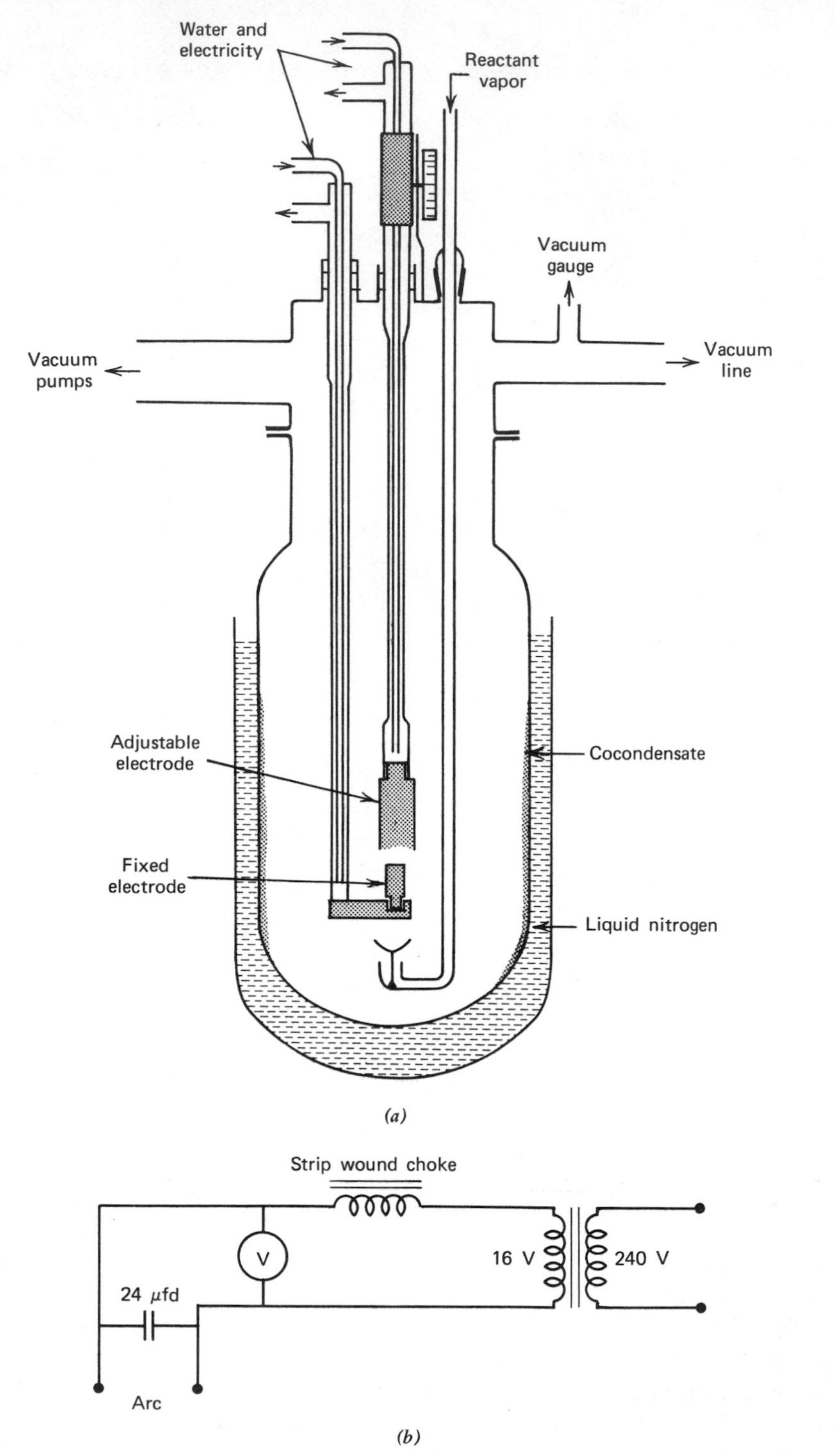

Figure 16. Apparatus for cryochemistry on carbon vapor.[47] (*a*) Apparatus, (*b*) electrical circuit to control carbon arc.

In an experiment, the reaction vessel is evacuated to about 10^{-5} torr through a pumping line with a speed of 8 liters/sec. The vessel is cooled, reactant compound addition begun, and then the arc is started by allowing the electrodes to touch with 16 V applied across them. The upper electrode is withdrawn until the arc appears to be 'stable,' although large fluctuations in the current may be expected: power dissipation of about 900 W is normal, for example, 12 V at 75 A. Rapid boil-off of liquid nitrogen from around the vessel is observed. In a typical run, 60 mmole of carbon are evaporated and condensed with 250 mmoles of a reactant vapor. Very much higher compound/carbon ratios have been used in some work.[2] The pressure in the reaction vessel remains low during runs except with compounds that are readily decomposed with evolution of a permanent gas. Both light and heat emission from the arc may contribute to the release of gases from sensitive compounds for example, trimethylborane or iron pentacarbonyl. The arcing is then best made intermittent to permit evolved permanent gases to be pumped away and the pressure to be kept below the level at which carbon vapor species can reach the cooled walls without aggregation.

The products are pumped out the reaction vessel at the end of a run through a series of traps on a conventional vacuum line. The apparatus shown in Figure 16 is not well suited to the recovery of involatile products as it is difficult to extract these products without exposing them to air.

As carbon vapor is a complex mixture of C_1, C_2, C_3, and higher homologs, with each species occuring in ground and excited electronic states, the yield of a single product from carbon atom reactions is seldom very high. However, as shown in Scheme 7, carbon vapor provides a unique route to some types of compounds. Much of the research on carbon vapor by Skell and his coworkers[2] has had the object of finding how the carbon vapor species react rather than using the species for the synthesis of previously unknown compounds.

Carbon can also be evaporated by resistance heating a graphite rod or filament. The composition of the vapor formed is slightly different from that formed in an arc.[2]

The other metalloids, boron, silicon, and germanium, cannot be vaporized satisfactorily from a simple arc system. Electron bombardment heating is the best method for evaporating boron.[53] Silicon has been evaporated by electron-bombardment heating or by resistance heating a silicon rod.[54] It is most readily evaporated from a small beryllia crucible heated by a coil of tungsten wire and mounted on the crucible holder in the apparatus of Figure 15. Germanium can be evaporated in large amounts from a resistance-heated graphite boat. The vapors of boron, silicon, and germanium have not yet found many uses in synthesizing compounds that are inaccessible by other routes.

Scheme 7. Some Synthetic Applications of Carbon Vapor Reactions at −196°C

$$C(s) \xrightarrow{\text{ac arc}} \underset{63-70\%}{C_1(g)} + \underset{25-30\%}{C{=}C(g)} + \underset{5-6\%}{C{=}C{=}C(g)} \quad \text{(Ref. 2)}$$

$$C{=}C{=}C + (CH_3)_2C{=}C(CH_3)_2 \longrightarrow \text{[bis(2,2,3,3-tetramethylcyclopropylidene)allene: } \triangleright C{=}C{=}C \triangleleft\text{]} \quad \text{(Ref. 1)}$$

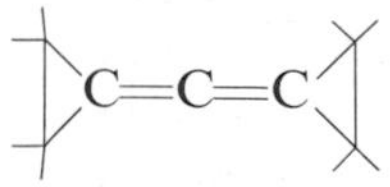

$$C_1(g) + CF_3CF{=}CF_2 \longrightarrow CF_3CF{=}C{=}CF_2 \text{ (20\% yield based on total carbon evaporated)} \quad \text{(Ref. 50)}$$

$$(CH_3)_3SiH \begin{cases} \xrightarrow{C=C} (CH_3)_3SiCH{=}CHSi(CH_3)_3 \\ \xrightarrow{C=C=C} (CH_3)_3SiCH{=}C{=}CHSi(CH_3)_3 \end{cases} \quad \text{(Ref. 49)}$$

$$\text{carbon vapor} + B_2F_4 \quad \underset{0.5\%}{C(BF_2)_4} + \underset{2\%}{(F_2B)_2C{=}C(BF_2)_2} +$$

$$(BF_2)_2C{=}C{=}C(BF_2)_2 \quad \text{(Ref. 47)}$$

3% yield based on total carbon evaporated

$$\text{carbon vapor} + SiCl_4 \longrightarrow Cl_3SiC{\equiv}CCl \text{ (low yield)} \quad \text{(Ref. 51)}$$

$$\text{carbon vapor} + GeCl_4 \longrightarrow Cl_3GeCCl{=}CCl_2 \quad \text{(Ref. 52)}$$

7. APPARATUS FOR LOW-TEMPERATURE REACTIONS OF METAL VAPORS

Only a limited range of examples are given in this section because of the work on metal atoms reviewed in Chapter 7. Apparatuses for metal-vapor synthesis can be classified into two groups, those using a static reaction vessel and those using a rotating reaction vessel.

7.1 Static Apparatus

Reactions in the Apparatus of Figure 15

As stated in Section 6, the apparatus of Figure 15 has a variety of uses in nonmetal and metal chemistry as it is equally suitable for the preparation and isolation of stable or unstable products of cryochemistry reactions.

The Preparation of B_2Cl_4 *Using Copper Atoms.* Copper vapor reacts on condensation with boron trichloride at −196°C to form diboron tetrachloride in yields that can range from 40 to 70% of that calculated from the equation below, depending on the Cu:BCl_3 ratio.[11]

$$2Cu(g) + 2BCl_3 \rightarrow 2CuCl + B_2Cl_4$$

The evaporator used for the copper can be identical to that described for preparing SiO (Section 6.2) except that it is not necessary to put a molybdenum cap over the crucible as molten copper evaporates smoothly. There is some advantage in having a double layer of molybdenum windings at the top of the alumina crucible as shown in Figure 3*b* to prevent condensation of the copper vapor near the top of the crucible. The crucible is charged with 15–20 g of copper (99.9% pure) that can be as a single lump but is preferably in the form of copper pellets. The vapor-inlet system is connected to a 4-liter gas burette that is itself connected to a cylinder of purified boron trichloride. It is convenient to monitor the rate of addition of BCl_3 during cocondensation by the rate of fall of pressure in the 4-liter volume and to fill up the volume with BCl_3 from the cylinder as required.

The reaction vessel is evacuated, the crucible is heated to just melt the copper, the vessel is cooled with liquid nitrogen, and a flow of BCl_3 is begun by opening a Teflon needle valve from the gas burette to give a flow of 20 mmole/min. The power input to the crucible is then raised to 350–400 W, which will evaporate the charge in about an hour: the exact power required has to be determined by a preliminary experiment with a particular crucible. Despite the rapid addition of BCl_3, the pressure in the vessel remains low, usually $<10^{-4}$ torr as BCl_3 condenses efficiently to form a glassy deposit on the walls. The upper parts of the reaction vessel, which are not cooled by liquid nitrogen, become coated with copper so that it becomes difficult to see into the vessel. After an hour or when most of the copper charge has evaporated, the crucible is allowed to cool, and the liquid nitrogen cooled reaction vessel is filled with gaseous nitrogen to enable the vaporization assembly to be lifted out and a blank stainless steel plate put in its place. The flask is reevacuated, and the liquid-nitrogen bath is replaced by a bath at −80°C (CO_2/acetone). This enables most of the BCl_3 to be pumped away from the B_2Cl_4, which stays in the reaction vessel. The pump-out process may last 2 hr. The vessel is then warmed to room temperature, and all remaining volatiles are pumped on to a vacuum line through a trap at −80°C: this collects the B_2Cl_4 and allows residual BCl_3 to pass through. From 15 g of copper about 7 g (43 mmole) of B_2Cl_4 is obtained. The residue in the reaction vessel can be safely washed out with water and acids after the vessel has been filled with nitrogen or carbon dioxide.

The Preparation of $Co_2(PF_3)_8$ *from Cobalt Atoms.* Many transition metal complexes of trifluorophosphine are conveniently prepared in good yield from metal atom reactions at −196°C.[55] The reaction of cobalt atoms with trifluorophosphine gives $Co_2(PF_3)_8$, a compound that readily abstracts hydrogen from other molecules to form $HCo(PF_3)_4$.

Cobalt is evaporated from the type of crucible shown in Figure 3*a*. Alumina cement is coated on a spiral of 1.2 mm diameter molybdenum wire and baked at 1400°C. The outside of the crucible is insulated with Kaowool. A crucible with a maximum diameter of 15 mm and a depth of 15 mm can be used to evaporate about 2 g of cobalt.

The apparatus of Figure 15 is assembled as described for the preparation of B_2Cl_4. The gas burette is filled with PF_3 that has been carefully purified to remove HCl and chlorofluorophosphines. The vessel is evacuated, and the crucible is heated to about 1300°C. It is advisable to allow the system to outgas very thoroughly at this stage, as traces of moisture react with the reaction product. The vessel is cooled by liquid nitrogen, and a flow of PF_3 of about 10 mmole/min is started. The crucible temperature is raised to about 1600°C, sufficient to vaporize the metal at the desired rate of one mmole/min. Cocondensation is continued for 30 min; then the crucible is allowed to cool, and the evaporator assembly is lifted out in a stream of gaseous nitrogen and is replaced by a blank plate. After reevacuation, the reaction vessel is allowed to warm to room temperature. Unreacted PF_3 is pumped out quickly, and at about −40°C a small amount of $HCo(PF_3)_4$ pumps out of the vessel. The desired product, $Co_2(PF_3)_8$, is only slightly volatile and can be pumped out of the vessel only over a period of 1–2 hr. However, the compound is so sensitive to addition of hydrogen from almost any source that it is better to handle it on a vacuum line than by solution techniques. The yield of dark-violet crystalline $Co_2(PF_3)_8$ is about 5 g (6 mmoles).

The residue in the reaction vessel is a buff-colored solid that contains PF_3 coordinated to cobalt, but its composition has not been studied. It is partly soluble in organic solvents, especially acetone.

Preparation of "Tetrakis(trimethylphosphine)iron", $[(Me_3P)_3H(CH_2PMe_2)] \cdot Fe$.[56–58] The problems in this preparation contrast sharply with those for preparing B_2Cl_4 or $Co_2(PF_3)_8$ by atom reactions. The reaction between iron atoms and trimethylphosphine is strongly affected by the rate of addition of the iron and by the ratio of ligand/metal on the cold surface. High rates of addition of iron and a low ligand/metal ratio may result in no product. Dilution of the trimethylphosphine with isopentane improves the yield.

Iron is evaporated from a small conical alumina crucible (12 mm diameter × 12 mm deep). The apparatus is assembled as in Figure 15 with a 2 *M* solution of trimethylphosphine in isopentane in the liquid burette. After the vessel has been evacuated, the crucible is heated to about 1300°C, and the vessel is cooled in liquid nitrogen. The flow of ligand solution is started (about 1.5 ml/min), and the crucible is heated to about 1500°C to evaporate the iron at a rate of about 0.2 mmole/min. A yellow-green condensate forms on the cold walls. At the end of 30 min, the crucible is cooled, and the evaporator assembly is replaced by a plate with a stoppered B29 glass socket cemented through it. Excess ligand and solvent is pumped out of the vessel as it warms to room temperature. A brown residue is left in the flask that is extracted with *n*-hexane (using three 20 ml portions) under a nitrogen atmosphere, and the solution is transferred to a Schlenk tube by means of a $\frac{1}{8}$ in. diameter stainless steel tube. The solution is filtered to remove any iron and is evaporated to dryness to give a yellow-brown oily residue. Yellow, crystalline product can be pumped out of this oil on to a liquid-nitrogen-cooled cold finger placed a short distance above it, over a 12 hr period. The oil must not be warmed above 25°C, otherwise it decompose quickly. About 0.3 mmole of product is collected (6% yield). The compound is exceptionally air sensitive, and the residues in the reaction flask are pyrophoric.

The Preparation of Tris(bicyclo 2.2.1 heptene)palladium(O).[56] Details of the preparation are very similar to those above, but the product, $Pd(C_7H_{10})_3$, is less thermally stable than the phosphine complex. An alumina crucible is charged with 1 g of palladium sponge that is melted and vaporized over a period of 1 hr using a similar electrical power input to that required for iron or cobalt. The ligand is added simultaneously as a 6 *M* solution in isopentane at a rate giving a 20:1 mole ratio of ligand/metal on the cold surface. At the end of the experiment, excess ligand and the isopentane is pumped off at −20°C leaving a dirty white solid residue containing the desired product and some free palladium. The product is washed out of the vessel with aliquots of *n*-hexane at 0°C, and the solution is filtered to remove metal. The hexane is pumped off at −30°C, leaving white crystalline $Pd(C_7H_{10})_3$. The compound decomposes slowly to palladium metal at 0°C. The yield is about 30% based on the palladium evaporated.

A Stainless Steel Apparatus[59]

The apparatus of Figure 17 is a design of atom reactor that has been superseded for most applications by the simple apparatus of Figure 15. Nevertheless, its stainless steel construction and the large diameter (50 mm) product pump-out line occasionally give this type of apparatus an advantage over that of Figure 15. The examples below illustrate this.

As shown in Figure 17, the apparatus is set up for subliming chromium from an open spiral of tungsten wire. The metal vapor is deflected downwards into the lower half of the reaction vessel by molybdenum shields. Vapors to be condensed with the chromium atoms on the cold walls of the

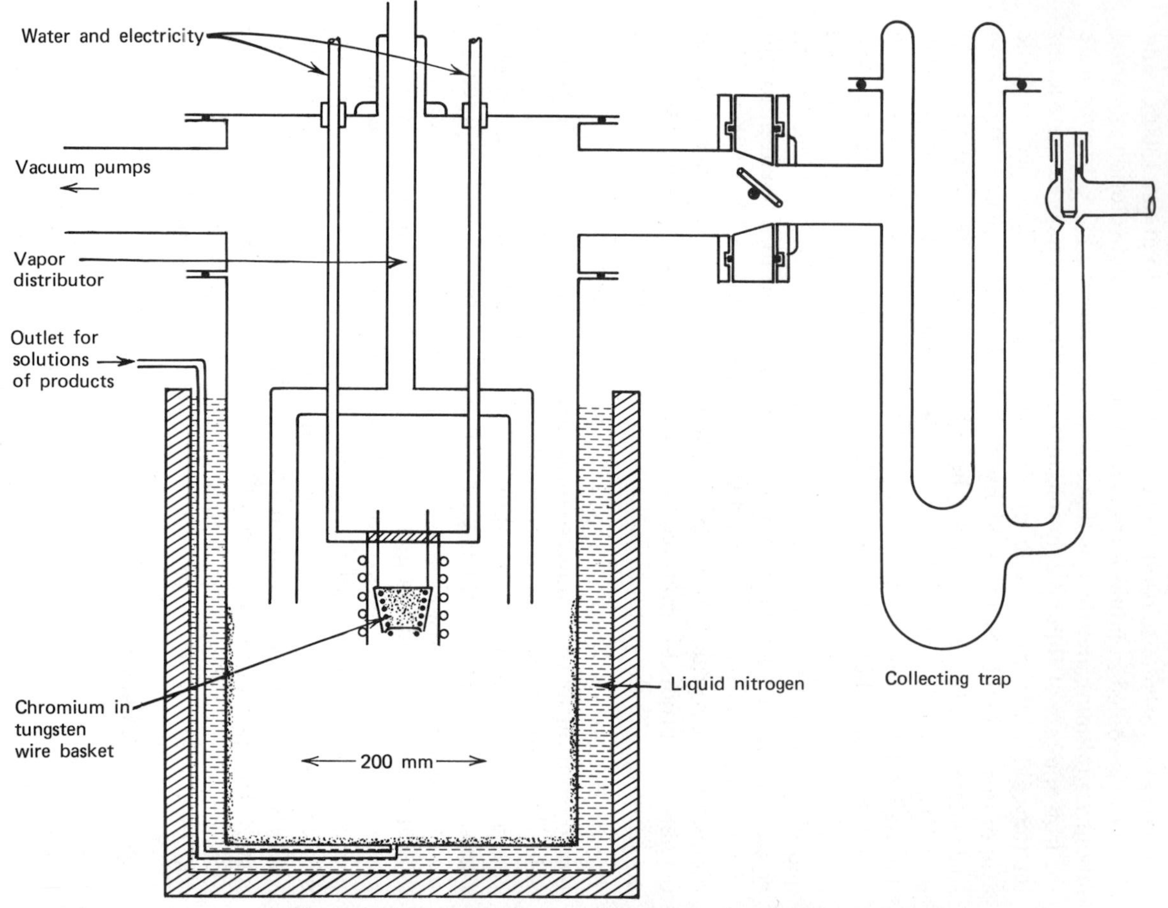

Figure 17. Apparatus for cryochemistry on chromium atoms.

vessel spray downwards from four tubes arranged around the chromium evaporator. Products can be either pumped out of the vessel through 50 mm tubing into a liquid-nitrogen-cooled trap or dissolved in solvents and the solution forced out through the stainless tube at the bottom of the reactor. This latter procedure is much more cumbersome than that in the reactors of Figure 15. The apparatus is evacuated by a diffusion-pumping system giving a speed of 15 liters/sec at the vessel.

Preparation of $Cr(C_6H_6)(C_6F_6)$. Condensation of chromium vapor and benzene at −196°C gives a 50% yield of dibenzene chromium.[5,10] When hexafluorobenzene is used in place of benzene, the condensate explodes on warming from −196°C. A mixture of benzene and hexafluorobenzene gives a small yield of the mixed complex $Cr(C_6H_6)(C_6F_6)$.[59,60]

The apparatus is assembled as in Figures 1 and 17. An 18 mm diameter, 20 mm deep spiral of 1 mm diameter tungsten wire is charged with about 2 g of chromium chips, and molybdenum shields are set to deflect the vapor downwards. The reaction vessel is evacuated and cooled in liquid nitrogen. An equimolar mixture of C_6H_6 and C_6F_6 is run into the vapor-distributor system through a Teflon needle valve, at the rate of 8 mmole/min for each compound. The chromium is evaporated at a rate of about 0.3 mmole/min using a current of 50–60 A and a total power input of 350–400 W. After an hour, the crucible is cooled, and the reaction vessel is allowed to warm to room temperature, continuously pumping through the product-collecting trap. The trap is initially cooled to −40°C, which allows most of the benzene and hexafluorobenzene vapors to pass through, but the trap is then cooled in liquid nitrogen to increase the cryogenic pumping effect and to collect the product. A small amount of a yellow solid is seen to collect in the trap over a period of 1–2 hr. The trap is isolated from the reaction vessel by closing the butterfly valve, and the condensate in the trap is transferred by cryogenic pumping to the central part of the trap. With the outer part of the trap at room temperature, the inner part is warmed to −40°C so that any free arene can be pumped off, and then to room temperature in a nitrogen atmosphere. The yellow deposit on the glass finger is washed off with chloroform in the air, and the chloroform solution is shaken with 0.2 *M* hydrogen peroxide to oxidize any $Cr(C_6H_6)_2$ present to the water-soluble $Cr(C_6H_6)_2^+$. The product, $Cr(C_6H_6)(C_6F_6)$, is recovered as yellow needles by evaporation of the chloroform solution: yield is about 250 mg (4%).

The residue in the reaction vessel will detonate on exposure to air. Its explosive activity can be destroyed by recooling the flask in liquid nitrogen and spraying 50 ml of ethyl bromide into the vapor. The vessel is allowed to warm to room temperature without pumping. After an hour, the ethyl bromide is pumped away, and the vessel is filled with nitrogen and can be safely opened to the air.

The Preparation of $Cr(\pi\text{-}C_5H_5N)(PF_3)_3$. Pyridine forms many complexes with transition metals in which it acts as a two-electron donor bonding through nitrogen. However, only one complex of pyridine, in which it acts as a six-electron donor, is known, and this has been prepared using chromium atoms.[59] Methylpyridines form π-complexes with chromium more readily.[61,62]

The experimental procedure is very similar to that described above. Over a period of an hour, 20 mmole of chromium is condensed with 50 mmole of pyridine and 300 mmole of trifluorophosphine in the liquid-nitrogen-cooled reaction vessel. The pyridine is bled through a Teflon needle valve and evaporated in the vapor-distributor system, and the PF_3 is added from a gas burette. At the end of the reaction the vessel is warmed to room temperature. The product trap is initially at a temperature that allows the PF_3 and some of the pyridine to pass through, and it is then cooled in liquid nitrogen to trap the products. The trap is isolated from the reaction vessel after about an hour. Warming the trap to 0°C with continuous pumping allows any PF_3, pyridine, and $Cr(PF_3)_6$ to be pumped away, leaving a residual yellow solid that is then sublimed on to the cold finger at the center of the trap. The condensate is washed off the finger with a few milliliters of hexane in air at room temperature. The hexane solution is poured on to a short silica gel column that has been previously treated with diethyl ether. On elution with hexane, small amounts of $Cr(PF_3)_6$ pass through, rapidly leaving a yellow band near the top of the column. Elution with an equimolar mixture of hexane and diethyl ether gives two fractions. The first to elute is $Cr(\sigma\text{-}C_5H_5N)(PF_3)_5$, yield 400 mg, 0.8 mmole (4%); the second fraction is the desired $Cr(\pi\text{-}C_5H_5N)(PF_3)_3$, yield 90 mg, 0.2 mmole (1%). The π-pyridine complex slowly decomposes on exposure to moist air. It readily dissolves in 11 *M* HCl to give a bright yellow solution.

The residue in the reaction vessel is partly soluble in organic solvents, although evaporation of the solutions leaves only intractable materials. The facility with which the sparingly volatile product can be pumped away from this intractable material greatly aids its isolation.

7.2 Rotary Apparatus

The possibility of using rotary rather than static reaction vessels for metal atom synthesis is discussed in Section 5.3. The rotary apparatuses lend themselves both to cocondensation and to experiments in which the metal atom is condensed into a cooled solution of the reactant compound.

A Rotary Cocondensation Apparatus Using Electron-Bombardment Evaporation

The reaction vessel of the apparatus built by Green et al.[18,63] is shown in Figure 18. An apparatus of similar design is now being sold by G. V. Planer

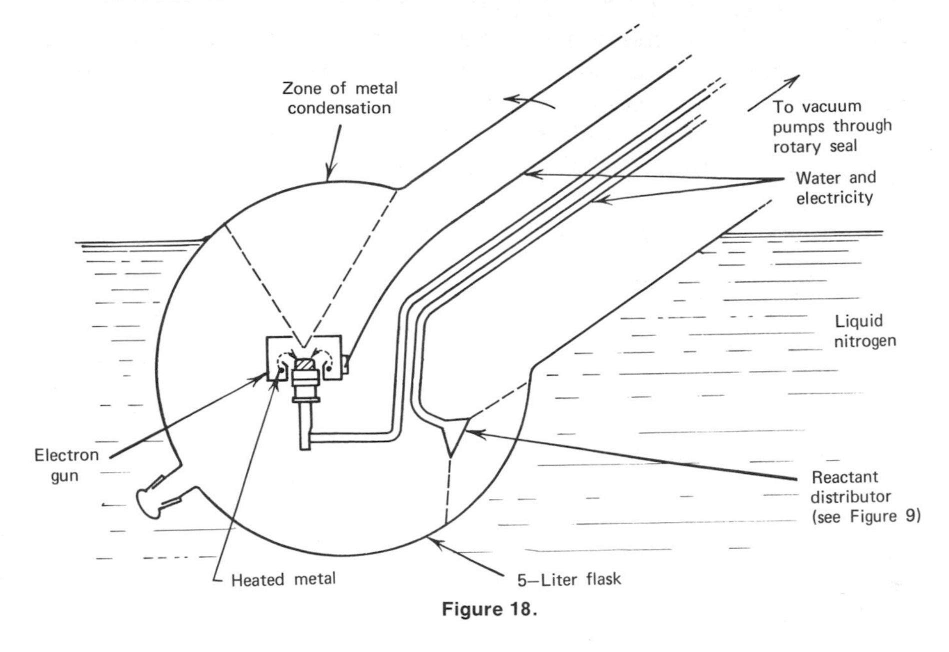

Figure 18.

Figure 21. The 5-liter reaction flask is rotated about an axis at 45° from the horizontal so that most of the flask can be immersed in liquid nitrogen. The neck of the flask is 100 mm diameter to accommodate the evaporation assembly, but the flask is pumped out through a 75 mm pumping line with an O-ring rotating vacuum seal. The pumping speed from the flask is about 40 liters/sec.

Metals are evaporated by electron bombardment (see Section 4.1): the gun is of the type shown in Figure 6*a* except that the polarity is reversed. Electrons are ejected from a heated molybdenum filament at a potential 0–7 kV negative with respect to an anode structure that focusses the electron beam into a small spot at the point where it strikes the target metal at ground potential. The metal is supported on a water-cooled copper pedestal so that there is no problem of finding a container for the hot metal. Up to 1.4 kW of electron beam power can be focused on a spot about 2 mm in diameter, giving intense local heating and vaporization of the metal. The gun can only be operated under vacuum, and the metal vapor emerges from the gun to condense on the walls of the rotating flask. Reactant vapors are sprayed into the lower part of the rotating flask so that a freshly deposited layer is continually being carried over and exposed to condensing metal.

Preparation of Dibenzene Titanium.[18] A piece of titanium rod (99.8% pure), 10 mm long and 12.5 mm diameter, is placed on the copper hearth of the electron gun. The reaction vessel is evacuated to better than 5×10^{-5} torr, cooled in liquid nitrogen, and rotated at about 60 r.p.m. Benzene is vaporized into the apparatus using the distributor device illustrated in Figure 9 or a simple funnel-ended tube, at the rate of 10 mmole/min. Titanium is vaporized at a rate of approximately 0.15 mmole/min with a power input of about 300 W (the rate has to be determined from preliminary experiments). Faster rates of titanium deposition have a very adverse effect on the yield of dibenzene titanium. The run is continued for 2 hr, then the electron gun is turned off, and the condensate is allowed to warm almost to room temperature under an atmosphere of argon. The ground joint at the bottom of the flask is removed, and a thin stainless tube is inserted through a rubber septum cap to suck out the solution of the product in excess benzene. The solution is filtered as quickly as possible and kept cool. Most of the benzene is evaporated under vacuum: addition of pentane then causes the red dibenzene titanium to crystallize from the solution at 0°C. The yield is about 1 g, 5 mmole (30% based on the Ti evaporated).

An Apparatus for Reacting Metal Atoms with Compounds in Solution

A simple but effective apparatus for reacting metal vapors with solutions of reactant compounds is shown in Figure 19. The apparatus is built around the drive mechanism of a commercial rotary evaporator assembly (Buchi). The rotary vacuum seal is provided by two stationary O-rings with vacuum pumping of the space between, which bear on to a rotating stainless steel tube. The end of the stainless steel tube is connected to a 40/45 glass joint through an O-ring flange. A 2-liter flask is normally attached to the 40/45 joint. The apparatus is connected via a liquid-nitrogen-cooled cold trap to a mercury diffusion pump. The pumping speed from the reaction flask is about 1 liter/sec. The pressure in the vacuum system is measured between the reaction flask and the diffusion pump: this gives only a crude indication of the pressure in the reaction flask, but it enables significant changes in pressure to be detected.

Preparation of Bis(cycloocta-1,5-diene)iron(O).[8] A conical alumina crucible is charged with 2–3 g of iron (mild steel, containing 0.5% carbon is very satisfactory), and the crucible is wrapped carefully with Kao-wool to reduce heat losses as much as possible. The apparatus is assembled as shown in Figure 19 with about 200 ml of a 0.5 *M* solution of cycloocta-1,5-diene in methylcyclohexane in the reaction flask. The apparatus is evacuated partially, and then the solution is cooled to −130°C in a bath of petroleum ether: the low temperature of the bath is maintained by addition of liquid nitrogen. The flask is rotated at 50–70 r.p.m., and full vacuum pumping is applied to

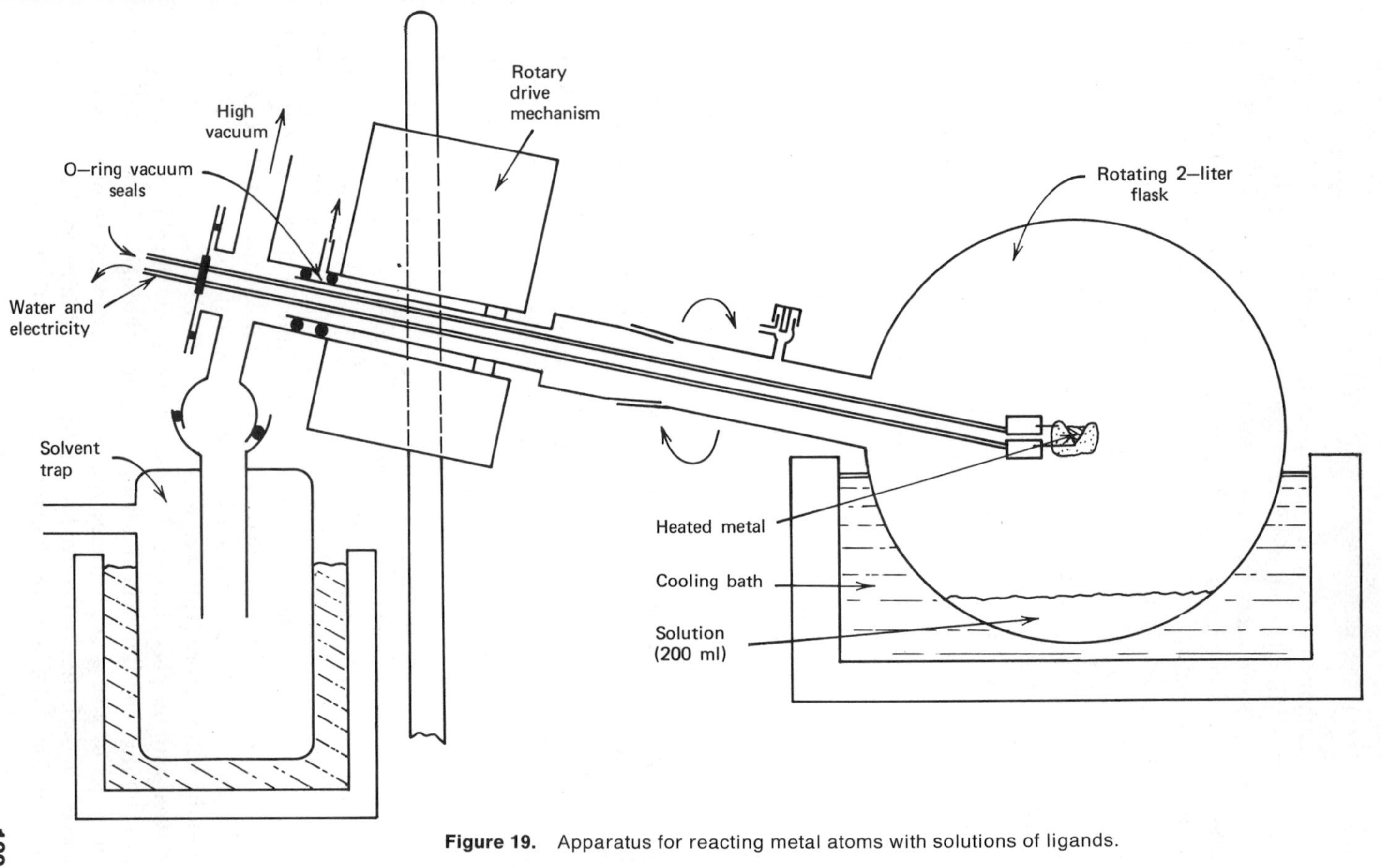

Figure 19. Apparatus for reacting metal atoms with solutions of ligands.

the flask. Although the vapor pressure of the solution at $-130°C$ is about 10^{-5} torr, small amounts of more volatile impurities and dissolved gases cause the pressure above the solution to be much higher initially. The rotation rate is chosen to give the maximum wetting of the surface of the flask. Rates higher than 100 r.p.m. cause the solution to form a narrow band around the equator of the flask. The crucible is heated to the evaporation temperature fairly slowly to allow outgassing to occur. The iron is evaporated at a rate of about 0.6 mmole/min. The solution in the flask turns a clear greenish brown color at first, and as the run proceeds a brown solid is seen to be suspended in the rotating solution. The iron vapor forms an opaque coating on the parts of the flask that are not continually wetted by fresh solution. After an hour, the crucible is cooled to room temperature, and the flask is warmed to $-20°C$ with continuing rotation so that the solvents can be pumped off and condensed in the liquid-nitrogen-cooled trap connected to the rotary assembly. Evaporation of the methylcyclohexane and cycloocta-1,5-diene may take up to an hour. The flask is filled with nitrogen and removed from the rotary apparatus while it is still cold. The product is washed out of the flask using several small portions of pentane at $-10°C$. The pentane solution is quickly filtered and cooled to $-80°C$. On standing, the green solution deposits brown crystals of $Fe(C_8H_{12})_2$: the yield is about 5 g, 20 mmoles (40% of theory). The isolation of $Fe(C_8H_{12})_2$ in the above yield requires good experimental techniques. The compound is very unstable both in the solid state and in solution above $-30°C$, and some losses during the work-up procedure are inevitable.

For a number of reactions of bis(cycloocta-1,5-diene) iron(O) it is not necessary to isolate the pure complex, but reactions with it can be carried out on the crude product prepared in methylcyclohexane solution. This is illustrated by the following experiment.

The Preparation of $(diphos)_2FeN_2$ [diphos $=(C_6H_5)_2PCH_2CH_2P(C_6H_5)_2$].[64] Iron atoms are reacted with cycloocta-1,5-diene in methylcyclohexane solution as described above. When the crucible has been allowed to cool, the cold reaction flask is filled with nitrogen and removed from the rotary apparatus. Using standard techniques, 8 mmole of diphos in 300 ml of diethyl ether at $-20°C$ is added to the solution in the reaction flask under a nitrogen atmosphere. Nitrogen is bubbled through the solution for 12 hr, and the temperature is allowed to rise to 0°C. Any $Fe(C_8H_{12})_2$ that has not reacted with the phosphine decomposes, leaving only the phosphine complex in solution. The liquid in the flask is filtered through a short column of silica gel to give a clear orange red solution. Evaporation of the solution leaves a red oil that is dissolved in diethyl ether, and isopentane is added to the solution at 0°C. On standing for some hours at 0°C the solution deposits fine crystals of $(diphos)_2FeN_2$: the yield is 2 g, 2.5 mmole, that is, about 60% of the diphos added is recovered in the product after reacting with excess $Fe(C_8H_{12})_2$.

8. OBSERVATIONS AND PREDICTIONS

The advent of commercially available apparatuses designed for metal atom synthesis will transform the whole field of preparative cryochemistry. At the time of writing two commercial apparatuses have been described. One is a simple static reactor similar to that in Figure 15, marketed by Kontes Martin of Evanston, Ill., (Figure 20) and the other is a rotary flask apparatus using electron-bombardment evaporation of metals similar to that described in Figure 19, which is being marketed by G. V. Planer of Sunbury, England (Figure 21). Both apparatuses will be capable of being adapted simply to carry out most of the chemistry of high-temperature species mentioned in this chapter, as the metal evaporator assembly could be replaced by a tube furnace for flash thermolysis or for gas-solid reactions. No doubt these latter items will become accessories for the metal atom apparatuses, to be plugged in when desired. The rotary apparatus by G. V. Planer can be adapted for reactions of metal atoms with solutions, although a simpler apparatus like that in Figure 19 is really more useful for this purpose; such an apparatus may be made available commercially. Thus, the preparation and reaction of high-temperature species will become something that every chemist can try using ready-made equipment. The greatest experimental problems will be faced in the recovery of products, especially when these are unstable below room temperature.

The last decade has been an exploratory phase for preparative cryochemistry, and most of the work has been on a rather small scale. There will be growing interest in using the technique on a larger scale. Difficulties of increasing the scale of working come from three sources. First, the ratio of the reactant compound to the high-temperature species condensed on a cold surface has to be large if reasonable yields are to be obtained. Scaling up the formation of the high-temperature species may require that very large quantities of the reactant compound be available. Most of the unreacted compound can be recovered, but in many cases this recovery is far from quantitative. Preparative cryochemistry is of little value at present for converting a rare or expensive reactant compound into a product by reaction with a high-temperature species except on an exploratory scale. It is doubtful if more than a 20% conversion could be achieved in most cases, even with recycling and repetition of the low-temperature reaction. Second, the rate of deposition of metal atoms and other high-temperature species on a cold surface seems to have a marked effect on the yield in some reactions.

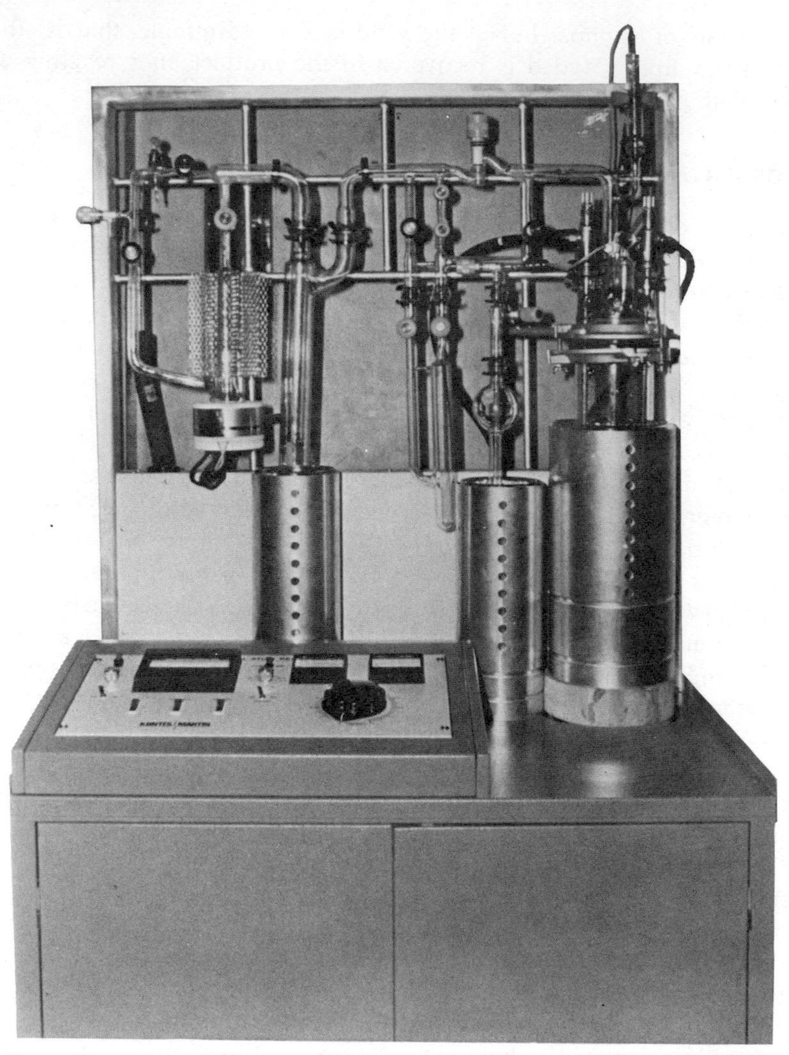

Figure 20. Commercial static apparatus using resistance evaporation of metals.

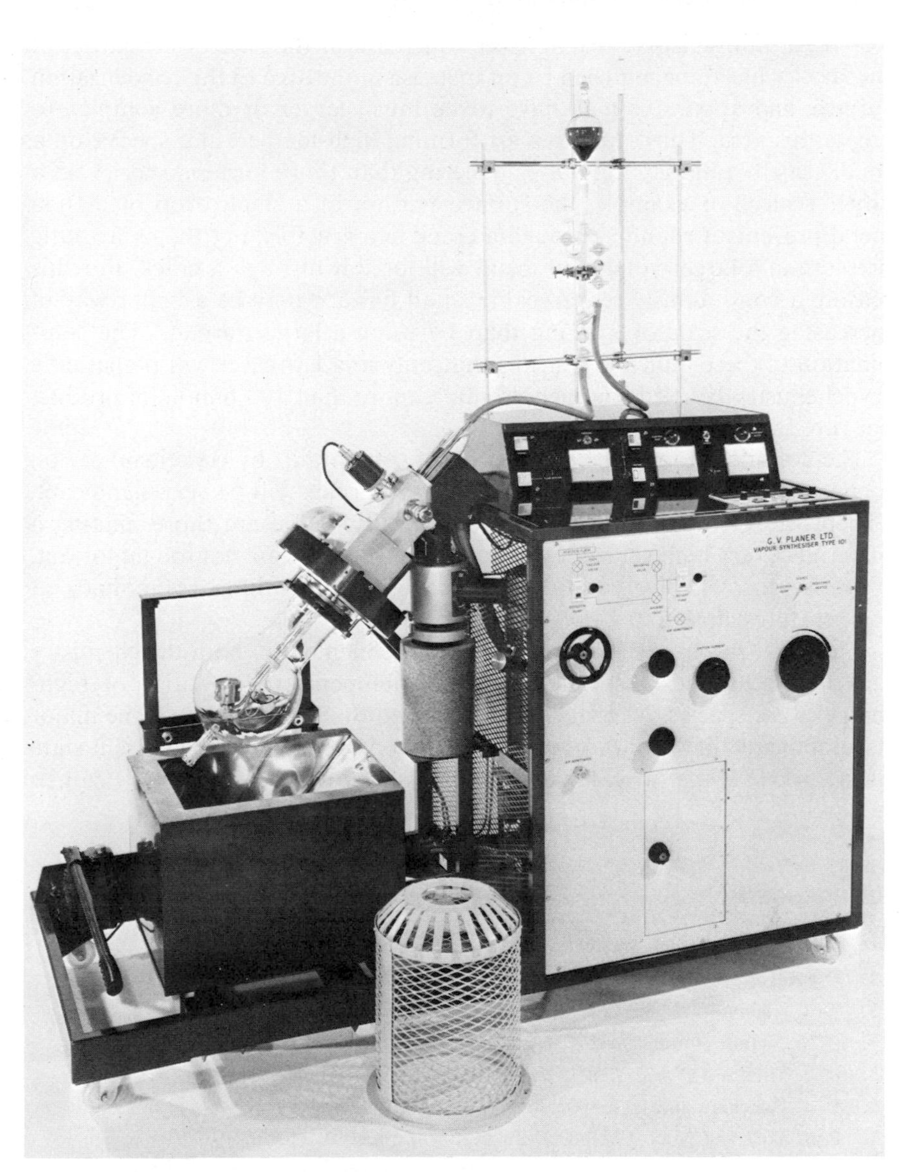

Figure 21. Commercial rotary-reactor apparatus using electron-beam evaporation of metals.

For these rate-sensitive reactions any increase in the rate of formation of the species has to be matched by an increase in the area of the condensation surface, and apparatus may have to be much larger or more complex to create this area. Third, furnaces for forming high-temperature species on a small scale require less careful engineering than those forming species on a larger scale. For example, the surface tension of a small drop of molten metal prevents it running through a crack in a crucible, but the hydrostatic pressure in a larger volume of metal will force it through a crack. For this reason, a continuously rechargeable small furnace may be a better way of increasing the scale of working than by using a larger furnace. The combination of these difficulties means that only in a few cases will preparative cryochemistry be used routinely to make more than 100 mmole of product per run.

The consideration of the production of compounds by cryochemistry on a commercial scale suggests that refrigeration costs will be very significant. The processes most likely to be economically viable are those making a material of very high value for which production costs are not too important, or those that involve reacting metals with high boiling compounds at temperatures close to 0°C.

The next few years should see a marked refinement in both the chemistry and the techniques associated with low-temperature reactions of high-temperature species. Nevertheless, it is hoped that there will be some major developments that cannot be predicted or considered from the present state of the art. It is the quest for developments such as these that adds zest to research.

REFERENCES

1. P. S. Skell and L. D. Westcott, *J. Amer. Chem. Soc.*, **85**, 1023 (1963).
2. J. J. Havel, M. J. McGlinchey, and P. S. Skell, *Account Chem. Res.*, **6**, 97 (1973).
3. K. J. Klabunde, *Account Chem. Res.*, **8**, 393 (1975).
4. P. L. Timms, *Advan. Inorg. Chem. Radiochem.*, **14**, 121 (1972).
5. P. L. Timms, *J. Chem. Educ.*, **49**, 782 (1972).
6. H. J. Hageman and U. E. Wiersum, *Chem. Brit.*, **9**, 206 (1973).
7. P. S. Skell and M. S. Cholod, *J. Amer. Chem. Soc.*, **91**, 6035 (1969).
8. R. Mackenzie and P. L. Timms, *J. Chem. Soc., Chem. Commun.*, 650 (1974).
9. H. Huber, E. P. Kündig, M. Moskovits, and G. A. Ozin, *J. Amer. Chem. Soc.*, **95**, 332 (1973).
10. P. L. Timms, *Chem. Commun.*, 1033 (1969).
11. P. L. Timms, *J. Chem. Soc. Dalton*, 830 (1972).
12. S. Dushman, in *Scientific Foundations of Vacuum Technique*, (J. M. Lafferty, Ed.), 2nd Edit., Wiley, New York, 1962.

13. L. I. Maissel and R. Glang, in *Handbook of Thin Film Technology*, McGraw-Hill, New York, 1970.
14. R. F. Bunshah (Ed.), *Techniques of Metal Research*, Vol. 1, Part 2, Techniques of Materials Preparation and Handling, Wiley-Interscience, New York, 1968.
15. L. O. Olsen, C. S. Smith, and E. C. Crittenden, *J. Appl. Phys.*, **16**, 425 (1945).
16. P. S. Skell, E. M. Van Dam, and M. P. Silvon, *J. Amer. Chem. Soc.*, **96**, 626 (1974).
17. M. J. D'Aniello and R. K. Burefield, *J. Organometal. Chem.*, **76**, C50 (1974).
18. F. W. S. Benfield, M. L. H. Green, J. S. Ogden, and D. Young, *J. Chem. Soc., Chem. Commun.*, 866 (1973).
19. E. K. von Gustorf, O. Jaenicke, and O. E. Polansky, *Angew. Chem.* (*Int. Edit.*), **11**, 532 (1972).
20. D. H. W. Carstens, J. F. Kozlowski, and D. M. Gruen, *High Temp. Sci.*, **4**, 301 (1972).
21. K. J. Klabunde, C. M. White, and H. F. Efner, *Inorg. Chem.*, **13**, 1778 (1974).
22. V. M. Akhmedov, M. T. Anthony, M. L. H. Green, and D. Young, *J. Chem. Soc., Chem. Commun.*, 777 (1974).
23. D. F. Shriver, *The Manipulation of Air-Sensitive Compounds*, McGraw-Hill, New York, 1969.
24. P. L. Timms, *Account Chem. Res.*, **6**, 118 (1973).
25. P. L. Timms, *J. Amer. Chem. Soc.*, **90**, 4585 (1968).
26. P. S. Maddren, Ph.D. thesis, University of Bristol, 1975.
27. P. L. Timms, *J. Amer. Chem. Soc.*, **89**, 1629 (1967).
28. R. W. Kirk, D. L. Smith, W. Airey, and P. L. Timms, *J. Chem. Soc., Dalton*, 1392 (1972).
29. D. L. Smith, Ph.D. thesis, University of Bristol, 1973.
30. P. L. Timms, *Inorg. Chem.*, **7**, 387 (1968).
31. J. L. Margrave and P. W. Wilson, *Account Chem. Res.*, **4**, 145 (1971).
32. P. L. Timms, D. D. Stump, R. A. Kent, and J. L. Margrave, *J. Amer. Chem. Soc.*, **88**, 940 (1967).
33. F. D. Catrett and J. L. Margrave, *J. Inorg. Nucl. Chem.*, **35**, 1087 (1973).
34. P. L. Timms, T. C. Ehlert, J. L. Margrave, F. E. Brinckmann, T. C. Farrar, and T. D. Coyle, *J. Amer. Chem. Soc.*, **87**, 3819 (1965).
35. J. L. Margrave, K. G. Sharp, and P. W. Wilson, *J. Inorg. Nucl. Chem.*, **32**, 1817 (1970).
36. D. Solan and P. L. Timms, *Inorg. Chem.*, **7**, 2157 (1968).
37. J. L. Margrave, K. G. Sharp, and P. W. Wilson, *J. Inorg. Nucl. Chem.*, **32**, 1813 (1970).
38. D. L. Smith, R. W. Kirk, and P. L. Timms, *Chem. Commun.*, 295 (1972).
39. R. W. Kirk and P. L. Timms, *Chem. Commun.*, 18 (1967).
40. R. W. Kirk, Ph.D. thesis, University of California, Berkeley, 1969.
41. D. Solan and P. L. Timms, *Chem. Commun.*, 1540 (1968).
42. J. F. King, P. de Mayo, C. L. McIntosh, K. Piers, and D. J. H. Smith, *Can. J. Chem.*, **48**, 3704 (1970).
43. E. Hedaya, *Account Chem. Res.*, **2**, 367 (1969).
44. R. D. Bush, C. M. Golino, G. D. Homer, and L. N. Somers, *J. Organometal. Chem.*, **80**, 37 (1974).
45. J. J. Havel and P. S. Skell, *J. Amer. Chem. Soc.*, **94**, 1799 (1972).
46. E. Schaschel, D. N. Gray, and P. L. Timms, *J. Organometal. Chem.*, **35**, 69 (1972).

47. J. E. Dobson, P. M. Tucker, R. Schaeffer, and F. G. A. Stone, *J. Chem. Soc.*, **A**, 1882 (1969).

48. P. S. Skell, L. D. Westcott, J. P. Golstein, and R. R. Engel, *J. Amer. Chem. Soc.*, **87**, 2829 (1965).

49. P. S. Skell and P. W. Owen, *J. Amer. Chem. Soc.*, **94**, 1578 (1972).

50. M. J. McGlinchey, T. Reynoldson, and F.G. A. Stone, *Chem. Commun.*, 1264 (1970).

51. J. Binenboym and R. Schaeffer, *Inorg. Chem.*, **9**, 1578 (1970).

52. M. J. McGlinchey, J. D. Odom, T. Reynoldson, and F. G. A. Stone, *J. Chem. Soc.*, *A*, 31 (1970).

53. P. L. Timms, *Chem. Commun.*, 258 (1968).

54. P. S. Skell and P. W. Owen, *J. Amer. Chem. Soc.*, **89**, 3933 (1972).

55. P. L. Timms, *J. Chem. Soc.*, *A*, 2526 (1970).

56. T. W. Turney and P. L. Timms, unpublished work.

57. J. W. Rathke and E. L. Muetterties, *J. Amer. Chem. Soc.*, **97**, 3272 (1975).

58. H. H. Karsch, H. F. Klein and H. Schmidbauer, *Angew. Chem. Internat. Edit.*, **14**, 637 (1975).

59. R. Middleton, Ph.D. thesis, University of Bristol, 1974.

60. R. Middleton, J. R. Hull, S. R. Simpson, C. H. Tomlinson, and P. L. Timms, *J. Chem. Soc., Dalton*, 120 (1973).

61. H. Biedermann, K. Ofele, N. Schuhbauer and J. Tajtelbrau, *Angew. Chem. Internat. Edit.*, **14**, 639 (1975).

62. L. H. Simons, P. E. Riley, R. E. Davis, and J. J. Lagowski, *J. Amer. Chem. Soc.*, **98**, 1044 (1976).

63. D. Young, Ph.D. thesis, Oxford University, 1974.

64. R. A. Cable, M. Green, R. E. Mackenzie, P. L. Timms, and T. W. Turney, *J. Chem. Soc. Chem. Commun.*, 270 (1976).

Organometallic and Organic Syntheses Using Main Group Elemental Vapors

4

Michael J. McGlinchey and Philip S. Skell

1. INTRODUCTION

The chemistry of high-boiling atomic species in the condensed phase started with the study of carbon vapor.[1] The techniques originally utilized have now been extended and improved so that a large number of elemental species have now been investigated.[2] The purpose of this chapter is to collect the published and unpublished data on the main group elemental vapors and to discuss not only their use in synthesis but also the chemistry they exhibit. The analogous reactions of the bulk material have been pointed out where it is appropriate to emphasize the different chemistry that is frequently observed in condensed phase processes.

Since this chapter is primarily concerned with metallic species, carbon has been but briefly mentioned to put it in perspective, and the elements of Groups V and VI, whose vapors are generally polyatomic and which are better produced by photochemical methods, have been omitted. Copper, silver, and gold are included here rather than in the transition metal chapter since their chemistry so far resembles more closely that of the main group elements. Furthermore, their *d* shells do not seem to become involved in the reactions so far investigated, and no transition metal type complexes have as yet been isolated.

2. ALKALI METALS

2.1 Lithium

Although alkali metal atoms have been shown to react with substrates such as halocarbons or oxygen in noble gas matrices,[3] the prime objective was a spectroscopic investigation, and as such this type of system is reviewed elsewhere in this book. However, a macro scale synthesis of lithiocarbons has been reported[4] in which lithium vapor, from a Knudsen cell, was codeposited on a cold (−196°C) finger with a deficit of halocarbon. The product was a grey-white solid, which was extremely air and moisture sensitive; indeed, addition of water led to an explosive reaction. However, addition of trimethylchlorosilane led to formation of tetrakis(trimethylsilyl)methane and lithium chloride. The intermediate tetralithiomethane also reacted with deuterium oxide at 0°C to yield CD_4, C_2D_4, and C_2D_2:

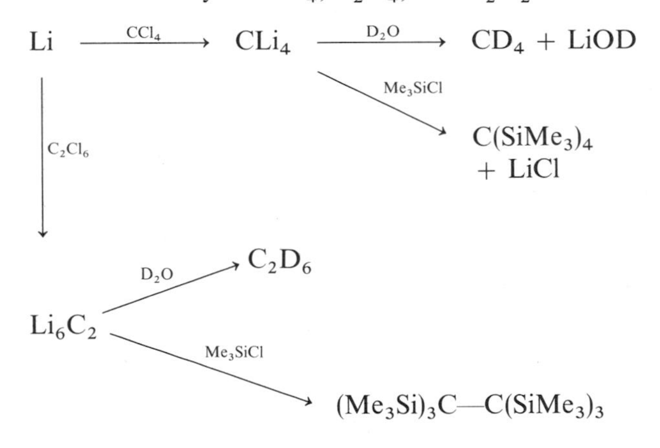

2.2 Sodium and Potassium

Alkali metals are well known to react with halocarbons under conventional reaction conditions to produce radicals in solution. However, radicals can be generated in the gas phase at temperatures between 260 and 280°C by reaction of sodium-potassium alloy vapor with halocompounds in a flow system and then trapping the products in liquid nitrogen.[5] Effectively this is an enormous scale-up of the type of experiment pioneered by Polanyi several decades ago. Although this does not involve the usual cocondensation apparatus, the comparisons are obvious. It is further possible by this technique to introduce other reactants into the system that they might interact with the short-lived species produced by the alkali-metal-vapor reaction. This technique of using a high-temperature reaction to produce a species that then reacts with an added substrate has been elegantly extended by Margrave, Timms, and their coworkers[6] to include such species as SiF_2, BF, and so forth, but these are reviewed elsewhere in this book.

The reaction of sodium-potassium vapor with dimethyldichlorosilane alone yielded mostly polymer but, in the presence of trimethylsilane, a 30% yield of pentamethyldisilane collected in the cold trap.[7]

$$Me_2SiCl_2 \xrightarrow{Na/K} Me_2Si: \xrightarrow{Me_3SiH} Me_3Si{-}SiHMe_2$$

This reaction must involve dimethysilene as an intermediate since a radical chain process involving $Me_2ClSi\cdot$ and $Me_3Si\cdot$ radicals would give a good yield of hexamethyldisilane. Potassium vapor and trimethylchlorosilane alone or in a mixture with methylene bromide produces hexamethyldisilane in good yield.

$$Me_3SiCl \xrightarrow{Na/K} Me_3Si{-}SiMe_3$$

These observations not only implicate a dimethylsilene intermediate, but also suggest it is in the singlet state. Spin conservation rules require that a triplet dimethylsilene react with trimethylsilane to produce a triplet radical pair for which geminate coupling is precluded. Nongeminate coupling would lead to hexamethyldisilane, pentamethyldisilane, and *sym*-tetramethyldisilane in 1:2:1 ratios:

$$Me_2\overset{\uparrow\uparrow}{Si} + Me_3SiH \longrightarrow Me_2\overset{\uparrow}{Si}H + Me_3\overset{\uparrow}{Si}$$

$$\downarrow$$

$$\underset{(25\%)}{Me_3Si{-}SiMe_3} + \underset{(50\%)}{Me_3Si{-}SiMe_2H} + \underset{(25\%)}{Me_2HSi{-}SiHMe_2}$$

However, experimentally the other disilane products summed to only 4% of the yield of pentamethyldisilane.

Until very recently the silacyclopropane system was considered to be too strained to be isolable, but Seyferth[8] demonstrated that substitution by bulky groups confers stability on the ring. Early evidence for the transitory existence of 1,1-dimethylsilacyclopropane was provided by the alkali metal vapor dehalogenation of suitable precursors.[9] In each case the final product was dimethylvinylsilane which could only have arisen via a cyclic intermediate:

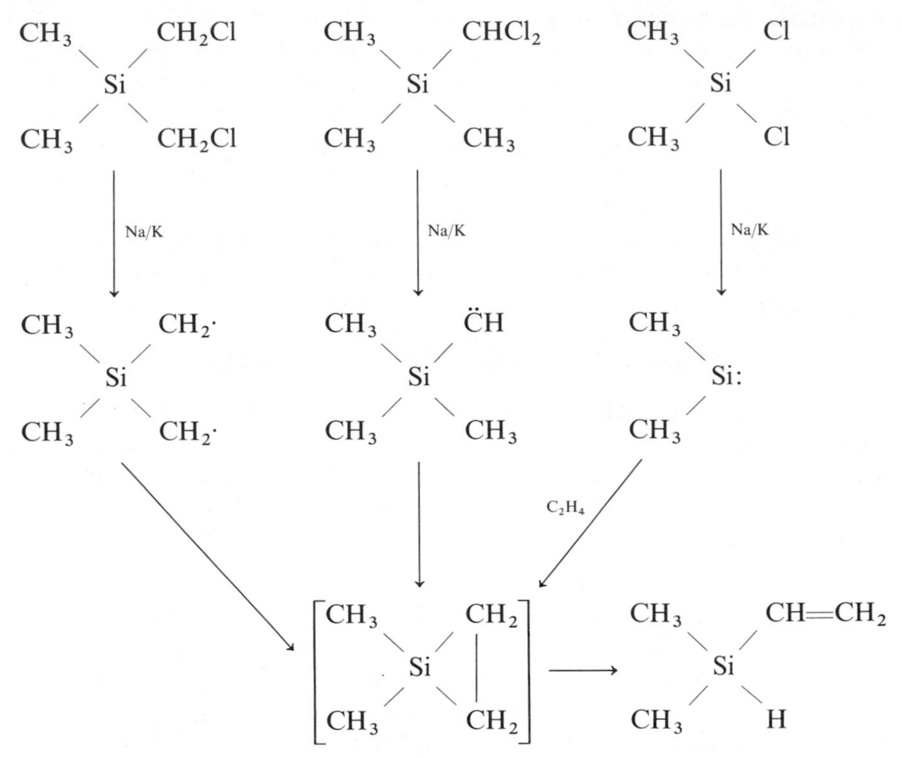

In like manner,[10] tetramethylcyclobutadiene, produced by the dehalogenation of 2,3-dichloro-1,2,3,4-tetramethylcyclobutene with Na/K vapor, yielded the dimers (I) and (II), as well as 1-methylene-2,3,4-trimethylcyclobutene (III), and cis-1,2,3,4-tetramethylcyclobutene (IV). In the presence of triplet methylene diradicals (from the Na/K dehalogenation of CH_2Br_2) tetramethylcyclobutadiene reacted to give 1,2-dimethylene-3,4-dimethylcyclobutene (V) and methane apparently via a transition state involving simultaneous transfer of two hydrogen atoms from the cyclobutadiene to methylene. These experiments support the hypothesis that free tetramethylcyclobutadiene possesses a triplet ground state as predicted theoretically.[11] Furthermore, (III) obtained from a mixture of perdeuterated and undeuterated starting materials indicated that this substance was formed in a bimolecular rather than a unimolecular reaction of tetramethylcyclobutadiene.

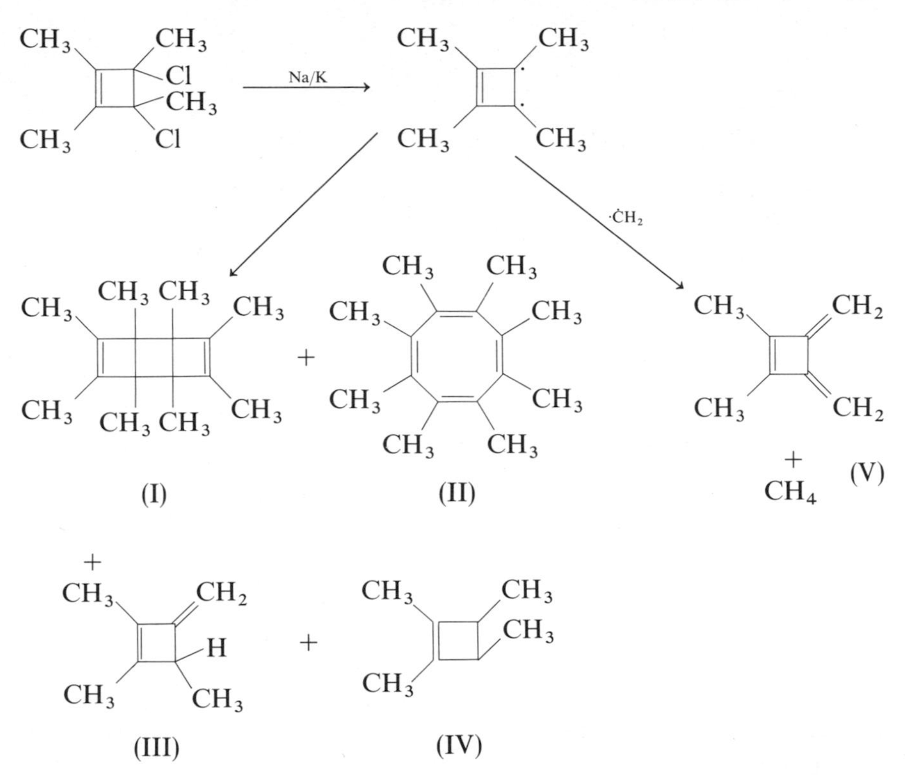

(III)

(IV)

Some work has also been done using Na/K vapor to dehalogenate α,α′-dihaloketones, thus producing cyclopropanones that decarbonylate under the reaction conditions yielding olefins:[12]

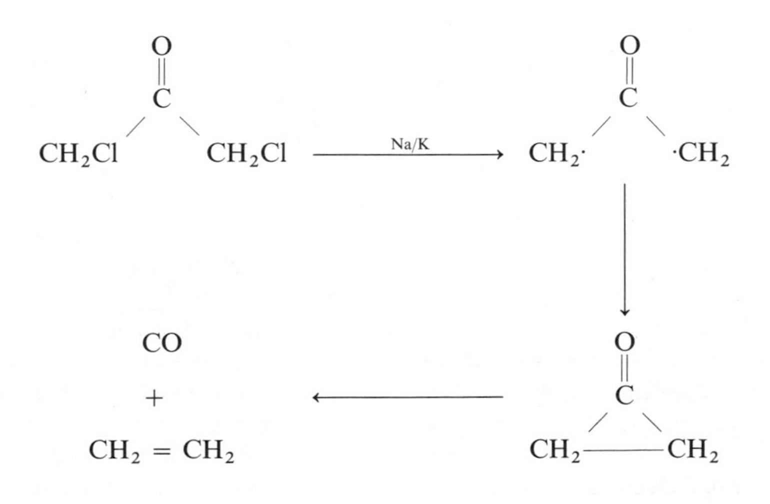

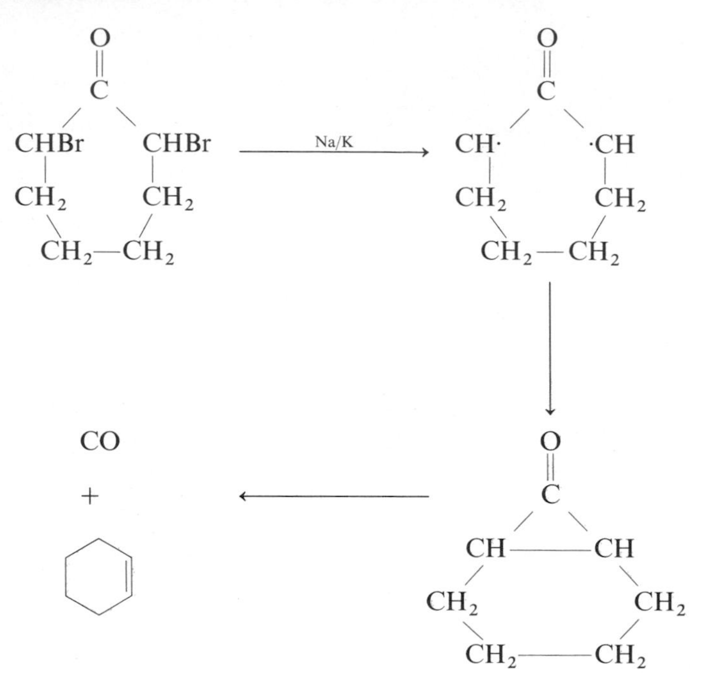

One must admit that there is little thermodynamic advantage to be gained by using alkali metal vapors in synthesis instead of the elements under conventional conditions since their heats of vaporization are low, and consequently the increase in reactivity is small. Nevertheless, the capability to perform clean reactions under controlled conditions and unencumbered by solvents has proven very useful in synthesis.

2.3 Reactions in Ammonia Matrices

Many spectroscopic studies have been made on the reactions of alkali metals in ammonia with a view to gaining information about solvated electrons. Alkali metal vapors when cocondensed with a large excess of ammonia give deep blue or blue-violet matrices.[13,14] During the codeposition very little hydrogen (2–3%, based on metal vaporized) is evolved, and, upon slow warming to ambient temperature, more hydrogen is released (~5%). Removal of the ammonia in vacuo leaves a residue of alkali metal that reacts with added heavy water to evolve deuterium almost quantitatively. It appears that all the alkali metals examined,[13] Li, Na, and K, react with ammonia in essentially the same way. Metal atoms are vaporized into the ammonia matrix; from centers well separated from their nearest neighbors, the atoms ionize, giving up their electron to the ammonia producing the characteristic blue color of solvated electrons in ammonia. Upon warming the matrix, the solvated electrons, having nothing else to react with, recombine with the electropositive metal ions producing a film of alkali metal which upon hydrolysis releases hydrogen quantitatively:

$$2M\cdot + 2D_2O \rightarrow 2MOD + D_2$$

3. ALKALINE EARTH METALS

3.1 The Electronic States of Magnesium

Although the chemistry of beryllium vapor has been largely ignored, some of the remaining Group II elements have been studied: indeed magnesium atoms have been the subject of a comprehensive study by Girard[13] in which not only the reactions but also the spin states of the reacting species have been elegantly elucidated.

Magnesium vapor has been generated in two ways; firstly by resistive heating of a ceramic-coated tungsten crucible, and secondly by striking a low-voltage arc (14 V and 30 A) between two magnesium electrodes in an identical apparatus to that originally designed for the production of carbon vapor. The reactions of thermally produced magnesium atoms and of those from arced magnesium frequently differ drastically, and this has been rationalized on the basis of their differing electronic states.[13]

When thermal magnesium is cocondensed with either water or hydrogen bromide at 77°K a white matrix is formed, and an almost quantitative yield of hydrogen is evolved during the deposition. In contrast, cocondensation of thermal magnesium with ammonia at 77°K produces a maroon matrix, and very little ($\sim 3\%$) hydrogen is evolved. Upon slow warming to ambient temperature, a 25% yield of hydrogen (based on magnesium vaporized) is released. Finally, removal of the ammonia and addition of D_2O releases the remaining 72% as D_2. By analogy with the alkali metals it appears that $\sim 25\%$ of the solvated electrons produced react with ammonia to release hydrogen during the warm-up procedure.

$$2NH_3 + 2e^- \rightarrow H_2 + 2NH_2^-$$

Arc-produced magnesium atoms react with ammonia at 77°K to produce a maroon-blue matrix, while H_2 (34%) is evolved during the deposition. Slight warming produces a deep blue matrix, and H_2 (45%) is released. Removal of the ammonia in vacuo followed by hydrolysis releases the remaining 21% H_2 and also yields over half a mole of ammonia (based on Mg vaporized).[13]

The thermally produced magnesium at $\sim 400°C$ must be almost entirely ^1S ground state atoms, whereas the high electron flux produced by the arcing procedure leads to a great excess of electronically excited state atoms over that expected from a simple Boltzmann distribution. The species most likely responsible for the characteristic arc-type reactions is the excited state,

^3P, of atomic magnesium that lies 65.8 Kcal above the ^1S ground state. It seems that the characteristic reaction of excited state ^3P magnesium is

$$\dot{\mathrm{M}}\mathrm{g}\cdot + \mathrm{NH_3} \xrightarrow{77^\circ \mathrm{K}} \mathrm{Mg\dot{N}H_2} + \mathrm{H}\cdot$$

$$\downarrow \mathrm{NH_3\ (warming)}$$

$$\mathrm{Mg(OH)_2} + 2\mathrm{NH_3} \xleftarrow{2\mathrm{H_2O}} \mathrm{Mg(NH_2)_2} + \mathrm{H}\cdot$$

These results indicate that between 65 and 75% of the magnesium produced in a low-voltage arc is in the ^3P excited state.[13]

Studies with thermally produced calcium atoms also support the concept of solvated electrons in ammonia.[13] Furthermore, similarities in the uv-visible absorption spectra of the potassium-ammonia and magnesium-ammonia matrices show that the two systems are analogous, that is, solvated electrons are produced by the reaction of magnesium atoms in a solid ammonia matrix.

3.2 Reactions of Magnesium with Acetylenes

Although nonconjugated, nonterminal olefins are stable to alkali metals in liquid ammonia, disubstituted acetylenes are readily reduced to form the corresponding *trans* olefin[15]—a useful contrast to catalytic hydrogenation that yields the *cis* isomers.

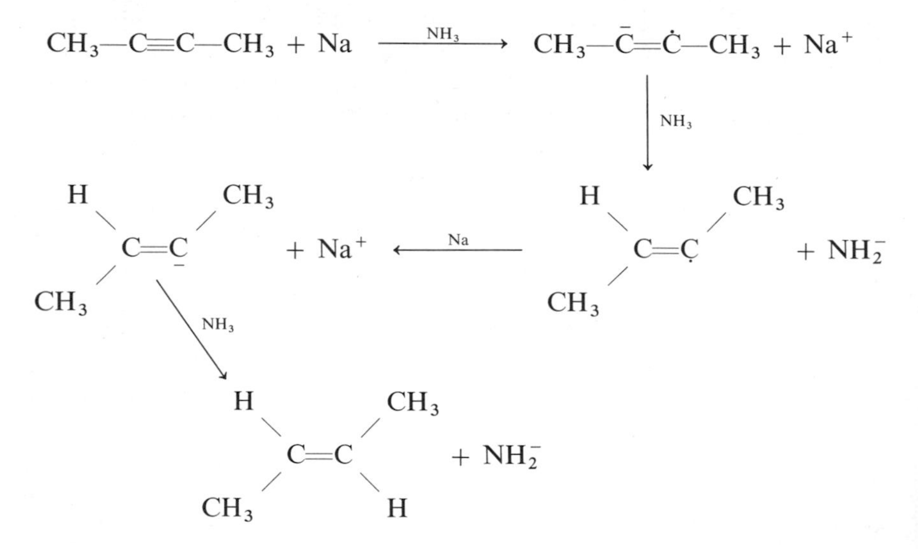

Preference for production of the *trans* radical on protonation of the radical anion could be considered analogous to the *trans* addition of electrophilic reagents or radicals to acetylenes. Presumably the further reduction of the radical to the anion and the formation of the olefinic product occurs more rapidly than inversion of the vinylic radical and carbanion.

The reaction of thermal (^1S) magnesium atoms with ammonia containing 2-butyne produces a maroon-colored matrix that, upon warming, gives an 87% yield of *trans*-2-butene. The mechanism postulated by Girard and Skell[13] is

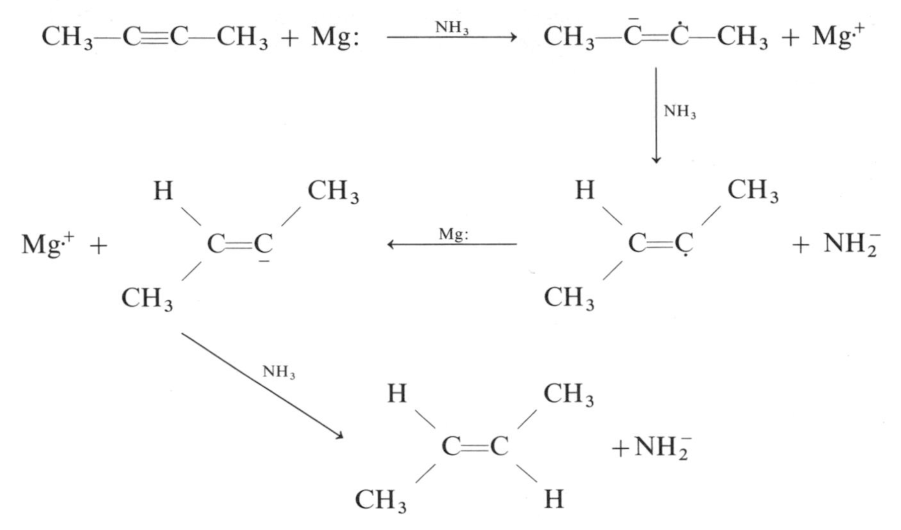

This mechanism would require that half a mole of hydrogen be released on hydrolysis per Mg atom reaction; also hydrolysis of the residue should release two molecules of NH_3 for every *trans*-2-butene molecule produced, and both of these criteria are realized experimentally.

It is not surprising that magnesium gives only one electron per atom to 2-butyne, since the first ionization potential for magnesium is 7.61 eV and the second is 14.96 eV. It is much easier for two atoms to give one electron each than for one atom to give both. A blank reaction revealed that deposition of an ammonia/2-butyne mixture upon a previously deposited magnesium film gave $<5\%$ reaction, demonstrating that the reaction does not occur to any appreciable extent with elemental Mg.

With ^3P, that is, arc-produced magnesium, the absolute yield of *trans* olefin is low (28%) since there is a competition between the ammonia and the 2-butyne for the electrons.

3.3 Reactions of Magnesium with Alkyl Halides

Attempts to produce nonsolvated Grignard reagents by pumping off the coordinated solvent has never proven entirely satisfactory, but more success has been achieved by Gault[16] who adsorbed alkyl halides onto a previously deposited magnesium film.

Recently, Girard has demonstrated that thermally produced magnesium atoms (i.e., ^1S ground state Mg) reacts with alkyl halides to provide a ready synthesis of nonsolvated Grignard reagents using the cocondensation procedure.[14] Thus, codeposition of *n*-propyl chloride and magnesium, after removal of the excess unreacted alkyl halide, produced a dry solid residue of nonsolvated propyl magnesium chloride; hydrolysis of this residue gave propane in 75% yield. Ground state magnesium atoms and 1,3-dichloropropane yield cyclopropane after hydrolysis; this could go either via a diabstraction process or through a mono Grignard reagent followed by elimination of $MgCl_2$.

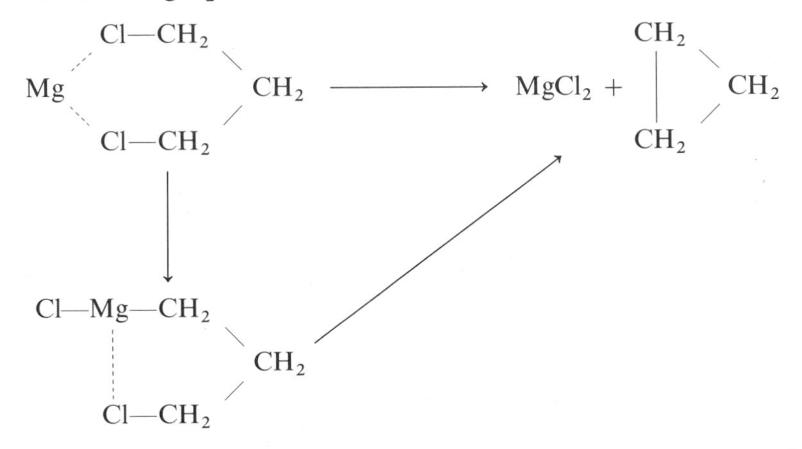

1,4-Dichlorobutane and magnesium give some di-Grignard intermediate since a 33% yield of butane is released on hydrolysis of the residue. On the other hand, ethylene (22%) and cyclobutane (8%) are formed before hydrolysis, indicating that the nonsolvated Grignard reagent has some free radical character as has been previously suggested,[17,18] or that diradical intermediates are formed. Low yields of benzene and ethylene were obtained after hydrolysis when the less reactive halides, chlorobenzene and vinyl chloride, were cocondensed with magnesium vapor.[13]

Interestingly, codeposition of *n*-propyl bromide and ^1S magnesium followed by adding methyl bromide to the matrix led, after hydrolysis, to ~90% methane, showing that nonsolvated Grignard reagent is not produced directly at −196°C, but only upon warming up the matrix. In a blank reaction, dry solid nonsolvated propyl Grignard reagent did not exchange with methyl bromide. Possibly, the initial jet-black matrix that appears on cocondensation of an alkyl halide with magnesium (as opposed to the white one from Mg and neopentane) is attributable to a charge-transfer complex that is formed at −196°C and reacts to give the nonsolvated Grignard on warming up.

The reactions of nonsolvated Grignard reagents do not always parallel the known chemistry of conventional Grignards in coordinating solvents.

Thus, although normal Grignards react with acetone to produce tertiary alcohols, nonsolvated Grignards prefer to react with acetone in its enol form to give alkanes.[14]

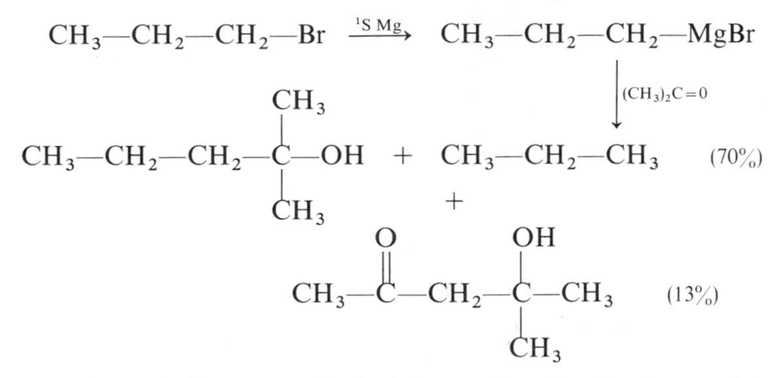

The formation of diacetone alcohol is readily visualized as arising via nucleophilic attack by the enolate anion of acetone upon the carbonyl function of acetone. To ensure that the hydrolysis was due to acetone-enol and not adventitious water, $(CD_3)_2CO$ was used and gave an 85% yield of monodeuteropropane.

The reaction of nonsolvated Grignard reagent with crotonaldehyde, which possesses no enolizable hydrogens, gives upon hydrolysis 3-pentene-2-ol rather than isovaleraldehyde (V), which might have been expected from a conventional 1,4 addition via a 6-membered transition state.

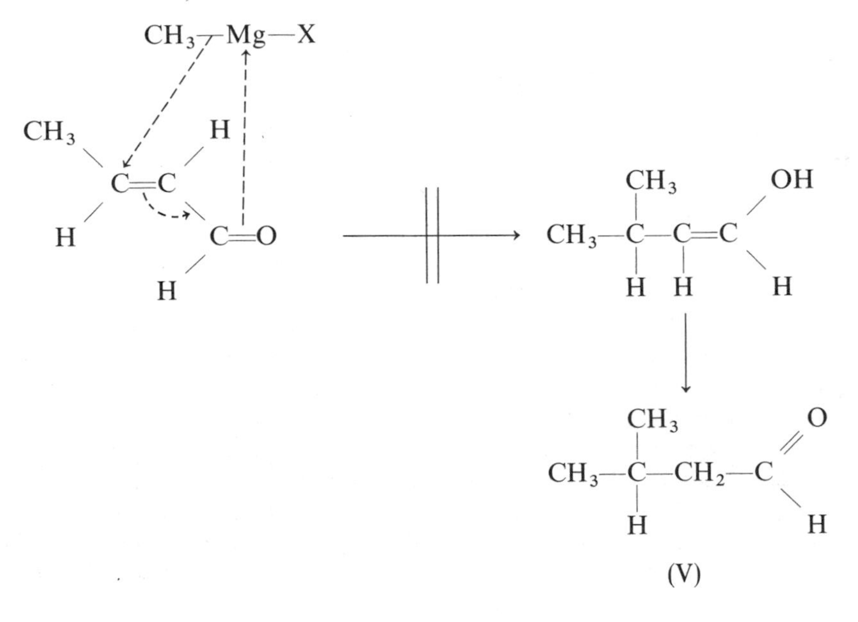

The formation of 3-pentene-2-ol is best interpreted as being a 1,2 addition involving two molecules of Grignard and so still goes via the favorable 6-membered transition state:

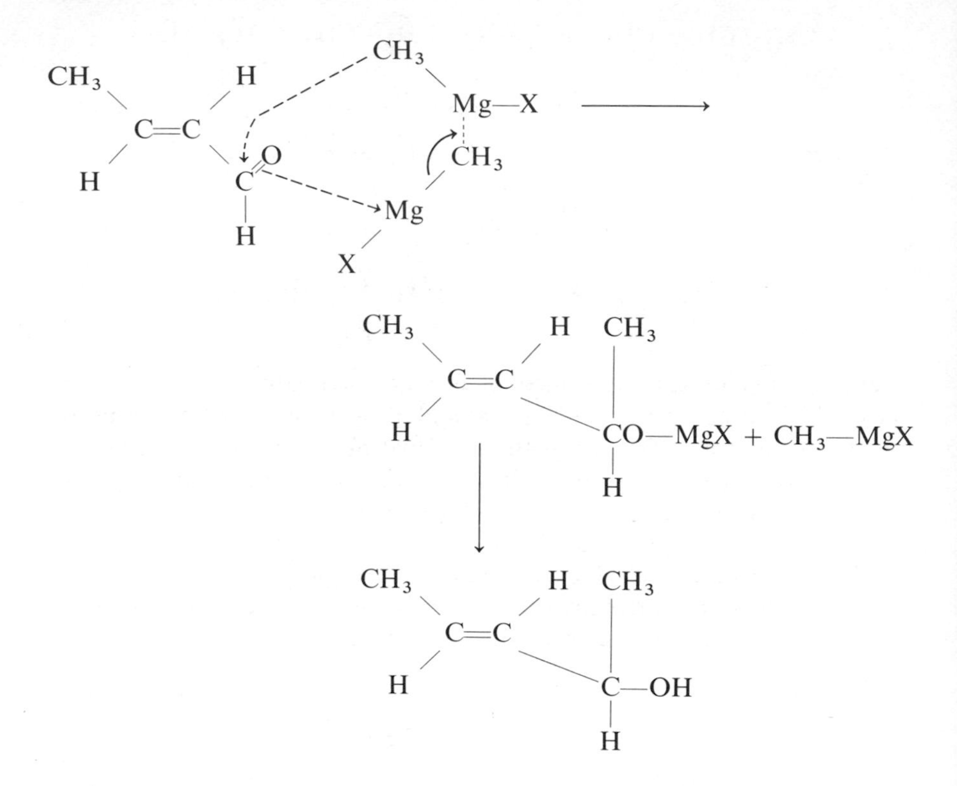

This evidence supports the contention that nonsolvated Grignard reagents are polymeric, or at least dimeric, that is, analogous to the alkylberyllium complexes.[19,20]

The reactions of arc-produced magnesium with alkyl halides, as one might expect, differ greatly from the products formed with thermal magnesium. Arced magnesium contains a high proportion of ^3P atoms, which have diradical character, and the products are rationalizable on the basis of radical intermediates. Typically, ^3P magnesium reacts with *iso*-propyl bromide to produce a white matrix (notice the difference from black matrix with ^1S Mg), and, upon warming to ambient temperature, propene (35%) and propane (34%) are produced along with traces of C_6 hydrocarbons. Hydrolysis of the residue produces 0.2% propane demonstrating that very little Grignard reagent is produced. In similar fashion, ^3P magnesium and *tert*-butyl bromide yield isobutane (40%) and isobutene (58%) before hydrolysis, and no paraffins are released when the residue is hydrolyzed. This, and much other evidence, points to a halogen-abstraction process by the diradical metal atoms and subsequent disproportionation of the intermediate radicals:

$$(CH_3)_2CH{-}Br + \dot{M}g\cdot \longrightarrow (CH_3)_2\dot{C}H + \dot{M}gBr \xrightarrow{RBr} MgBr_2 + R\cdot$$

$$(CH_3)_2\dot{C}H \longrightarrow CH_3{-}CH_2{-}CH_3 + CH_3{-}CH{=}CH_2 + [(CH_3)_2CH]_2$$

Radicals such as neopentyl that cannot lose a β-hydrogen but that can β-eliminate only an alkyl group show a higher tendency to yield coupling products.

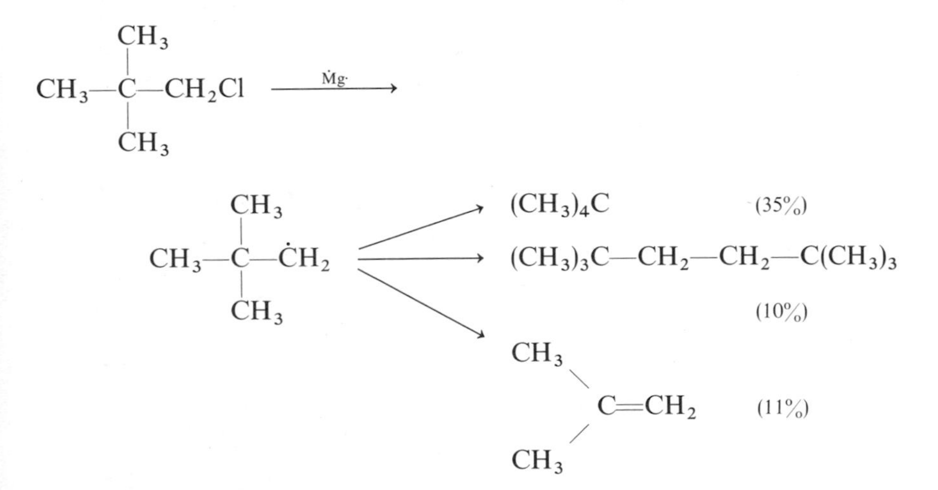

Methyl radicals, which can neither disproportionate nor β-eliminate, yield 44% of ethane the coupling product.

Arc-produced magnesium reacts with vicinal alkylene dihalides to produce, as expected, the corresponding olefin in high yield. Also, 1,3-dihaloalkanes yield the corresponding 1,3-diradicals producing cyclopropane, but, as the halogen atoms are moved further apart on the alkyl chain, the yield of products coming from the diabstraction of halogens drops. With 1,4-dichlorobutane, abstraction of only one halogen is the major process.[13]

$$Cl{-}(CH_2)_4{-}Cl \xrightarrow{\dot{M}g\cdot} Cl{-}CH_2{-}CH_2{-}CH{=}CH_2 \quad (15\%)$$
$$+$$
$$Cl{-}CH_2{-}CH_2{-}CH_2{-}CH_3 \quad (31\%)$$
$$+$$
$$C_2H_4 \quad (9\%)$$

^3P Magnesium reacts with *d,l*-2,3-dibromobutane and with *meso*-2,3-dibromobutane in separate experiments to give very similar product yields. This points to the fact that there is a common intermediate in the reaction and that the abstraction is not a concerted elimination. Under these conditions *cis*-2-pentene in the matrix is also isomerized, implicating a bromine atom as the common intermediate.

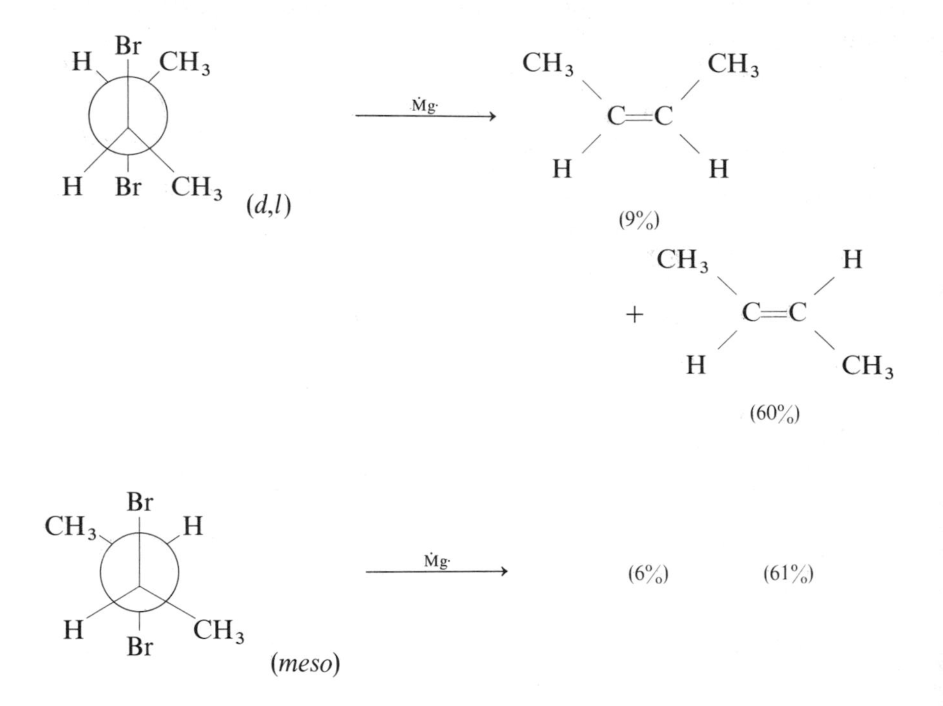

When the reactions of *d,l*- and *meso*-2,3-dibromobutane were carried out in the presence of 1,3-pentadiene (a radical trap) the reactions showed high stereospecificity, and this is rationalizable on the basis of a bridged free-radical intermediate[21] that is debrominated by 1,3-pentadiene prior to isomerization.

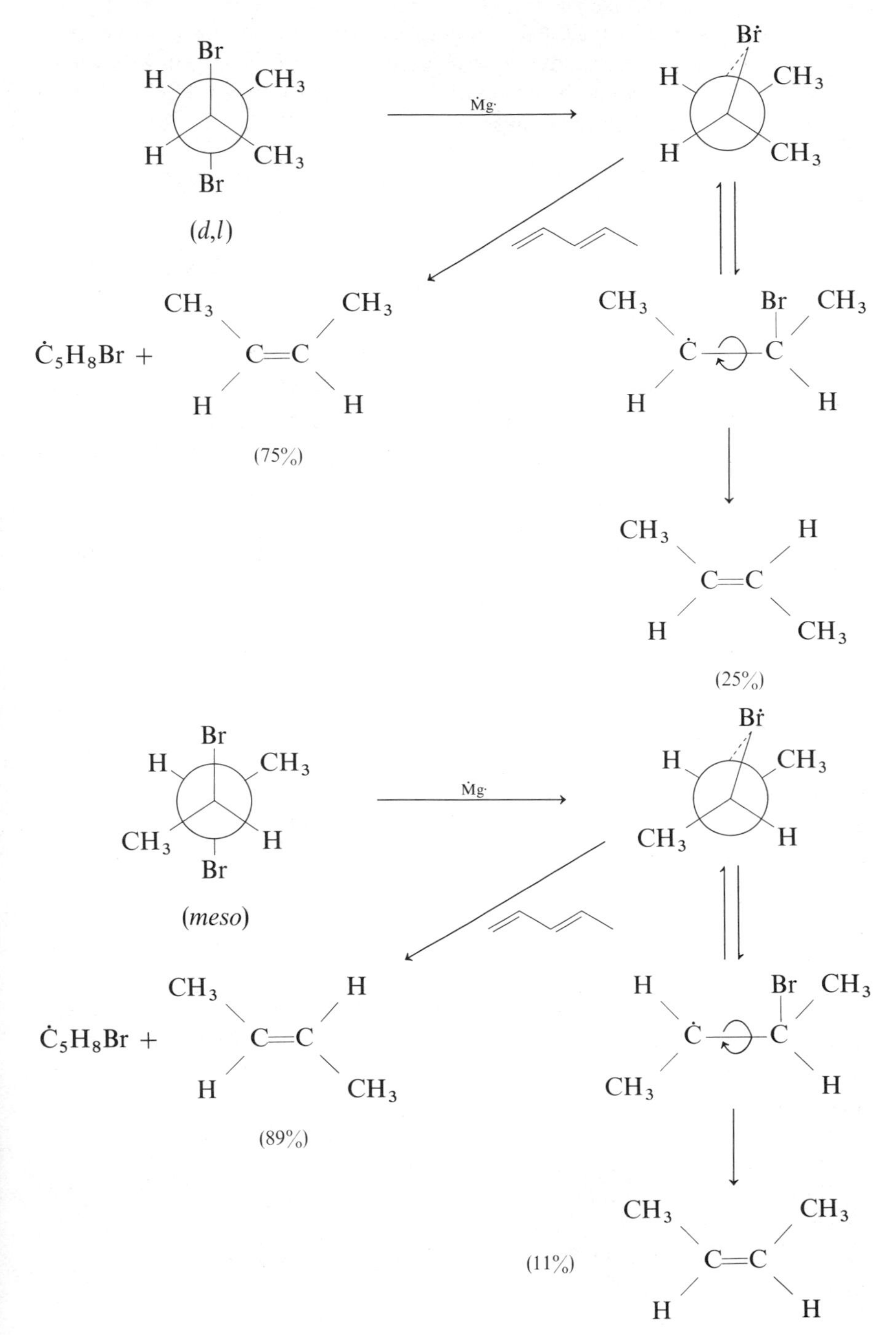
Br
H
CH3
H
CH3
Br
(d,l)
Mg
Br
H
CH3
H
CH3
CH3
CH3
C5H8Br +
C=C
H
H
(75%)
CH3
Br
CH3
C
C
H
H
CH3
H
C=C
H
CH3
(25%)
Br
H
CH3
CH3
H
Br
(meso)
Mg
Br
H
CH3
CH3
H
CH3
H
C5H8Br +
C=C
H
CH3
(89%)
H
Br
CH3
C
C
CH3
H
CH3
CH3
(11%)
C=C
H
H

It is also possible to produce carbenes by the debromination of geminal dihalides with ^3P magnesium. Methylene produced this way inserts into a C—Br bond of methylene bromide and also adds to *cis*-2-butene to give a mixture of dimethylcyclopropanes.[13]

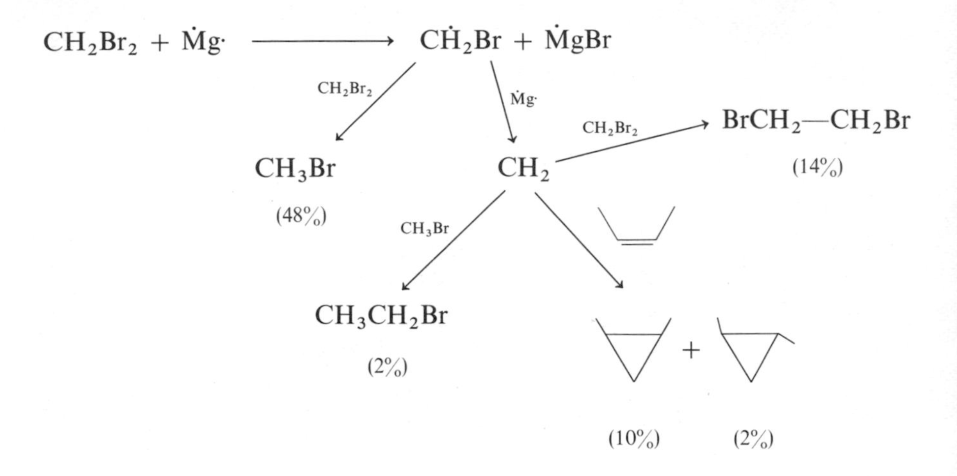

3.4 Calcium

Although calcium atoms in ammonia matrices have been shown to give rise to solvated electrons (see Section 3.1), their use in synthesis has been but little investigated. However, Klabunde and his coworkers[22] have demonstrated that in certain systems calcium atoms behave as moderately efficient defluorinating agents. Thus perfluoro-2-butene is defluorinated to perfluoro-2-butyne. Perfluoropropene or perfluorocyclobutene, on the other hand, yielded only polymeric products, and fluorines in saturated environments, for example, perfluoropropane or benzotrifluoride, failed to react at all.

Attempts to eliminate fluorine from hexafluorobenzene led to polymer formation that may have been produced by the polymerization of tetrafluorobenzyne; but this supposed intermediate could not be trapped with benzene as the barrellene, that is, its Diels–Alder adduct. Nevertheless, initial insertion of a calcium atom into a carbon-fluorine bond does occur since the intermediate organocalcium fluoride can be trapped if water-vapor-saturated C_6F_6 is used as the substrate. In this case an 88% yield of pentafluorobenzene is isolable.

Since these defluorinations with calcium vapor only occur with vinyl or aromatic fluorines one might postulate an initial coordination of the calcium atom to the π system; subsequent fluorine migrations could then lead to the observed products:

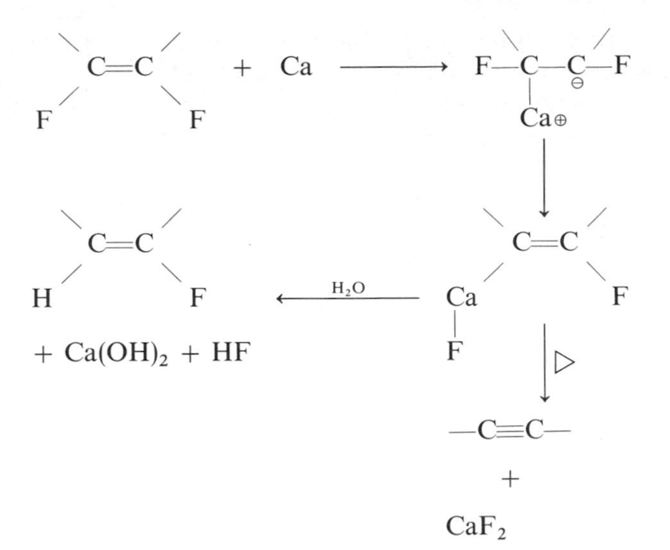

This may turn out to be a useful synthetic technique since, although the lower halogens are readily removed, carbon-fluorine bonds are notoriously difficult to break.

4. GROUP III ELEMENTS

4.1 Boron

Boron atoms have been produced by electron bombardment of a boron rod such that it is heated to 2500–2800°C; the boron produced is believed to be monoatomic and in the ^2P electronic state.[23] One of the biggest problems with boron atoms is that they are odd-electron species and so form free radicals that promote the polymerization of many reactants. Thus boron has not been widely utilized in chemical synthesis except for the preparations of boron monofluoride[24] and to a much lesser extent, B_2O_2,[2] HBS,[25] and FBS.[2]

Boron atoms react with HC1 and HBr to give hydrogen and also BX_3 and HBX_2 (X = Cl or Br), whereas reaction with boron or phosphorus trihalides yields small quantities of the catenated tetrahalides, B_2X_4 and P_2X_4.[23] Reaction of B atoms with benzene or ammonia led to polymeric solids, but the reaction with CO_2 caused explosive reduction to CO.[2]

Boron atoms and water yield boric acid and hydrogen, and the overall stoichiometry follows the equation,

$$B + 3H_2O \rightarrow B(OH)_3 + \tfrac{3}{2}H_2$$

The reaction probably involves an intermediate with a boron-hydrogen bond since deposition of acetone on top of the B/H_2O matrix leads to isopropanol and a reduction in the hydrogen yield. A possible intermediate is dihydroxyborane, and this gains support from the fact that the reaction of B atoms with methanol yields some dimethoxyborane. There is no evidence to suggest the presence of monohydroxyborane, since if this were to result from the reaction of boron atoms and water, one would expect on further reaction (warm up) 2 moles of hydrogen per mole of boron, and this is not realized experimentally.[26]

A plausible route involves the initial insertion of a ^2P boron atom into an O—H bond followed by a homolytic displacement reaction to give dihydroxyborane and a hydrogen radical.

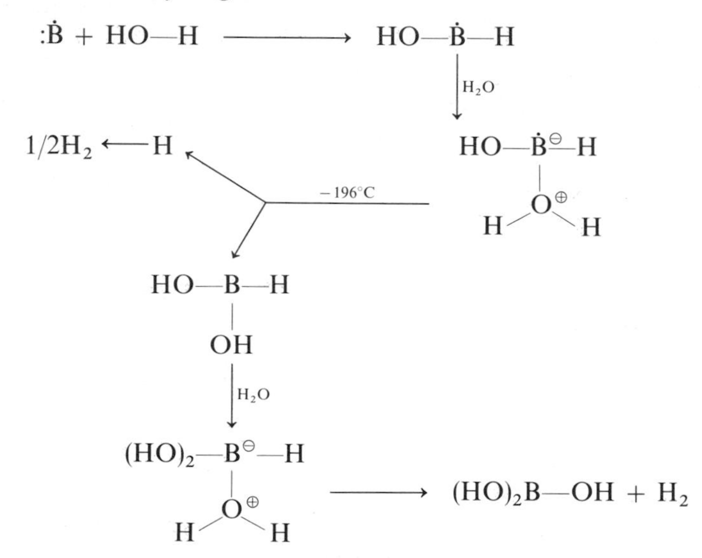

Such a homolytic displacement process is thermodynamically much more favored than an abstraction by a boron radical of a hydrogen from a hydroxyl linkage.

Similarly, the course of reaction of boron atoms with alcohols would be expected to give dialkoxyboranes that upon alcoholysis produce borate esters, and this is what is observed.

$$:B\cdot + CH_3OH \rightarrow CH_3\text{—}O\text{—}\dot{B}\text{—}H \xrightarrow{CH_3OH} (CH_3O)_2BH + H\cdot$$

$$(CH_3O)_3B + H_2 \xleftarrow{\quad CH_3OH \quad} (CH_3O)_2BH$$

Analogous reactions also occur with ethers and either alkyl or aryl halides to give alkyldialkoxyboranes and alkylborondihalides, respectively.

$$CH_3{-}O{-}CH_3 + :B\cdot \rightarrow CH_3O{-}\dot{B}{-}CH_3 \xrightarrow{Me_2O} (CH_3O)_2BCH_3 + \dot{C}H_3$$

$$PhBr + :B\cdot \rightarrow Ph{-}\dot{B}{-}Br \xrightarrow{PhBr} Ph{-}BBr_2 + Ph$$

In view of the known ability of boron vapor to deoxygenate carbon dioxide to carbon monoxide, an attempt was made to deoxygenate ketones, thus producing carbenes whose rearrangement products could be observed. However, using 2-butanone, less than 2% carbene formation was observed,[26] and the major reaction was the reductive coupling of the ketone to form the cyclic ester, 2-(2-butoxy)-4,5-diethyl-4,5-dimethyl-1,3,2-dioxaborolane (VI). This reaction was especially facile in the presence of 2-butanol, which may serve as a source of hydrogen radicals.

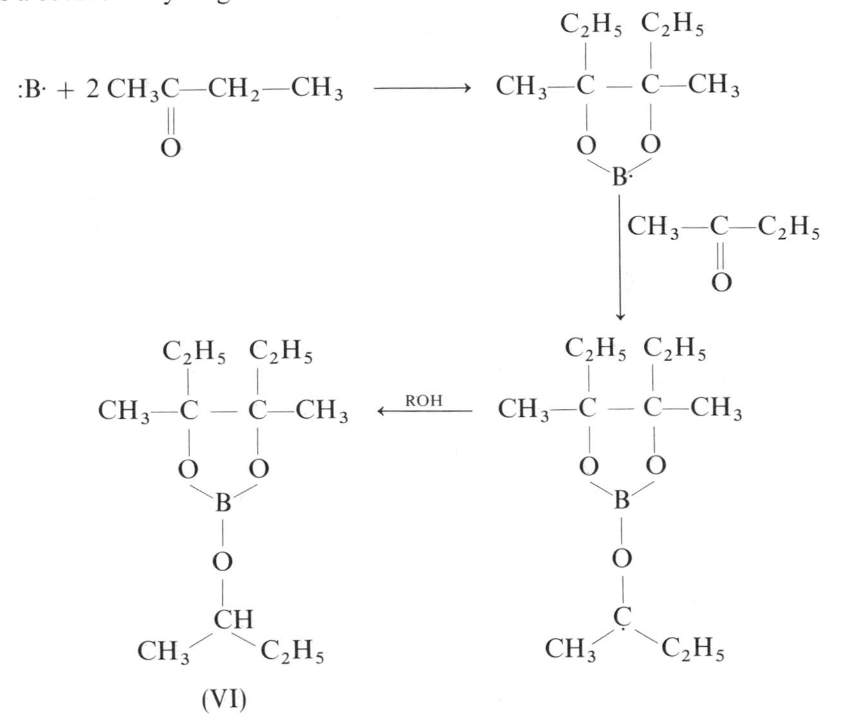

4.2 Aluminum

Aluminum atoms, like boron, of course, are odd-electron species and thus give rise to many polymeric products. For this reason much of the chemistry has to be inferred by hydrolyzing the primary reaction products and examining the hydrolysis products.

Typically, the direct cocondensation of aluminum atoms with propene gave no isolable simple products, and so the yellow viscous residue remaining after the removal of excess unreacted propene was analyzed by hydrolysis with D_2O to mark each carbon-aluminum bond by converting it to a carbon-deuterium bond. The major products upon hydrolysis were propane, 2,3-dimethyl butane, 2-methylpentane, and traces of *n*-hexane.[27] One can readily visualize the formation of such products via intermediates in which secondary radicals are formed preferentially over primary radicals; coupling of such radicals would lead to the observed products.

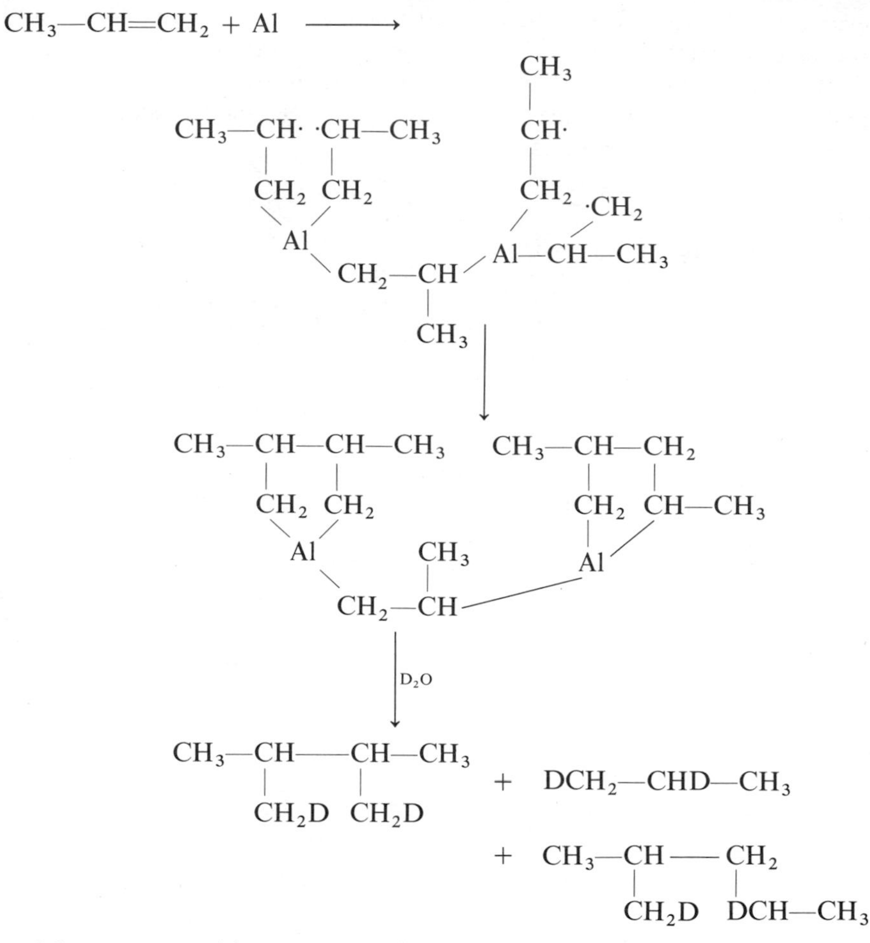

Of course, a multicomponent mixture of organoaluminum compounds is produced, and the one shown above merely represents one route to the major products. However, to go back one further stage in the mechanism, one must consider the fact that at the temperature of vaporization of the aluminum it exists predominantly in its ground state ^2P electronic configuration and is not a triradical. Therefore, the question arises as to the possible intermediacy of an aluminocyclopropane that opens up to yield a diradical and follows the scheme outlined above. A preliminary indication is the reaction of 1,3-butadiene with aluminum atoms since, upon hydrolysis, almost exclusively 1-butene is formed and very little 2-butene.[27] This militates for an aluminocyclopropane, but such evidence is equivocal, and more compelling data are necessary.

The reaction of aluminum atoms with either *cis*- or *trans*-2-butene to produce, after deuterolysis, a mixture of *meso*- and *d,l*,-2,3-dideuterobutanes shows that the reaction is not stereospecific.[27] If, for example, a stable *cis*-dimethylaluminocyclopropane were produced, one would have to postulate two different modes of deuterolysis involving either retention (or inversion) at each ring carbon atom to produce the *meso*-2,3-dideuterobutane, or retention at one carbon atom and inversion at the other to get the *d,l*-isomer. Thus, if an aluminocyclopropane is initially formed it must be thermally unstable so that ring opening and bond rotation are very facile. This latter postulate gains considerable support from the experiment whereby aluminum vapor, *cis*-2-butene, and a Lewis base, such as dimethyl ether or trimethylamine, were cocondensed and then after D_2O treatment gave 95% of the *meso* product.[28] Similarly, use of *trans*-2-butene led to 95% of the *d,l* isomer. These results tell us that the initially formed aluminocyclopropane system is stabilized by the Lewis base, and hence the rate of ring opening is greatly reduced, leading to 95% stereospecificity.

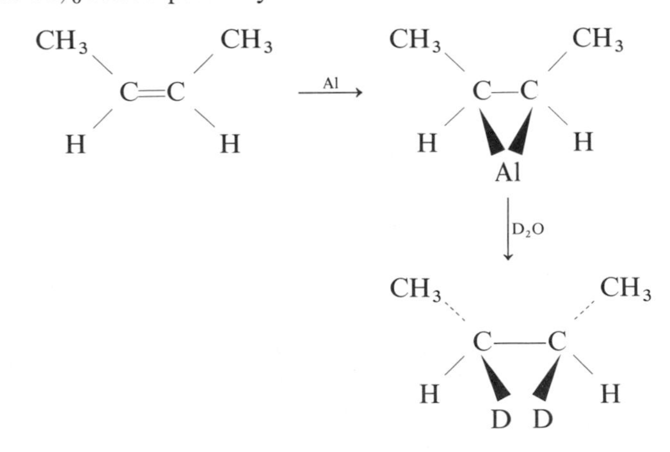

Formally, one could regard the ^2P aluminum atoms as being somewhat analogous to singlet carbenes, and so the stereospecificity of the reactions can be rationalized on this basis. Furthermore, it is difficult to explain the retention of stereochemistry on the basis of two aluminum atoms per double bond.

A further interesting observation is that in the aluminum-propene reaction a consideration of the number of deuterium atoms incorporated per mole of metal reacted accounts for about 2.7 of the 3 valences of Al.[27] By contrast, the stoichiometry derived from the hydrolysis of the aluminum/2-butene reaction corresponds to 2 aluminums/1 H_2/2 butanes. Also, use of D_2O instead of H_2O yields only D_2, and the presence of Al—H bonds would surely have produced HD. A possible explanation is the presence of aluminum-aluminum bonds, a previously unreported phenomenon in organoaluminum chemistry. Although it has not yet proven possible to isolate a molecule such as bis(2,3-dimethylaluminocyclopropane), it is possible to titrate out with iodine that portion of the reactivity that gives D_2. Thus, addition of one molar equivalent of I_2 removes the ability to produce D_2 on hydrolysis, yet leaves the aluminum/butane ratio unchanged.

Aluminum atoms react with acetylenes mainly by addition to make 1,2-dialuminoalkenes and upon deuterolysis yield mainly dideuterated alkenes. There is little stereospecificity in these reactions since 2-butyne yields a 2:1 mixture of *cis*- and *trans*-2-butenes. Propyne gives mostly propene, but also propyne monodeuterated in either the methyl or ethynyl positions, indicating the intermediacy of $AlCH_2—C{\equiv}CH$ as well as $CH_3—C{\equiv}CAl$.[27]

Aluminum atoms react readily with alkyl halides to produce the corresponding alkyl aluminum sesquihalides. Thus ethyl bromide and aluminum react to form an amber involatile liquid that upon deuterolysis yields 147% d_1-ethane based on aluminum vaporized. A small amount of d_1-butane is also isolable, presumably formed by the well-known growth reaction of organoaluminums and ethylene produced by thermal decomposition of the ethyl bromide.

$$2Al + 3RX \rightarrow R_3Al_2X_3 \xrightarrow{3D_2O} 3RD + Al_2O_3 + 3DX$$

Bulk aluminum metal yields sesquihalides with methyl or ethyl halides, but the higher homologues are unstable.[29]

Aluminum atoms also react with ammonia to form the amide at low temperature and release hydrogen. Upon warm-up, however, ammonia is lost, and the polymeric nitride is formed:[27]

$$Al + 3NH_3 \rightarrow Al(NH_2)_3 + \tfrac{3}{2}H_2$$

$$Al(NH_2)_3 \xrightarrow{\Delta} AlN + 2NH_3$$

The intense black color of the initially produced Al/NH_3 matrix suggested that solvated electrons might be present, but addition of 2-butyne to the matrix gave only small amounts of 2-butene.[27] This contrasts with magnesium, where stereospecific reduction of 2-butyne to *trans*-2-butene has been observed.[13]

Finally, the reaction of aluminum atoms with water yields hydrogen as expected, but the interesting point is that some hydrogen is released during codeposition and some during warm up of the matrix. The isotopic composition of the hydrogen released (when using an H_2O/D_2O mixture) differs greatly for these two periods of hydrogen evolution and suggests that two different mechanisms of hydrogen evolution are utilized.[28] Further work is needed to clarify the position, and it may well turn out that solvated electrons are involved as has been demonstrated in the magnesium vapor reactions.

5. GROUP IV ELEMENTS

5.1 Carbon

Although carbon clearly does not qualify as a main group metal, a very brief review is presented to place the field in historical perspective. Carbon vapor was the first elemental species to be studied in considerable detail using the cocondensation procedure,[1] and as such it served as a prototype for the study of the remainder of the periodic table. The chemistry of carbon vapor has been fully reviewed elsewhere.[30]

Carbon vapor provides an interesting and complex problem in that unlike most of the other elements studied it does not consist solely of monoatomic species in their electronic ground state. Carbon vapor contains C_1, C_2, C_3, C_4, and other molecular species in diminishing amounts. The problem of assigning any given product to a particular carbon precursor was overcome by enriching the carbon to be arced or resistively heated with radioactive carbon-14. This provided an easily identifiable marker to identify the molecular carbon species involved. Thus, the acetylene produced by an insertion-fragmentation process in alkyl halides is derived from C_1,[31] whereas the acetylene derived from the reaction of carbon vapor with alcohols or ketones is entirely arc derived.[32]

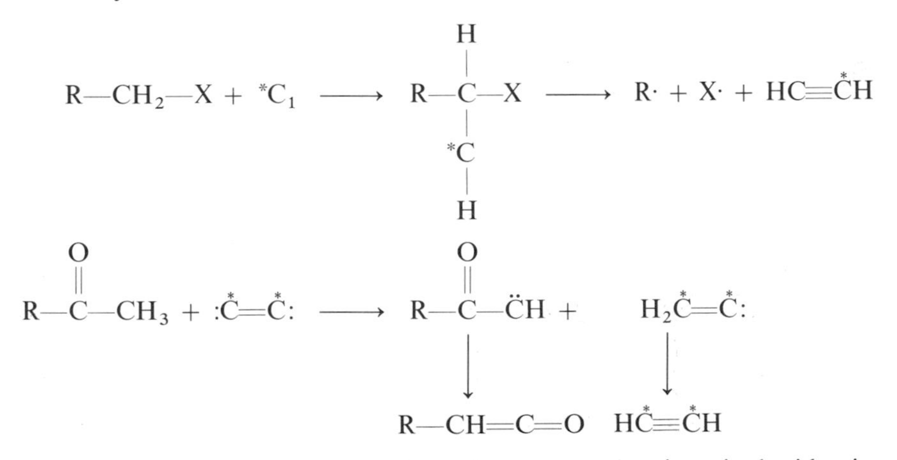

The other interesting problem in carbon vapor chemistry is the identity of the electronic state of the molecular carbon species. Thus, C_3 has a singlet ground state that reacts by two successive stereospecific additions to an olefin to produce a single bisethanoallene.[1]

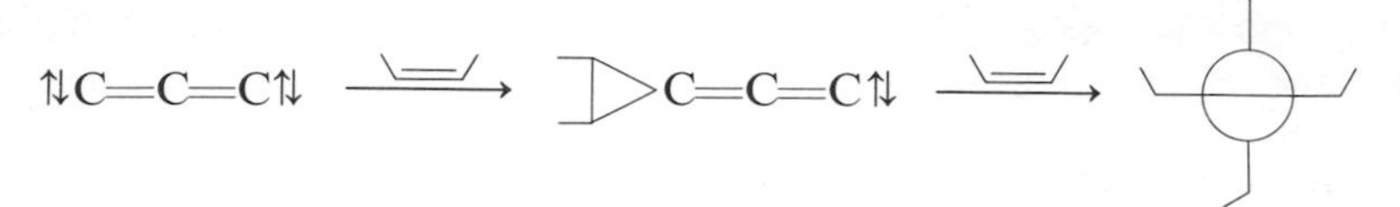

On the other hand, the metastable excited-state triplet C_3, produced by electronic bombardment in the arc, adds nonstereospecifically in its second addition step to produce two isomeric bisethanoallenes.[33]

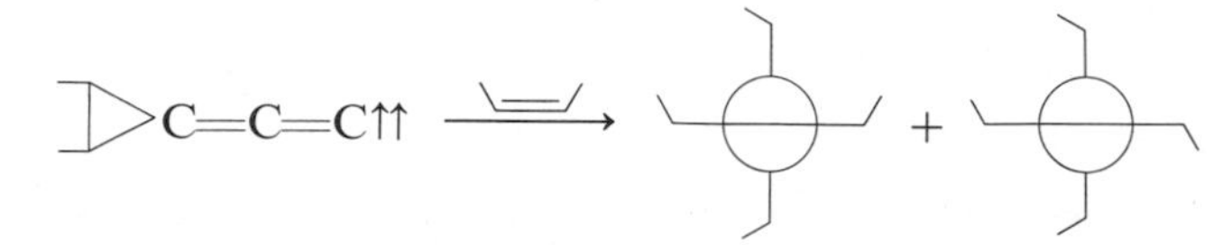

Carbon atoms have also been utilized for purely synthetic purposes and can lead directly to products that are either difficult or impossible to synthesize by conventional procedures.[34–36]

$$CF_3{-}CF{=}CF_2 \xrightarrow{C} CF_3{-}CF{=}C{=}CF_2$$

$$B_2Cl_4 \xrightarrow{C} C(BCl_2)_4$$

$$SiCl_4 \xrightarrow{C} Cl_3Si{-}C{\equiv}C{-}Cl$$

The success of carbon-vapor chemistry from both a mechanistic and synthetic point of view undoubtedly stimulated further interest in the chemistry of high-boiling atomic and molecular species.

5.2 Silicon and Germanium

Silicon differs from carbon in that the monoatomic species predominates in the vapor phase, and so this is the only silicon species to have been investigated. Skell and Owen[37] vaporized silicon from a resistively heated silicon rod and cocondensed the vapor with trimethylsilane to produce 1,1,1,3,3,3-hexamethyltrisilane in 27% yield. This reaction apparently proceeds via successive silene insertions into Si—H bonds.

$$Me_3Si{-}H + Si \rightarrow \underset{\text{(VII)}}{Me_3Si{-}\ddot{Si}{-}H} \xrightarrow{Me_3SiH} Me_3Si{-}SiH_2{-}SiMe_3$$

In contrast to the thermal production of silicon atoms, electron bombardment of silicon produces (VII) in 14% yield as well as an involatile residue and 10 other products including hexamethyldisilane and hexamethylsiloxane. These additional products were shown to be formed in a blank reaction in which no silicon was vaporized, but the substrate was subjected to electron bombardment, which produced silenes and hence the additional products.[38]

Similarly, other silanes reacted with silicon atoms via successive silene insertions into Si—H bonds, and in each case electron bombardment was the cause of much independent chemistry as revealed by appropriate blank reactions.[38]

$$SiH_3{-}SiH_2{-}SiH_2{-}SiH_2{-}SiH_3 \xleftarrow{Si_2H_6} Si \xrightarrow{Me_2SiH_2} (Me_2SiH)_2SiH_2 \ (30\%)$$

$$Si \xrightarrow{MeSiH_3} (MeSiH_2)_2SiH_2 \ (34\%)$$

Silicon tetrafluoride reacts with hot silicon to produce the very interesting species SiF_2, whose chemistry has been reviewed previously.[39,40] Silicon difluoride reacts with boron trifluoride to give a variety of silicon-boron fluorides, including the long-sought-after parent compound, $SiF_3{-}BF_2$.[41] Alternative routes to these compounds via the reaction of boron atoms with SiF_4 or silicon atoms with BF_3 were generally unsuccessful,[42] but an interesting product, namely, $(BF_2)_3SiF$, was obtained from silicon atoms and B_2F_4. Timms[42] postulates initial insertion by silicon into a B—B bond and then B—F insertion by the bis(difluoroboryl)silene to yield a product unstable with respect to loss of boron monofluoride polymer:

$$B_2F_4 + Si \longrightarrow BF_2{-}Si{-}BF_2 \xrightarrow{BF_2BF{-}F} (BF_2)_2Si(F)(BF{-}BF_2)$$

$$(BF_2)_2Si(F)(BF{-}BF_2) \xrightarrow{-BF_3} \text{polymers}$$

$$(BF_2)_2Si(F)(BF{-}BF_2) \longrightarrow (BF_2)_3SiF + (BF)_n$$

However, one could equally well postulate loss of $(BF)_n$ from tetrakis(difluoroboryl)silane, which would be formed by successive B—B insertions and would be analogous to the known $(BF_2)_4C$, which is also unstable.[35]

Germanium vapor is still relatively unexplored as a synthetic reagent, but it does react with some halides of Groups IV and V to give the corresponding trichlorogermyl derivatives.[43]

$$CHCl_3 + Ge \rightarrow CHCl_2{-}GeCl_3$$

$$SiCl_4 + Ge \rightarrow SiCl_3{-}GeCl_3$$

$$PCl_3 + Ge \rightarrow PCl_2{-}GeCl_3$$

No double insertion products are observed, suggesting that germanium is reacting in its ^3P electronic ground state whereby an initial insertion is followed by two radical abstractions:

$$\text{R—X} + {\uparrow\downarrow}\text{Ge}{\uparrow} \rightarrow \text{R—}\overset{\uparrow\uparrow}{\text{Ge}}\text{—X} \xrightarrow{\text{RX}} \text{R—}\overset{\uparrow}{\text{Ge}}\text{—X}_2 + \text{R}\cdot \xrightarrow{\text{R—X}} \text{R—GeX}_3 + \text{R}\cdot \quad (2\,\text{R}\cdot \rightarrow \text{R}_2)$$

It has also been reported that tin vapor reacts with some alkyl halides to give alkyltin halides at low temperatures.[2]

6. DEHALOGENATIONS USING COPPER, SILVER, AND GOLD

Elemental copper has long been used as a reagent to couple halides, and the Ullman biaryl synthesis is a classic example:

$$\text{ArX} + \text{Ar}^1\text{X} \xrightarrow[\Delta]{\text{Cu}} \text{Ar—Ar}^1$$

Copper vapor is monoatomic, and Timms has demonstrated that copper atoms function very well as dechlorinating agents. Typically, condensation of BCl_3 and Cu in the molar ratio 6:1 gives a 40% yield of B_2Cl_4; with a mole ratio of 20:1 the yield increases to 70%.[44] This has now been developed to the point where a preparation of 10 g of B_2Cl_4 is now routine,[45] and this technique is much superior to the commonly used discharge methods.[46] Dechlorinations with copper have also been extended to organochloroboranes[44] and also to phosphorus halides:[2]

$$\text{MeBCl}_2 \xrightarrow{\text{Cu}} \text{MeBCl—BClMe}$$

$$\text{PCl}_3 \xrightarrow{\text{Cu}} \text{P}_2\text{Cl}_4$$

Interestingly, thermally vaporized copper does not react with Si—Cl compounds, yet copper atoms produced by electron bombardment react with $SiCl_4$ to give perchlorosilanes.[2] This may be an analogous case to magnesium (see Section 3.2), where different methods of production yield metal atoms in different electronic states and consequently exhibit different chemistry.

Timms also reported that copper dehalogenated alkyl halides to give alkyl radicals that either dimerize or disproportionate to the corresponding alkane and alkene:[2]

$$\text{C}_2\text{H}_5\text{Br} \xrightarrow{\text{Cu}} \text{CuBr} + \text{C}_2\text{H}_5^{\cdot}$$

$$\text{C}_2\text{H}_5^{\cdot} \rightarrow n\text{—C}_4\text{H}_{10} \quad \text{or} \quad \text{C}_2\text{H}_4 + \text{C}_2\text{H}_6$$

Wolf[28] has investigated these reactions in more detail and showed that the ratio of dimerization to disproportionation products increased dramatically as the ratio of substrate to metal was lowered from 100:1 to 10:1. The high yield of coupling product suggested the possible intermediacy of alkyl copper intermediates, and attempts were made to detect methyl copper in the reaction of copper atoms with methyl bromide. However, much ethane was evolved even at $-78°C$, at which temperature methyl copper should be stable. This suggests that the major process was the abstraction of bromine to form methyl radicals which coupled. The formation of methyl copper would require either the coupling of a copper atom with a methyl radical or the reaction of two copper atoms with one bromomethane. Both processes are unfavorable because of the presence of a large excess of substrate.

Copper has been used to dehalogenate acetyl chloride to give biacetyl.[28] Silver and gold are also capable of debrominating ethyl bromide to give *n*-butane, ethane, and ethylene, but coupling is much more efficient using copper. Silver vapor gives poor yields of B_2Cl_4 from BCl_3 but is quite efficient for the production of P_2Cl_4 from PCl_3.[2]

7. ZINC

Organozinc compounds have been known for some time,[47] but cocondensation of zinc atoms with perfluoroalkyl iodides leads to nonsolvated fluoro-organo-zinc compounds directly.[48] Typically, CF_3I and Zn atoms reacted to produce a solid, which decomposed vigorously at $-80°C$, liberating C_2F_6 and C_2F_4. When the CF_3I substrate was saturated with water vapor before deposition, fluoroform was produced as well as the previous two products. This indicates that the fluoroorgano zinc complex is unstable at well below room temperature and is hydrolyzable by water while still cold. Even more striking is the fact that C_3F_7ZnI hydrolyzes to C_3F_7H in the cold, while the same compound made by conventional means (and presumably solvated) is stable and not easily hydrolyzed.[49] The results have been interpreted in terms of radical decompositions of the nonsolvated organo zinc iodides, and trapping experiments support this rationale. The isolation of C_2F_4 suggested the intermediacy of difluorocarbene, and this was confirmed by a trapping experiment with isobutene:[48]

$$CF_3I + Zn \longrightarrow CF_3ZnI$$

$$CF_3ZnI \xrightarrow{-80°C} :CF_2 + ZnFI$$

$$CF_3ZnI \xrightarrow[-100°C]{H_2O} CF_3H + Zn(OH)I$$

$$:CF_2 \xrightarrow{\text{isobutene}} \text{1,1-difluoro-2,2-dimethylcyclopropane (F F)}$$

8. SUMMARY

Although a wide variety of main group elemental species have now been investigated, the coverage is by no means even. Thus, magnesium vapor exhibits many reactions, and in some cases the identity of the electronic state of the reactive species has been elucidated. One might mention here the relevance to this work of the recent reports of active metals by Klabunde[50] and by Rieke.[51] However, the heavier elements of Groups II, III, and IV have been little studied, and these may provide fruitful areas of investigation over the next few years.

REFERENCES

1. P. S. Skell and L. D. Wescott, Jr., *J. Amer. Chem. Soc.*, **85**, 1023 (1963).
2. P. L. Timms, *Advan. Inorg. Chem. Radiochem.*, **14**, 121 (1972).
3. L. Andrews and G. C. Pimentel, *J. Chem. Phys.*, **44**, 2527 (1966).
4. C. Chung and R. J. Lagow, *Chem. Commun.*, 1078 (1972); ibid., 303 (1975).
5. P. S. Skell, E. J. Goldstein, R. J. Petersen, and G. L. Tingey, *Chem. Ber.*, **100**, 1442 (1967).
6. P. L. Timms, R. A. Kent, T. C. Ehlert, and J. L. Margrave, *J. Amer. Chem. Soc.*, **87**, 2824 (1965).
7. P. S. Skell and E. J. Goldstein, *J. Amer. Chem. Soc.*, **86**, 1442 (1964).
8. R. L. Lambert, Jr., and D. Seyferth, *J. Amer. Chem. Soc.*, **94**, 9246 (1972).
9. P. S. Skell and E. J. Goldstein, *J. Amer. Chem. Soc.*, **86**, 1442 (1964).
10. P. S. Skell and R. J. Petersen, *J. Amer. Chem. Soc.*, **86**, 2530 (1964).
11. J. D. Roberts, A. Streitwieser, Jr., and C. M. Regan, *J. Amer. Chem. Soc.*, **74**, 4579 (1952).
12. R. G. Doerr and P. S. Skell, *J. Amer. Chem. Soc.*, **89**, 4684 (1967).
13. J. E. Girard, Ph.D. thesis, The Pennsylvania State University, University Park, Pa., 1971.
14. P. S. Skell and J. E. Girard, *J. Amer. Chem. Soc.*, **94**, 5518 (1972).
15. A. Farkas and L. Farkas, *Trans. Faraday Soc.*, **33**, 337 (1937).
16. M. Lefrancois and Y. Gault, *J. Organometal. Chem.*, **10**, 493 (1972).
17. D. Bryce-Smith, *J. Chem. Soc.*, 1603 (1956).
18. L. I. Zakharkin, O. Yu. Okhlobystin, and B. N. Strunin, *J. Organometal. Chem.*, **4**, 349 (1965).
19. A. I. Snow and R. E. Rundle, *Acta Crystallogr.*, **4**, 348 (1951).
20. H. Kleinfeller, *Berichte.*, **62**, 2736 (1929).
21. P. S. Skell and K. J. Shea, *Isr. J. Chem.*, **10**, 493 (1972).
22. K. J. Klabunde, J. Y. F. Low, and M. S. Key, *J. Fluorine Chem.*, **2**, 207 (1972).
23. P. L. Timms, *Chem. Commun.*, 258 (1968).
24. P. L. Timms, *J. Amer. Chem. Soc.*, **89**, 1629 (1967).
25. R. Kirk and P. L. Timms, *Chem. Commun.*, 18 (1967).
26. W. N. Brent, Ph.D. thesis, The Pennsylvania State University, University Park, Pa., 1974.
27. P. S. Skell and L. R. Wolf, *J. Amer. Chem. Soc.*, **94**, 7919 (1972).

28. L. R. Wolf, Ph.D. thesis, The Pennsylvania State University, University Park, Pa., 1974.
29. H. Adkins and C. Scanly, *J. Amer. Chem. Soc.*, **73**, 2854 (1951).
30. P. S. Skell, J. J. Havel, and M. J. McGlinchey, *Account. Chem. Res.*, **6**, 97 (1973).
31. L. Eng, Ph.D. thesis, The Pennsylvania State University, University Park, Pa., 1970.
32. P. S. Skell and R. F. Harris, *J. Amer. Chem. Soc.*, **88**, 5933 (1966).
33. P. S. Skell, L. D. Wescott, Jr., J. P. Golstein, and R. R. Engel, *J. Amer. Chem. Soc.*, **87**, 2829 (1965).
34. M. J. McGlinchey, T. Reynoldson, and F. G. A. Stone, *Chem. Commun.*, 1264 (1970).
35. J. E. Dobson, P. M. Tucker, R. Schaeffer, and F. G. A. Stone, *J. Chem. Soc., A*, 1882 (1969).
36. J. Binenboym and R. Schaeffer, *Inorg. Chem.*, **9**, 1578 (1970).
37. P. S. Skell and P. W. Owen, *J. Amer. Chem. Soc.*, **89**, 3933 (1967).
38. P. S. Skell and P. W. Owen, *J, Amer. Chem. Soc.*, **94**, 5434 (1972).
39. J. L. Margrave and P. W. Wilson, *Account. Chem. Res.*, **4**, 145 (1971).
40. P. L. Timms, *Account. Chem. Res.*, **6**, 118 (1973).
41. D. L. Smith, R. W. Kirk, and P. L. Timms, *Chem. Commun.*, 295 (1972).
42. R. W. Kirk and P. L. Timms, *J. Amer. Chem. Soc.*, **91**, 6315 (1969).
43. M. J. McGlinchey and T. S. Tan, *Inorg. Chem.*, **14**, 1209 (1975).
44. P. L. Timms, *Chem. Commun.*, 1525 (1968).
45. P. L. Timms, *J. Chem. Soc., A*, 830 (1972).
46. A. G. Massey, *Advan. Inorg. Chem. Radiochem.*, **10**, 1 (1967).
47. G. E. Coates, M. L. H. Green, P, Powell, and K. Wade, *Principles of Organometal. Chem.*, p. 65, Methuen, London, 1971.
48. K. J. Klabunde, M. S. Key, and J. Y. F. Low, *J. Amer. Chem. Soc.*, **94**, 999 (1972).
49. R. D. Chambers, W. K. R. Musgrave, and J. Savory, *J. Chem. Soc.*, 1993 (1962).
50. K. J. Klabunde, H. F. Efner, L. Satek, and W. Donley, *J. Organometal. Chem.*, **71**, 309 (1974).
51. R. D. Rieke and S. E. Bales, *J. Chem. Soc., Chem. Commun.*, 879 (1973).

Organometallic and Organic Syntheses Involving Transition Metal Vapors

5

Michael J. McGlinchey and Philip S. Skell

1. INTRODUCTION

The use of high-temperature species as synthetic reagents has been developed over the past decade,[1] and some areas within this field have been reviewed.[2,3] The use of transition metal atoms is of much more recent vintage and has allowed the direct synthesis of a number of novel organometallic complexes, as well as many known compounds, previous routes to which were cumbersome. Furthermore, it has proved possible to isolate at low temperatures interesting and unstable compounds whose existence had only been inferred through mechanistic studies.

The general technique in this area is to vaporize a transition metal and to cocondense this metal vapor at $-196°C$ with that of one or more substrates. Throughout the codeposition process the pressure is kept as low as possible to minimize gas phase reactions and ensure that the reactions studied are those occurring on the cold surface.

For the purposes of a logical presentation of this chapter it is preferable to classify the reaction types into three categories (which may occasionally overlap).

1. After removal of the excess unreacted substrate(s) a less volatile organometallic compound remains, and this is isolated either by a pump-out procedure or by the less attractive method of solvent extraction. Sometimes the products are sufficiently air stable to allow them to be scraped from the reaction flask and purified by conventional methods.
2. Removal of the excess substrate leaves an involatile residue the nature of which is investigated by treatment with a variety of reagents, for example, hydrogen, deuterium oxide, carbon monoxide, trifluorophosphine, and so forth.
3. Removal of the substrate reveals that a sizable percentage of it has been catalytically transformed via an organometallic intermediate: typical examples are isomerization or dimerization.

2. DIRECT SYNTHESIS OF ORGANO-TRANSITION METAL COMPLEXES

2.1 Sandwich Compounds

The first published synthesis of organo-transition metal complexes using transition metal vapors was by Timms[4] in 1969, who demonstrated that cocondensation of iron vapor and cyclopentadiene monomer produced ferrocene. This was interesting, but not dramatic synthetically, since ferrocene is readily available by a number of routes. However, the direct synthesis of cyclopentadienyl sandwich compounds has been extended from ferrocene to include those of titanium,[5] chromium,[2,6] cobalt,[2] nickel,[4] and molybdenum[7] (see Figure 1). As in their more conventional syntheses[8] the cyclopentadienyl complexes of Cr and Mo have their usual 16- and 18-electron systems, respectively, that is, Cp_2Cr and Cp_2MoH_2. No evidence was observed for the formation of molybdenocene, Cp_2Mo, such as has been claimed by Brintzinger[9] in a different system. It is also noteworthy that the yields of cobaltocene and nickelocene (19- and 20-electron systems, respectively) are rather low and that the major products from the reactions of cyclopentadiene with cobalt vapor and with nickel vapor are (π-cyclopentadienyl)cobalt(cyclopentadiene) and (π-cyclopentadienyl)(π-cyclopente-

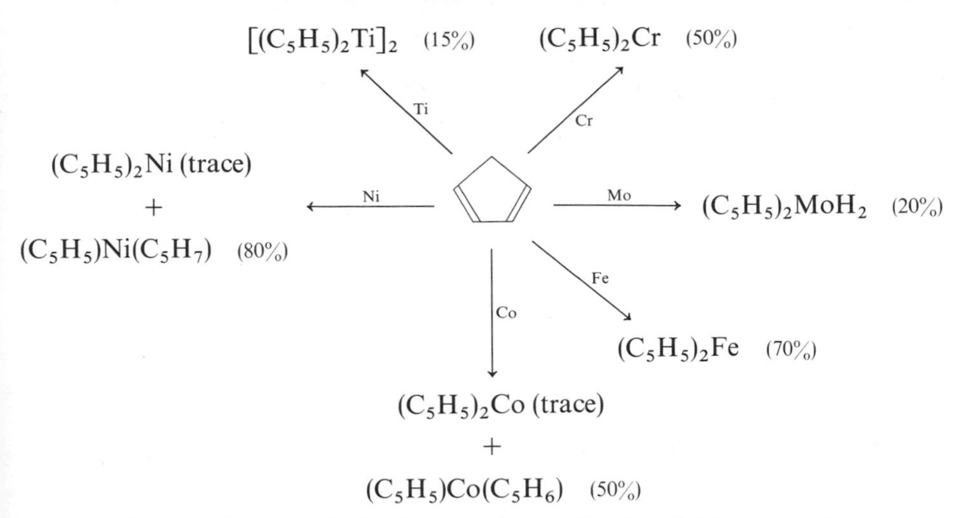

Figure 1. Direct synthesis of some cyclopentadienyl sandwich compounds.

nyl)nickel, which are both 18-electron systems. The nickel reaction probably involves the intermediate, bis(cyclopentadiene)nickel (I), which rearranges by hydrogen migration—possibly involving a nickel hydride.

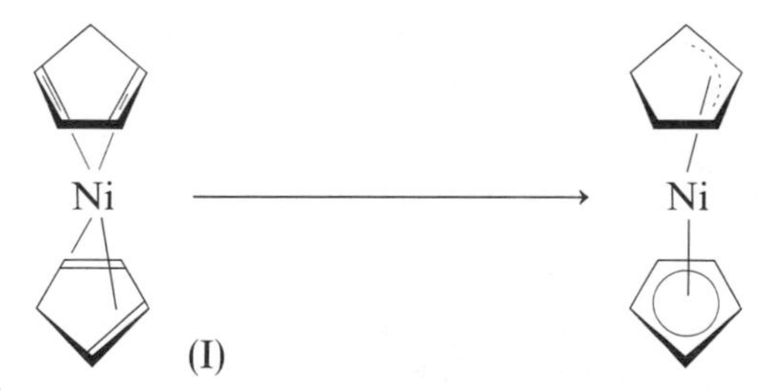

Such a postulate is supported by the nickel-atom-catalyzed disproportionation[10] of 1,3-cyclohexadiene to cyclohexene and benzene, which is readily visualized as involving a bis(diene) nickel intermediate that rapidly undergoes hydrogen migration to give the observed product (see also Section 4.3).

An extremely useful synthetic reaction is the direct formation of bis(arene)metal complexes by the simultaneous codeposition of metal vapor and the appropriate arene.[4]

$$C_6H_6 \xrightarrow[\text{vapor}]{\text{Cr}} (C_6H_6)_2Cr\ (60\%)$$

This synthesis of bis(benzene)chromium provided a simple route to a compound previously obtainable only by a difficult two-stage process pioneered by Fischer and Hafner:[11]

$$CrCl_3 + C_6H_6 + AlCl_3 + \text{Al powder}$$

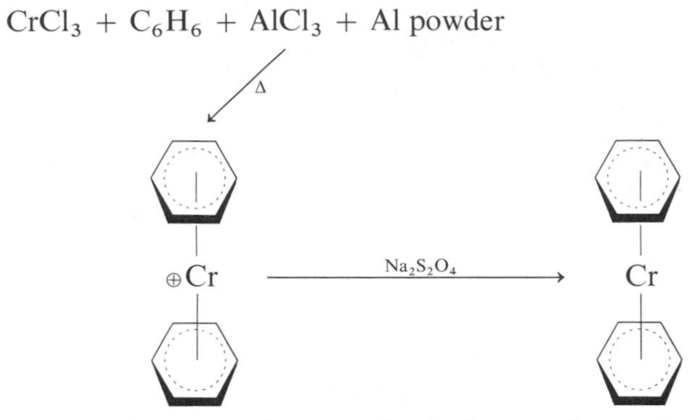

Furthermore, the Fischer–Hafner synthesis frequently produced bis(arene)chromiums of doubtful purity, since, with alkylbenzenes, the Friedel–Crafts conditions employed led to rearrangement of alkyl groups. Thus, 'bis(cumene)chromium' was reported to be a liquid,[12] but this in fact contains a complex mixture including bis(toluene)-, bis(xylene)-, and bis(diisopropylbenzene)chromium. Bis(cumene)chromium prepared by the cocondensation method is a dark yellow solid, m.p. 46°C.[13]

Another limitation of the conventional bis(arene)chromium synthesis is that it fails when used with arenes having substituents possessing lone pairs of electrons, since they complex with the aluminum trichloride catalyst. However, this limitation is removed when the cocondensation method is employed, and many bis(arene)chromium complexes have been prepared with substituents such as fluorine,[6] chlorine,[6] or methoxide.[14] Attempts to prepare bromoarene complexes by this route have so far failed due to abstraction of bromine from the ring to produce $CrBr_3$.[14] Some of the corresponding sandwich compounds of molybdenum and tungsten have also been prepared by evaporating the metal from a filament (see Table 1).[15–17]

By use of electron-beam evaporation of some of the early transition metals it has been possible to synthesize directly bis(benzene)titanium:[16]

Table 1. Mo and W Sandwich Compounds Prepared via Direct Synthesis

Mo	W
$(C_5H_5)_2MoH_2$	$(C_5H_5)_2WH_2$
$(C_6H_6)_2Mo$	$(C_6H_6)_2W$
$(C_6H_5{\cdot}Me)_2Mo$	$(o\text{-}C_6H_4{\cdot}Me_2)_2W$
$(C_6H_5F)_2Mo$	$(C_6H_5F)_2W$
$(C_6H_5{\cdot}OMe)_2Mo$	$(C_6H_5{\cdot}OMe)_2W$
$(C_6H_5{\cdot}NMe_2)_2Mo$	$(C_7H_7)W(C_7H_9)$
$(C_6H_5{\cdot}CO_2Me)_2Mo$	

bis(benzene)vanadium has also been claimed.[18] The titanium compound is diamagnetic as is to be expected by analogy to the isoelectronic (π-cycloheptatrienyl)(π-cyclopentadienyl)titanium.[19] Iron atoms and benzene may produce the very unstable bis(benzene)iron at low temperatures (but see Section 3.1).[4]

The ready availability of these complexes has stimulated their study by a variety of spectroscopic techniques with a view to further elucidating their metal-ring bonding.

A ^{19}F nmr study on a series of substituted bis(fluorobenzene)chromium complexes possessing an extra substituent in the *ortho*, *meta*, or *para* position such as CH_3, CF_3, Cl, or F indicated a similarity between the π-complexed fluorobenzene ring and the uncomplexed pentafluorobenzene ring.[14] Thus the π-bonded chromium atom clearly withdraws considerable electron density from the ring. Further studies using both X-ray and ultraviolet photoelectron spectroscopy have revealed that there are noticeable changes in the carbon 1*s* binding energies on complexation.[20] A series of bis(arene)molybdenum complexes have been examined by ^{13}C nmr spectroscopy and show large upfield chemical shifts for the arene carbons relative to the free-ligand values.[21]

The overall picture indicated is that the arene is relatively poorer in electron density when it is complexed to chromium and should thus be susceptible to nucleophilic attack, but it would require the presence of a suitable leaving group on the arene ring to enable substitution to occur. This prediction has now been justified since methoxide ion displaces fluoride from bis(fluorobenzene)chromium to produce bis(anisole)chromium identical to that formed by the direct reaction of anisole with chromium vapor.[22] Furthermore, the mixed arene-chromium complex, $(C_6H_6)Cr(C_6F_6)$, has been shown to undergo very facile substitution of fluorines using alkyl- or aryllithium reagents.[22] Such advances should open up the chemistry of the chromium arene sandwich complexes, which has been relatively untouched for nearly twenty years!

Hexafluorobenzene complexes of several first-row metals have been claimed, but apart from those of vanadium and to a lesser extent nickel they have low thermal stability.[23] However, as a general rule, it has been clearly shown that introduction of electron-withdrawing substituents (particularly trifluoromethyl groups) leads to greatly enhanced stability to oxidation.[14,24] This is thought to be a kinetic effect whereby the electron-withdrawing groups retard the one electron oxidation of the chromium. However, infrared data on these complexes suggest that there is less C=C bond weakening (via back-donation to the ring by the chromium) in the CF_3-substituted complexes than in bis(benzene)chromium itself. This in turn implies a slightly weaker metal-ring bond in the former (CF_3 substituted) complexes.[24] Thus there must be a compromise between the kinetic advantage and the thermodynamic disadvantage of substituting the rings with electronegative groups. Timms has investigated this point in a competition reaction using an equimolar mixture of C_6H_6, C_6H_5F, 1,4-$C_6H_4F_2$, 1,3,5-$C_6H_3F_3$, 1,2,3,5-$C_6H_2F_4$, C_6HF_5, and C_6F_6. Along with excess PF_3 he cocondensed this mixture with chromium vapor.[25] The yield of (arene)Cr$(PF_3)_3$ was a maximum with $C_6H_4F_2$ and decreased to a minimum with C_6F_6. Timms suggests that an increasing tendency for defluorination results in the decreasing yield from the difluoro- to the hexafluorobenzene.[25]

Manganese vapor has also been used in an attempt to prepare sandwich compounds directly. Although this does not provide a route to manganocene, the mixed arene(cyclopentadienyl)manganese(I) systems can be prepared.[25,26] Fluorine-substituted bis(arene)vanadium complexes have also been reported very recently.[24]

Sandwich compounds of larger rings have also been prepared: a tentative claim has been made for bis(cycloheptatriene)chromium,[6] but evidence has also been presented to show that (1-4-η-cycloheptadiene)(η-cycloheptatri-

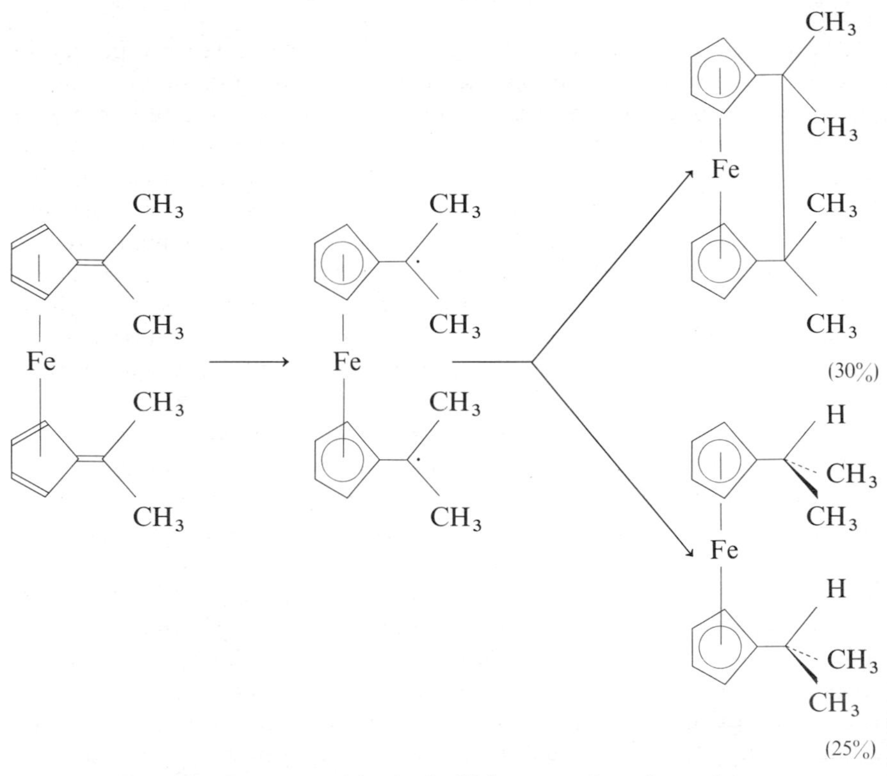

Figure 2. Template synthesis of a [2]-ferrocenophane from a fulvene.

enyl)chromium(−1) is also isolable from the product mixture.[26] The reaction of iron atoms with cycloheptatriene is much cleaner and gives good yields of $(C_7H_9)Fe(C_7H_7)$.[26] Furthermore, it is possible to prepare bis(cyclooctatetraene)iron directly,[27] and the synthesis of uranocene from cyclooctatetraene and uranium vapor has also been claimed.[28]

Finally, it is reported to be possible to synthesize bridged sandwich compounds via the cocondensation procedure. Thus, cocondensation of 6,6-dimethylfulvene with iron vapor leads to 1,1,2,2-tetramethyl-[2]-ferrocenophane,[29] probably via the mechanism depicted in Figure 2.

2.2 Olefin Complexes

The reactions of a typical olefin such as propene with a wide range of transition metals is considered in detail in Section 3 since the major product after removal of the excess olefin is an involatile residue. However, a recent claim has been made that codeposition of nickel atoms with ethylene leads directly to tris(ethylene)nickel,[30] which was first synthesized more conventionally by Wilke.[31]

Nonconjugated dienes such as 1,5-cyclooctadiene react with nickel[10] or platinum[32] vapor to produce bis(1,5-cyclooctadiene)nickel (or platinum) in low yields.

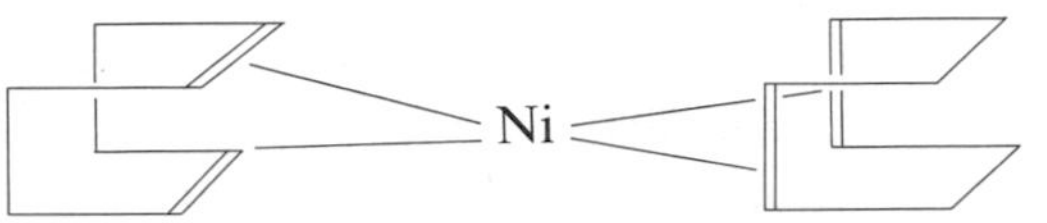

However, this route is unlikely to oust the more conventional method, which utilizes ethoxydiethylaluminum to reduce nickel acetylacetonate in the presence of 1,5-cyclooctadiene and which is amenable to relatively large-scale work.[33] In contrast, nickel vapor and norbornadiene[10] yielded no isolable organometallic product, but gave predominantly the *exo-trans-exo* dimer of norbornadiene (II).[34]

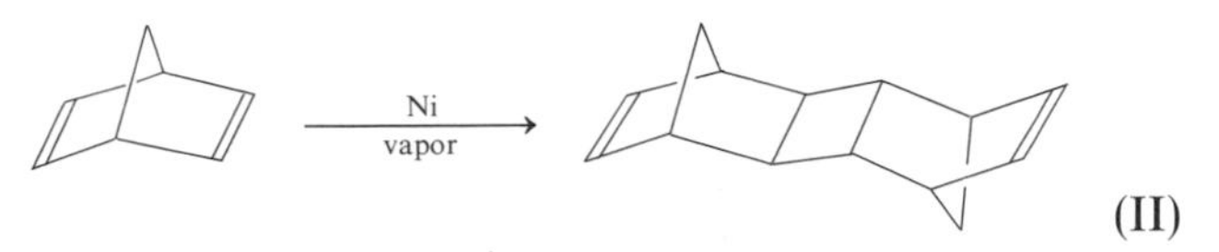

An exciting new discovery in this area has been the isolation by Timms[35] of bis(cyclooctadiene)iron, a 16-electron system that shows high reactivity toward other ligands.[25] The geometry of this molecule has not yet been determined, but its lack of an nmr signal might be interpreted as suggesting a paramagnetic tetrahedral configuration.

Linear nonconjugated dienes, for example, 1,4-pentadiene, are catalytically isomerized and also undergo disproportionation when they are cocondensed with cobalt vapor, but a small yield of bis(1,3-pentadiene)cobalt hydride is also formed.[18] Further discussion of this is delayed until Section 4.1, but the synthetic aspects are of interest as the closest analog is probably bis(1,3-butadiene)rhodium chloride prepared by the sealed-tube reaction of 1,3-butadiene with rhodium trichloride:[36] this is known to possess C_{2v} symmetry in the crystalline state, and bis(1,3-butadiene)cobalt hydride may well have the same structure.

Very novel diene complexes have recently been reported in which tungsten or molybdenum react in the condensed phase with 1,3-butadiene to yield tris(butadiene)tungsten(O) or the corresponding molybdenum analog.[37] The only closely comparable previously known system is tris(methylvinylketone)tungsten,[38] which adopts the relatively unusual trigonal prismatic structure:[39] the X-ray structural analysis of the tris(butadiene)molybdenum complex shows that the central molybdenum atom is equidistant from all the carbon atoms. The butadienes are all cisoid and in a trigonal prismatic arrangement in a girdle around the Mo such that they are all aligned in the same direction. There is a small distortion in the planarity of the butadiene system, and the C_2—C_3 bond is longer than in free 1,3-butadiene, giving the impression that the butadiene ligands have been pulled in around the central metal atom.[40,41]

One of the most important catalytic reactions of zero-valent nickel complexes is the cyclotrimerization of 1,3-butadiene to 1,5,9-cyclododecatriene, a major precursor of $HOOC(CH_2)_{10}COOH$, which is used in the manufacture of nylon-12 and space vehicle lubricants. The activity of nickel in the dimerization and trimerization reaction has been attributed by Wilke[42] to the ability of nickel to alternate between 16- and 18-electron systems, and several intermediates in the catalytic system have been isolated either as stable compounds or as their phosphine adducts:[43]

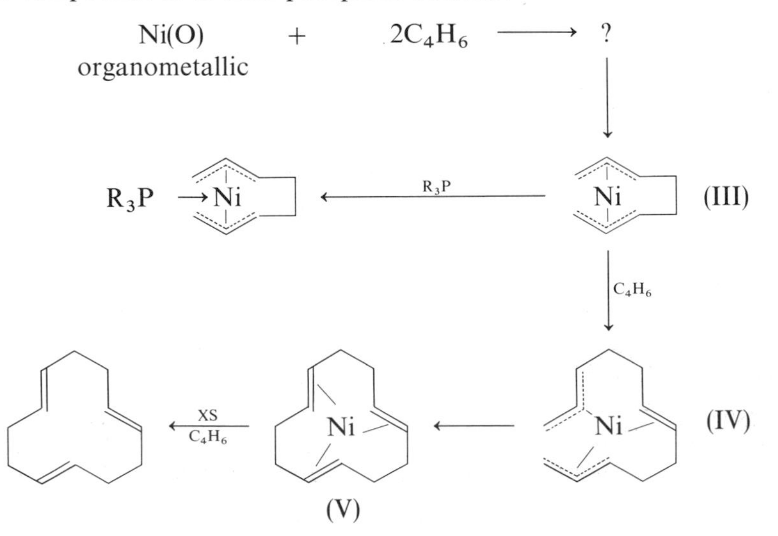

A possible intermediate early in the catalytic scheme might be bis(butadiene)nickel, but this was not isolated by Wilke. The reaction of nickel vapor with 1,3-butadiene gives as the major product a grey residue from which a volatile, oily, yellow compound (ca. 2% yield) can be pumped away at room temperature. Hydrogenation of this compound yields *n*-butane, linear butenes, and metallic nickel, and this together with its mass spectrum led to a tentative assignment as bis(butadiene)nickel.[10] Treatment of this yellow complex with excess liquid butadiene at −20°C leads to formation of blood-red 2,6,10-dodecatriene-1,12-diyl nickel (IV),[43] identified by its nmr, mass spectrum, and hydrogenation to yield *n*-dodecane. However, the process is further complicated by the presence of bis(crotyl)nickel,which has also been isolated from this system[44] and is known to be capable of promoting the trimerization of 1,3-butadiene.[45] The major product from the cocondensation of nickel and 1,3-butadiene is the grey nonvolatile material that remains when the excess diene is removed at −78°C. Treatment of this residue with D_2O at room temperature liberates 2 moles of undeuterated butadiene per mole of nickel vaporized. When the whole nickel-butadiene cocondensate is stirred in excess diene at −20°C, the acyclic nickel complex (IV) is produced in 65% yield based on nickel vaporized.[10]

The complex butadienes in the grey residue are exchangeable since removal of excess butadiene at −78°C, addition of a large excess of isoprene, and 3 hr stirring at 0°C produced red organonickel complexes that upon hydrogenation produced acyclic hydrocarbons in 63% yield based on total nickel vaporized. The composition of these hydrocarbons was $C_{13}H_{28}$, 3%; $C_{14}H_{30}$, 42%; and $C_{15}H_{32}$, 55%; no C_{12} hydrocarbons were found.[18] The butadienes are thus exchangeable with isoprene, yet the involatility of the material suggests that it is polymeric. It may be that the butadiene ligands bridge two nickel atoms. A number of complexes of butadiene are known in which the diene bonds to nickel[46] or palladium[47] through only one double bond, and furthermore a binuclear rhodium complex has been reported in which one butadiene is π-bonded to both rhodium atoms.[48]

2.3 Phosphine Complexes

Trifluorophosphine is a weak base but a good π-acid and was thus expected to form a series of low-valent metal complexes analogous to the metal carbonyls. Many PF_3 complexes were initially prepared by Kruck[49] using high-pressure techniques, but an attractive alternative on the laboratory scale is the cocondensation procedure.[4] Thus, chromium, nickel, or palladium reacted with trifluorophosphine to give good yields (ca. 60%) of $Cr(PF_3)_6$, $Ni(PF_3)_4$, or $Pd(PF_3)_4$ respectively, identical to the products previously described by Kruck. (See Figure 3).[49] No isolable products were obtained with either manganese or copper vapor despite a noticeable gettering action in the latter case

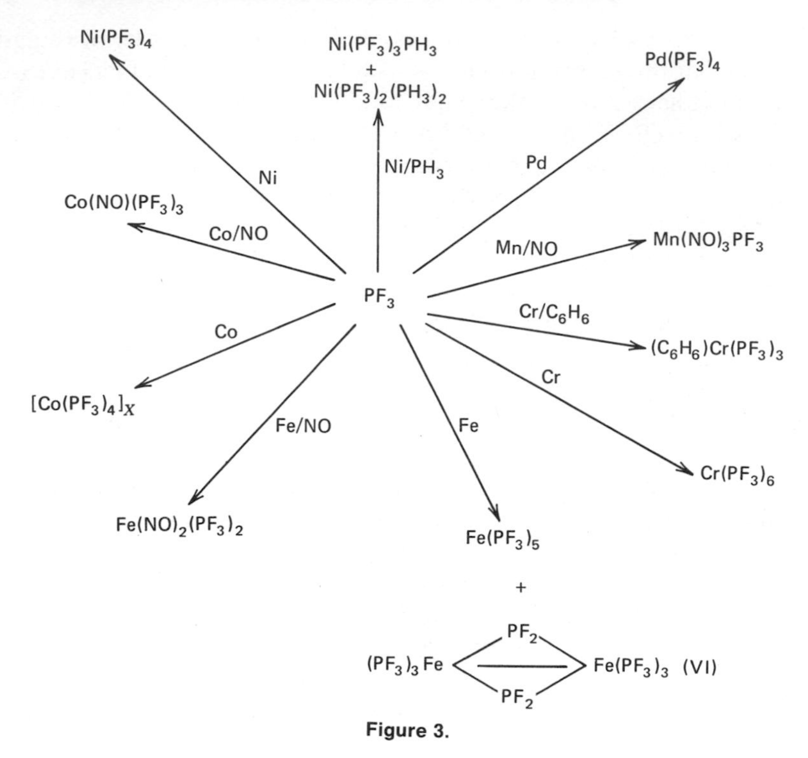

Figure 3.

during the codeposition;[50] thus any compounds formed must have very low thermal stability. The reaction with iron vapor was more interesting in that not only was the expected pentakis(trifluorophosphine)iron(O) produced, but also the $-PF_2$- bridged compound, $(PF_3)_3Fe(PF_2)_2Fe(PF_3)_3$ (VI), which ia analogous to the known $(PF_3)_3Co(PF_2)_2Co(PF_3)_3$ (VII) previously prepared by Kruck and Lang[51] by a high-pressure method. Surprisingly, cobalt vapor did not defluorinate PF_3 to give thc known cobalt compound (VII), but instead produced $[Co(PF_3)_4]_x$ in $\sim 50\%$ yield.[50] This moderately volatile, violet, crystalline solid seemed identical to the one previously reported by Kruck and Lang:[52] it was very readily decomposed by water to give tetrakis(trifluorophosphine)cobalt hydride, and indeed small quantities of this were always obtained, presumably due to traces of adsorbed water on the walls of the reaction flask. Although it is not understood why defluorination did not occur with cobalt as it did with iron, it is typical of reactions at $-196°C$ that a tiny difference in activation energy or other kinetic parameter can radically affect the course of two apparently similar reactions.[50]

Even the ligand chlorodifluorophosphine which would be expected to be a poorer π-acid than trifluorophosphine has been observed to react directly with nickel vapor to yield tetrakis(chlorodifluorophosphine)nickel(O). The identification of this compound relied mostly on mass spectral evidence that showed a parent peak and successive losses of Cl and PF_2Cl. The ^{19}F nmr spectrum showed a doublet, approximately 1300 Hz wide centred at +12.8 p.p.m. relative to $CFCl_3$: this peak was completely symmetrical making it very unlikely that any of the possible isomers such as $Ni(PF_2Cl)_2(PFCl_2)(PF_3)$ were present.[50]

2.4 Use of Two Ligands Simultaneously

Mixed Phosphines

The use of more than one ligand, while obviously increasing the possibility of forming multiple products and thus complicating separation procedures, yet has a great advantage: it is possible to prepare complexes in which one ligand is strong, capable of supplying a large part of the electronic requirements of the metal, while the other ligand is too weak or thermally unstable to compete with the first in conventional preparative procedures.

This is well exemplified by the reaction, first reported by Timms,[50] whereby nickel atoms were cocondensed with an equimolar mixture of trifluorophosphine (a good π-acid) and phosphine (a notoriously weak ligand). The three products isolated were tetrakis(trifluorophosphine)nickel(O) in low yield, phosphine-tris(trifluorophosphine)nickel(O) in 10% yield, and bis(phosphine)bis(trifluorophosphine)nickel(O) in 15% yield. All the compounds were characterised by ir, nmr, and mass spectra. Furthermore, $(PH_3)_2(PF_3)_2Ni$ reacted with excess PF_3 at 5 atm pressure to give $(PH_3)(PF_3)_3Ni$ and PH_3. In contrast, nickel atoms and phosphine alone yielded no organonickel complexes. The mixed phosphine reaction demonstrates the nonequilibrium nature of reactions on the cold surface in which $(PH_3)_2(PF_3)_2Ni$ is the most abundant product, despite the fact that not only is it less stable than $(PH_3)(PF_3)_3Ni$ and considerably less stable than $(PF_3)_4Ni$, but also that unreacted PF_3 was present on the cold surface.

Phosphines with Arenes

Arene chromium tricarbonyl complexes have been well established for many years, but the corresponding trifluorophosphine complexes are relatively new; in fact, Kruck reported $(C_6H_6)Cr(PF_3)_3$ only in 1967.[49] However, cocondensation of chromium vapor with a six fold excess of both trifluorophosphine and arene produces (arene)tris(trifluorophosphine)chromium(O) complexes in approximately 15% yield. Arenes used include benzene, cumene, mesitylene, and even hexafluorobenzene.[13] As C_6F_6 would not be expected to be a good π-electron donor (it is well known to be highly susceptible to nucleophilic attack) it is clear that the trifluorophosphine ligands must supply most of the electronic requirements of the zero-valent metal. Surprisingly, benzene itself can apparently stabilize the (π-C_6F_6)Cr system and the complex, (π-benzene)(π-hexafluorobenzene)chromium(O), has been reported to be air stable.[13] A variety of other novel hydrocarbon-metal-trifluorophosphine complexes have also been prepared, but the technique used was to condense PF_3 onto the cold hydrocarbon-metal matrix rather than by cocondensation: such reactions are further discussed in Section 3.1.

Nitric Oxide with Other Ligands

Nitric oxide has a vapor pressure of approximately 5×10^{-3} torr at $-196°C$ and is thus difficult to use in cocondensation reactions where a high vacuum has to be maintained. However, boron trifluoride, which is relatively inert towards metal atoms, complexes weakly with nitric oxide at low temperatures and is thus useful in reducing the effective vapor pressure of NO.

Application of the effective atomic number rule would predict the existence of the series of compounds, $Cr(NO)_4$, $Mn(NO)_3PF_3$, $Fe(NO)_2(PF_3)_2$, and $Co(NO)(PF_3)_3$. Kruck had previously prepared the mixed nitrosyl trifluorophosphine complexes of cobalt[53] and iron,[54] but the manganese complex was unknown before 1970 when Timms prepared it by the cocondensation of manganese vapor with PF_3, NO, and BF_3:[13] $Mn(NO)_3PF_3$ is a green liquid that reacts readily with carbon monoxide to produce the known nitrosyl carbonyl complex, $Mn(NO)_3CO$.[55] $Fe(NO)_2(PF_3)_2$ and $Co(NO)(PF_3)_3$ can both be prepared by the cocondensation technique,[13] whereas tetranitrosylchromium(O) cannot despite the fact that it has been prepared more conventionally by passing nitric oxide through an irradiated solution of $Cr(CO)_6$.[56] This again demonstrates that minor differences in activation energy under these conditions of low temperature can drastically affect the course of a reaction. Attempts were also made to synthesize $Co(NO)_3$ (previously prepared by Burg[57] under photolytic conditions) and also $Co(C_6H_6)NO$, which would be analogous to $Ni(C_5H_5)NO$, but they were unsuccessful.

It is noteworthy that reactions involving the cocondensation of nitric oxide occasionally exploded at low temperatures, and the use of BF_3 did not always prevent this, so workers in this field must continue to exercise great care.

2.5 Halocarbons

It is well established that low-valent transition metal complexes may undergo oxidative-addition reactions whereby they increase their coordination number and formal oxidation state. Archetypical of such reactions is the addition of alkyl halides to Vaska's compound, $(Ph_3P)_2Ir(CO)Cl$, whereby the central metal undergoes oxidative addition to a square planar d^8 environment producing an octahedral d^6 system.[58] In similar fashion we would anticipate the oxidative addition of suitable addenda to zero-valent metals in the form of free atoms.

This was first realized by Piper and Timms,[59] who showed that π-allyl nickel halides are efficiently produced (ca. 60% yield) by the cocondensation of nickel atoms and allyl chloride or bromide. The presumed mechanism would involve initial oxidative addition of the alkenyl halide to nickel, rearrangement of the σ-allyl to a π-allyl system, and subsequent dimerization via halogen bridges:

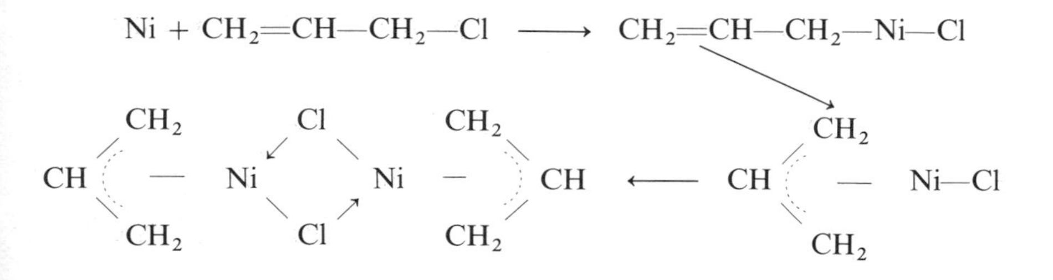

This route certainly provides a useful laboratory synthesis that avoids the toxicity hazards of nickel carbonyl inherent in the conventional syntheses of such compounds; subsequently, this was extended to include the palladium analogs.[60] The reaction of nickel atoms with 1,4-dichloro-2-butene gave bis-(1-chloromethyl)-π-allyl-di-μ-chlorodinickel.[18]

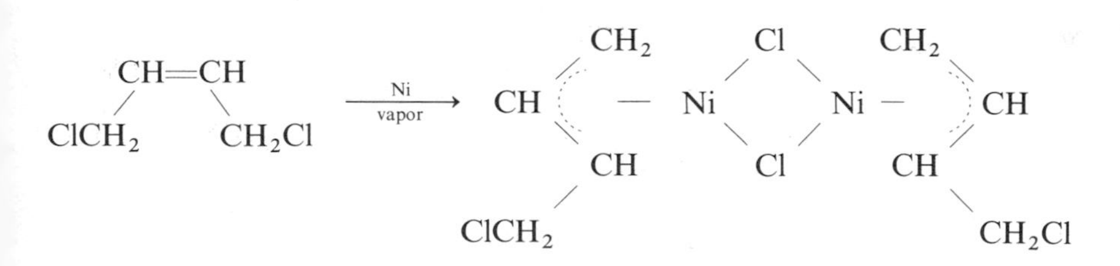

Platinum vapor[32] reacts with allyl chloride to produce a yellow solid (VIII) that reacts with triphenylphosphine to yield the known complex, $Ph_3P(\pi\text{-}C_3H_5)PtCl$. On the basis of its hydrolysis to propene and platinum, mass spectrum, and melting point,[18] (VIII) is thought to be the previously described tetrameric complex $[Pt(C_3H_5)Cl]_4$ first synthesized by the reaction of anhydrous hydrogen chloride with bis(π-allyl)platinum.[45] X-ray analysis of this tetramer showed it to contain both bridging chlorines and bridging allyl groups that are π-bonded to one platinum atom and σ-bonded to another.[61]

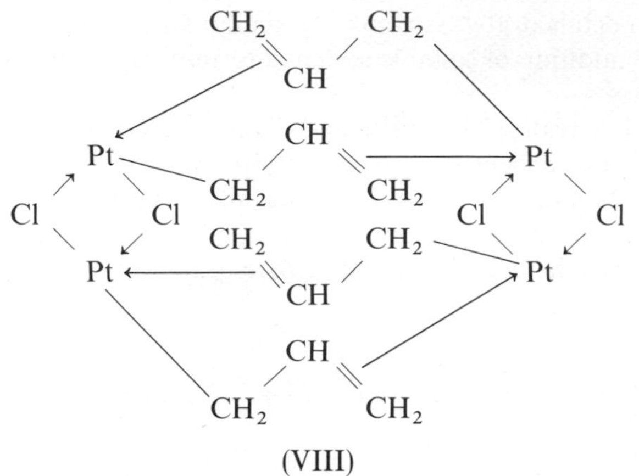

(VIII)

The oxidative addition of aryl halides to nickel or palladium atoms has been elegantly demonstrated by Klabunde and Low:[62] they showed that pentafluorobromobenzene and palladium atoms yield a poly(arylpalladium bromide) that reacts with phosphines to give the known bis(phosphine)aryl palladium bromide complexes:

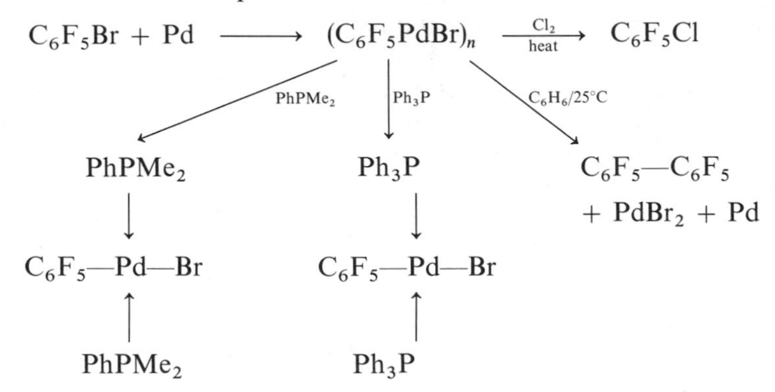

The polymeric species originally produced decomposes homolytically in benzene at room temperature to give perfluorobiphenyl. In contrast, palladium atoms and bromobenzene yield biphenyl below room temperature, and this clearly reflects the well-known stability of fluorosubstituted alkyl and aryl transition metal complexes.[63] Nickel atoms react with either bromobenzene or bromopentafluorobenzene to give the corresponding biphenyls at room temperature, but addition of triphenylphosphine to the latter at low temperature yields bis(triphenylphosphine)pentafluorophenyl nickel bromide.[62]

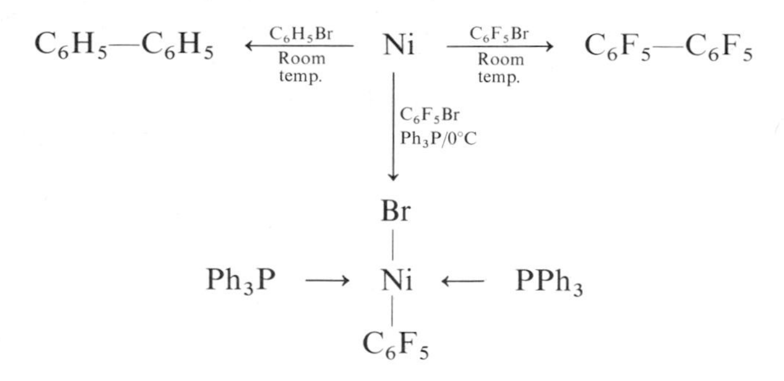

Klabunde has extended this series of reactions to alkyl, fluoroalkyl, and fluoroacyl systems and has also used Pd atoms as the inserting reagent.[64–66] The current evidence points to the conclusions that (a) RMX is formed initially at low temperatures and (b) phosphine trapping always yields *trans* complexes so that this latter addition presumably occurs in a stepwise process to give the stereochemically most favored products.

A particularly interesting recent development has been reported whereby palladium atoms and benzyl chloride react to produce a novel η^3-benzyl palladium chloride dimer.[67]

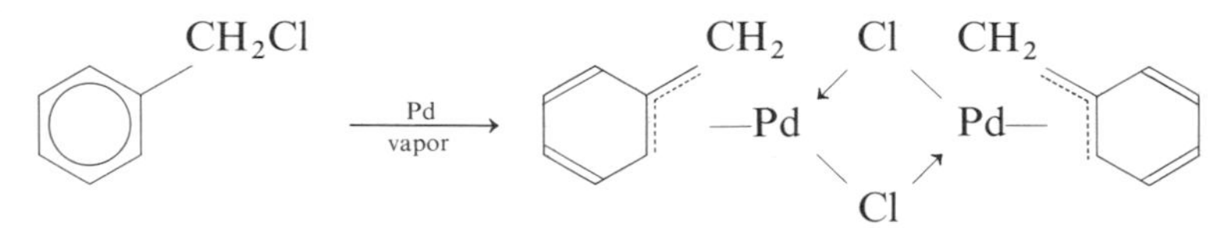

Such an intermediate had previously been postulated by Heck[68] in the palladium-acetate-catalyzed reaction of benzyl chloride with methyl acrylate. Thus, once again a metal-atom reaction provides convincing support for a mechanism operative in a conventional organometallic procedure.

2.6 Miscellaneous

The synthesis of anhydrous metal acetylacetonates has been a difficult problem that is now resolvable by the cocondensation procedure.[26] Thus, 2,4-pentanedione reacts directly with many transition metal vapors to give modest yields of the anhydrous acetylacetonate complexes—even the sensitive tin(II) compound may be prepared in this way.

Other hydroxylic reagents that can be utilized in this way are alcohols and carboxylic acids. Typically, one can synthesize $Cr(OCH_3)_3$ and $Cr(OCOCH_3)_n$ (n = 2 or 3) from methanol or acetic acid, respectively.[26]

Finally, reaction of iron vapor with dimethylmaleic anhydride produces a paramagnetic material that contains two ligand molecules. Hydrolysis with hydrochloric acid yields the original ligand and also an interesting dimer.[26]

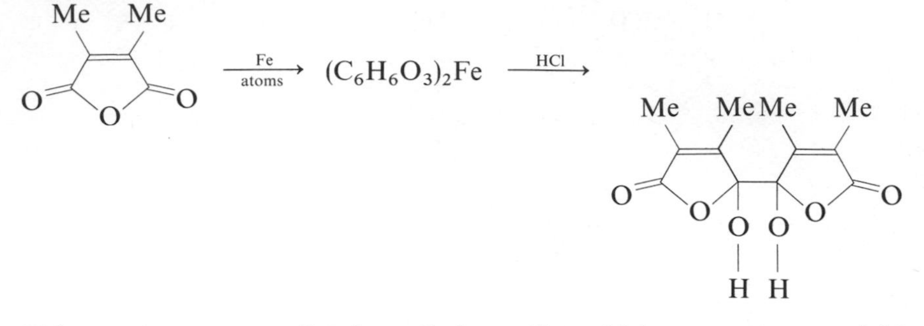

This reaction may parallel the radical coupling of fulvenes on iron to yield [2]-ferrocenophanes.[29]

3. INDIRECT SYNTHESIS VIA TREATMENT OF INVOLATILE RESIDUES WITH ADDITIONAL LIGANDS

3.1 Addition of PF_3, CO, or dienes

In his early work on the direct synthesis of sandwich compounds, Timms[4] reported the possible formation of bis(benzene)iron at low temperature, but was unable to isolate the compound presumably because of its thermal instability. However, addition of hydrogen to the cold matrix led to the formation of some cyclohexane. Williams-Smith, Wolf, and Skell[69] first demonstrated clearly the existence of a π-bonded arene iron complex at low temperature by treating the residue remaining after codeposition of iron and toluene with trifluorophosphine to obtain $(\pi\text{-}C_6H_5\cdot CH_3)Fe(PF_3)_2$. The parent compound, (benzene)bis(trifluorophosphine)iron(O), was reported subsequently,[13] but the analogous $(C_6H_6)Fe(CO)_2$ could not be prepared by this method. Interestingly, Timms[25] has shown that the reaction of iron vapor with mesitylene yields the (mesitylene)iron(trimethylcyclohexadiene) complex directly rather than bis(mesitylene)iron, which would be a 20-electron system.

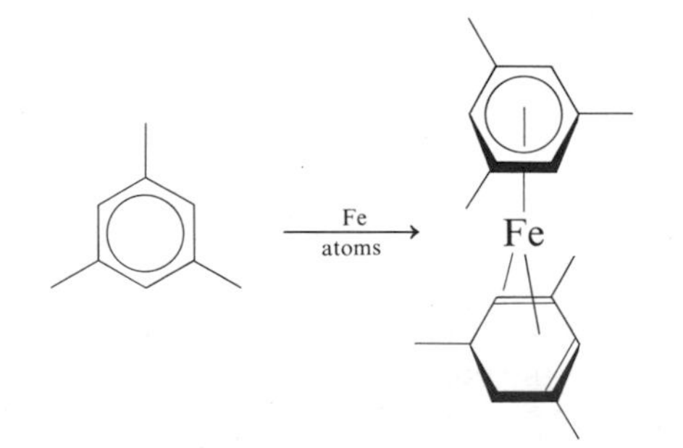

It is also possible to use 1,3-butadiene as the stabilizing agent, and by this means a 20% yield of toluene(1,3-butadiene)iron(O) has been obtained and unambiguously identified by mass spectrometry and nmr.[69] The high field absorptions are entirely compatible with a π-bonded butadiene moiety and compare favorably with those reported for butadiene iron tricarbonyl.[70]

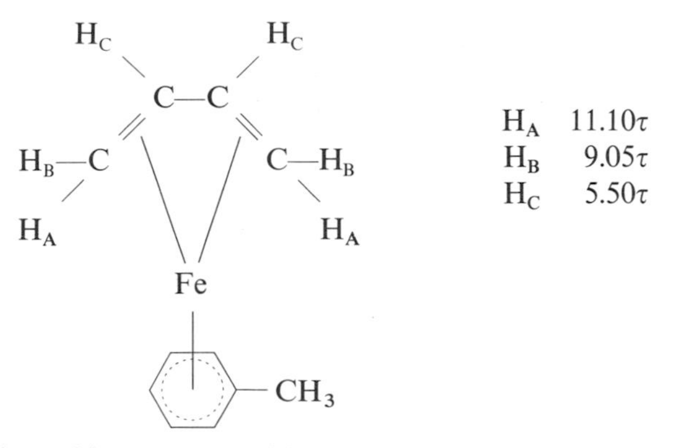

The reaction of iron atoms with 1,3-butadiene leaves, upon removal of excess unreacted butadiene, a residue believed to contain bridging butadiene ligands. Treatment of this residue with trifluorophosphine or carbon monoxide yields bis(butadiene)trifluorophosphine-iron(O) or bis(butadiene)carbonyl-iron(O), respectively.[69] These complexes were independently prepared by E. K. von Gustorf and his coworkers,[71] who used a laser to evaporate the metal. Furthermore, because of the similarities in the ^{1}H chemical shifts in both the PF_3 and CO complexes, it is postulated that the structure of the PF_3 complex is (IX), as has been previously demonstrated for the monocarbonyl analog.[72,73] Furthermore, treatment of the chromium-1,3-butadiene residue with carbon monoxide yields the highly labile complex butadienetetracarbonylchromium(O), whose infrared spectrum indicates structure (X).[71]

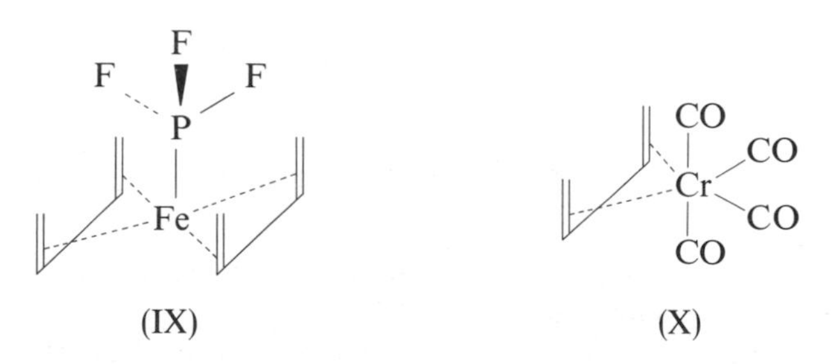

(IX) (X)

Clearly, this technique of treating the involatile and perhaps polymeric residue of a reaction with a strong ligand such as PF_3, CO, or a diene should

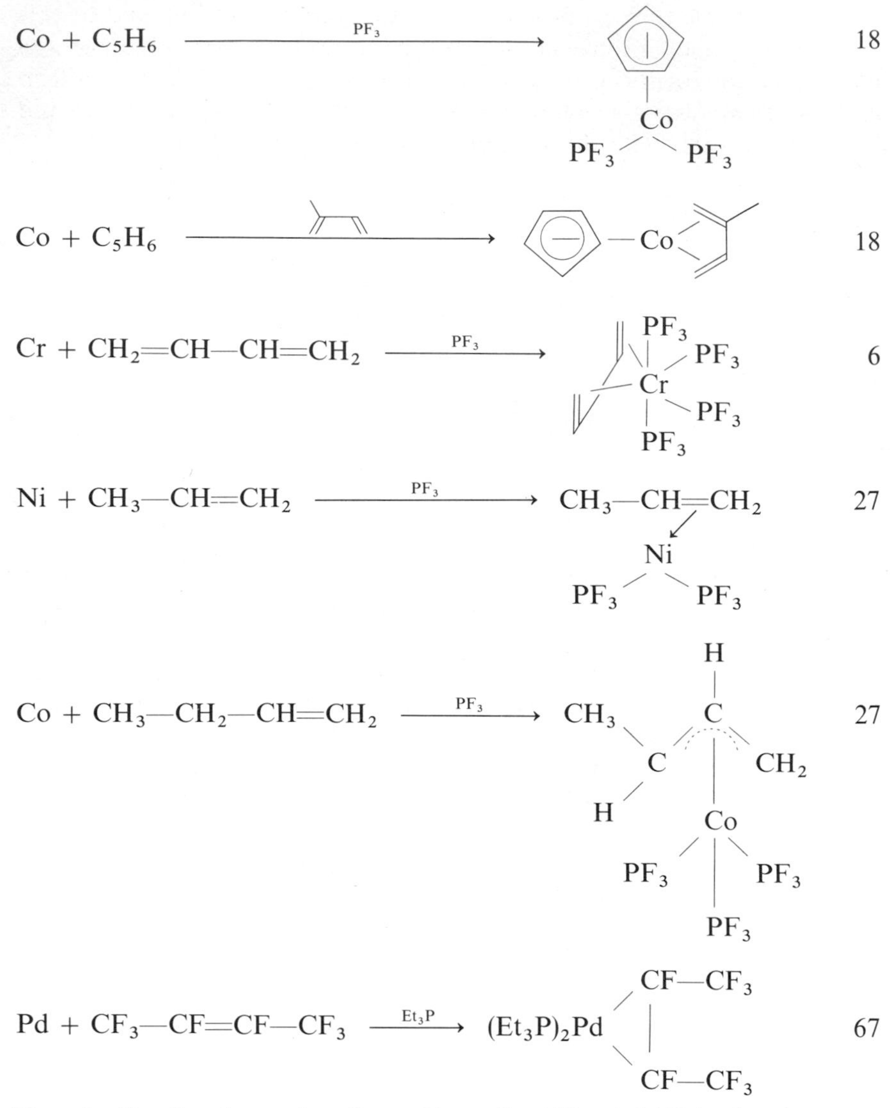

Figure 4. Reaction of cocondensation residues with additional stabilizing ligands.

lead to many novel complexes in the future. Typical reactions which have been performed are shown in Figure 4.

These complexes are not just chemical curiosities, however, and von Gustorf has demonstrated that the bis(diene)monocarbonyl iron complexes are very useful polymerization catalysts.[26] Other elegant work from von Gustorf's group has demonstrated that it is possible to trap out, as its PF_3

adduct, the intermediate in the rearrangement of 1,3- to 1,5-cyclooctadiene on chromium. The X-ray crystal structure confirms that cocondensation of chromium, PF_3, and either 1,3- or 1,5-cyclooctadiene leads to (1-5-η-cyclooctadienyl)hydridotris(trifluorophosphine)chromium.[26]

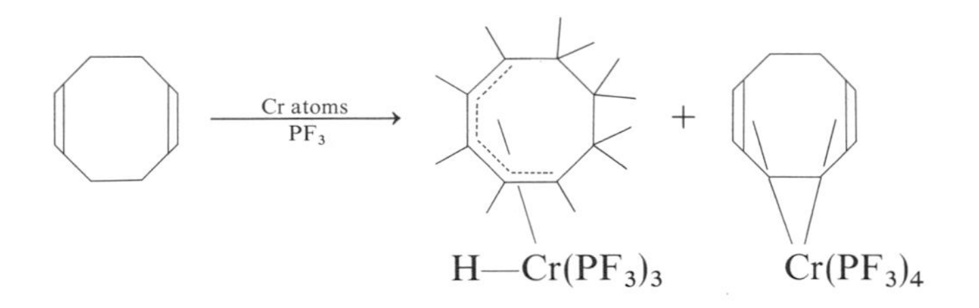

A final example is provided by the reaction of chromium atoms with PF_3 and pyridine. One observes not only the expected σ-bonded pyridine complex but also $(C_5H_5N)Cr(PF_3)_3$, which appears to be the first authentic π-pyridine complex.[25]

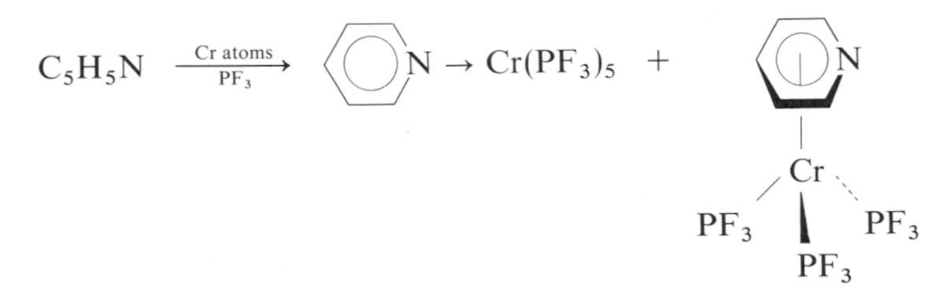

3.2 Hydrolysis Reactions

A most informative series of reactions was the cocondensation of propene with most of the first-row transition elements and subsequent hydrolysis of the residues. In each case the metal and propene were cocondensed at $-196°C$, allowed to warm to $-78°C$, and the excess unreacted propene was removed *in vacuo*. Deuterium oxide was then cocondensed onto the organometallic residue, and the products were analyzed by gas chromatography and mass spectrometry. The products were propene, propane, and the isomeric hexanes. Although the central transition elements, chromium,[6] manganese,[27] and iron,[27] were not very specific in their behavior, the elements at each end of the series showed noticeable differences. The early transition metals, titanium[5] and vanadium,[5] showed appreciable incorporation of deuterium in the propane and hexanes, whereas cobalt,[27] nickel,[10] palladium,[18] and platinum[18] showed very little deuterium incorporation. This may be rationalized on the basis of a gradual change from σ to π character of the bonding in metal olefin complexes as one crosses the periodic table from left to right. Thus, at one extreme, aluminum[74] produces almost quantitative yields of dideuterated propane and hexanes, whereas nickel,[10] palladium,[18] and platinum[18] yield for the most part undeuterated propene. The aluminum reaction is thought to involve radical intermediates whereby hydrolysis with D_2O marks every metal-carbon bond with a deuterium. The possible intermediacy of an aluminocyclopropane is discussed in more detail in the chapter on main-group elemental species.

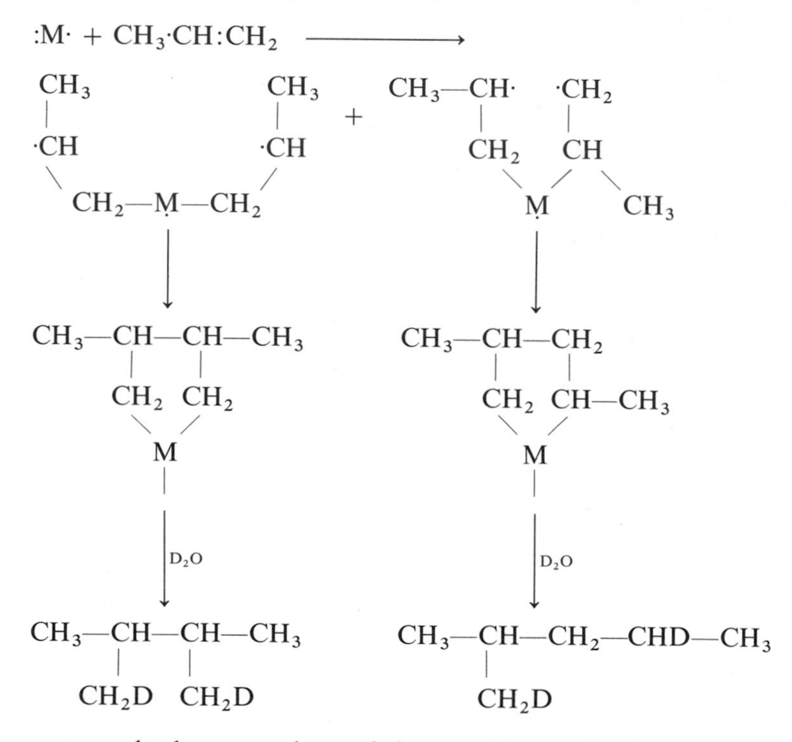

In contrast the later members of the transition series (Co, Ni, Pd, Pt) apparently form π-complexes that release predominantly undeuterated propene upon hydrolysis. There is no doubt of the intermediacy of hydrocarbon-metal complexes, since addition of suitable ligands such as CO or PF_3 leads to isolable products (see Section 3.1). Furthermore, the isolation of $(C_2H_4)_nM$ complexes (where $M = $ Co, Ni, Cu, $n = 1, 2, 3$; $M = $ Ag, Au, $n = 1$) have recently been claimed.[30,84]

However, these structures with olefins π-complexed to metals do not appear to be static complexes. They probably involve highly mobile systems in which there is rapid interconversion of π-complex and π-allyl metal hydride, as has been previously demonstrated by the elegant low-temperature nmr studies of Bönnemann[75] showing exchange between (π-propene)trifluorophosphine nickel and (π-allyl)trifluorophosphine nickel hydride. The π-allyl metal hydride may then be the intermediate responsible for deuterium incorporation (46% C_3H_5D) into propene when the propene-chromium residue is treated with deuterium oxide.[6] The protons of stable metal carbonyl hydrides are readily exchangeable[76] so the hydridic proton of the π-allyl metal hydride could exchange with D_2O and so deuterium would be incorporated into the organic ligand by conversion of the π-allyl metal deuteride to the π-propene complex.

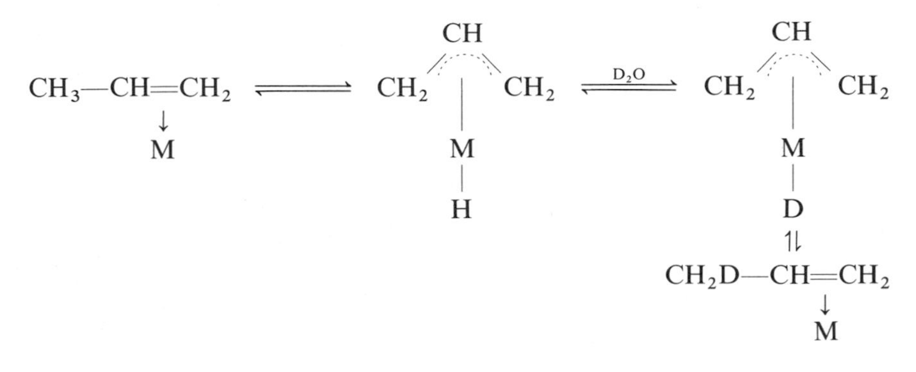

In support of this equilibrium whereby hydrogen transfer may be achieved, cocondensation of a mixture of C_3H_6 and C_3D_6 leads to scrambling.[10]

It is of interest to note that the reactions of lanthanide vapors (erbium, dysprosium, terbium) yield products that, superficially at least, resemble those from aluminum, that is, they give good yields of propanes and hexanes. However, further examination reveals that there is very little correspondence in deuterium incorporation:[77] aluminum reactions produce approximately 90% dideuterated products,[74] but with lanthanides three and even four deuterium atoms are commonly found in the resulting paraffins. Preliminary investigations with uranium vapor follow a similar pattern.[27] Participation by *f* orbitals seems to be very unlikely since yttrium resembles the lanthanides in its behavior.[77]

4. CATALYTIC PROCESSES

4.1 Isomerization Reactions

During the initial survey of the reactions of transition metal atoms with unsaturated hydrocarbons it rapidly became clear that atoms of several of the later transition metals, notably iron, cobalt, and nickel, were efficient isomerization catalysts. Typically, 1 mmole of cobalt vapor catalyzed the isomerization of 600 mmoles of 1-butene to a mixture of *cis*- and *trans*-2-butenes when the whole cocondensate was stirred at $-30°C$.[78] Following the pattern of the C_3H_6–C_3D_6 scrambling experiment and also of the incorporation of deuterium into propene after deuterolysis of the metal-hydrocarbon residues, it seemed reasonable to postulate a 1,3 hydrogen shift via a π-allyl metal hydride intermediate:

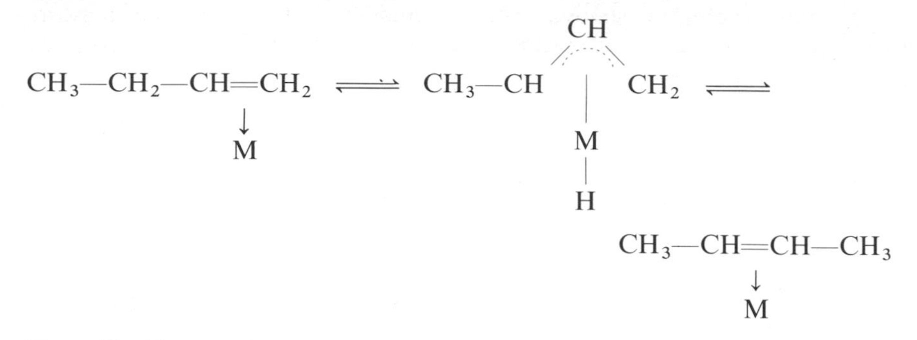

To verify this postulate of a 1,3 hydrogen shift, nickel vapor was cocondensed with 1-deuteropropene, 2-deuteropropene, and 3-deuteropropene in three separate experiments.[41] 1-Deuteropropene reacted to produce a mixture of 1- and 3-deuteropropenes as well as dideutero and unlabeled propane; 3-deuteropropene did likewise, but 2-deuteropropene remained apparently unchanged since 1,3 shifts in such molecules scramble only the terminal hydrogens and leaves the deuterium marker unchanged. The formation of d_2- and d_0-propene demonstrates that this hydrogen shift is intermolecular and not solely intramolecular.

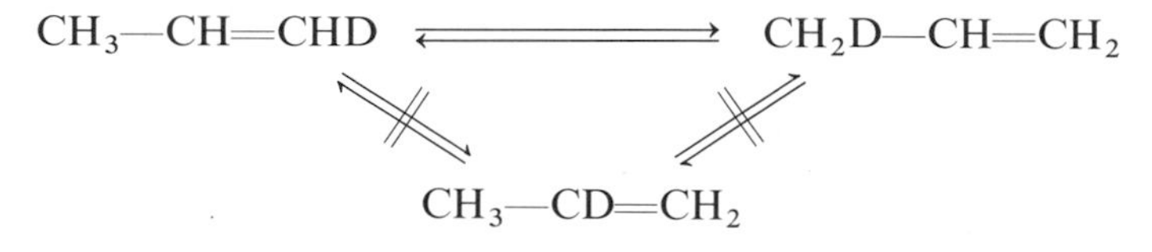

A number of other transition metal vapors such as rhodium or rhenium bring about isomerization of terminal to internal olefins; however, these very high-boiling metals necessitated the use of a magnetically focussed electron beam, and considerable modification of the substrate was observed due to electron bombardment.[5] It is thus difficult to separate the transformations attributable solely to reaction with the metal vapor. Clearly it is crucial to the understanding of many of these metal vapor experiments to perform the appropriate blank reactions such as were done for the production of silicon vapor using an electron gun.[79] Furthermore, it is necessary to determine whether the catalytic reactions discussed in Section 4 are also brought about by finely divided metals. In several cases, for example, disproportionation of cyclohexadienes or isomerization of nonconjugated to conjugated dienes, powdered metals or metal carbonyls are also capable of functioning as catalysts.

Platinum vapor, produced by electrically heating Pt wire in a tungsten spiral, fails to isomerize 1-butene, and furthermore the reaction of D_2O with the propene-platinum cocondensate yields only 10% monodeuteropropene.[18] This indicates that the equilibrium between π-olefin platinum and π-allyl platinum hydride is considerably less facile than with nickel.

The successful isomerization of terminal to internal olefins prompted attempts to rearrange nonconjugated dienes to their thermodynamically more stable conjugated isomers. However, isomerization yields were low since the conjugated diene-metal complexes, the supposedly catalytic intermediates, were sufficiently stable to poison the catalyst and stop the process. Typically, cobalt vapor and 1,4-pentadiene yield unchanged 1,4-pentadiene (93%), *cis*-1,3-pentadiene (1%), and *trans*-2-pentenes together with some polymer. Thus cobalt atoms isomerize 1,4-pentadiene to 1,3-pentadiene and also bring about disproportionation reaction that produces linear pentenes and some C_5H_6 compounds that polymerize. A low yield of bright-red, volatile organocobalt complex was identified on the basis of mass spectral, nmr, and infrared data ($\nu_{Co—H}$ at 1997 cm^{-1}; cf. $HCo(CO)_4$, $\nu_{Co—H}$ at 1934 cm^{-1}) as bis(*trans*-1,3-pentadiene)cobalt hydride. Piperylene (1,3-pentadiene) reacts with cobalt vapor to give what seems to be the identical product.[18]

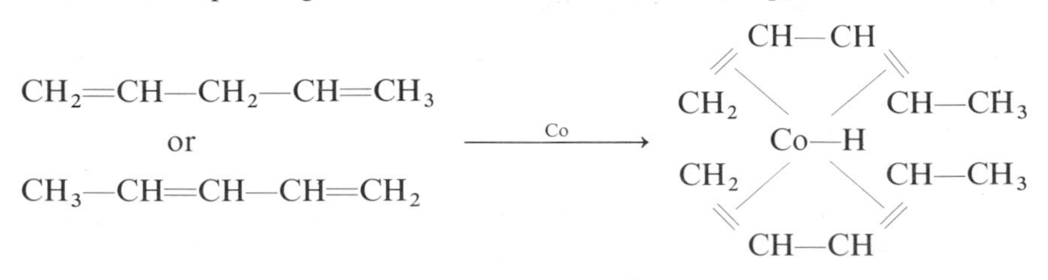

The ^1H nmr data are similar to those reported by Pettit for *trans*-1,3-pentadiene iron tricarbonyl prepared from 1,4-pentadiene and iron pentacarbonyl.[70]

In an attempt to increase the yields of these novel bis(diene)cobalt hydride complexes, 1,3-butadiene and cobalt were cocondensed in the presence of isobutane, which possesses a readily abstractable tertiary hydrogen atom. A 2% yield of yellow volatile bis(1,3-butadiene)cobalt hydride was obtained.[18] The closest known analog of these interesting bis(diene)metal hydrides is probably bis(butadiene)rhodium chloride, which possesses C_{2v} symmetry in the crystalline state.[36]

$$C_4H_6 + RhCl_3 \longrightarrow$$

Cl

Rh

4.2 Hydrogenations

The cocondensates of several of the later transition metals with olefins, upon treatment with hydrogen, have been shown to be good catalytic hydrogenation systems. Thus, when palladium vapor was cocondensed with olefins, allowed to warm to −78°C, and then treated with hydrogen, catalytic conversion to the corresponding alkanes occurred. Use of deuterium instead of hydrogen led to scrambled products, and typically palladium, cyclopentadiene, and deuterium produced cyclopentane (60%), a small percentage of which was completely deuterated: piperylene was also obtained in 30% yield, indicating that considerable ring opening had occurred.[80]

It is of interest to note that treatment of the platinum-butadiene cocondensate at −78°C with hydrogen, equimolar with the quantity of 1,3-butadiene used, led to a product distribution containing 80% butane (based on the butadiene converted) with lesser amounts of the butenes.[18] In contrast, Adam's catalyst (finely divided Pt in ethanol) yielded a mixture of butenes (60%) and butane (40%).[81] Thus the catalyst produced by the cocondensation route hydrogenates butenes preferentially over butadiene.

4.3 Disproportionation Reactions

The disproportionation of 1,3- or 1,4-cyclohexadiene to benzene and cyclohexene can be accomplished by numerous metal carbonyls and also by a film of palladium metal.[82] This catalytic transformation can be brought about by cocondensation of 1,3- or 1,4-cyclohexadiene with any of a number of metal vapors, for example, Cr,[6] Mn,[27] Fe,[69] Co,[18] or Ni,[10] when the total cocondensate is stirred at 0°C. It is tempting to predicate the formation of a bis(diene)metal complex that then undergoes hydrogen migration possibly via a metal hydride intermediate.

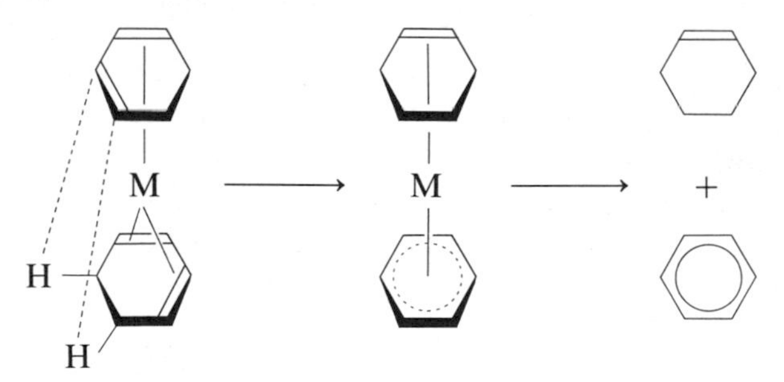

The obvious metals with which to hope to isolate such an intermediate are nickel, palladium, or platinum, but these bring about catalytic disproportionation, and no organometallics are isolable. With iron vapor,[69] however, it is possible to isolate (π-benzene)(1,3-cyclohexadiene)iron, which is analogous to (toluene)(1,3-butadiene)iron previously mentioned in Section 3.1.

4.4 Acetylene Trimerizations

The trimerization of acetylenes by nickel[10] or chromium[6] atoms has been reported. 2-Butyne is trimerized to hexamethylbenzene, and terminal acetylenes such as 1-butyne (or 1-pentyne) yield isomeric mixtures of 1,2,4- and 1,3,5-triethyl- (or tripropyl-) benzenes. These yields, however, are not catalytic, and dark green involatile residues remain in the reaction flask after sublimation or solvent extraction of the benzenes. Nuclear magnetic resonance studies of these residues gave no evidence of the presence of bis-(arene)chromium complexes that might be postulated as intermediates in these metal-promoted trimerizations, perhaps implying that no more than one arene ring is bonded to the chromium at any time. Interestingly, manganese vapor and 2-butyne give excellent yields of an elastomeric polymer.[27]

5. SCOPE AND LIMITATIONS OF THE METHOD

The technique of codeposition of transition metal atoms and organic or inorganic reactants has been applied to a wide variety of systems, and its general applicability to organic and organometallic synthesis is now evident. There are numerous systems in which cocondensation is the only known route to synthesis and others in which it provides a viable alternative to conventional methods. The very fact of proving a compound's existence is frequently a stimulus to other workers to develop different routes that may be more applicable to routine laboratory synthesis. In the cases where parallel synthetic routes are already known, these have usually been mentioned so that the comparisons are readily drawn.

The advantages of the method are now so sufficiently widely accepted that a number of laboratories on both sides of the Atlantic are actively involved in investigations of these fascinating metal vapor reactions. There are still difficulties in vaporizing large quantities of the very high-boiling metals, since the very energetic vaporization methods such as the electron gun or the laser can damage the substrate; but improvements in technique[25,26] are gradually resolving this problem. In fact, industrial concerns are already showing interest in large-scale reactors.[83]

Suffice to say that the hopes and expectations of the pioneers of this field are being rapidly realized, and that the cocondensation technique should soon graduate from the position of laboratory curiosity to that of an accepted and viable synthetic method.

REFERENCES

1. P. S. Skell, L. D. Wescott, Jr., J. -P. Golstein, and R. R. Engel, *J. Amer. Chem. Soc.*, **87**, 2829 (1965).
2. P. L. Timms, *Advan. Inorg. Chem. Radiochem.*, **14**, 121 (1972).

3. P. S. Skell, J. J. Havel, and M. J. McGlinchey, *Account Chem. Res.*, **6**, 97 (1973).
4. P. L. Timms, *Chems. Commun.*, 1033 (1969).
5. P. S. Skell and K. Hess, unpublished results.
6. P. S. Skell, D. L. Williams-Smith, and M. J. McGlinchey, *J. Amer. Chem. Soc.*, **95**, 3337 (1973).
7. P. S. Skell and W. N. Brent, unpublished results.
8. R. B. King, *Transition Metal Organometal. Chem.*, Academic, New York, 1969.
9. J. L. Thomas and H. H. Brintzinger, *J. Amer. Chem. Soc.*, **94**, 1386 (1972).
10. P. S. Skell, J. J. Havel, D. L. Williams-Smith, and M. J. McGlinchey, *Chem. Commun.*, 1098 (1972).
11. E. O. Fischer and W. Hafner, *Z. Naturforsch.*, *B*, **10**, 665 (1955).
12. Gmelin, *Handbuch der anorganische Chemie, erganzungwerk, Chrom organisch Verbindungen*, 1971.
13. R. Middleton, J. R. Hull, S. R. Simpson, C. H. Tomlinson, and P. L. Timms, *J. Chem. Soc.*, *A*, 120 (1973).
14. M. J. McGlinchey and T. S. Tan, *Can. J. Chem.*, **52**, 2439 (1974).
15. M. P. Silvon, E. M. Van Dam, and P. S. Skell, *J. Amer. Chem. Soc.*, **96**, 1945 (1974).
16. F. W. S. Benfield, M. L. H. Green, J. S. Ogden, and D. Young, *Chem. Commun.*, 866 (1973).
17. M. J. D'Aniello and E. K. Barefield, *J. Organometal. Chem.*, **76**, C50 (1974).
18. J. J. Havel, Ph.D. thesis, The Pennsylvania State University, University Park, Pa., 1972.
19. H. D. von Oven and H. J. deLiefde Meijer, *J. Organometal. Chem.*, **23**, 159 (1970).
20. T. Birchall, personal communication.
21. P. S. Skell, L. M. Jackman, and E. M. Van Dam, unpublished results.
22. M. J. McGlinchey and T. S. Tan, *J. Amer. Chem. Soc.*, **98**, (1976).
23. K. J. Klabunde and H. F. Efner, *J. Fluorine Chem.*, **4**, 115 (1974).
24. K. J. Klabunde and H. F. Efner, *Inorg. Chem.*, **14**, 789 (1975).
25. P. L. Timms, *Angew. Chem.* (*Int. Edit.*), **14**, 273 (1975).
26. E. A. Koerner von Gustorf, O. Jaenicke, O. Wolfbeis, and C. R. Eady, *Angew. Chem.* (*Int. Edit.*), **14**, 278 (1975).
27. P. S. Skell and D. L. Williams-Smith, unpublished results.
28. V. Graves and J. J. Lagowski, 165th National Meeting of *A.C.S.*, Abstract Number, Inorg. 52, Dallas, Texas, April 1973.
29. J. L. Fletcher, T. S. Tan, and M. J. McGlinchey, *J. Chem. Soc.*, *Chem. Commun.*, 771, 1975
30. M. Moskovits and G. A. Ozin, communicated at Merck Symposium "Metal Atoms in Chemical Synthesis," Darmstadt, Germany, May 1974; H. Huber, G. A. Ozin and W. Power, *J. Amer. Chem. Soc.*, 1976 (in press).
31. K. Fischer, K. Jonas, and G. Wilke, *Angew. Chem.* (*Int. Edit.*), **12**, 565 (1973).
32. P. S. Skell and J. J. Havel, *J. Amer. Chem. Soc.*, **93**, 6687 (1971).
33. B. Bogdanovic, M. Kroner, and G. Wilke, *Justus Liebigs Ann. Chem.*, **699**, 1 (1966).
34. D. R. Arnold, D. J. Trecker, and E. B. Whipple, *J. Amer. Chem. Soc.*, **87**, 2596 (1965).
35. R. MacKenzie and P. L. Timms, *J. Chem. Soc.*, *Chem. Commun.*, 650 (1974).

36. L. Porri, A. Lionetti, G. Allegra, and A. Immerzi, *Chem. Commun.*, 336 (1965).
37. P. S. Skell, E. M. Van Dam, and M. P. Silvon, *J. Amer. Chem. Soc.*, **96**, 627 (1974).
38. R. B. King and A. Fronzaglia, *Inorg. Chem.*, **5**, 1837 (1966).
39. R. E. Moriarty, R. D. Ernst, and R. Bau, *Chem. Commun.*, 1242 (1972).
40. M. M. Yevitz and P. S. Skell, unpublished results quoted in Ref. 41.
41. P. S. Skell and M. J. McGlinchey, *Angew Chem.* (*Int. Edit.*), **14**, 195 (1975).
42. G. Wilke, *Angew. Chem.* (*Int. Edit.*), **2**, 105 (1963).
43. G. Wilke, M. Kroner, and B. Bogdanovic, *Angew. Chem.*, **73**, 755 (1961).
44. P. S. Skell and D. V. Howe, unpublished results.
45. G. Wilke, *Angew. Chem.* (*Int. Edit.*), **5**, 151 (1966).
46. P. W. Jolly, I. Tkatchenko, and G. Wilke, *Angew. Chem.* (*Int. Edit.*), **10**, 328 (1971).
47. M. Donati and F. Conti, *Tetrahedron Lett.*, 1219 (1966).
48. J. Powell and B. L. Shaw, *J. Chem. Soc., A*, 597 (1970).
49. Th. Kruck, *Angew. Chem.* (*Int. Edit.*), **6**, 53 (1967).
50. P. L. Timms, *J. Chem. Soc., A*, 2526 (1970).
51. Th. Kruck and W. Lang, *Angew. Chem.* (*Int. Edit.*), **6**, 454 (1967).
52. Th. Kruck and W. Lang, *Z. Anorg. Chem.*, **343**, 181 (1966).
53. Th. Kruck and W. Lang, *Chem. Ber.*, **99**, 3794 (1966).
54. Th. Kruck and W. Lang, *Angew. Chem.* (*Int. Edit.*), **3**, 700 (1964).
55. C. G. Barraclough and J. Lewis, *J. Chem. Soc.*, 4842 (1960).
56. M. Herberhold and A. Razavi, *Angew. Chem.* (*Int. Edit.*), **11**, 1092 (1972).
57. I. H. Sabherwal and A. B. Burg, *Chem. Commun.*, 1001 (1970).
58. L. Vaska, *Account. Chem. Res.*, **1**, 335 (1968).
59. M. Piper and P. L. Timms, *Chem. Commun.*, 52 (1972).
60. P. L. Timms, personal communciation.
61. G. Raper and W. S. McDonald, *Chem. Commun.*, 655 (1970).
62. K. J. Klabunde and J. Y. F. Low, *J. Organometal. Chem.*, **51**, C33 (1973).
63. P. M. Treichel and F. G. A. Stone, *Advan. Organometal. Chem.*, **1**, 143 (1964).
64. K. J. Klabunde, J. Y. F. Low, and H. F. Efner, *J. Amer. Chem. Soc.*, **96**, 1984 (1974).
65. K. J. Klabunde and J. Y. F. Low, *J. Amer. Chem. Soc.*, **96**, 7674 (1974).
66. J. S. Roberts and K. J. Klabunde, *J. Organometal. Chem.*, **85**, C13 (1975).
67. K. J. Klabunde, *Angew. Chem.* (*Int. Edit.*), **14**, 287 (1975).
68. R. F. Heck and J. P. Nolley, Jr., *J. Org. Chem.*, **37**, 2320 (1972).
69. D. L. Williams-Smith, L. R. Wolf, and P. S. Skell, *J. Amer. Chem. Soc.*, **94**, 4042 (1972).
70. G. F. Emerson, J. E. Mahler, R. Kochar, and R. Pettit, *J. Org. Chem.*, **29**, 3620 (1964).
71. E. K. von Gustorf, O. Jaenicke, and O. E. Polansky, *Angew. Chem.* (*Int. Edit.*), **11**, 532 (1972).
72. E. K. von Gustorf, J. Buchkremer, A. Pfaifer, and F. -W. Grevels, *Angew. Chem.* (*Int. Edit.*), **10**, 260 (1971).
73. E. K. von Gustorf, Z. Pfaifer, and F. -W. Grevels, *Z. Naturforsch., B*, 66 (1971).
74. P. S. Skell and L. R. Wolf, *J. Amer. Chem. Soc.*, **94**, 7919 (1972).
75. H. Boenneman, *Angew. Chem.* (*Int. Edit.*), **9**, 736 (1970).

76. F. A. Cotton and G. Wilkinson, *Advanced Inorganic Chemistry*, 3rd Edit., Wiley-Interscience, New York, 1972.
77. P. S. Skell and M. J. McGlinchey, unpublished results.
78. P. S. Skell and F. A. Fagone, unpublished results.
79. P. S. Skell and P. W. Owen, *J. Amer. Chem. Soc.*, **94**, 5434 (1972).
80. P. S. Skell, S. Lander, and M. J. McGlinchey, unpublished results.
81. W. G. Young, *J. Amer. Chem. Soc.*, **69**, 2046 (1947).
82. V. M. Gryaznov and V. D. Yagodovski, *Kinet. Katal.*, **4**, 404 (1963).
83. W. Reichelt, *Angew. Chem.* (*Int. Edit.*), **14**, 218 (1975).
84. H. Huber, D. McIntosh and G. A. Ozin, *J. Organometallic Chem.*, **112**, C50 (1976), and *J. Amer. Chem. Soc.*, 1976 (in press).

Infrared and Raman Spectroscopic Studies of Alkali-Metal-Atom Matrix-Reaction Products

6

Lester Andrews

1. INTRODUCTION

Historically, alkali metal atom reactions have been used to produce reactive chemical species since the early work of Polanyi and coworkers.[1] Applications of the "sodium flame" technique to the study of free-radical reactions have been discussed by Steacie.[2] In more recent work, crossed-molecular-beam reactions of alkali metal atoms and halogen compounds have provided a means of studying reaction dynamics: the early crossed-beam work has been reviewed by Herschbach.[3] The alkali-metal-atom matrix-reaction technique was developed by Andrews and Pimentel[4] in a study of the lithium atom-nitric oxide matrix cocondensation reaction. Subsequent use of alkali-metal-atom matrix reactions by this author has produced a series of free radicals[5] by halogen abstraction

$$M + R\text{—}X \rightarrow MX + R \tag{1}$$

and new alkali metal species[6] by direct synthesis

$$M + O_2 \rightarrow M^+O_2^- \tag{2}$$

which were matrix trapped for spectral study.

In recent review articles, this author has described matrix infrared spectra of free radicals[7] and infrared and Raman spectra of unique matrix-isolated molecules.[8] The application of infrared and Raman matrix techniques to problems in inorganic chemistry has been reviewed by Ozin and Vander Voet.[9]

Although experimental methods in matrix-isolation spectroscopy are described in Chapter 2 and references,[7,8] the alkali-metal-atom matrix reaction technique, in particular, is discussed briefly here.

2. EXPERIMENTAL

The experimental goal is to bring alkali metal atoms and reactive molecules together long enough for primary reactions to take place and then to quickly trap the primary reaction products for spectroscopic study. This is accomplished by depositing a vapor stream of alkali metal atoms from a Knudsen cell along with some reactive molecule at high dilution in argon together on a substrate cooled to 15°K. Figure 1 illustrates the reaction geometry. During the condensation process, collisions between alkali atoms and reactive molecules take place, which in many cases produce reaction products that are quickly trapped and prevented from further reaction by the matrix host after sample solidification. In the infrared transmission experimental apparatus, shown in Figure 1, the substrate is a CsI plate. Sample deposition is governed by the desire to maximize the yield of reaction products, while at the same time maintain infrared transmission and good spectral conditions of the sample. For the Raman scattering experiment, a tilted metal wedge is used to condense the gas sample streams from which the laser beam can be scattered. Here a maximum yield of products is counterbalanced by the need to keep the sample light in color to maximize light scattering and minimize light absorption by the sample. The most critical aspect of both infrared and Raman matrix experiments is sample preparation.

This article is concerned with infrared and Raman studies of the products of specific alkali-metal-atom matrix reactions. First, halogen abstraction reactions yielding free radicals are described, and second, the synthesis of new alkali- and alkaline-earth-metal atom species are presented.

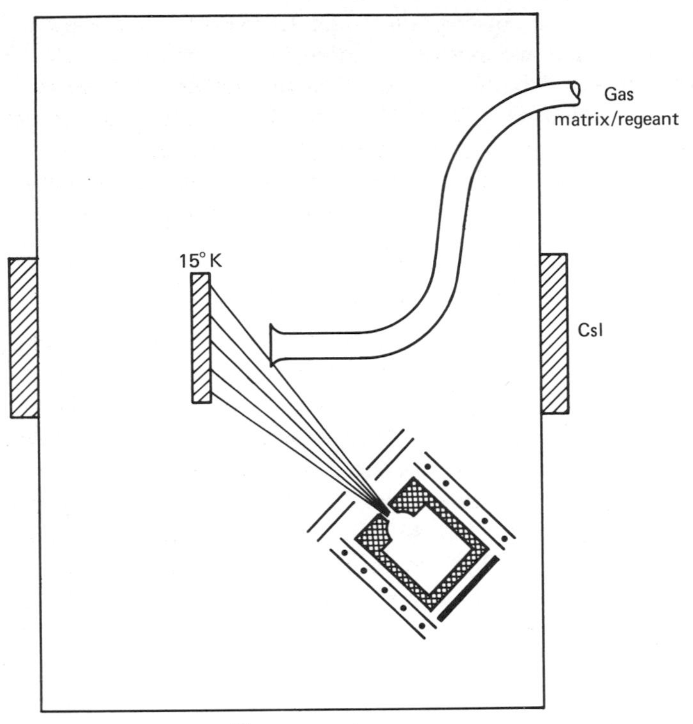

Figure 1. Cross-section of the vacuum vessel base showing reaction geometry, Knudsen cell alkali-metal-atom source, and gas-deposition line.

3. FREE RADICALS AND CHEMICAL INTERMEDIATES

3.1 Methyl and Trihalomethyl Radicals

The first attempt at producing a free radical by the alkali-metal-atom matrix reaction technique was the lithium atom-methyl iodide study of Andrews and Pimentel.[10] Infrared spectra of the reaction products exhibited prominent features due to LiI, a sharp, intense feature at 730 cm^{-1} and a moderately-intense feature at 616 cm^{-1} coincident with a frequently appearing window impurity. The 730 cm^{-1} band exhibited deuterium substitution counterparts appropriate for a harmonic vibration of a species containing three equivalent hydrogen atoms: the deuterium counterpart of the 616 cm^{-1} feature was obscured by lithium iodide absorptions. The 730 cm^{-1} band was originally assigned to ν_2 of the methyl radical.[10] In subsequent vacuum-ultraviolet photolysis studies of CH_4, Milligan and Jacox[11] attributed a feature at 616 cm^{-1} to ν_2 of CH_3: the deuterium counterparts exhibited considerable anharmonicity in the out-of-plane mode. In the following work of Tan and Pimentel,[12] sodium and potassium atom matrix reactions with CH_3I produced the 616 cm^{-1} band unshifted, but the 730 cm^{-1} band in Li—CH_3I experiments shifted to 696 cm^{-1} with Na—CH_3I and 680 cm^{-1} with K—CH_3I. Clearly, the ν_2 assignment of Milligan and Jacox[11] to matrix-isolated CH_3 is correct: the 730 cm^{-1} feature is due to the methyl radical-alkali halide complex, CH_3—LiI,[12] that is the product of reaction (3) which was trapped in the same matrix site.

$$CH_3I + Li \rightarrow CH_3 + LiI \quad \Delta H = -28 \text{ kcal/mole} \tag{3}$$

Note that these reactions are highly exothermic, and this energy, which appears largely as vibrational excitation of the new bond formed, is instrumental in effecting separation of the LiI and CH_3 products during condensation of the matrix sample. Apparently, sufficiently rapid quenching occurs to trap some CH_3 and LiI in the same site; however, some of the products separate before sample solidification, producing isolated CH_3 and isolated LiI. The need to produce a free radical with two different metal reagents, such that isolated radical and MX-perturbed radical can be identified, is clear.

The trichloromethyl radical was first observed by Andrews[13] as the primary reaction product of alkali-metal matrix reactions with carbon tetrachloride.

$$M + CCl_4 \rightarrow MCl + CCl_3 \tag{4}$$

Li, Na, and K atom matrix reactions with CCl_4 produced an intense sharp band at 898 cm^{-1}, which was assigned to ν_3 of CCl_3, the primary reaction product: a weak band observed in the Li—CCl_4 reaction at 674 cm^{-1} was attributed to ν_1 of CCl_3.[5] However, Rogers and coworkers[14] using reduced alkali-metal concentrations did not observe the 674 cm^{-1} band. Extraordinarily large yields of CCl_3 have been isolated following proton beam irradiation of argon-CCl_4 samples during deposition: the 674 cm^{-1} feature was not present in the final spectrum.[15]

Figure 2 illustrates spectra for recent matrix reactions of Li, Na, and K atoms with CCl_4.[16] Notice the intense sharp peak at 898 cm^{-1} in the inset spectrum showing the chlorine isotopic splitting representative of a doubly degenerate chlorine stretching mode.[5] The yield of CCl_3 in the Li atom reactions is sufficient to observe the naturally occurring (1.1%) carbon-13 species at 869 cm^{-1}. In the Na and K reactions satellite bands labeled C are observed near the intense A_1 band. These C bands are attributed to the radical complexed to NaCl (or KCl), that is the CCl_3 and NaCl (or KCl) reaction products trapped in the same matrix site.

The effect of heavier metals on the disputed A_2 band is indicated in Figure 2. The A_2 feature at 674 cm^{-1} with the lithium atom reagent shifted to 635 cm^{-1} with sodium and 608 cm^{-1} with potassium. Clearly, the alkali

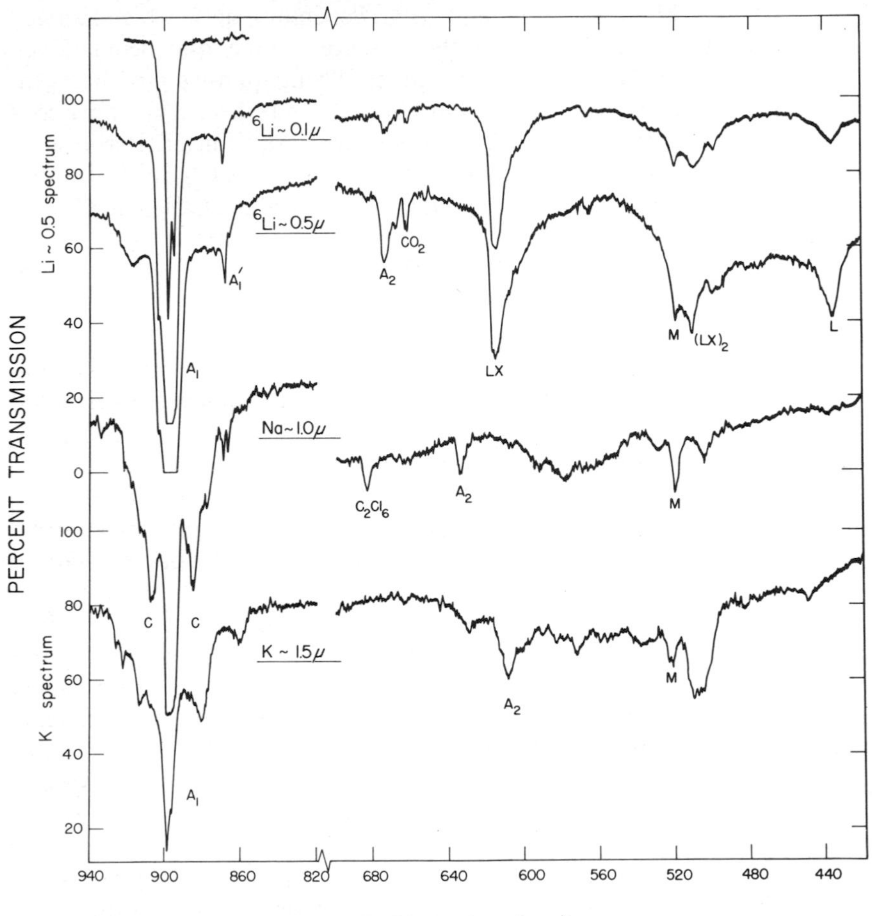

Figure 2. Infrared spectra in the 420–940 cm^{-1} region following matrix reactions of alkali metal atoms with carbon tetrachloride (Ar/CCl_4 = 100), deposited at 15°K. (Top) ^{6}Li, 0.1 μ vapor pressure, insert shows earlier scan of the A_1 band; (second) ^{6}Li, 0.5 μ vapor pressure; (third) Na, 1.0 μ vapor pressure; (bottom) K, 1.5 μ vapor pressure.

metal contributes to this vibration, which is probably due to the carbene, CCl_2, perturbed by the MCl reaction product.[16]

The identification of ν_3 of CCl_3 and CBr_3 was confirmed in the spectra of the CCl_2Br and $CClBr_2$ species of lower symmetry.[5] The doubly degenerate ν_3 mode of CCl_3 at 898 cm^{-1} was split into two bands at 888 and 835 cm^{-1} for CCl_2Br and two bands at 856 and 783 cm^{-1} for $CClBr_2$; a single ν_3 band was observed at 773 cm^{-1} for CBr_3. Each of the radicals, CCl_3, CCl_2Br, $CClBr_2$, and CBr_3, was produced from two and only two of the precursors,

CCl_4, CCl_3Br, CCl_2Br_2, $CClBr_3$, and CBr_4, depending upon which halogen was abstracted: this clearly indicates the presence of three equivalent halogen atoms in these primary reaction products. Using pyrolysis of halogen-substituted organo-mercury compounds, Maltsev and coworkers[17] have confirmed the above vibrational assignments to the CX_3 radicals. Pyrolysis of $C_6H_5HgCCl_2Br$ produced a small amount of CCl_3 and CCl_2Br: the lithium atom-CCl_3Br matrix reaction produced large yields of the CCl_3 and CCl_2Br free radicals.[5]

$$CCl_3 + M \rightarrow MCl + CCl_2 \tag{5}$$

$$CCl_3 + M \rightarrow M^+CCl_3^- \tag{6}$$

Secondary reactions (5) and (6) of alkali atoms with the primary reaction product CCl_3 yielded the dichlorocarbene and the trihalomethyl alkali metal species.

The observation of resolved chlorine isotopic splittings led to the assignments of an intense 745.7 cm^{-1} band to ν_3 and a weak 719.5 cm^{-1} band to ν_1 of dichlorocarbene, since these two modes contain slightly different chlorine atomic mass participation.[18] In the very recent pyrolysis work of Maltsev and coworkers[17], which produced better chlorine isotopic resolution, these bands were observed at 745.8 and 719.5 cm^{-1}. In the CBr_4 reaction, an intense band at 640.5 cm^{-1} and a weak band at 595.0 cm^{-1} were assigned to ν_3 and ν_1, respectively, of dibromocarbene.[19] The dibromocarbene assignments have been confirmed by very recent proton beam matrix reaction experiments conducted in this laboratory.[15] Evidence for the trapping of both products of reaction (5) in the same matrix site forming an MCl-perturbed CCl_2 species was found in bands at 674, 635, and 608 cm^{-1} in the Li, Na, and K reactions of Figure 2.

The trichloromethyl lithium secondary reaction product produced a 521 cm^{-1} band, labeled M in Figure 2, which was assigned to ν_3 of the CCl_3^- part of $CCl_3^-Li^+$.[20] The bromine counterpart was observed at 462 cm^{-1}. Intermediate bands for the mixed chlorine-bromine species indicated the presence of three equivalent halogen atoms. The low frequency was presumably due to repulsions of the extra electron and interaction between the CCl_3^- anion and the Li^+ cation.

The isotopic data for ν_3 of CCl_2 provide ample basis for calculating the Cl—C—Cl valence angle. The carbon-12 and carbon-13 isotopic ν_3 frequencies,[18] 745.7 and 723.2 cm^{-1}, provide a lower limit of $100 \pm 9°$ for the bond angle: the chlorine-35 and chlorine-37 ν_3 frequencies,[17] 745.8 and 741.7 cm^{-1} give a $111 \pm 7°$ upper limit. The upper limit-lower limit average, 106°, provides a good measure of the bond angle for isotopes whose anharmonicities are nearly the same. It is of interest to point out that the bond angle of CF_2 is 105°.[21]

Analogous matrix-reaction studies were done by Smith and Andrews[22] for CI_4 with Li and Na atoms. An intense band at 693 cm^{-1} was assigned to ν_3 of the Cl_3 radical: the same 693 cm^{-1} feature was also produced by pyrolysis of CI_4 near 350°C. An additional feature at 437 cm^{-1} in Li and Na experiments was attributed to ν_3 of CI_3^- in the $CI_3^- M^+$ species.

Alkali metal matrix reactions with the less reactive CF_4 species were not attempted. Owing to the high C—F bond energy, the activation energy for fluorine atom abstraction by an alkali atom was believed to be excessive for appreciable product formation.

3.2 Dihalomethyl Radicals

Carver and Andrews[23–25] have studied the HCF_2, $HCCl_2$, and $HCBr_2$ free radicals and their deuterium counterparts using matrix reactions of Li and Na atoms with HCF_2Br, DCF_2Br, $HCCl_3$, $DCCl_3$, $HCBr_3$, $DCBr_3$. Subsequently, Smith and Andrews[22] observed HCI_2 and DCI_2 from the analogous iodoform-matrix reactions. The spectra were characterized by a sharp band assigned to ν_5, the antisymmetric H—C—X bending mode, and an intense band for ν_6, the antisymmetric X—C—X stretching mode. Table 1 contrasts the frequencies for dihalomethyl radicals. Notice the unusual shift of ν_6 of difluoromethyl to a higher frequency upon deuteration, which is caused by interaction between the ν_5 and ν_6 antisymmetric modes.[23] This interaction also caused an intensity sharing between modes for the $DCCl_2$ species.[24] The progressive decrease of ν_5 and ν_6 of the HCX_2 species clearly follows the increase in atomic weight of the halogen atom.

Table 1. Antisymmetric H—C—X Bending (ν_5) and Antisymmetric C—X Stretching (ν_6) Frequencies (cm^{-1}) for the HCX_2 and DCX_2 Free Radicals

Radical	ν_5	ν_6
HCF_2	1317	1175
DCF_2	934	1217
$HCCl_2$	1226	900
$DCCl_2$	974	816
$HCBr_2$	1166	786
$DCBr_2$	898	725
HCI_2	1106	716
DCI_2	850	653

In the haloform matrix reactions with Li and Na atoms, sharp bands independent of the alkali reagent were assigned to the isolated free radical. Nearby satellite bands were attributed to the same vibrational mode of the radical perturbed by the alkali halide molecule in a radical-alkali halide complex.[25]

3.3 Monohalomethyl Radicals

The four monohalomethyl radicals, CH_2X, have been produced by appropriate matrix reactions and stabilized by the argon matrix.[26–29] The CH_2F radical produced an intense C—F stretching mode, whereas C—F stretching and symmetric D—C—D bending modes were observed for CD_2F.[25]

The spectrum of CH_2Cl is representative of the CH_2Br and CH_2I species[28,29] as well, so it is discussed in some detail. Figure 3 contrasts the synthesis of CH_2Cl as the product of ^7Li atom matrix reactions with CH_2ClF, CH_2Cl_2, CH_2ClBr, and CH_2ClI.[27] The CH_2Cl free-radical absorptions are labeled A_1, A_2, and A_3, and the C_1, C_2, and C_3 absorptions are due to the CH_2Cl—LiX species. Note the progressive decrease in the C_3 band frequency as the perturbing species changes from LiF, LiCl, LiBr, to LiI. The A_1 band at 1391 cm^{-1} is due to the symmetric H—C—H bending mode, the A_2 chlorine-isotopic doublet 826.3, 820.2 cm^{-1} is due to the C—Cl stretching mode, and the intense A_3 band is assigned to the out-of-plane bending mode of the CH_2Cl species. The sense of the anharmonicity in this vibration is indicative of a planar radical. The relatively high C—Cl stretching fundamental for the CH_2Cl radical suggests π-bonding as a means of stabilizing the radical and strengthing the C—Cl bond.[27]

3.4 The OF and ClO Free Radicals

The oxygen fluoride free radical was first observed by Arkell and coworkers[30] following the mercury-arc photolysis of argon-OF_2 samples at 4°K. The matrix reactions of OF_2 and Li, Na, K, and Mg atoms were studied by Andrews and Raymond.[31] The intense, sharp OF fundamental was observed at 1028.6 ± 0.3 cm^{-1}, independent of the metal reagent: metal fluoride molecules were observed for all reagent metals. In the lithium-OF_2 experiments, a secondary reaction product identified as the ionic species, Li^+OF^-, was characterized by two infrared bands.[31]

The sodium-OF_2 matrix reaction was examined in this laboratory using the matrix Raman technique: a weak band at 1028 cm^{-1} was observed. However, it was noticed that the 1028 cm^{-1} band grew during the laser-illumination period. The Raman spectrum of argon-matrix-isolated OF_2 without alkali metal atoms is illustrated in Figure 4*a*: the sharp OF_2 fundamental bands, ν_2 and ν_3, are noted at 464 and 825 cm^{-1} along with the

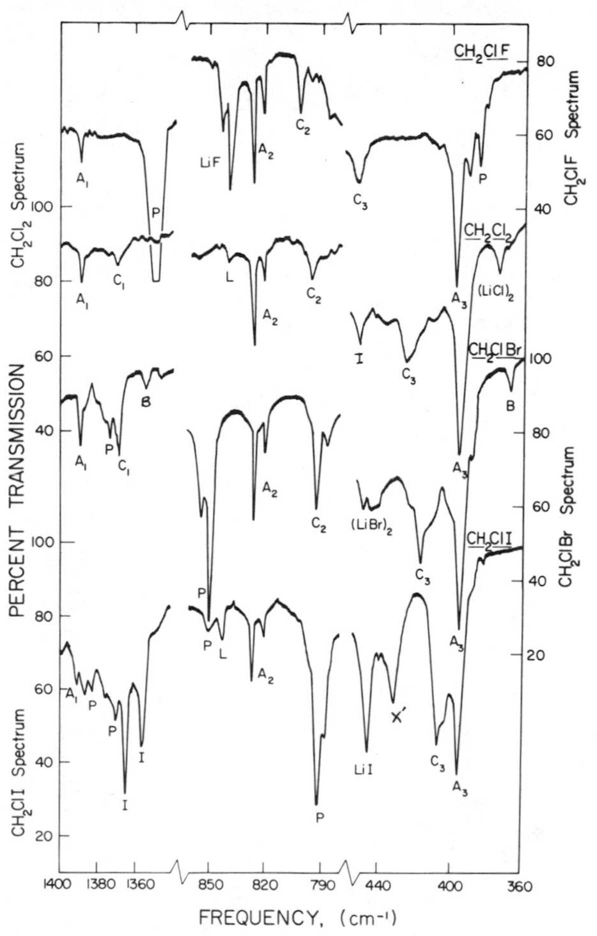

Figure 3. Matrix reactions of CH_2ClX precursors with lithium atoms producing the CH_2Cl free radical whose infrared absorptions are labeled A_1, A_2, and A_3. (Ref. 27, used with permission.)

Fermi doublet $\nu_1, 2\nu_2$ at 920, 931 cm^{-1}.[32] These frequencies are within 1 cm^{-1} of the infrared matrix assignments. Figure 4*a* also shows the laser photolytic production of the OF species. Note in the first trace, no signal was detected at 1029 cm^{-1} after 8 min of exposure to 700 mW of 4880 Å excitation; however, after 25 min of laser illumination on the same spot of sample, a small signal appeared at 1029 cm^{-1} that grew to a strong signal after 2 hr of laser photolysis. Clearly, the 1029 cm^{-1} signal increased as a function of laser illumination time. In addition to the OF photolysis product, O_2 and

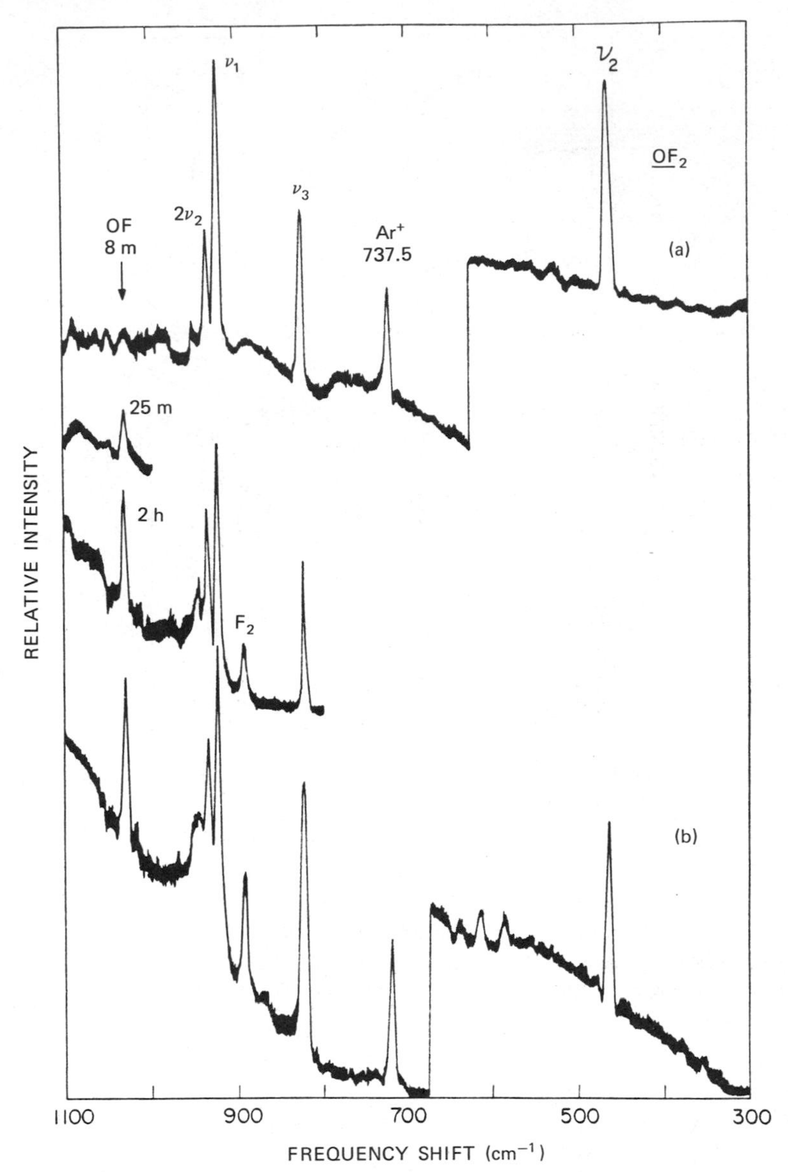

Figure 4. Raman spectra from 300 to 1100 cm^{-1} for oxygen difluoride in solid argon at 16°K using 4880 Å excitation ($Ar/OF_2 = 100$). Spectrum (*a*), initial spectrum recorded, arrows denote region of spectrum scanned after indicated time of exposure to the 4880 Å laser line. Spectrum (*b*), sample deposited for 4 hr with simultaneous 4880 Å photolysis for the last 3 hr, initial spectrum recorded. (Ref. 32, used with permission.)

F_2 were also observed at 1552 and 892 cm^{-1}, respectively. Accurate measurements of the OF fundamental frequency using argon emission lines superimposed on the actual scan yielded 1028.9 $\pm$ 0.5 cm^{-1}, which is in excellent agreement with the infrared observations.

Figure 4*b* illustrates the Raman spectrum of an Ar/OF_2 = 100 sample deposited for 4 hr with simultaneous 4880 Å photolysis for the last 3 hr.[32] Note the strong signals at 1029 cm^{-1} and 892 cm^{-1} for photolysis products of OF_2 and the strong signals for unphotolysed OF_2. The matrix apparently moderates the laser photodecomposition of OF_2 such that excellent Raman spectra of percursor and photolysis products can be obtained.

The advantage of laser photolysis is demostrated for OF_2. The electronic transition for OF_2 photodecomposition increases in intensity with decreasing wavelength: the extinction coefficient[40] is 75×10^{-5} at 2100 Å. Accordingly, the OF species was first observed by Arkell and coworkers following mercury arc photolysis of matrix isolated OF_2.[30] However, the OF_2 absorption tails out into the visible region with an extinction coefficient of 0.14×10^{-5} at 4880 Å. The high power density of the laser beam compensates for the low extinction coefficient at 4800 Å such that photolysis products of OF_2 are produced by laser illumination, and they can be identified by Raman spectra.

Andrews and Raymond[33] studied matrix reactions of Cl_2O with Li, Na, and K atoms: the lithium reaction spectra are shown in Figure 5. First, notice the very intense ^6LiCl and ^7LiCl product bands. Two new ClO species were observed at 995 and 850 cm^{-1}, labeled R and C, which were independent of the metal reagent. The broad band labeled M in Figure 5 clearly shows a lithium isotopic shift. In very recent laser Raman studies of matrix-isolated Cl_2O, Chi and Andrews[34] observed the photoproduction of a weak band at 850 cm^{-1} and no signal at 995 cm^{-1}. This observation, and the hot band spacings of 845 $\pm$ 4 cm^{-1} in electronic spectra of ClO reported by Basco and Morse,[35] indicate that the band labeled C at 850 cm^{-1} is due to the matrix-isolated ClO free radical.[34] The M band is attributed to the MX perturbed species, ClO—LiCl. The R band at 995 cm^{-1} was produced in greater yields following the ultraviolet photolysis of $Ar/Cl_2O/O_3$ samples: it is believed to be due to a Cl=O—Cl=O species.[34] The L band in Figure 5, which shows a large lithium isotopic shift, is probably due to the ionic species Li^+ClO^-.[33]

4. NEW ALAKALI-METAL-CONTAINING SPECIES

We now focus our attention on a number of infrared and Raman studies of new chemical species that require the alkali-metal-atom matrix reaction technique for synthesis and stabilization.

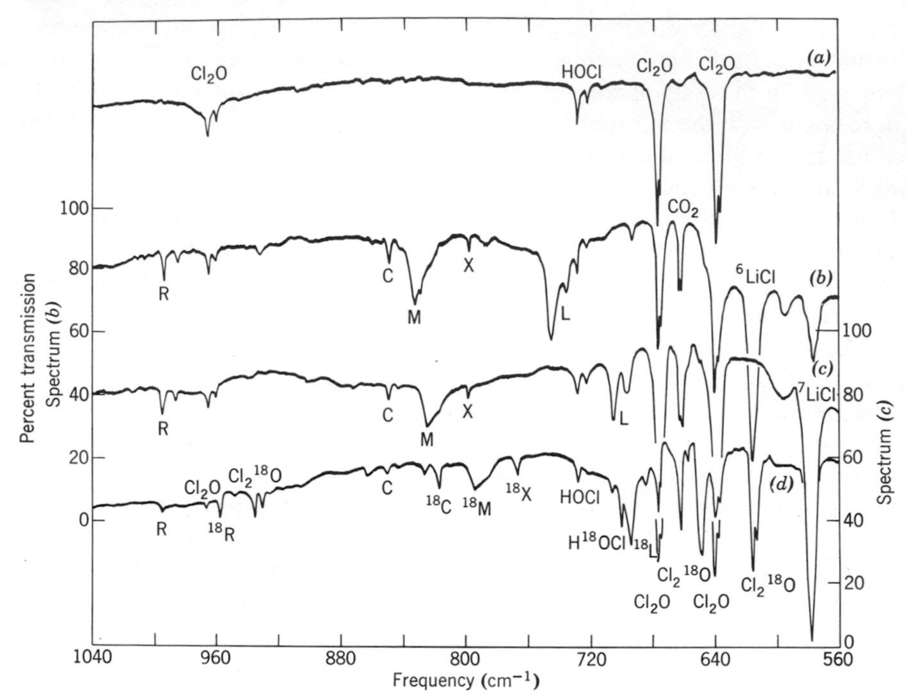

Figure 5. Infrared spectra in the 560–1040 cm^{-1} region for samples of dichlorine monoxide in argon (M/R = 100) deposited at 15°K. (*a*) 0.50 mmoles of Cl_2O deposited without alkali metal, (*b*) codeposition with 2μ beam of ^{6}Li, (*c*) codeposition with 2μ beam of ^{7}Li, (*d*) 0.96 mmoles of 70% ^{18}O-enriched Cl_2O deposited with 1μ beam of ^{7}Li. (Ref. 33, used with permission.)

4.1 Matrix Reactions with NO

The first new species synthesized using matrix reactions of alkali metal atoms was LiON.[4]

$$\text{Li} + \text{NO} \rightarrow \text{LiON} \tag{7}$$

In this work, Andrews and Pimentel assigned intense bands at 651 and 1352 cm^{-1} to the Li—O stretch and the N—O stretch of a presumably bent species, LiON: the use of ^{6}Li and ^{7}Li atoms and ^{15}N- and ^{18}O-enriched NO reagents were important for characterizing the vibrational modes. In subsequent studies, Milligan and Jacox also observed the 651 and 1352 cm^{-1} bands; however, mercury arc photolysis markedly decreased the 651 cm^{-1} feature while the 1352 cm^{-1} signal remained.[36] These workers attributed the 1352 cm^{-1} band to $(NO)^-$ and the 651 cm^{-1} feature to some other lithium-nitric oxide species. In the most recent study, aimed at comparison of the $M^+O_2^-$ and M^+NO^- species, Tevault and Andrews[37] prepared matrix samples of LiON and studied their mercury arc photolysis behavior. The decrease in intensity of the Li—O mode at 651 cm^{-1} was accompanied by the growth of a new Li—O mode at 447 cm^{-1}, while the $(NO)^-$ mode at 1352 cm^{-1} remained. The photolysis behavior was explained as the photoisomerism of the triangular ionic species, $Li^+(ON)^-$, to a bent form.

$$(Li)^+\begin{pmatrix} O \\ | \\ N \end{pmatrix}^- \xrightarrow{h\nu} (Li)^+(O\diagdown N)^- \tag{8}$$

Here the frequency of the interionic $Li^+ \leftrightarrow NO^-$ mode was changed by a rearrangement in cation-anion structure, whereas the intraionic mode was not measurably affected.

Tevault and Andrews[37] observed three bands that showed isotopic splittings appropriate to a species containing two equivalent Li atoms and one NO molecule. These bands were assigned to the two antisymmetric interionic modes and the intraionic $(N—O)^{2-}$ mode in the secondary reaction product, $Li^+(NO)^{2-}Li^+$.

$$Li^+(ON)^- + Li \rightarrow Li^+(NO)^{2-}Li^+ \tag{9}$$

In studies involving all of the heavier alkali metal reagents, Tevault and Andrews observed the M^+NO^- species for all of the alkali metal reagents.[38] The M^+NO^- spectra were characterized by two bands, an $(N \leftrightarrow O)^-$ intraionic mode in the 1352 to 1374 cm^{-1} region and a $M^+ \leftrightarrow (NO)^-$ interionic mode between 692 and 219 cm^{-1}, depending upon the alkali counterion. Table 2 lists the M^+NO^- frequencies. Note the significant alkali-metal mass effect on ν_2, the $M^+ \leftrightarrow NO^-$ mode. Note also the small reverse effect on

Table 2. Fundamental Frequencies (cm^{-1}) Assigned to the ν_1 Intraionic and the ν_2 Interionic Modes of the $M^+(NO)^-$ Species

Metal	ν_1	ν_2
^6Li	1353.5	692.3
^7Li	1352.5	651.8
Na	1358.0	361.0
K	1372.0	280.4
Rb	1373.0	235.0
Cs	1374.0	219.0

ν_1, the $(N—O)^-$ mode. Here, the larger, more polarizable alkali cation accommodates more of the antibonding anion electron, which results in the removal of a small amount of antibonding charge density from NO^- and a slight increase in the $(N—O)^-$ frequency. The M^+ effect on the $(N—O)^-$ frequency is relatively small compared to the difference between $(N—O)^-$ frequencies as a function of M^+, 1352–1374 cm^{-1}, and NO, 1875 cm^{-1}.

Recently, alkaline earth metal matrix reactions with NO have been studied in this laboratory.[39] Ca, Sr, and Ba produced an A^+NO^- species characterized by an intense intraionic $(N \leftrightarrow O)^-$ mode at 1356.8 cm^{-1} for Ca, 1360.5 cm^{-1} for Sr, and 1364.0 cm^{-1} for Ba. Note that these are A^{+1} species, as determined by the NO^{-1} frequency. The trend of increasing NO^- frequency with increasing A^+ ion size and polarizability follows the alkali metal series.

$$A + NO \rightarrow A^{+1}(NO)^{-1} \tag{10}$$

4.2 Matrix Reactions with O_2

Alkali-metal-atom–oxygen-molecule matrix reactions have been extensively studied in this laboratory using infrared and Raman techniques. The primary reaction yields the superoxide species, which is discussed in detail.

$$M + O_2 \rightarrow M^+O_2^- \tag{2}$$

Two secondary reactions yield the peroxide species

$$M^+O_2^- + M \rightarrow M^+O_2^{2-}M^+ \tag{11}$$

and the disuperoxide species.

$$M^+O_2^- + O_2 \rightarrow M^+O_4^- \tag{12}$$

In the first study of $M—O_2$ matrix reactions, this author reported a new chemical species, lithium superoxide ($Li^+O_2^-$), with ionic binding between Li^+ and O_2^-, and an isosceles triangular structure.[6] These conclusions were based upon the near agreement between the O—O mode for LiO_2 and the O_2^- fundamental and the observation of sharp triplet bands in the reaction of $^{16}O_2/^{16}O^{18}O/^{18}O_2$ with ^7Li atoms at high dilution in argon. Figure 6 contrasts the infrared spectrum in this critical experiment with the Raman spectrum of a similar argon-matrix sample.[40]

First, examine the O_2^- region near 1100 cm^{-1}. The 1096.1 cm^{-1} Raman band observed at 1096.6 cm^{-1} in the infrared is due to $^7Li^+$ $^{16}O_2^-$; the 1065.7 cm^{-1} Raman band recorded at 1066.5 cm^{-1} in the infrared is due to $^7Li^+$ $^{16}O^{18}O^-$; the 1034.6 cm^{-1} Raman signal measured at 1035.2 cm^{-1} in the infrared is due to $^7Li^+$ $^{18}O_2^-$. The large 61 cm^{-1} oxygen isotopic shift indicates the pure oxygen character of this vibration, which is identified as ν_1, the intraionic $(O \leftrightarrow O)^-$ mode. The central feature reveals the molecular

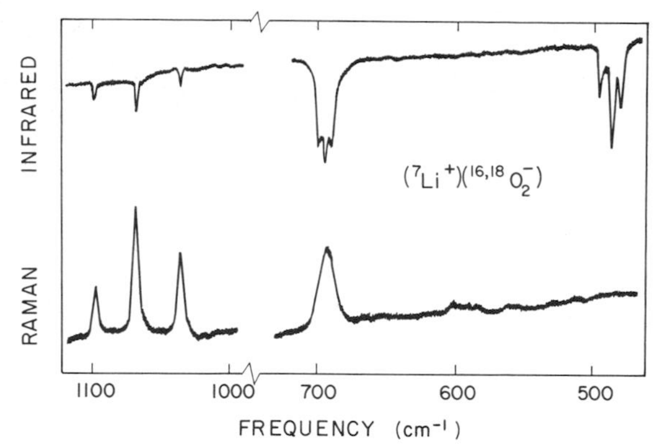

Figure 6. Infrared and Raman spectra of lithium superoxide, $Li^+O_2^-$ using lithium—7 and 30% $^{18}O_2$, 50% $^{16}O^{18}O$, 20% $^{16}O_2$ ($Ar/O_2 = 100$). Raman spectrum recorded using 200 mW of 4880 Å excitation and long-wavelength-pass dielectric filter in 1000 cm^{-1} region.

structure. In both infrared and Raman experiments, the single 16–18 isotopic band was sharp, just as sharp as the 16–16 and 18–18 isotopic bands. This indicates that the 16–18 isotope produces a single molecular arrangement, M(16|18), with equivalent oxygen atoms, in contrast to two possible M—16—18, M—18—16 arrangements that would be expected to produce a split central band as was observed for $H^{16}O^{18}O$ and $H^{18}O^{16}O$.[41] In the infrared experiments, this band was very sharp, 1.0 cm^{-1} half-width. It appears that 0.2 cm^{-1} is a reasonable upper limit on the inequivalence of the O atoms in $Li^+O_2^-$ in the $Li^+\ ^{16}O^{18}O^-$ isotopic band.

Second, in the Li—O stretching region, two intense resolved triplet bands were observed in the infrared, and a single broader, weaker feature was observed in the Raman spectrum. The large lithium isotopic shift and smaller oxygen isotopic shift indicate that the band near 700 cm^{-1} is the symmetric interionic mode ν_2, $Li^+ \leftrightarrow O_2^-$, and the smaller lithium isotopic shift and larger oxygen isotopic shift show that the band near 500 cm^{-1} is ν_3, the antisymmetric interionic mode. The G-matrix elements for these normal modes weight the participation of Li and O atoms differently.[6]

Comparison of band intensities in Figure 6 also reflects on the bonding in the $Li^+O_2^-$ species. The infrared spectrum shows a weak intraionic mode, ν_1, and very intense interionic modes, ν_2 and ν_3, as would be expected for an ionic model. The Raman spectrum contains a very intense intraionic mode ν_1, a moderately intense symmetric interionic mode ν_2, and the antisymmetric interionic mode ν_3 is absent. The ionic model for LiO_2 is supported by the complementary Raman spectrum.

The heavier alkali metals, Na, K, Rb, and Cs, react with oxygen to produce the superoxide $M^+O_2^-$ species.[42–44] Raman spectra for the Na, K, Rb, and Cs atom reactions with O_2 show a strong band near 1110 cm^{-1} for the O_2^- vibration.[45,46] The infrared spectra are typified by Figure 7, which shows Cs atom cocondensation spectra with four isotopic oxygen samples. The very sharp, weak infrared band at 1115.6 cm^{-1} has a strong Raman counterpart at 1114 cm^{-1}. Isotopic substitution shows that this is the ν_1 O_2^- intraionic fundamental in a $Cs^+O_2^-$ species with two equivalent oxygen atoms. The very intense bands near 230 cm^{-1} and weak bands at 260 cm^{-1} are the interionic ν_2 and ν_3 modes, respectively, of $Cs^+O_2^-$.

The $Cs^+O_2^-$ vibrational data are contrasted in Table 3 with frequencies for the other $M^+O_2^-$ molecules. Note the increase in ν_1 with increasing cation

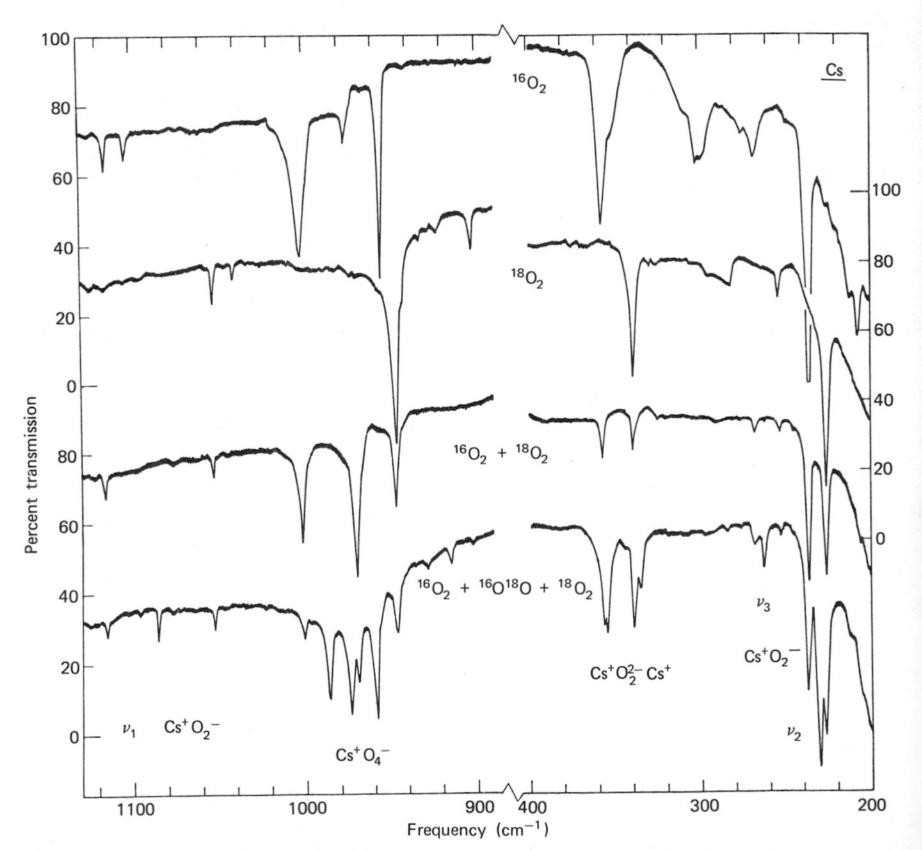

Figure 7. Infrared spectra of cesium-oxygen species produced by the cesium-atom–O_2 matrix reaction (Ar/O_2 = 100). Fundamentals of $Cs^+O_2^-$ and bands due to $Cs^+O_4^-$ and $Cs^+O_2^{2-}Cs^+$ are labeled. (Ref. 44, used with permission.)

Table 3. Fundamental Frequencies (cm^{-1}) Assigned to the ν_1 Intraionic and ν_2 and ν_3 Interionic Modes of the C_{2v} Alkali Metal Superoxide Molecules in Solid Argon at 15°K

Molecule	ν_1	ν_2	ν_3
$^6LiO_2$	1097.4	743.8	507.3
$^7LiO_2$	1096.9	698.8	492.4
NaO_2	1094	390.7	332.8
KO_2	1108	307.5	—
RbO_2	1111.3	255.0	282.5
CsO_2	1115.6	236.5	268.6

size. This trend has been rationalized[40] as follows: increasing the cation polarizability increases the induced dipole moment on the cation opposite in sense to the ion dipole; accordingly, a small amount of antibonding electron density is removed from O_2^-, increasing the fundamental in the direction of the O_2 value (1552 cm^{-1}). Also note the larger decrease in ν_2 and smaller decline in ν_3 with increasing alkali atomic weight, which is required by the G-matrix elements for these normal modes.

The peroxide species, $M^+O_2^{2-}M^+$, has been observed for the five alkali metals. Intense antisymmetric Li—O stretching bands at 796 and 446 cm^{-1} for the ^7Li species showed isotopic splittings appropriate for a species containing *two* equivalent lithium atoms and *two* equivalent oxygen atoms. A rhombus structure with an O—O bond was suggested for the $M^+O_2^{2-}M^+$ molecule.[6] In the heavier species, the sodium frequency counterparts were observed at 525 and 254 cm^{-1}.[42] The upper band only was observed for the K, Rb, and Cs species at 433, 389, and 357 cm^{-1}.[43,44]

The symmetric $(O—O)^{2-}$ intraionic mode is infrared inactive, and accordingly, it was not observed in the infrared studies. However, Huber and Ozin[47] observed a new Raman band at 802 cm^{-1} following the temperature cycling to 30°K of a highly concentrated lithium-atom–*pure*-oxygen matrix sample deposited at 4°K: this operation destroyed the 1097 cm^{-1} $Li^+O_2^-$ band. In spite of the fact that this assignment was not supported by isotopic substitution, it does fall near the crystal O_2^{2-} mode.[48] However, in the temperature cycling of 14°K lithium-oxygen matrix samples, infrared studies in this laboratory revealed the growth of $(LiO_2)_2$ bands and *no* appearance of LiO_2Li interionic modes.[6] It is difficult to conceive of unreacted Li atoms in an oxygen matrix to carry out secondary reaction (11) upon diffusion. However, it is possible that LiO_2 dimer formed during temperature cycling could eliminate O_2 to form the peroxide species.

$$2LiO_2 \rightarrow (LiO_2)_2 \rightarrow LiO_2Li + O_2 \quad (13)$$

On the other hand, Li_2 could react with LiO_2 upon diffusion to produce the peroxide species.

$$LiO_2 + Li_2 \rightarrow LiO_2Li + Li \quad (14)$$

The disuperoxide species, $M^+O_4^-$, have been observed for the heaviest alkali metals. The small lithium ion apparently is not large enough to stabilize the large O_4^- anion. The new MO_4 species was first observed in K and Rb studies that produced extraordinarily intense, very sharp, bands at 993.4 and 991.7 cm^{-1}, respectively.[43] The cesium counterpart at 1002.5 cm^{-1} in Figure 7 typifies the isotopic behavior of the new species: the intense new band shifted to 946.5 cm^{-1} in the $^{18}O_2$ reaction. The stoichiometry of the new species is revealed by the mixed isotopic reactions: the $^{16}O_2$ + $^{18}O_2$ sample produced a symmetric triplet at 1002.0, 970.2, and 946.5 cm^{-1}, which indicates the presence of two equivalent O_2 molecules in the new species. A well-resolved sextet was produced in the scrambled oxygen isotopic experiment, which is illustrated in Figure 7. The explicit interpretation of the sextet indicates a species with equivalent O_2 units with equivalent O atoms in each unit. Accordingly, a D_{2d} structure was first proposed by this author for the O_2KO_2 species.[43] Subsequently, Jacox and Milligan pointed out the existence of the O_4^- ion in ion-molecule reactions and suggested the O_4^- arrangement.[49] Possible O_4^- anion arrangements and M^+ cation positions have been explored by CNINDO calculations; a "puckered five-membered ring" $M^+O_4^-$ structure was suggested in which the two O_2 parts of O_4^- are equivalent, but the inequivalence in atomic positions for each O_2 unit, which would be spectroscopically small, was not resolved in the infrared spectrum.[44] The O_4^- anion was suggested to contain two "superoxide" bonds and a weak intermolecular bond between the two O_2 units. A very strong Raman band near 300 cm^{-1} for the $K^+O_4^-$, $Rb^+O_4^-$, and $Cs^+O_4^-$ species was assigned to this symmetric intermolecular mode, $(O_2 \leftrightarrow O_2)^-$: interpretation of this low-frequency Raman band requires a very weak oxygen bond that is provided by the O_4^- anion in the $M^+O_4^-$ species.[46] The presence of the cation certainly influences the geometry of the O_4^- anion in the $M^+O_4^-$ ion pair.

Alkaline-earth-metal-atom–O_2 matrix reactions were first studied by Abramowitz and Aquista[50] for the Ba—O_2 system in argon matrices. These workers observed a prominent band at 570 cm^{-1}, which was complicated by aggregation effects and a very weak, poorly defined band at 1066 cm^{-1} that were assigned, respectively, to an interionic mode $Ba^+ \leftrightarrow O_2^-$ and the O_2^- intraionic mode. Subsequent work in this laboratory has confirmed the intense $Ba^+O_2^-$ band at 571.3 cm^{-1} and identified this feature as the ν_2

symmetric interionic mode, but the weak 1066 cm^{-1} band did not show an $^{18}O_2$ shift.[51] However, a sharp band was observed at 1120 cm^{-1}, which is appropriate for the ν_1 intraionic mode for $Ba^+O_2^-$.[52] Note the near agreement of the ν_1 modes of $Ba^+O_2^-$ and $Cs^+O_2^-$ (1115.6 cm^{-1}). We believe $Ba^+O_2^-$ is a symmetrical triangular species like $Cs^+O_2^-$.

Ault and Andrews have very recently investigated the strontium and calcium atom matrix reactions with O_2.[52] These atoms are less reactive with O_2, which made possible the observation of only the more intense interionic vibrational modes of the $Ca^+O_2^-$ and $Sr^+O_2^-$ species.

4.3 Matrix Reactions with Ozone

Extensive infrared and Raman studies of matrix-isolated isotopic ozone molecules have been conducted by Andrews and Spiker.[53] Isotopic ozones ($^{16}O_3$, $^{18}O_3$, and $^{16,18}O_3$) were synthesized by Tesla coil discharge of O_2 gas in a Pyrex finger immersed in liquid nitrogen. Oxygen was outgassed from the blue, liquid ozone sample by evacuating the sample at 77°K. Argon/ozone samples were prepared using standard manometric techniques in a stainless steel vacuum system.

Matrix reactions of alkali metal atoms and ozone have also been studied using infrared and Raman techniques. The infrared spectra[54] were characterized by very intense bands near 800 cm^{-1} depending upon the alkali atom and weaker bands near 600 cm^{-1} for the heavier alkali metal reactions. The intense 800 cm^{-1} bands were assigned to ν_3, and the weak 600 cm^{-1} bands were attributed to ν_2 of O_3^- in the $M^+O_3^-$ species. Splittings observed in the intense ν_3 band are represented by the $Na^+O_3^-$ spectra in Figure 8: the intense bands were resolved into doublets with 3 wavenumber splittings. Note the large oxygen-18 isotopic shift, from 807 to 762 cm^{-1} and the resolved sextet of isotopic bands in the $^{16,18}O_3$ spectrum. The relative intensities of the $Na^+\ ^{16,18}O_3^-$ isotopic ν_3 bands are identical to the analogous O_3 bands, as Figure 8 illustrates. The ozonide ion valence angle in the $Na^+O_3^-$ species was calculated from the ν_3 frequencies for the C_{2v} isotopic bands in Figure 8. The data produced an upper limit of $116 \pm 1°$ and a lower limit of $106 \pm 1°$ for the ozonide valence angle. The average value, 111°, represents a reliable determination of the ozonide valence angle.

The cesium-atom–ozone reaction produced intense ν_3 bands at 802 and 757 cm^{-1} for $Cs^+\ ^{16}O_3^-$ and $Cs^+\ ^{18}O_3^-$, respectively, and weaker ν_2 bands at 600 and 567 for the same respective isotopic species. In addition, oxygen atom abstraction was evidenced by observation of Cs_2O at 457 cm^{-1} and CsO at 322 cm^{-1}.[54]

Sample preparation for Raman scattering experiments was closely monitored using white light: formation of the $M^+O_3^-$ species was evidenced by

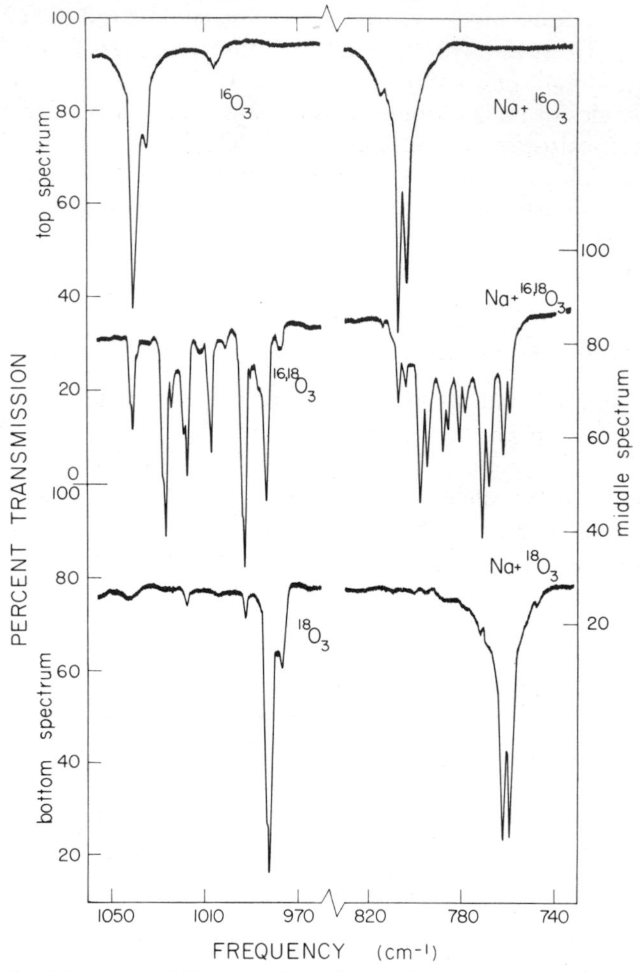

Figure 8. Infrared spectra of the ν_3 regions of isotopic ozone and argon-matrix-isolated sodium ozonide (Ar/O_3 = 200). (Ref. 54, used with permission.)

the appearance of the orange ozonide color.[55] Great care must be taken to prevent excess alkali metal, which produces darker and thus poorer scattering samples. Figure 9 illustrates the Raman spectra of Na—O_3 matrix reactions[56] and provides a complement to the infrared, Figure 8. A very intense Raman band was observed at 1011 cm^{-1} with a site splitting at 1024 cm^{-1} in the $Na^{+16}O_3^-$ spectrum. Large oxygen-18 shifts are indicated by the $Na^+ \, ^{18}O_3^-$ spectrum: the intense band and site splitting shifted to 956 and 970 cm^{-1}, respectively. The $Na^+ \, ^{16,18}O_3^-$ Raman spectrum shows five

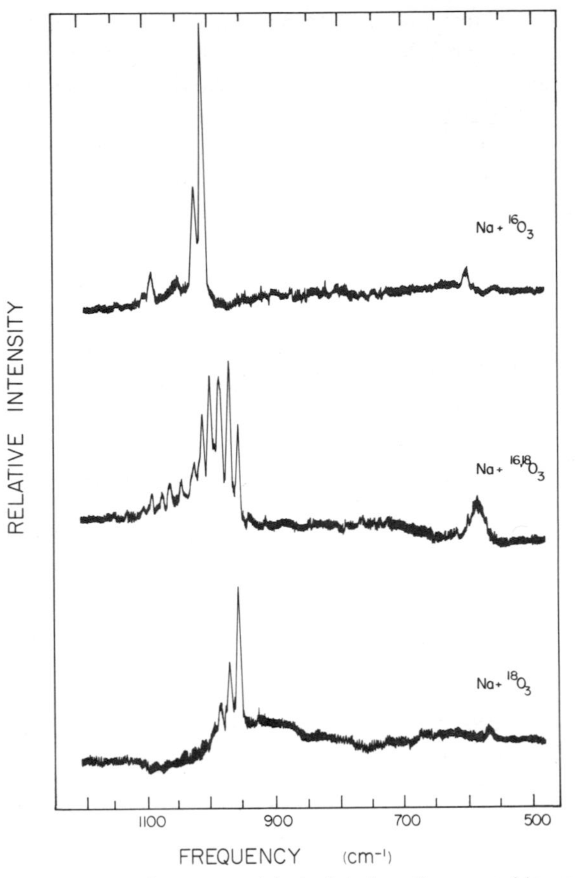

Figure 9. Raman spectrum of argon matrix-isolated sodium ozonide using 200 mW of 5145 Å excitation. (Ref. 56, used with permission.)

well-resolved components, the broader central component contains both 16—18—16 and 18—16—18 isotopic species. A five-component multiplet with the same relative intensities has been observed for ν_1 of matrix-isolated 55% ^{18}O-enriched ozone.[53] As was the case for ν_3 of O_3 and O_3^- in the infrared spectra, ν_1 of O_3 and O_3^- produced similar isotopic bands in the Raman spectra.

The weak band at 599 cm^{-1} in the $Na^{+16}O_3^-$ Raman spectrum has an ozone-18 counterpart at 567 cm^{-1}: individual isotopic components were not resolved in the $^{16,18}O_3$—Na reaction. These features are in good agreement with the infrared assignments to ν_2 of O_3^- in the cesium species.[54] Accordingly the weak Raman bands were assigned to ν_2 of O_3^- in the sodium ozonide molecule.[56]

The very intense fundamental Raman bands for the ozonide species and the deep orange sample color prompted a search for overtones of the ν_1 fundamental. An electronic band with vibrational fine structure in the 5100–3700 Å region has been reported by Jacox and Milligan[57] for the O_3^- species. In the case of $Cs^+O_3^-$, a regular progression of fundamental and overtone bands at 1018, 2028, 3024, and 4014 cm^{-1} was observed with regularly decreasing intensity. The $Cs^+ \, {}^{18}O_3^-$ species produced a progression out to $5\nu_1$: bands were observed at 962, 1915, 2859, 3795, and 4724 cm^{-1}.[56] The regularly decreasing intensity pattern for an overtone progression is characteristic of the resonance Raman effect.

Figure 10 contrasts the use of six different laser exciting lines on the Raman spectrum of $Na^+O_3^-$. The 6471 Å line produced a weak fundamental at 1011 cm^{-1} on a steep fluorescence background. 5682 Å excitation gave an intense fundamental at 1011 cm^{-1} and a weak first overtone at 2013 cm^{-1}. The 5309 Å line produced an intense fundamental at 1012 cm^{-1}, a first overtone at 2013 cm^{-1}, and a weak second overtone at 3001 cm^{-1}. 5145 Å excitation yielded the intense 1011 cm^{-1} fundamental, intense first and second overtones at 2013 and 3002 cm^{-1}, and a weak third overtone at 3977 cm^{-1}: the 4880 Å line produced the intense fundamental and two intense overtones at the same frequencies. Figure 10 concludes with the 4579 Å exciting line that gave the intense fundamental and two moderately intense overtones.[56] Clearly, the overtone intensity is enhanced as the excitation wavelength approaches the electronic absorption band maximum. This increase in overtone intensity relative to fundamental intensity as the exciting wavelength enters the electronic band is characteristic of resonance Raman spectra.

Analysis of the overtone progressions for the $Cs^+ \, {}^{16}O_3^-$ species produced the vibrational constants, $\omega_1 = 1028.2 \pm 1.0$ cm^{-1} and $x_{11} = 4.95 \pm 0.25$ cm^{-1}. The harmonic and anharmonic vibrational constants provide a measure of the heat of atomization, $O_3^- \rightarrow 2O + O^-$, of the ozonide ion using a linear Birge–Spooner extrapolation to the dissociation limit. The spectroscopic heat of atomization, 153 ± 3 kcal/mole, agrees well with the thermodynamic value, and it provides an independent check on the electron affinity of ozone.[56]

The present Raman spectra for the $M^+O_3^-$ species demonstrate the usefulness of the matrix reaction technique to produce and stabilize observable quantities of unstable species for laser examination. The ozonide Raman fundamentals near 1010 cm^{-1} in the matrix isolation study agree with He—Ne observations on polycrystalline ozonides.[58] The matrix host moderates the photodecomposition of the $M^+O_3^-$ molecule, quenches fluorescence, and allows the resonance Raman spectrum of O_3^- to be observed using argon ion excitation.

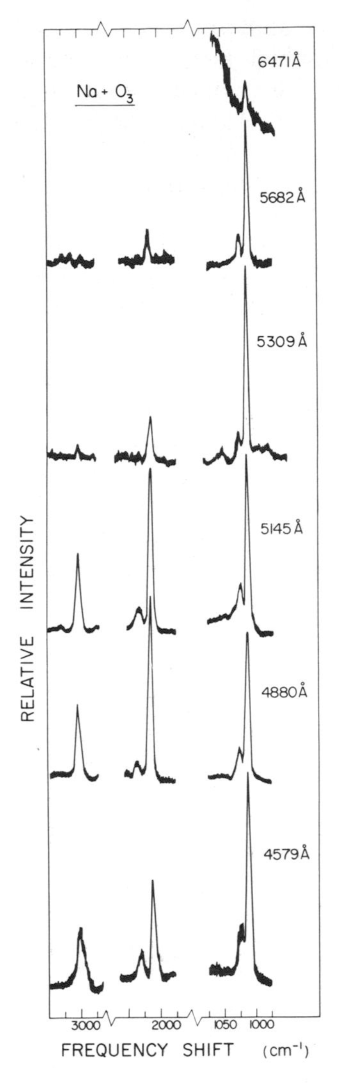

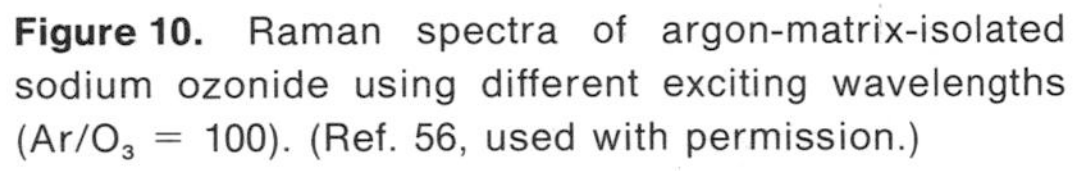
Figure 10. Raman spectra of argon-matrix-isolated sodium ozonide using different exciting wavelengths (Ar/O_3 = 100). (Ref. 56, used with permission.)

Alkaline earth metal atom reactions with ozone have been examined in the infrared spectral region using both argon[51] and nitrogen matrices.[52] The two primary reactions yielding the alkaline earth ozonide and oxide species were sought.

$$A + O_3 \rightarrow A^+O_3^- \tag{15}$$

$$A + O_3 \rightarrow AO + O_2 \tag{16}$$

The alkaline ozonide species is of interest for comparison to the alkali metal ozonides. The alkaline oxide molecules are high-temperature species that are difficult to study using conventional Knudsen effusion techniques. Hence, matrix reactions provide a unique source of new alkaline earth oxide species.

Barium-ozone matrix reactions conducted by Thomas and Andrews[51] yielded ν_3 of the ozonide ion in $Ba^+O_3^-$ at 804.3 cm^{-1}, ν_2 of $Ba^+O_2^-$ at 571.3 cm^{-1}, BaO at 634.3 cm^{-1}, and $(BaO)_2$ at 460 cm^{-1} in solid argon. These features and their oxygen-18 substituted counterparts are illustrated in Figure 11, which contrasts the barium atom reaction with $^{16}O_3$, $^{18}O_3$, and scrambled $^{16,18}O_3$. Notice how the oxygen atom stoichiometry of the molecular species is revealed by the $^{16,18}O_3$ experiment. The sextet clearly denotes a symmetrical ozonide species, $Ba^+O_3^-$: the doublet identifies BaO and the triplet is due to symmetrical $Ba^+O_2^-$.

The ozonide feature was observed at 804.2 cm^{-1} in strontium experiments. The Ca reactions produced ν_3 of O_3^- in $Ca^+O_3^-$ at 804.5 cm^{-1}, $(CaO)_2$ bands at 635.7 and 575.5 cm^{-1}, and possibly $Ca^+O_2^-$ at 593 and 582 cm^{-1}. The ozonide frequencies in the $A^+O_3^-$ species agree within a few cm^{-1} with

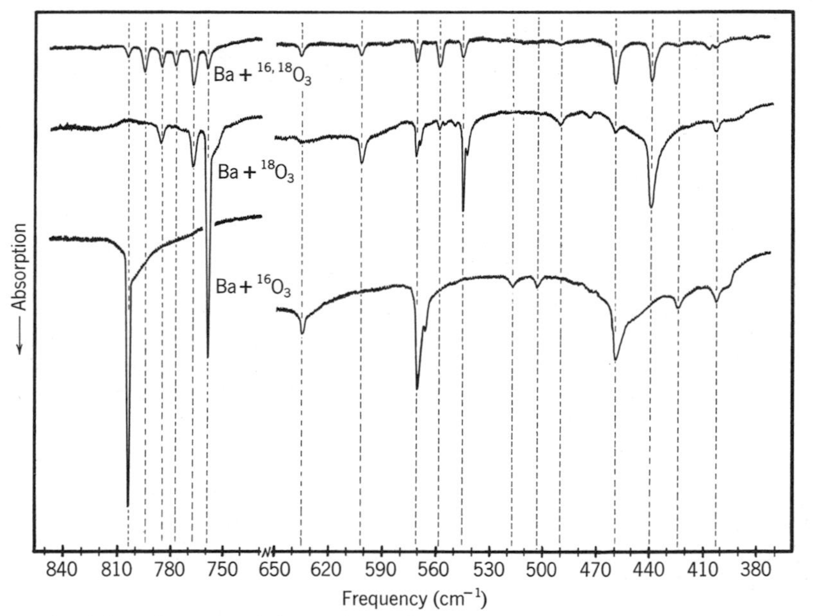

Figure 11. Infrared spectra of the products of barium-atom–ozone argon-matrix reactions (Ar/O_3 = 200). (top) $^{16,18}O_3$, 55% oxygen–18, (middle) $^{18}O_3$, 93% oxygen–18, (bottom) $^{16}O_3$, natural isotope.

the alkali metal species; hence a single ionization of the alkaline atom has been produced. The BaO produced by reaction (16) is in excellent agreement with Linevsky's argon matrix 634 cm^{-1} assignment to BaO produced by evaporation from BaO solid at 1650°C.[59]

Ault and Andrews have performed Ba, Sr, and Ca matrix reactions with ozone in nitrogen matrices.[52] These reactions are characterized by a broad, intense band at 800 cm^{-1}, which is due to the ozonide species. Of most interest are very intense, sharp bands at 612.4 and 581.5 cm^{-1} produced from Ba-$^{16}O_3$ and $^{18}O_3$ reactions, respectively, which are the nitrogen matrix counterparts of BaO and $Ba^{18}O$, respectively. In the analogous strontium reactions, intense sharp bands at 620.0 and 589.9 cm^{-1} are appropriate for assignment to SrO and $Sr^{18}O$. The calcium-ozone reaction produced a sharp, intense band at 707.0 cm^{-1}, 3 cm^{-1} above the ν_2 band of O_3: this feature shifted to 678.2 cm^{-1} in the $^{18}O_3$ reaction, whereas ν_2 of $^{18}O_3$ appeared at 664.5 cm^{-1}. A scrambled $^{16,18}O_3$ reaction produced the sharp, intense 707.0 and 678.2 cm^{-1} bands prominent above the weaker sextet of bands due to ν_2 of the six $^{16,18}O_3$ isotopes. The harmonic ratio for the $Ca^{18}O/Ca^{16}O$ frequencies is 0.95942: the observed ratio is 0.95926 for the two intense nitrogen matrix bands. This excellent agreement supports the identification of CaO in solid nitrogen. The enhanced reactivity of O_3 in primary reaction (16) in nitrogen matrices has also been noted for N_2O[60,61]. Apparently, the nitrogen matrix interacts more strongly with the A^+O^- species, thus contributing a small aid to the decomposition of a $A^+O_3^-$ "collision complex" to A^+O^- and O_2.

Table 4 contrasts gas phase fundamentals of CaO, SrO, and BaO in the ground $^1\Sigma$ electronic state as determined from gas-phase electronic band spectra[62] and the nitrogen matrix frequencies.[52] Notice the increase in the gas-to-nitrogen matrix shift of 15 cm^{-1} for CaO, 25 cm^{-1} for SrO, to 54 cm^{-1} for BaO as the size and polarizability of the alkaline cation increases. This trend confirms the CaO assignment and that the $^1\Sigma$ state is the electronic ground state for CaO, which has thermodynamic significance.[63]

Table 4. Vibrational Frequencies (cm^{-1}) of Alkaline Earth Oxides in the Gas Phase and in Nitrogen Matrices

Molecule	Gas	$M^{16}O$	$M^{18}O$
CaO	722	707.0	678.2
SrO	645	620.0	589.9
BaO	666	612.4	581.5

4.4 Matrix Reactions with Cl_2

Cl_2^- is another species of limited stability that requires matrix synthesis and stabilization. Hass and Griscom have attributed a Raman band at 265 cm^{-1} in γ-irradiated alkali-halide–alkali-borate glasses to the Cl_2^- fundamental.[64] Howard and Andrews have extensively studied matrix reactions of Cl_2 and alkali metal atoms using Raman and infrared techniques.[65] The Raman spectra were characterized by intense bands near 250 cm^{-1} depending upon the alkali reagent: the fundamental was observed at 246 cm^{-1} with Li, 225 cm^{-1} with Na, 264 cm^{-1} with K, 260 cm^{-1} with Rb, and 259 $\pm$ 1 cm^{-1} with Cs. These frequency differences indicate the interaction of the interionic mode ν_2, $M^+ \leftrightarrow Cl_2^-$, with the observed frequencies that were assigned to ν_1, the symmetric intraionic vibration $(Cl \leftrightarrow Cl)^-$, in the $M^+Cl_2^-$ species.

$$M + Cl_2 \rightarrow M^+Cl_2^- \tag{17}$$

Owing to the yellowish-orange color of the $M^+Cl_2^-$ sample and the great intensity of the fundamental Cl_2^- band, the higher frequency region was searched for overtones. The optical absorption of Cl_2^- hole-type centers tails out above 5000 Å:[64] argon ion excitation falls within the electronic band. Figure 12 shows the regular progression of overtones observed out to 9ν for Cl_2^- in $Li^+Cl_2^-$ using either 4880 or 5145 Å excitation. A similar regular progression of overtones was observed for the $Cs^+Cl_2^-$ species using 4579 Å excitation. The fundamental and overtone bands are nicely fit by the well-known equation, $\nu(v) = v\omega_e - \omega_e x_e(v + 1)(v)$. The regularly decreasing intensity of the overtones with increasing vibrational quantum number suggests that these spectra are due to the resonance Raman effect.[65]

Infrared spectra of lithium-6 and -7 isotopic reactions with Cl_2 in this laboratory produced intense absorptions at 616.9 and 579.5 cm^{-1}, which are due to ^6LiCl and ^7LiCl and bands at 519.6, 511.4, and 375.7 cm^{-1} and 489.2,

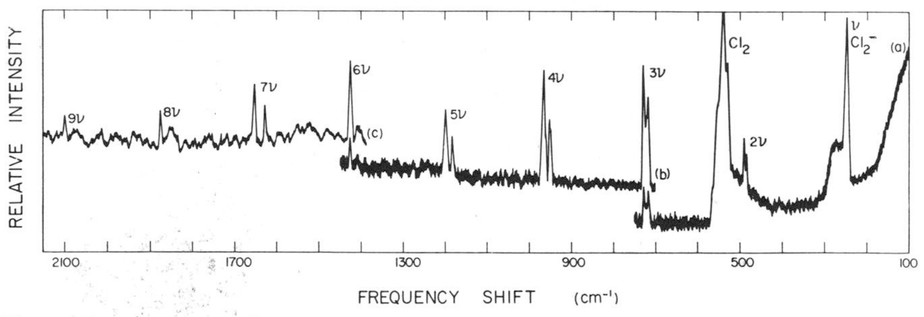

Figure 12. Resonance Raman spectrum of Cl_2^- in the matrix-isolated $Li^+Cl_2^-$ species using approximately 50 mW of 5145 Å excitation (Ar/Cl_2 = 100). Amplification range (*a*) 1 × 10^{-9} A, (*b*) 0.3 × 10^{-9} A, (*c*) 0.1 × 10^{-9} A. (Ref. 65, used with permission.)

481.1, and 352.6 cm^{-1}, which are due, respectively, to $(^6LiCl)_2$ and $(^7LiCl)_2$. The LiCl monomer bands have an isotopic frequency ratio of 0.9394, in excellent agreement with the calculated 0.9372 ratio with the expected small anharmonic deviation: the dimer assignments also produce the correct isotopic frequency ratio. In addition, infrared bands were observed at 501.3 cm^{-1} in the lithium-6 reaction and at 469.6 cm^{-1} with lithium-7, which are appropriate for assignment to the ν_2 interionic mode $Li^+ \leftrightarrow Cl_2^-$. The LiCl monomer and dimer bands and the ν_2 $Li^+Cl_2^-$ band were not observed in the Raman spectrum due to the low polarizability change for interionic modes.

4.5 Matrix Reactions with ClO_2

Chlorine dioxide was first studied in argon matrices by Arkell and Schwager.[66] These workers observed the ν_3 and ν_1 bands of ClO_2 split into quartets at 1100 and 940 cm^{-1}. The bending mode ν_2 was observed at 448 cm^{-1}. In situ photolysis using near ultraviolet radiation effected a photoisomerization of OClO to ClOO, which was identified by fundamental frequencies at 1441, 407, and 373 cm^{-1}. In mixed oxygen isotopic studies, the O—O stretch of ClOO was resolved into a quartet indicating inequivalent oxygen atoms and the C_s structure for ClOO.

Matrix isolated ClO_2 was extensively examined by Chi and Andrews,[67] using argon ion laser excitation. The 4579 Å excitation spectrum produced an intense resonance Raman progression in ν_1 of ClO_2 with regularly decreasing intensities out to $6\nu_1$. Four different progressions of ClO_2 fluorescence bands were also observed.

Tevault, Chi, and Andrews recorded the infrared spectra of alkali-metal-atom–ClO_2 matrix reaction products.[68] The spectra were characterized by intense chlorine isotopic doublets at 820 and 790 cm^{-1} and a band in the 400–440 cm^{-1} spectral region depending upon the alkali reagent. The matrix infrared bands are in excellent agreement with the Raman bands of $NaClO_2$ in solution and solid phases, which confirms the matrix assignments to ClO_2^- in the $M^+ClO_2^-$ species. The matrix reaction involved a simple electron-transfer process resulting in a bound ion pair.

$$M + ClO_2 \rightarrow M^+ClO_2^- \tag{18}$$

4.6 Matrix Reactions with N_2

There has been a lot of recent research activity on metal-atom–dinitrogen matrix reactions with lithium atoms and transition metal atoms. Spiker and coworkers observed sharp bands at 1800 and 1535 cm^{-1} and a broad band at 2300 cm^{-1} following the deposition of lithium atoms into pure nitrogen matrices.[69] The intense 1535 cm^{-1} band is illustrated in Figure 13 along

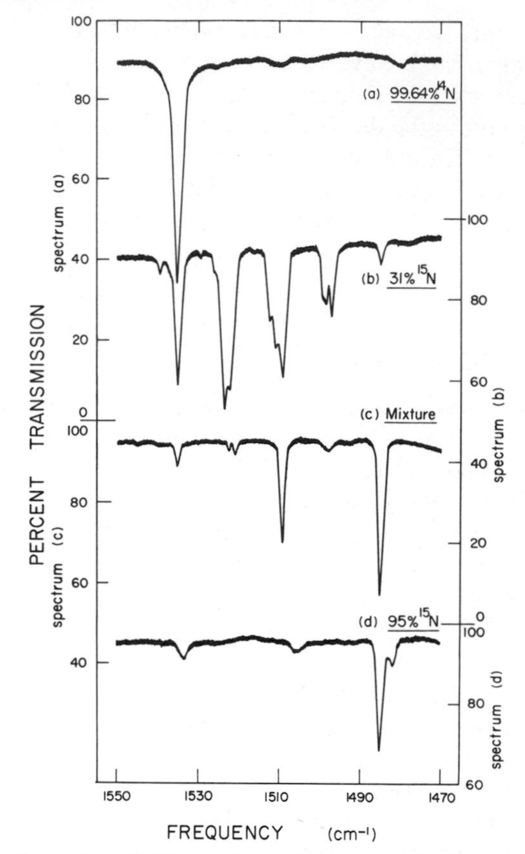

Figure 13. Infrared spectra of lithium-atom–nitrogen matrix cocondensation products. (*a*) Natural isotopic N_2, (*b*) 9% $^{15}N_2$, 43% $^{15}N^{14}N$, 48% $^{14}N_2$, (*c*) 74% $^{15}N_2$, 8% $^{15}N^{14}N$, 18% $^{14}N_2$, (*d*) 90% $^{15}N_2$, 10% $^{15}N^{14}N$. (Ref. 69, used with permission.)

with spectra using nitrogen-15 enriched matrices. With a $^{15}N_2$ matrix the broad band shifted to 2222 cm^{-1}, and the sharp bands shifted respectively to 1740 and 1487 cm^{-1}: these large isotopic shifts indicate pure nitrogen stretching modes.

A mixed $^{14}N_2/^{15}N_2$ matrix sample produced an intense central component at 1509 cm^{-1} in addition to the 1535 and 1487 cm^{-1} bands, indicating the presence of two equivalent N_2 units in this molecular species. Using a scrambled $^{14,15}N_2$ isotopic matrix, the 1537 cm^{-1} band was split into a multiplet of nine components, which showed that the N_2 molecules were bonded "end on" (i.e., inequivalent N atoms) to the lithium reagent. It was suggested that this new species, with the lowest reported N—N frequency, has the structure, N=N(Li_2)N=N. Owing to the proximity of the 1800 cm^{-1}

nitrogen frequency to N_2^- Raman bands in alkali halide lattices,[57] the 1800 cm^{-1} band was assigned to the intraionic (N—N) mode in $Li^+N_2^-$. The broad feature at 2300 cm^{-1} just below the fundamental of diatomic N_2 was attributed to a nitrogen molecular mode perturbed by lithium, $Li_x \cdots N_2$.

The broad 2300 cm^{-1} nitrogen matrix band has also been observed in Ca, Sr, and Ba nitrogen matrix codeposition reactions in this laboratory.[51,52] The Ba—N_2 reaction also produced a broad band at 1770 cm^{-1}, which is probably due to $Ba^+N_2^-$.

4.7 Matrix Reactions with N_2O

Spiker and Andrews have studied alkali metal matrix reactions with nitrous oxide for the purpose of producing alkali metal oxides.[60,61]

$$N_2O + M \rightarrow MO \tag{19}$$

$$MO + M \rightarrow MOM \tag{20}$$

Infrared spectra of ^7Li reactions with N_2O in nitrogen matrices produced sharp, strong bands at 945.5, 804.5, and 700.0 cm^{-1}.[60] These bands showed ^6Li and ^{18}O isotopic shifts appropriate for assignment to ν_3 of the linear molecule Li—O—Li, the Li—O mode of a new unsymmetrical species Li_2O, and the LiO diatomic, respectively. The analogous argon matrix reactions produced only an 875, 871 cm^{-1} doublet that exhibited ^6Li and ^{18}O shifts appropriate for the Li_2O species. The large argon-to-nitrogen matrix shift is expected for ionic alkali species: a similar shift has been found for LiF. No Li—O—Li or LiO was produced in the argon matrix reaction. The nitrogen matrix clearly plays an important role in the oxygen atom abstraction reaction.

Reactions of Na, K, Rb, and Cs with N_2O in nitrogen matrices were examined by Spiker and Andrews.[61] The sodium reaction yielded no products: the rubidium reaction produced a single sharp band at 472.7 cm^{-1} with an ^{18}O counterpart at 448.1 cm^{-1}, which were assigned to ν_3 of Rb—O—Rb. The K and Cs reactions are typified by the cesium study illustrated in Figure 14. The sharp, intense band at 455.2 cm^{-1} shifted to 431.0 cm^{-1} in the $N_2^{18}O$ experiment: the mixed isotopic experiment produced a doublet indicating a single-oxygen atom absorber. Mixed K—Cs atom reactions with N_2O produced an intense triplet at 455, 486, and 502 cm^{-1}; the latter band was prominent in K—N_2O experiments; the central component, which was absent in pure K or pure Cs experiments, was assigned to K—O—Cs. The triplet indicates a two-metal atom absorber, which, therefore, must be the M—O—M species. Bond angle calculations for Cs—O—Cs place the angle lower limit at 125° and suggest that the Cs—O—Cs angle lies in the 130–140° range. Cs—O—Cs is clearly a bent molecule, in contrast to Li—O—Li, which is linear.

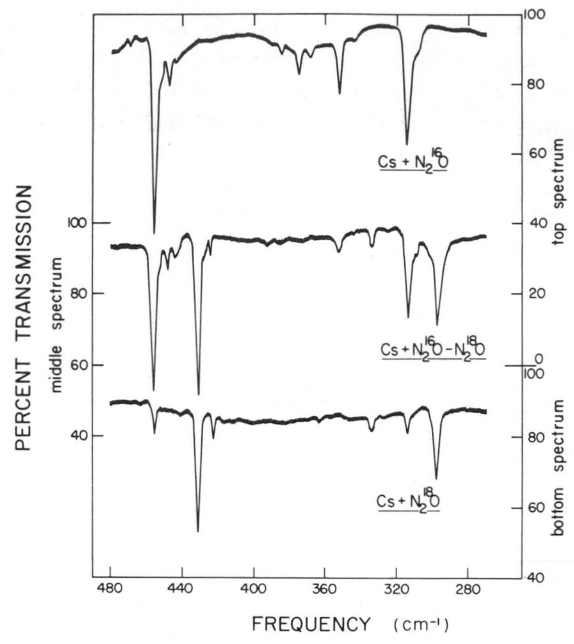

Figure 14. Infrared spectra of nitrogen matrix reactions of cesium atoms with isotopic nitrous oxide molecules (N_2/N_2O = 200/1). 86.5% ^{18}O-enriched $N_2^{18}O$ used in last spectrum. (Ref. 61, used with permission.)

The sharp lower-frequency band in Figure 14 at 314.0 cm^{-1} shows an oxygen-18 shift to 297.3 cm^{-1}. The mixed oxygen isotopic experiment produced a doublet, again denoting a single oxygen atom. The mixed metal experiment showed two bands at 314 and 384 cm^{-1}; the latter band is the potassium counterpart. Accordingly, the 314 cm^{-1} feature was assigned to the diatomic CsO. It is interesting to note that argon matrix counterparts for Cs_2O and CsO were observed from Cs + O_3 reactions[54] at 457 and 322 cm^{-1}, respectively, which showed relatively small nitrogen-to-argon matrix shifts.

The Ba atom–N_2O nitrogen matrix reaction[51] produced an intense 612 cm^{-1} band due to BaO, in agreement with ozone reactions. The Sr—N_2O reaction produced a medium intensity band at 620 cm^{-1} for SrO, also in agreement with the ozone reaction. The calcium-atom–N_2O-nitrogen-matrix reaction produced a weak 707 cm^{-1} band that could have come from reaction with O_2 impurity: the 707 cm^{-1} feature was assigned to CaO from the ozone-nitrogen-matrix reaction with Ca atoms.[52]

Milligan and Jacox[36,57] have studied several argon matrix reactions involving alkali atoms, N_2O, and some other small molecule, such as O_2, where the deposited reagents were photolysed using a medium-pressure mercury arc. When $Ar/O_2/N_2O$ samples were deposited with Cs atoms, an intense 1001 cm^{-1} absorption and two moderately intense bands at 802 and 789 cm^{-1} appeared in the initial spectrum. Upon photolysis of the deposited sample, the 1001 cm^{-1} $Cs^+O_4^-$ band changed little, but the 802 and 789 cm^{-1} absorptions grew to about 0.2 OD (optical density units). The latter two bands were produced directly in reactions of Cs atoms with ozone[54] at 802.4 and 788.8 cm^{-1} with significantly greater intensity (1.1 OD for the 802.4 cm^{-1} band). Clearly, these two bands are due to ν_3 of O_3^- in the $Cs^+O_3^-$ species. The interesting point is the mechanism of formation of the $Cs^+O_3^-$ species in the Jacox and Milligan work.[57] These workers proposed that $M + N_2O$ produced a complex that photolysed to give "O^- which is free to migrate through the argon lattice" and find reaction partners such as O_2.

$$M + N_2O \rightarrow M\cdots N_2O \xrightarrow{h\nu} M^+ + N_2 + O^- \tag{21}$$

They indicated ozonide formation by the simple reaction (22) without any comment on the role or position of the M^+ cation.

$$O^- + O_2 \rightarrow O_3^- \tag{22}$$

We do not believe that reaction mechanism (21) is responsible for the production of ozonide for the following fundamental reason: O^- anion cannot diffuse away from the the M^+ cation responsible for the production of O^-, owing to the strong electrostatic attraction between unlike charged ions. However, some $Cs^+O_3^-$ is produced upon sample deposition so photolysis is not required for the reaction, but photolysis clearly does enhance the yield. We propose that two better mechanisms for this reaction exist: one involves the direct formation of Cs^+O^- from the N_2O reaction and the subsequent reaction of Cs^+O^- with O_2 molecules during sample deposition.

$$N_2O + Cs \rightarrow Cs^+O^- + N_2 \qquad \Delta H \approx -30 \text{ kcal/mole} \tag{23}$$

$$Cs^+O^- + O_2 \rightarrow Cs^+O_3^- \tag{24}$$

Reaction (23) produced a sizable yield of the 322 cm^{-1} CsO band in solid argon:[70] noteworthy is the fact that no NaO could be produced in a similar reaction.[60] This nicely explains the fact that some $Cs^+O_3^-$ but no $Na^+O_3^-$ was observed during sample deposition of matrix samples of N_2O and O_2 with sodium.

Very recent studies of the Ca—O_2 nitrogen matrix reaction[52] provided evidence for the metal oxide mechanism. CaO, produced by reactions (25), reacts further with O_2 to yield $Ca^+O_3^-$ without photolysis [reaction (26)].

$$Ca + O_2 \rightarrow CaO_2;\ CaO_2 + Ca \rightarrow CaO_2Ca \rightarrow 2CaO \tag{25}$$

$$CaO + O_2 \rightarrow Ca^+O_3^- \tag{26}$$

Reaction sequence (25), which is exothermic by approximately 110 kcal/mole, is probably a more productive source of CaO than mechanism (27) suggested by Brewer and Yang,[63] which is exothermic by approximately 20 kcal/mole.

$$Ca + 2O_2 \rightarrow CaO + O_3 \tag{27}$$

The latter mechanism could not be confirmed in our experiments.[52] It is also important to point out that sample warming experiments cause CaO to disappear while $Ca^+O_3^-$ grows strongly. Therefore, reaction (26), which proceeds with little activation energy, is the major mechanism for the production of calcium ozonide in O_2 experiments.

The photolytic mechanism most likely responsible for the ozonide production is the photolysis of N_2O in the presence of a *near neighbor*, $Cs^+O_2^-$, which effects an oxygen atom transfer, reaction (28).

$$N_2O + Cs^+O_2^- \xrightarrow{h\nu} N_2 + Cs^+O_3^- \tag{28}$$

In the reaction of N_2O with Cs atoms in argon matrices in this laboratory, $Cs^+O_2^-$ was observed in the initial sample deposit; the $Cs^+O_2^-$ absorptions grew upon photolysis, suggesting reaction (29).

$$N_2O + Cs^+O^- \xrightarrow{h\nu} N_2 + Cs^+O_2^- \tag{29}$$

It appears that a chain reaction between the several cesium-oxygen species and photolytically produced oxygen atoms is responsible for the observed synthesis.[70]

In similar reactions with $Ar/NO/N_2O/Na$, Na^+NO^- was present in the initial sample deposit.[36,70] Upon mercury-arc irradiation of the sample, the Na^+NO^- bands[38] decreased in intensity, while bands due to $Na^+NO_2^-$ appeared.[36] This observation, and its confirmation using the analogous reaction in this laboratory,[70] support the O abstraction from N_2O mechanism by Na^+NO^- in a nearby matrix site.

$$Na^+NO^- + ON_2 \xrightarrow{h\nu} Na^+NO_2^- + N_2 \tag{30}$$

The role of the alkali metal is in forming an oxyanion that abstracts an oxygen atom from N_2O with the aid of photolysis, forming a more stable oxyanion (NO_2^- in $Na^+NO_2^-$, as compared to NO^- in Na^+NO^-).

4.8 Matrix Reactions with NO_2

The nitrite anion was first observed by Milligan and coworkers[71] following vacuum-ultraviolet photolysis, electron bombardment, and alkali metal matrix reactions with NO_2. The three techniques produced an identical 1244 cm^{-1} band with a 1247 cm^{-1} splitting; however, the alkali reactions produced the 1244 cm^{-1} band with a threefold increase in intensity along with additional bands in the 1200–1230 cm^{-1} region. On the basis of the observation of the 1244 cm^{-1} band unshifted between electron bombardment, photoionization, and alkali-metal-atom electron sources, these workers assigned the 1244 cm^{-1} band to NO_2^- surrounded by argon atoms, in spite of the fact that the 1244 cm^{-1} signal was destroyed by photolysis in electron impact and photoionization experiments, but not in the alkali metal reaction experiments. In a subsequent more-detailed alkali-metal-reaction study, Milligan and Jacox[36] characterized the 1244 cm^{-1} band as isolated NO_2^- (completely surrounded by argon atoms) and the 1200–1230 cm^{-1} bands as $M_x^+NO_2^-$ (NO_2^- perturbed by M^+ or M_2^+). These workers explained the formation of isolated NO_2^- in alkali metal experiments by "gas phase charge transfer interaction during deposition and separation of the ion-pair before the products are trapped in the solid argon lattice."

Following additional experiments in this laboratory,[39] we concluded that in alkali-metal–NO_2 matrix reactions, the 1244 cm^{-1} band is also due to $M^+NO_2^-$. It happens that NO_2^- is a very stable anion, and that the M^+ perturbation on the 1244 cm^{-1} band is not observable, perhaps due to a different $M^+NO_2^-$ structural arrangement such as a pyramidal species with M^+ above the NO_2^- plane. The $Ca^+O_3^-$, $Sr^+O_3^-$, and $Ba^+O_3^-$ species produced ν_3 ozonide bands at 804.4 $\pm$ 0.2 cm^{-1}; the lack of a cation effect on the anion vibration does not require the absence of the cation.[51]

On the other hand, energetics do require the presence of the cation, coulombically bonded to the anion, after charge transfer between the alkali atom and NO_2. All of the experimental observations[36] can, we believe, be best explained by assigning the 1244 cm^{-1} feature in alkali metal matrix reactions to $M^+NO_2^-$ and the 1200–1230 cm^{-1} bands to different matrix sites or structural isomers of $M^+NO_2^-$ and/or to $M_2^+NO_2^-$.

An important point must be made here. Matrix reactions of alkali metal atoms with small molecules that produce anions also produce alkali cations, and these two ions must be coulombically bound together. Energetics require the intimate association of anion and cation. Since the metal ionization potential exceeds the electron affinity of the small molecule, coulombic attraction is necessary for an exothermic process. The curve crossing point, r_c,[3] can be calculated for a given alkali metal-molecule reaction. The significance for matrix reactions is that once the alkali atom and reactive molecule come within r_c and electron transfer—cation and anion formation—takes place, the ion pair *must* be drawn together owing to the very strong coulombic attractive force between oppositely charged ions. It is, therefore, not possible for isolated anions to be produced by alkali metal matrix reactions. Once the ion pair is formed, it cannot diffuse apart: it must be drawn together, resulting in a cation-anion pair.

ACKNOWLEDGMENTS

The author gratefully acknowledges research support from the National Science Foundation and an Alfred P. Sloan Research Fellowship during the writing of this chapter.

REFERENCES

1. H. Beutler, S. V. Bogdandy, and M. Polanyi, *Naturwissenschaften* **13**, 711 (1925); ibid., **14**, 164 (1926).
2. E. W. R. Steacie, *Atomic and Free Radical Reactions*, 2nd Edit., Reinhold, New York, 1954.
3. D. R. Herschbach, *Appl. Opt., Suppl. 2, Chem. Lasers*, 128 (1965).
4. W. L. S. Andrews and G. C. Pimentel, *J. Chem. Phys.*, **44**, 2361 (1966).
5. L. Andrews, *J. Chem. Phys.*, **48**, 972 (1968).
6. L. Andrews, *J. Chem. Phys.*, **50**, 4288 (1969).
7. L. Andrews, *Annu. Rev. Phys. Chem.*, **22**, 109 (1971).
8. L. Andrews, "Infrared and raman spectra of unique matrix-isolated molecules," in *Vibrational Spectra and Structure*, (J. R. Durig, Ed.), Elsevier Scientific Publishing Co., Amsterdam, 1975.
9. G. A. Ozin and A. Vander Voet, *Progr. Inorg. Chem.*, **19**, 105 (1974).
10. L. Andrews and G. C. Pimentel, *J. Chem. Phys.*, **47**, 3637 (1967).
11. D. E. Milligan and M. E. Jacox, *J. Chem. Phys.*, **47**, 5146 (1967).
12. L. Y. Tan and G. C. Pimentel, *J. Chem. Phys.*, **48**, 5202 (1968).
13. L. Andrews, *J. Phys. Chem.*, **71**, 2761 (1967).
14. E. E. Rogers, S. Abramowitz, D. E. Milligan, and M. E. Jacox, *J. Chem. Phys.*, **52**, 2198 (1970).
15. L. Andrews, R. O. Allen, and J. M. Grzybowski, *J. Chem. Phys.*, **61**, 2156 (1974); *J. Phys. Chem.* **79**, 904 (1975).
16. D. A. Hatzenbühler, L. Andrews, and F. A. Carey, *J. Amer. Chem. Soc.*, **97**, 187 (1975).
17. A. K. Maltsev, R. H. Hauge, and J. L. Margrave, *J. Phys. Chem.*, **75**, 3984 (1971).
18. L. Andrews, *J. Chem. Phys.*, **48**, 979 (1968).
19. L. Andrews and T. G. Carver, *J. Chem. Phys.*, **49**, 896 (1968).
20. L. Andrews and T. G. Carver, *J. Phys. Chem.*, **72**, 1743 (1968).
21. F. X. Powell and D. R. Lide, Jr., *J. Chem. Phys.*, **45**, 1067 (1966).
22. D. W. Smith and L. Andrews, *J. Phys. Chem.*, **76**, 2718 (1972).
23. T. G. Carver and L. Andrews, *J. Chem. Phys.*, **50**, 5100 (1969).
24. T. G. Carver and L. Andrews, *J. Chem. Phys.*, **50**, 4235 (1969).
25. T. G. Carver and L. Andrews, *J. Chem. Phys.*, **50**, 4223 (1969).
26. J. I. Raymond and L. Andrews, *J. Phys. Chem.*, **75**, 3235 (1971).
27. L. Andrews and D. W. Smith, *J. Chem. Phys.*, **53**, 2956 (1970).
28. D. W. Smith and L. Andrews, *J. Chem. Phys.*, **55**, 5295 (1971).
29. D. W. Smith and L. Andrews, *J. Chem. Phys.*, **58**, 5222 (1973).
30. A. Arkell, R. R. Reinhard, and L. P. Larson, *J. Amer. Chem. Soc.*, **87**, 1016 (1965).
31. L. Andrews and J. I. Raymond, *J. Chem. Phys.*, **55**, 3078 (1971).

32. L. Andrews, *J. Chem. Phys.*, **57**, 51 (1972).
33. L. Andrews and J. I. Raymond, *J. Chem. Phys.*, **55**, 3087 (1971).
34. F. K. Chi and L. Andrews, *J. Phys. Chem.*, **77**, 3062 (1973).
35. M. Basco and R. D. Morse, *J. Mol. Spectrosc.*, **45**, 35 (1973).
36. D. E. Milligan and M. E. Jacox, *J. Chem. Phys.*, **55**, 3404 (1971).
37. D. E. Tevault and L. Andrews, *J. Phys. Chem.*, **77**, 1640 (1973).
38. D. E. Tevault and L. Andrews, *J. Phys. Chem.*, **77**, 1646 (1973).
39. D. E. Tevault and L. Andrews, unpublished results, 1974.
40. L. Andrews and R. R. Smardzewski, *J. Chem. Phys.*, **58**, 2258 (1973).
41. D. W. Smith and L. Andrews, *J. Chem. Phys.*, **60**, 81 (1974).
42. L. Andrews, *J. Phys. Chem.*, **73**, 3922 (1969).
43. L. Andrews, *J. Chem. Phys.*, **54**, 4935 (1971).
44. L. Andrews, J. -T. Hwang, and C. Trindle, *J. Phys. Chem.*, **77**, 1065 (1973).
45. R. R. Smardzewski and L. Andrews, *J. Chem. Phys.*, **57**, 1327 (1972).
46. R. R. Smardzewski and L. Andrews, *J. Phys. Chem.*, **77**, 801 (1973).
47. H. Huber and G. A. Ozin, *J. Mol. Spectrosc.*, **41**, 595 (1972).
48. J. C. Evans, *Chem. Commun.*, 682 (1969).
49. M. E. Jacox and D. E. Milligan, *Chem. Phys. Lett.*, **14**, 518 (1972).
50. S. Abramowitz and N. Acquista, *J. Res. Nat. Bur. Stand.*, *A* **75**, 23 (1971).
51. D. M. Thomas and L. Andrews, *J. Mol. Spectrosc.* **49**, 220 (1974).
52. B. S. Ault and L. Andrews, *J. Chem. Phys.* **62**, 2312, 2320 (1975).
53. L. Andrews and R. C. Spiker, Jr., *J. Phys. Chem.*, **76**, 3208 (1972).
54. R. C. Spiker, Jr., and L. Andrews, *J. Chem. Phys.*, **59**, 1851 (1973).
55. L. Andrews, *J. Amer. Chem. Soc.*, **95**, 4487 (1973).
56. L. Andrews and R. C. Spiker, Jr., *J. Chem. Phys.*, **59**, 1863 (1973).
57. M. E. Jacox and D. E. Milligan, *J. Mol. Spectrosc.*, **43**, 148 (1972).
58. J. B. Bates, M. H. Brooker, and G. E. Boyd, *Chem. Phys. Lett.*, **16**, 391 (1972).
59. M. J. Linevsky, Technical Report RADC-TR-70-212, General Electric Co., 1970.
60. R. C. Spiker, Jr. and L. Andrews, *J. Chem. Phys.*, **58**, 702 (1973).
61. R. C. Spiker, Jr. and L. Andrews, *J. Chem. Phys.*, **58**, 713 (1973).
62. B. Rosen, *Spectroscopic Data Relative to Diatomic Molecules*, Pergamon, New York, 1970.
63. L. Brewer and J. L. -F. Wang, *J. Chem. Phys.*, **56**, 4305 (1972).
64. M. Hass and D. L. Griscom, *J. Chem. Phys.*, **51**, 5185 (1969).
65. W. F. Howard, Jr., and L. Andrews, *J. Amer. Chem. Soc.*, **95**, 2056 (1973); *Inorg. Chem.*, **14**, 767 (1975).
66. A. Arkell and I. Schwager, *J. Amer. Chem. Soc.*, **89**, 5999 (1967).
67. F. K. Chi and L. Andrews, *J. Mol. Spectrosc.*, **52**, 82 (1974).
68. D. E. Tevault, F. K. Chi, and L. Andrews, *J. Mol. Spectrosc.*, **51**, 450 (1974).
69. R. C. Spiker, Jr., L. Andrews, and C. Trindle, *J. Amer. Chem. Soc.*, **94**, 2401 (1972).
70. L. Andrews and D. E. Tevault, *J. Mol. Spectrosc.*, **55**, 452 (1975).
71. D. E. Milligan, M. E. Jacox, and W. A. Guillory, *J. Chem. Phys.*, **52**, 3864 (1970).

Matrix-Isolation Studies Involving Main Group Metals

7

J. S. Ogden

1. INTRODUCTION

In this chapter we are primarily concerned with describing some of the ir and Raman studies that have been carried out involving the remaining main group metals, and this survey can be regarded as a sequel to the alkali metal systems described earlier. The greater part of the discussion is concerned with traditionally metallic elements, such as tin or aluminum, but there are certain advantages to be gained by referring to related matrix work on metalloids such as germanium.

Infrared spectroscopy continues to be the most popular method of investigation, but the growing importance of Raman spectroscopy is self-evident, particularly for the heavier elements.

The majority of papers published in this field fall into two distinct classes that may conveniently be described as (a) studies on high-temperature molecules trapped in inert matrices and (b) studies on metal atom cocondensation reactions, and the kind of apparatus required for this work is identical to that described earlier. However, before discussing the information obtained from the numerous systems that have now been studied,[1] it is convenient to summarize the ways in which vibrational spectroscopy may be used, first, to establish the identity of a particular species and, second, to obtain structural information.

One of the principal advantages of matrix isolation is that it allows one to resolve vibrational transitions separated by as little as 2 or 3 cm^{-1}. This means that one is in a position to resolve isotope fine structure for several main group metals such as Mg, Ca, Sr, Ga, or Sn; and the combination of this data with the larger frequency shifts associated with ^{18}O or ^{34}S enrichment turns out to be extremely useful. If all the atoms in a particular molecule can undergo isotopic substitution, it is sometimes possible to identify the molecule unequivocally simply from isotope intensity patterns, without necessarily performing any numerical calculations. The most obvious examples here are diatomics such as GeO or SnS, but the method may be extended to triatomics without too much complexity. Figure 1 shows part of the ir spectrum obtained[2] when tin atoms are deposited with a krypton matrix containing $^{16}O_2$, $^{16}O^{18}O$, and $^{18}O_2$ in the ratio, 1:2:1. The basic triplet indicates that two equivalent atoms of oxygen are involved in the vibration, and the fine structure on each of these bands reflects the natural isotopic abundance of tin. This particular vibration is, therefore, assigned to molecular SnO_2, and only two point groups are possible, $D_{\infty h}$ or C_{2v}. This method of characterization requires that all the atoms be involved in the vibration, and it is preferable that the mode in question should be singly degenerate. Other molecules that may be identified in this way include $MgCl_2$ and $ZnCl_2$.

Once the stoichiometry is established, the problems of molecular symmetry and geometry can be handled by traditional methods such as ir or Raman selection rules, and isotope product or sum rule calculations. However, in the case of matrix-isolation studies, it is not always easy to establish the number of vibrational fundamentals associated with a particular species, and confusion between monomer and dimer absorptions has led to a number of incorrect band assignments. Secondly, one has the problems associated with site effects, and finally, matrix isolation studies rarely provide information on the anharmonicity of molecular vibrations. Any form of normal coordinate analysis that attempts to extract, for example, bond angles from frequency data, must take account of this. One way of minimizing this particular problem for triatomics, XY_2, is to use double-isotopic substitution. This procedure was originally employed[3] by Mann et al. to determine the apex angle in SO_2 from ^{18}O and ^{34}S isotope shifts, and has since been applied to a considerable number of main group metal oxides and halides. For most

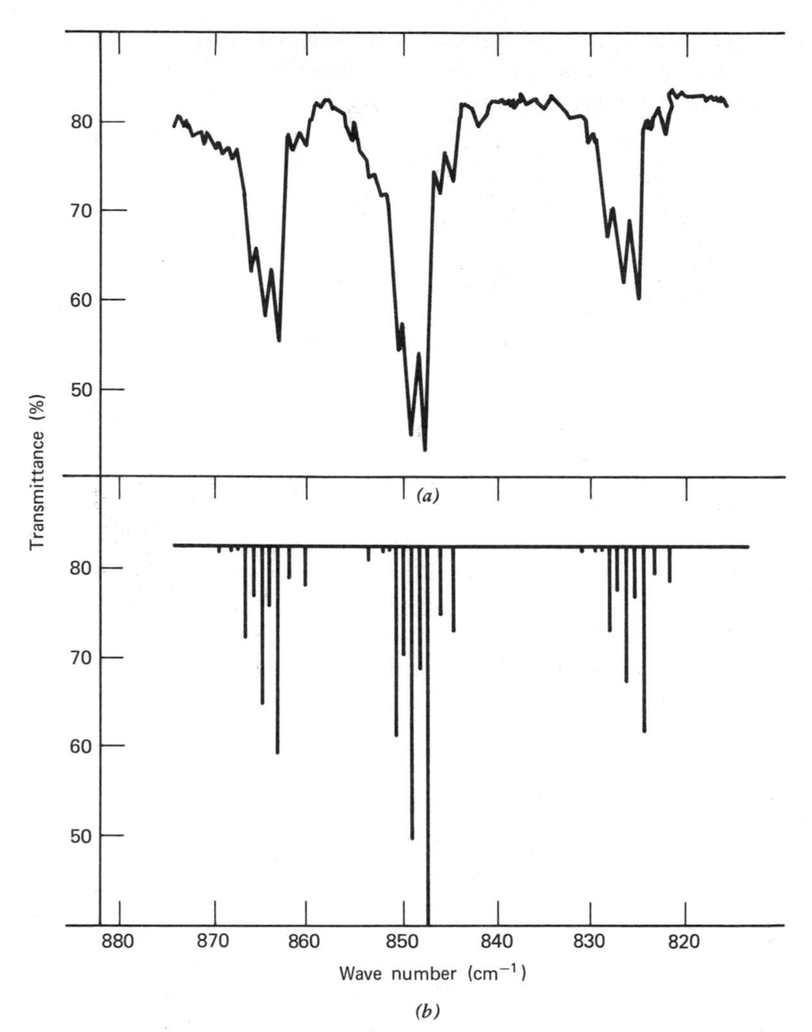

Figure 1. Identification of molecular SnO_2 formed in the matrix reaction of tin atoms with a mixture of $^{16}O_2$, $^{16}O^{18}O$, and $^{18}O_2$. (*a*) Infrared spectrum in the region of the antisymmetric stretching mode showing the effect of tin and oxygen isotopes, (*b*) calculated spectrum assuming a $D_{\infty h}$ structure. Reprinted with permission from A. Bos and J. S. Ogden, *J. Phys. Chem.*, **77**, 1513 (1973). Copyright by the American Chemical Society.

triatomics, the antisymmetric stretching mode is the most intense ir absorption, and for a general triatomic XY_2 with angle YXY $= 2\theta$, a value of $\sin\theta$ is obtainable directly from the equation

$$\sin^2\theta = \frac{M_X M_X^i[M_Y(\omega_3)^2 - M_Y^i(\omega_3^i)^2]}{2M_Y M_Y^i[M_X^i(\omega_3^i)^2 - M_X(\omega_3)^2]} \tag{1}$$

if one uses frequency data from two different isotopic molecules. Mann et al.[3] showed that if the anharmonicity constant is positive and follows the Darling–Dennison relationship, then the value of $\sin\theta$ obtained from the experimental frequencies is less than the true value if isotopic substitution takes place at atom X, but that an upper limit for $\sin\theta$ is obtained as a result of substitution at Y. When this procedure is applied to SnO_2,[2] the upper and lower limits found for $\sin\theta$ are 1.02 and $\sim$1.000. This clearly indicates a linear $D_{\infty h}$ geometry, and the extent to which this strategy has been applied may be gauged from the bond angle data listed in Tables 1–4. Where isotope data indicates a C_{2v} geometry, the direct observation of an ir active symmetric stretch provides additional confirmation of a bent structure.

The methods used to obtain structural data for other molecular geometries, such as the planar four-membered D_{2h} ring, usually rely on the validity of secular equations derived from force fields in which some or all of the interaction constants are neglected. In particular, the two equations[4]

$$^{\lambda}B_{2u} = 2\left(\frac{1}{M_X} + \frac{1}{M_Y}\right)\left(k_r\cos^2\alpha + 2\frac{k_\gamma}{r^2}\sin^2\alpha\right) \tag{2}$$

$$^{\lambda}B_{3u} = 2\left(\frac{1}{M_X} + \frac{1}{M_Y}\right)\left(k_r\sin^2\alpha + 2\frac{k_\gamma}{r^2}\cos^2\alpha\right) \tag{3}$$

for the ir active stretching modes of the X_2Y_2 D_{2h} ring have been used to estimate the angle at atom X (2α) and the stretching constant, k_r. The bending constant, k_γ, is assumed to be small. It is difficult to estimate how serious these approximations are, but the data now available for these ring structures is probably sufficiently accurate to form the basis for simple bonding models.

Finally, it must be acknowledged that some prior information on the nature of the vapor species is an important guide as to what may ultimately be observed in the matrix. This data is most easily obtainable from high-temperature mass spectrometry,[5] but it is often misleading to place too much reliance upon the relative proportions of species present in the vapor. Extensive chemical reaction can take place during matrix deposition, and although this usually has the effect of increasing the proportion of known polymers at the expense of monomer species, it may sometimes result in entirely novel products, which could arise, for example, from atom-molecule reactions. It has been suggested[6] that the confusion over the interpretation of the aluminum/oxygen system may have arisen in this way.

For reasons of space, it is not possible to mention all the matrix-isolation studies that have been reported in this field, and in the discussion that follows, attention is concentrated on the identification of molecular species produced either from the condensation of high-temperature vapors or from metal atom reactions that take place during cocondensation.

2. MATRIX-ISOLATION STUDIES ON GROUP II ELEMENTS

Matrix isolation studies involving Be, Mg, Ca, Sr, Ba, Zn, Cd and Hg have been primarily concerned with the vaporization and condensation of the dihalide species. The combined techniques of ir spectroscopy, electron diffraction, and electric deflection have now provided definitive experimental data on the geometries of many of these molecules.[7] At the same time, there is a growing interest in metal atom cocondensations involving these elements, and studies reported to date involving oxygen and hydrogen indicate that in addition to obtaining structural information, matrix isolation is a particularly useful technique for studying the initial products of metal atom reactions.

2.1 Dihalides of Groups IIA and IIB

The elements, Be–Ba and Zn–Hg, both form the complete range of anhydrous dihalides, and on vaporization, mass spectrometric studies[5] have shown that one may expect the monomer species, MX_2, and smaller amounts of dimers, M_2X_4, as discrete molecules. There seems little doubt that the Group IIB dihalides are linear $D_{\infty h}$ both in the gas phase and in low-temperature matrices, and, in particular, Schnepp et al.[8] have examined all the halides of Zn, Cd, and Hg in krypton matrices over the spectral range, 800–35 cm^{-1}. Double-oven vaporization was used to distinguish between monomer and dimer species, and their results are summarized in Table 1. Evidence for linear geometry in these studies comes both from isotope data and from the nonobservation of the totally symmetric stretching mode. Resolvable isotope shifts are reported for the halides of zinc and cadmium, and for all the dichlorides. Metal isotope data alone would be expected to yield a lower limit for the apex angle, and anharmonicity effects may be responsible for the nonlinear geometry ($2\theta \simeq 165°$) deduced for ZnF_2 in an independent study by Margrave.[9] Where metal isotope shifts indicate a rather wide angle, and where there is no positive evidence for the symmetric stretch in the ir, one cannot dismiss the possibility that the species may be linear. Molecule $ZnCl_2$ provides an excellent example of double-isotope labeling, and this is shown in Figure 2.

Table 1. Vibration Frequencies (cm^{-1}) and Derived Bond Angles for Matrix-Isolated Group II Metal Dihalides

Species	Matrix	ν_1	ν_2	ν_3	Angle $\widehat{XMX}$	Ref.
BeF_2	(Ne)	(680)	345	1555	180°	53, 54
$BeCl_2$	(Ne)	(390)	250	1135	180°	53, 54
$BeBr_2$	(Ne)	(230)	220	1010	180°	54
BeI_2	(Ne)	(160)	(175)	873	180°	54
$^{24}MgF_2$	(Ar)	(552)	247	851	180°	12
$^{24}MgCl_2$	(Kr)	(298.2)	87.7	590.0	180°	10
$^{40}CaF_2$	(Kr)	484.75	163.36	553.66	140°	11
$^{40}CaCl_2$	(Kr)	(242.6)	63.6	402.3	180°	10
$^{88}SrF_2$	(Kr)	440.95	82.0	442.2	108°	11
$^{88}SrCl_2$	(Kr)	269.3	43.7	299.5	120°	10
	(Ar)	275.0	—	308.0	130°	55
BaF_2	(Kr)	413.22	(64)	389.58	100°	11
$BaCl_2$	(Kr)	255.2	(36)	260.0	100°	10
	(Ne)	262.0	62	268.0	120°	55
$^{64}ZnF_2$	(Kr)	(600)	150.5	758.0	180°	8
	(Ar)	—	(157)	763.2	160–170°	9
$^{64}ZnCl_2$	(Kr)	(351)	102	508.5	180°	8
$^{64}ZnBr_2$	(Kr)	(217)	74	404.0	180°	8
$^{64}ZnI_2$	(Kr)	(155)	62	346.0	180°	8
CdF_2	(Kr)	(572)	123.0	662.0	180°	8
$CdCl_2$	(Kr)	(327)	88.5	419.0	180°	8
$CdBr_2$	(Kr)	(205)	62.0	319.0	180°	8
CdI_2	(Kr)	(149)	(50.0)	269.5	180°	8
HgF_2	(Kr)	(588)	171.5	641.5	180°	8
$HgCl_2$	(Kr)	(348)	107.0	407.0	180°	8
$HgBr_2$	(Kr)	(219)	73.0	294.0	180°	8
HgI_2	(Kr)	(158)	63.0	237.5	180°	8

The position regarding the dihalides of Group IIA is interesting. All the halides of beryllium appear to be linear, while those of barium have all been shown to be bent.[7] For the intermediate elements, the pattern that seems to be emerging is that the fluorides of calcium and strontium are bent, but linear structures become increasingly favored as the halogen mass increases. Thus recent studies by White et al.[10] have established that $MgCl_2$ and $CaCl_2$ are linear, but that $SrCl_2$ is bent. The situation regarding MgF_2 is that gas-phase electric deflection experiments indicate a linear structure; and although this conclusion was countered by a matrix study claiming a C_{2v}

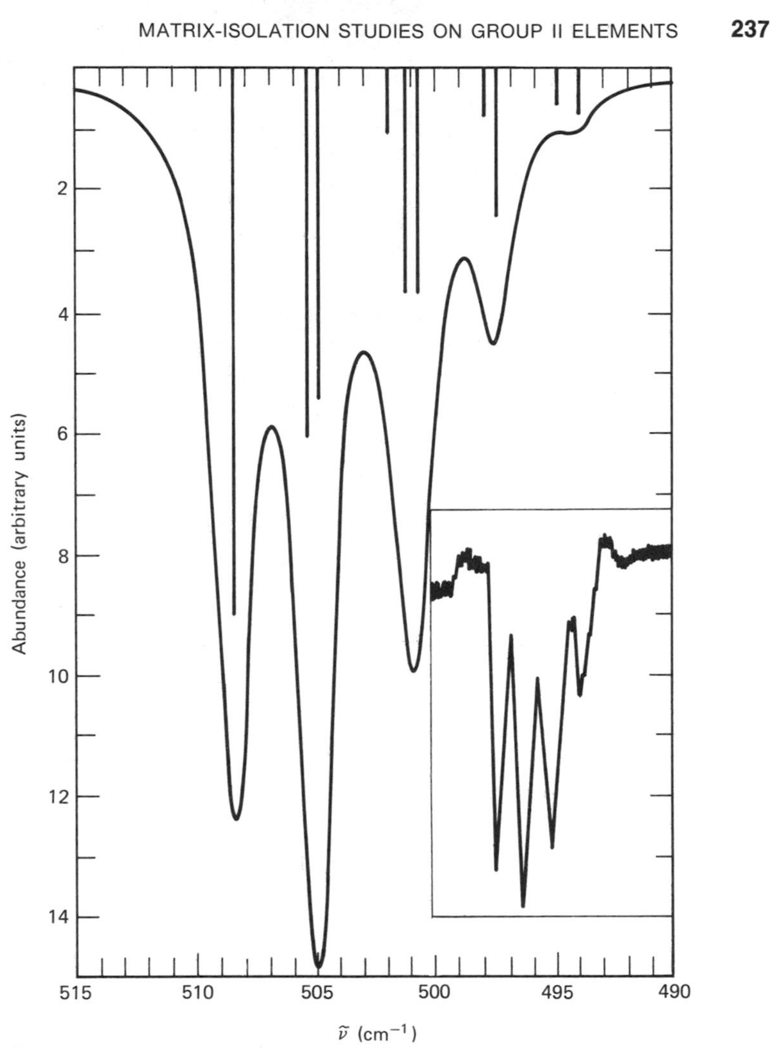

Figure 2. Synthetic spectrum of ν_3 $ZnCl_2$ calculated assuming a linear geometry and a natural abundance of Zn and Cl isotopes. Insert: observed spectrum (different frequency scale). Reproduced from A. Loewenschuss, A. Ron, and O. Schnepp, *J. Chem. Phys.*, **49**, 272 (1968).

geometry (with angle $\widehat{FMgF} \sim 160°$),[11] a recent paper by Margrave[12] demonstrates that some of the earlier band assignments were incorrect, and that MgF_2 should also be considered linear in a low-temperature matrix.

The dihalides of this group provide an interesting study in valence theory, as some are linear and others bent. Our present knowledge is summarized below, where l = linear and b = bent.

	F	Cl	Br	I
Be	1	1	1	1
Mg	1	1	1	1
Ca	b	1	1	1
Sr	b	b	1	1
Ba	b	b	b	b

Coulson[62] has discussed these results using first, an electrostatic model, involving coulomb and polarization terms, and second, a covalent model which explores the possibility of significant *sd* hybridization. The electrostatic model is found to be unsatisfactory in its predictions of molecular shape, but Coulson points out that if pure *sd* hybrids were formed, one would anticipate a 90° geometry,[63] in which each *sd* hybrid overlaps with a $p\sigma$ orbital on the halogen.

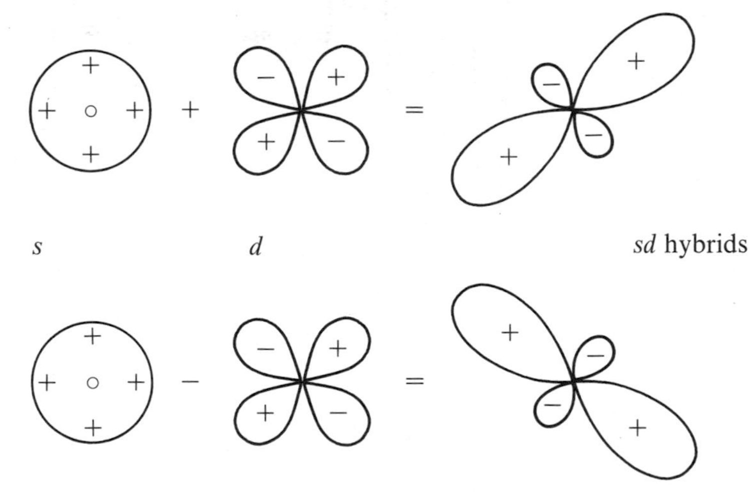

The importance of this type of hybridization in comparison with more familiar linear *sp* hybrids may be inferred from the relative promotion energies $ns^2 \rightarrow ns^1np^1$, $ns^2 \rightarrow ns^1nd^1$, and $ns^2 \rightarrow ns^1(n-1)d^1$, where n is the principal quantum number. Experimentally, it is found that the promotion energies $ns^2 \rightarrow ns^1np^1$ are lower than $ns^2 \rightarrow ns^1nd^1$ for all Group II atoms, but for the elements, Ca, Sr, and Ba, where the promotion $ns^2 \rightarrow ns^1(n-1)d^1$ provides an alternative route to *sd* hybrids, there is a distinct probability that nonlinear geometries will result, particularly if the metal atom carries an effective positive charge. The table of promotion energies below shows that this is particularly favorable for the barium halides and satisfactorily accounts for the pattern of nonlinear geometries found for these Group IIA halides.

	$ns \rightarrow (n-1)d$ (cm^{-1})	$ns \rightarrow np$ (cm^{-1})
Ca	20,727	17,334
Sr	18,712	16,354
Ba	9,811	14,119
Ca^+	13,681	25,303
Sr^+	14,696	24,116
Ba^+	5,274	21,107

For the remaining elements in Group II (Be, Mg, Zn, Cd, Hg) the promotion $ns^2 \rightarrow ns^1(n-1)d^1$ is not possible, either because the $(n-1)d$ orbitals are not available, or because they are already filled. Since $ns^2 \rightarrow ns^1np^1$ promotion is favored over $ns^2 \rightarrow ns^1nd^1$, linear structures are to be expected for the dihalides of all these metals, and no exceptions have yet been found.

From an experimental point of view, it is clear that one of the principal difficulties has been the occurrence of polymer species, and for the majority of systems it has been essential to use both single- and double-oven vaporizations and to carry out controlled-diffusion experiments. The dimer species, such as Mg_2Cl_4 and Mg_2Br_4, that have been identified in this way[13] are believed to possess the D_{2h} structure

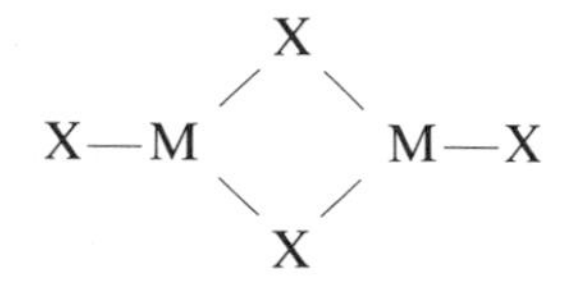

A series of experiments has also been carried out in which the solid difluorides of Mg, Ca, Sr, and Ba were mixed with either boron or aluminum, and the products of vaporization were examined using epr spectroscopy.[14] Hyperfine coupling constants and g values were obtained for MgF, CaF, SrF, and BaF.

2.2 Cocondensation Reactions of Group II Metal Atoms

Matrix-isolation cocondensations involving Group II metal atoms have so far been confined to reactions with hydrogen, oxygen, ozone, and nitrous oxide, and two of the earliest papers in this field are concerned with the characterization of the diatomic hydrides, MH (M = Mg, Ca, Sr, Ba, Zn, Cd, Hg). These species were generated in argon matrices at 4°K by cocondensing the metal vapors with atomic hydrogen, and were detected by epr spectroscopy.[15,16]

The first ir study was by Abramowitz and Acquista,[17] who assigned bands at 570 and 1066 cm^{-1} to molecular BaO_2 formed as a reaction product in the cocondensation of barium atoms in an argon matrix. This molecule was shown to have a C_{2v} ring structure on the basis of ^{18}O-enrichment studies, and similar experiments have since been reported by Brewer and Wang,[18] Spoliti et al.,[19] and Thomas and Andrews.[20] These initial experiments were concerned with generating metal oxide molecules in low-temperature matrices in a variety of ways and may conveniently be discussed together.

Magnesium

Spoliti et al.[19] have cocondensed magnesium vapor with pure oxygen and with oxygen/ozone mixtures at 20°K and report that there is no reaction product. However, similar experiments using discharged oxygen (containing O_2, O_3, and O atoms) produced a number of new features in addition to the ir bands of matrix-isolated O_3. In krypton matrices, these new bands were observed at 839.5, 819.5, and 592 cm^{-1}, and the corresponding argon matrix frequencies are 844, 823, and 592 cm^{-1}. A constant intensity ratio was observed for the lower pair of bands, and they are tentatively assigned to a D_{3h} cyclic molecule, Mg_3O_3, on the basis of ^{18}O enrichment. The third band was attributed to a matrix effect.

Thomas and Andrews[20] carried out a similar study in which magnesium was cocondensed with ozone in an argon matrix, and they report three very weak features at 824.6, 799.6, and 795.7 cm^{-1}. The doublet at 799.6/795.7 cm^{-1} was assigned to the ozonide ion, O_3^- (see below), and it seems reasonable to assume that the 824.6 cm^{-1} band arises from the same molecule as the 823 cm^{-1} feature observed by Spoliti.

Calcium

The initial paper on the Ca + O_2 cocondensation reaction was by Brewer and Wang,[18] who identified ozone as one of the reaction products, and assigned a prominent ir band at 794 cm^{-1} to diatomic CaO. This system was reinvestigated briefly by Thomas and Andrews,[20] who reported a broad feature at ~800 cm^{-1} and weaker bands at 637, 594, 582, and 575 cm^{-1}. The 800 cm^{-1} band appears to correspond to the earlier 794 cm^{-1} feature, but these authors suggest that it is in fact due to O_3^- in $Ca^+O_3^-$. Support for this assignment comes from related experiments on the cocondensation of calcium atoms with ozone in an argon matrix, where a strong band at 804.5 cm^{-1} is shown conclusively to be the antisymmetric stretching mode of the C_{2v} ion, O_3^-. Several additional bands were also observed in the Ca/O_3 system, and these are assigned to CaO_2 (593.0 and 582.2 cm^{-1}) and Ca_2O_2 (635.7 and 575.5 cm^{-1}) on the basis of ^{18}O frequency shifts. However, the bands in this lower frequency region were badly overlapped in the ^{18}O work, and these assignments should be regarded as tentative.

Strontium and Barium

The Sr/O_3 and Ba/O_3 systems were also studied by Thomas and Andrews,[20] and the principal reaction product in the cocondensation of Sr atoms with O_3/Ar was shown to be $Sr^+O_3^-$. The ozonide ion was also prominent in the corresponding $Ba—O_3$ matrix reaction, but in this system, additional features were also observed at 634.3 cm^{-1}, 571.3, and 459.6 cm^{-1}, which were assigned to BaO, BaO_2, and Ba_2O_2, respectively. The band at 571.3 cm^{-1} is very close to that obtained previously by Abramowitz and Acquista,[17] and the identification of BaO and Ba_2O_2 is consistent with earlier work by Linevsky on the matrix isolation of molecular barium oxides and is supported by ^{18}O frequency shifts. BaO is also produced in the cocondensation reaction between barium atoms and nitrous oxide.

Table 2 summarizes the results so far obtained from M/O_2 and M/O_3 cocondensation reactions (M = Mg, Ca, Sr, Ba). Although many of the assignments are rather tentative at the present time, one may anticipate that our understanding of these systems will increase as more experiments are carried out and, particularly, if metal isotope data becomes available.

Table 2. Vibration Frequencies and Band Assignments for Some Matrix-Isolated Group II Metal Oxides

Species	Matrix	Vibration Frequencies (cm^{-1})	Ref.
Mg_3O_3	(Ar)	823, 592	19
CaO_2	(Ar)	593.0, 582.2	20
Ca_2O_2	(Ar)	635.7, 575.5	20
BaO	(Ar)	634.3	20
BaO_2	(Ar)	570, 1066	17
Ba_2O_2	(Ar)	459.6	20
$Mg^+O_3^-$(?)	(Ar)	799.6/795.7	20
$Ca^+O_3^-$	(Ar)	804.5	20
$Sr^+O_3^-$	(Ar)	804.2	20
$Ba^+O_3^-$	(Ar)	804.3	20

3. MATRIX-ISOLATION STUDIES ON THE GROUP III METALS AL, GA, IN, AND TL

3.1 Oxides

High-temperature mass spectrometric studies have shown that the principal metal oxide species present in the vapor above heated M_2O_3 or M_2O_3/M mixtures (M = Al, Ga, In, Tl) are the suboxide molecules M_2O, and all four systems have been the subject of several electron-diffraction[21] or matrix-isolation[22,23] studies. However, it has been surprisingly difficult to achieve a consensus of opinion on the interpretation of both the electron-diffraction results and of the matrix-isolation results. It is not possible here to give a comprehensive account of all the aspects of this saga, but the present position seems to be as follows.

As regards the matrix-isolation ir work, only the antisymmetric stretching modes, ν_3, for the species, Al_2O, Ga_2O, In_2O, and Tl_2O, have been identified with reasonable confidence, and these absorptions are listed in Table 3. The minor frequency differences may be attributed to matrix effects, and all

Table 3. Vibration Frequencies (cm^{-1}) and Band Assignments for some Matrix Isolated Group III Metal Oxides and Halides

Species	Matrix	Vibrational Frequencies	Shape	Ref.
Al_2O	(Ar)	991.7 (ν_3)	—	59, 22
	(Ar)	994 (ν_3)	$\widehat{AlOAl} \geq 145°$	60
	—	—	$\widehat{AlOAl} \sim 144°$	21
Ga_2O	(N_2)	808.1 (ν_3)	—	59, 22
	(N_2)	809.4 (ν_3), 472 (ν_1?)	$\widehat{GaOGa} \geq 142°$	23
	—	—	$\widehat{GaOGa} \sim 140°$	21
In_2O	(N_2)	721.8 (ν_3)	—	59, 22
	(N_2)	722.4 (ν_3), 442 (ν_1?)	$\widehat{InOIn} \geq 135°$	23
	—	—	$\widehat{InOIn} \sim 144°$	21
Tl_2O	(N_2)	626.1 (ν_3)	—	59
	(N_2)	625.3 (ν_3)	$\widehat{TlOTl} \geq 130°$	61
	—	—	$\widehat{TlOTl} \sim 133°$	21
Al_4O_2(?)	(Ar)	714.8, 503(?)	—	22
Ga_4O_2	(N_2)	590.9, 423.9	—	23
In_4O_2	(N_2)	550.2, 412.2	—	23
Tl_4O_2	(Ar)	509.0, 383.0	—	25
AlF_3	(Ne)	960(E′), 284(A_2''), 252(E′)	D_{3h}	29
$AlCl_3$	(Ar)	618.8(E′), 393.5(A_1'), 183(A_2''), 151(E′)	D_{3h}	28
GaF_3	(Ar)	745(E′), 193, 188	D_{3h}	30
$GaCl_3$	(Ar)	470.3(E′), 136.2(A_2''), 132.1(E′), 386.2(A_1')	D_{3h}	67, 69
$GaBr_3$	(Ar)	354.8(E′), 107.0(A_2''), 88.6(E′)	D_{3h}	67
GaI_3	(Ar)	292.7(E′)	D_{3h}	67
$InCl_3$	(Ar)	400.5(E′), 359.0(A_1'), 119(E′)	D_{3h}	69
Ga_2F_2	(Ar)	575, 547, 416	—	30
Tl_2F_2	(Ar)	316.0(B_{2u}), 297.0(B_{1g}), 257.5(B_{3u}) 226.0(A_g), 93(A_g), 81(B_{1u})	D_{2h}	27
Tl_2Cl_2	(Ar)	190.4(B_{1g}), 189.6(B_{2u}), 170.6(B_{3u}) 131.0(A_g), 61(A_g)	D_{2h}	27
OAlF	(Ar)	1022, 386	$C_{\infty v}$	32

the papers on these systems agree that this mode is relatively intense. ^{18}O frequency shifts have been measured for all four M_2O species and for InOGa, and this data allows one to calculate a lower limit for the $\widehat{MOM}$ apex angle. The results obtained are included in Table 3, and they agree surprisingly well with the most recent electron-diffraction results.[21] However, both sets of values are unlikely to be accurate. In the first place, the ir data is uncorrected for anharmonicity, and it has been shown[23] that the inclusion of only a modest anharmonic correction for Ga_2O could result in an apex angle much closer to 180°. Secondly, the electron diffraction angles as presented[21] in Table 3 are uncorrected for shrinkage produced as a result of low-frequency bending vibrations. An appropriate correction for shrinkage can, however, be made if the bending frequency is known, and it is shown[21] that if the bending modes in these species lie below 200 cm^{-1}, then the apex angles are probably in the range, 150–180°. Numerous attempts have been made to locate the bending frequencies in these suboxides, and it is now fairly certain that they must lie below 200 cm^{-1}. If this is true, one might anticipate that these molecules would have much wider apex angles and may even be linear,[24] as has been shown for Li_2O. This would at least account for the largely unsuccessful search for ν_1 in the infrared.

The additional bands observed in these M/O systems, some of which were originally assigned as fundamentals of M_2O, are now believed to arise from either polymerization reactions or metal atom reactions which take place during sample deposition, and papers by Hinchcliffe and Ogden,[23] and Carlson et al.,[22] are significant here. These papers show that certain bands previously assigned[59] to the totally symmetric vibrations of In_2O, Ga_2O, and Al_2O in fact produce triplet patterns on ^{18}O enrichment, and most probably arise from a species containing two equivalent atoms of oxygen. These bands have also been observed to increase in intensity on diffusion at the expense of M_2O, and have tentatively been assigned (Table 3) to dimer molecules, while in the Tl/O system, Franzen et al.[25] have found similar evidence for Tl_4O_2. Figure 3 shows the diffusion plot obtained for matrix-isolated aluminum oxides. The structures of these dimer species are unknown, but it has been suggested[23] that they may contain a four-membered M_2O_2 ring similar to that found in related organometallic systems.

3.2 Halides and Oxyhalides

The existence of more than one formal oxidation state and lack of useful isotope data for the metals Al, In, and Tl make characterization by vibrational spectroscopy quite difficult for even the simplest halides of these elements, and it is only comparatively recently that matrix-isolation ir and Raman studies have been combined with earlier gas phase data to solve the basic problems of molecular structure. One of the earliest systems studied[26] was

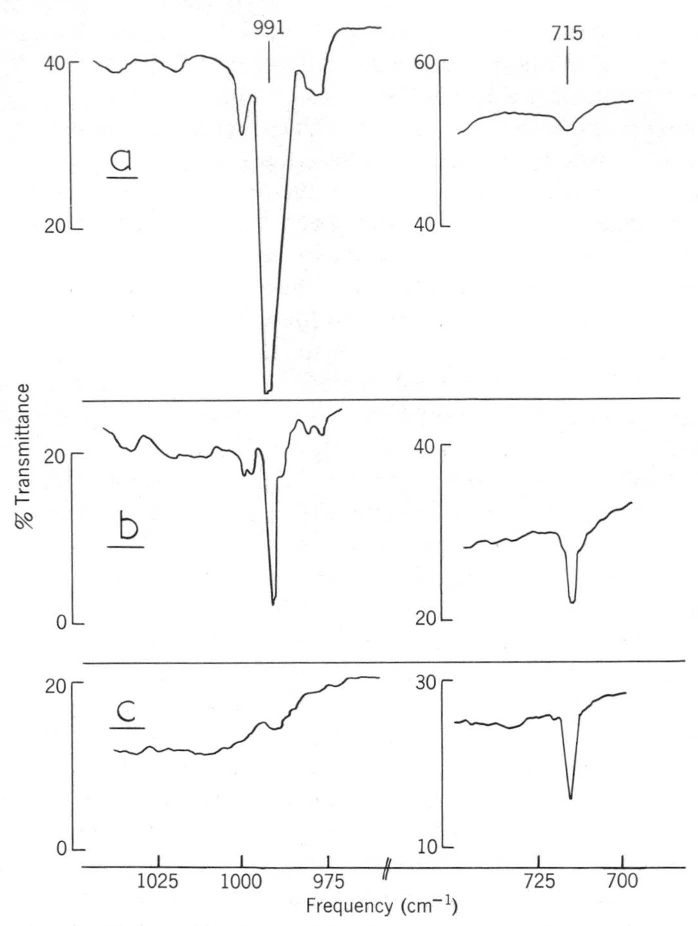

Figure 3. Infrared spectra of Al_2O isolated in an argon matrix at M/R = 500. (*a*) for the freshly prepared matrix, (*b*) after two diffusion warmings, (*c*) after four diffusion warmings. The feature at 991 cm^{-1} is assigned to monomer, while that at 715 cm^{-1} is assigned to a polymer. Reprinted with permission from D. A. Lynch, Jr., M. J. Zehe, and K. D. Carlson, *J. Phys. Chem.*, **78**, 236 (1974). Copyright by the American Chemical Society.

the vaporization of thallium (I) halides in which ir bands were observed for matrix-isolated Tl_2F_2 and Tl_2Cl_2 in addition to the complete range of monomer species. These dimer absorptions were assigned to a linear model X-Tl-Tl-X that was believed to be the correct structure at the time. However, in a subsequent paper describing both ir and Raman matrix-isolation studies, Lesiecki and Nibler[27] present convincing evidence that Tl_2F_2 and Tl_2Cl_2 have planar ring structures.

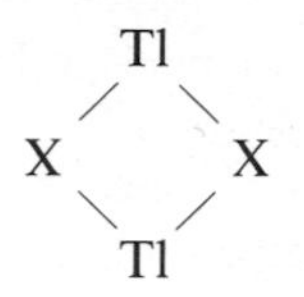

and the frequencies listed in Table 3 for these molecules are assigned on this basis.

In view of the structural differences discussed earlier for MX_2 monomers (linear or bent), one might anticipate a similar divergence in structure between the monomeric trihalides of Group IIIA (Sc-La) and Group IIIB (Ga-Tl). Here, involvement of the $(n - 1)d$ orbitals might be expected to lead to *pyramidal* structures for some of the trihalides of Sc, Y, or La, but all the Group IIIB trihalides and AlX_3 should adopt the D_{3h} structures found for the boron halides. From the experimental point of view, Shirk et al.[67] have demonstrated that $GaCl_3$, $GaBr_3$, and GaI_3 are all planar, and the controversy[28] regarding the shape of $AlCl_3$ has also been resolved in favor of the planar model. The range of trifluorides so far studied is not extensive, but there is evidence that LaF_3 is pyramidal,[68] and the fluorides of Al and Ga have also been investigated. The vaporization of AlF_3 and AlF_3/Al mixtures has led to the identification[29] of monomeric AlF_3 and AlF, but these systems also produce additional bands that may be assigned to Al_2F_6 and AlF polymers.

The gallium fluoride system is also complex,[30] and although it has been possible to use gallium isotopes to identify GaF_3 and GaF, numerous additional bands have been observed, which might be attributed to Ga_2F_6 and GaF polymers formed during deposition. Monomeric AlF_3 and GaF_3 are believed to have D_{3h} symmetry, and it seems likely that their dimers are isostructural with Al_2Cl_6, and thus contain both terminal and bridging fluorine atoms. These two kinds of metal-fluorine bond are also believed to be present in molecular $LiAlF_4$ and $NaAlF_4$, which have been identified[31] in low-temperature matrices as a product of the vaporization of AlF_3/LiF or AlF_3/NaF mixtures. These molecules are chemically very interesting as they are believed to possess C_{2v} symmetry with structural type (II) shown below:

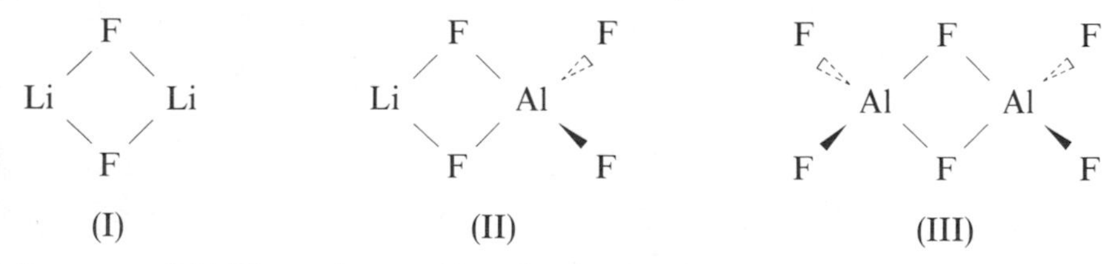

In essence $LiAlF_4$ can be considered to be the expected intermediate between the Li_2F_2 (I) and Al_2F_6 (III) dimers found in the gaseous and matrix phases.

The most prominent matrix frequencies of $LiAlF_4$ are similar to those found in the gas phase.

Our knowledge of vapor phase oxyhalides OMX (X = halogen) is confined almost exclusively to boron compounds[64,65] where mass spectrometric studies have indicated the existence of both monomers and trimers. A convenient source of monomeric OBF was obtained[64] from the Knudsen cell reaction

$$B_2O_3(s) + MgF_2(s) \xrightarrow[\text{Pt oven}]{700^\circ C} OBF$$

Using $^{10}B_2O_3/MgF_2$ and $^{11}B_2O_3/MgF_2$ isotopically enriched samples, the product was characterized as a species containing a single boron and a single oxygen atom. The linear (*sp* hybridized) O=B—F structure was suggested, rather than B—O—F, from the frequency of the boron-oxygen stretching mode ($\sim 2100\ cm^{-1}$) and by analogy with other compounds containing linear —B=O groups, for example,

Other oxyhalides of boron have been characterized[65] by passing the halide vapor BX_3 over heated B_2O_3, and the force constants derived for the boron-oxygen bond show an interesting trend:

Molecular species	K_{OB} (mdyne/Å)
OBF	12.69
OBCl	12.87
OBBr	12.94
OB	13.49

The decrease in the —B=O bond order on passing from OBBr to OBF has been rationalized by noting the absence of antibonding electrons in the molecular orbitals of the parent B=O molecule. Therefore, as the electronegativity of X increases, the extent of B—X $p\pi$—$p\pi$ bonding will increase,

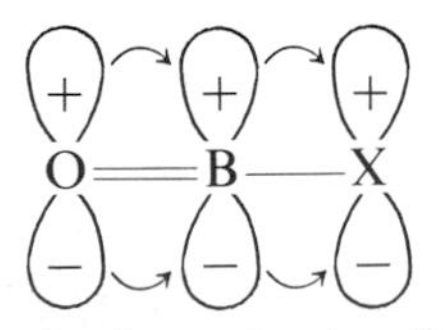

causing greater electron transfer from the *bonding* π orbitals of the O=B— residue. Hence the order, $k_{BO}^{F} < k_{BO}^{Cl} < k_{NO}^{Br}$. These ideas receive some support from the trend in the —N=O bond order observed for the molecules O=N—X,[66] where $k_{NO}^{F} > k_{NO}^{Cl} > k_{NO}^{Br}$, consistent with the decreased withdrawl of *antibonding* charge densities from the —N=O residue as the electronegativity of X decreases.

For the remaining elements, Snelson has shown that vaporization of AlF_3/Al_2O_3 mixtures[32] produces not only AlF_3 and other previously identified Al/F species, but also several additional new bands that must be assigned to oxyfluorides, AlO_xF_y. In particular, two of these at 1022 and 386 cm^{-1} are identified as the Al=O stretch and the O—Al—F bend in the linear molecule OAlF by analogy with the related boron oxyhalides. This type of experiment could obviously be extended to encompass the other metals in the group, and it would be particularly interesting if polymer species such as $(OAlF)_3$ could be identified, since they would presumably contain metal/oxygen rings similar to those proposed in Mg_3O_3 and Si_3O_3.[19]

3.3 Cocondensation Reactions of Aluminum and Gallium

Only brief accounts have so far appeared describing a few Group III metal vapor cocondensations, but it seems probable that both aluminum and gallium atoms have potentially rather interesting chemistries. Marino and White have indicated[6] that aluminum atoms produced in the vaporization of Al_2O_3/Al mixtures can react with O_2/Ar matrices to produce a species containing two equivalent atoms of oxygen. A band is observed for this molecule at 496 cm^{-1}, and the authors suggest that it may be a cyclic peroxide, Al_2O_2. The cocondensation of gallium vapor with argon matrices containing molecular fluorine has been reported by Margrave et al.[30] In addition to identifying GaF_3 and GaF, they observe a prominent feature at 547 cm^{-1}, which is assigned to Ga_2F_2, and suggest that this may be produced by the direct reaction of Ga_2 molecules.

Preliminary experiments have also been carried out on the cocondensation of aluminum[33] and gallium[34] vapor with carbon monoxide, and these have led to the first characterization of Group III metal carbonyls. When aluminum vapor is codeposited with krypton containing ~3% $C^{16}O$, prominent ir bands are produced at 1988 and 1890 cm^{-1}, and these are assigned to a nonlinear dicarbonyl of aluminum, $Al_x(CO)_2$, on the basis of the characteristic frequency and intensity patterns observed on $C^{18}O$ enrichment. The ir spectra obtained are reproduced in Figure 4. Analagous results have also been obtained using gallium vapor, and ir bands at 2006.0 and 1912.0 cm^{-1} (xenon matrix) are assigned to $Ga_x(CO)_2$. These results indicate that metal carbonyls, therefore, exist for both pretransition and posttransition elements, and it is interesting that the carbonyl vibration frequencies in these compounds do not appear to be significantly different from those found in the more familiar transition metal carbonyls (see Chapter 8 for further details).

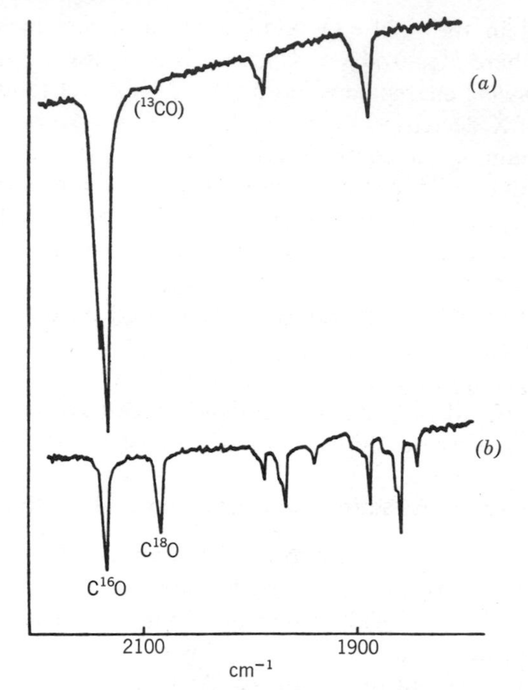

Figure 4. Infrared spectra obtained after cocondensing aluminum atoms in a krypton matrix containing (*a*) 3% $C^{16}O$, and (*b*) 1.7% $C^{16}O$ + 1.3% $C^{18}O$. Reproduced from A. J. Hinchcliffe, J. S. Ogden, and D. D. Oswald, *J. Chem. Soc., Chem. Commun.* 338 (1972).

4. MATRIX-ISOLATION STUDIES ON GROUP IV METALS

4.1 Vaporization of Germanium, Tin, and Lead Halides

Many of the tetrahalides of the elements in Group IV are well-characterized covalent molecules that can be studied by conventional techniques, and the main impact of matrix isolation has been in the characterization of the molecular dihalides, MX_2. These species exist in the vapor above the heated anhydrous solids, and both electron diffraction and electric deflection experiments show that they have C_{2v} geometries.[7] Until quite recently, the majority of matrix experiments on these systems have relied solely on ir spectroscopy, and although it has usually been possible to observe all the vibrational fundamentals, it was not until parallel Raman experiments were carried out[35] using polarized light that one could make unequivocal assignments for the symmetric (A_1) and antisymmetric (B_2) stretching modes for some of these species. The reason for this is that the apex angles in many of these triatomics are found to lie in the range, 90–100°, and if stretch-stretch interaction constants are small, these two stretching modes have similar

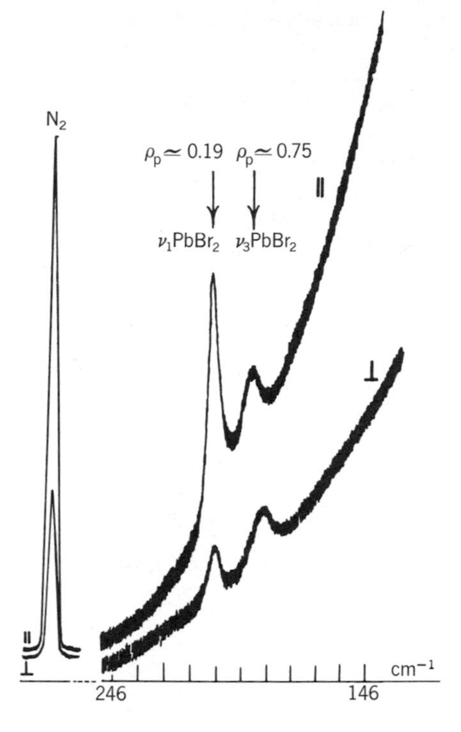

Figure 5. The matrix Raman spectrum with polarization data for monomeric $PbBr_2$ in N_2 at 4.2°K. Reproduced from G. A. Ozin and A. Vander Voet, *J. Chem. Phys.*, **56**, 4768 (1972).

frequencies and ir intensities, and cannot easily be distinguished solely from ir studies. Figure 5 shows how the totally symmetric stretching mode in $PbBr_2$ may be identified from polarization data in the Raman.[35]

In addition to straightforward studies on the pure dihalides, a number of experiments have been carried out using a mixture of solids that on vaporization yield species such as PbClBr,[35] or $KSnF_3$.[36] The triatomics have, as expected, C_s symmetry, but the products of the alkali halide + MX_2 systems, such as $KSnF_3$, appear to have C_{2v} symmetry with both bridge and terminal metal-halogen vibrations. The tendency for fluorine and chlorine to form bridge bonds in these systems has also been observed in dimer species such as $(GeF_2)_2$ and $(PbCl_2)_2$. These molecules have sometimes been identified in mass spectrometric studies, and matrix ir and Raman data indicate[37] that they probably have C_{2h} structures.

The matrix studies on these MX_2 systems compare very favorably with gas-phase Raman studies on the high-temperature vapors, and Table 4 compares the fundamental frequencies and bond angles obtained for these molecules under various conditions. A number of gas-phase ir studies have also been reported, but the location of the band centers is not sufficiently

Table 4. Vibration Frequencies (cm^{-1}) and Derived Bond Angles for Some Group IV Metal Dihalides

Species	Study	ν_1	ν_2	ν_3	Angle X-M-X	Ref.
GeF_2	(Ne)^a	685	263	655	94° ± 3°	38
	$(N_2)^b$	653	—	702	—	37
SnF_2	(Ar)^a	592.7	197	570.9	94° ± 5°	56
PbF_2	(Ar)^a	531.2	165	507.2	~90°	56
$GeCl_2$	(Ar)^a	399.7	—	374.5	90°–100°	57
	$(N_2)^b$	390	163	362	—	35
	—^{c}	399	159	—	—	58
$SnCl_2$	(Ar)^a	354	—	334	90°–100°	57
	$(N_2)^b$	341	124	320	—	35
	—^{c}	352	120	—	—	58
$PbCl_2$	(Ne)^a	326	81	304	96° ± 3°	55
	(Ar)^a	322	103	300	—	35
	—^{c}	314	99	—	—	58
$SnBr_2$	(Ar)^b	244	82	231	—	35
	—^{c}	—	80	—	—	58
SnClBr	$(N_2)^b$	328	—	228	—	35
	—^{c}	—	100	240	—	58
PbClBr	$(N_2)^b$	295	—	200	—	35

^{a} Infrared matrix.
^{b} Raman matrix.
^{c} Raman gas phase.

precise for a critical comparison to be made. Bond-angle data for the chlorides have been obtained from the chlorine isotope-frequency shifts and, therefore, represent upper limits, and the apex angles derived for GeF_2[38] and SnF_2[36] conversely represent lower limits. A much more precise value for GeF_2 has recently been obtained[39] from the microwave spectrum (97°10′), and this compares very favorably with Margrave's ir study.[38]

4.2 Oxides, Sulfides, Selenides, and Tellurides

High-temperature mass spectrometry has shown that the principal vapor species over the heated solids, M_nX_m (M = Ge, Sn, Pb; X = O, S, Se, Te), are the diatomic molecules, MX, and that varying amounts of dimers M_2X_2, trimers, and higher polymers are also to be expected.[40] The combination of this information with a knowledge of the ground state vibration frequencies of the monomers (from electronic spectroscopy) has led to the characterization of several new matrix-isolated species, and the facility for controlled diffusion presents a valuable opportunity for following polymerization reactions in these systems.

All the chalcogenides of germanium have now been studied in low-temperature matrices, and for tin and lead, corresponding data is lacking only for the tellurides.

Germanium

Matrix-isolation studies on germanium oxides have been reported by Ogden and Ricks.[41] Vaporizations from three solid systems (GeO_2, "amorphous" GeO, and solid GeO_2/Ge mixtures) were studied, and the products of reaction between O_2 gas and hot Ge metal were also investigated. Four different species were identified on the basis of isotope patterns and controlled-diffusion experiments, and the most prominent bands were assigned to GeO, Ge_2O_2, Ge_3O_3, and Ge_4O_4. Only a small matrix shift was observed for GeO (973.4 cm^{-1} in a nitrogen matrix compared with 977.1 cm^{-1} for the gas-phase molecule), and strong bands at 667.0 and 599.0 cm^{-1} were assigned to Ge_2O_2. The spectrum of this molecule was interpreted assuming a planar four-membered ring structure (D_{2h} symmetry) with OGeO = 83°, and ring structures were also proposed for the trimer and tetramer. Approximate force constants and bond angles were calculated for the dimer and trimer rings using a simple valence force field, and the experimental frequencies and derived parameters are listed in Table 5. Diffusion studies in this system were interpreted on the basis of a stepwise polymerization

$$GeO \xrightarrow{GeO} Ge_2O_2 \xrightarrow{GeO} Ge_3O_3 \xrightarrow{GeO} Ge_4O_4$$

Analogous studies on the sulfides, selenides, and tellurides of germanium have been reported by Marino et al.[42] The vibration frequencies of the diatomics, GeX, are shifted only slightly from the vapor-phase values, and the natural abundance of germanium, sulfur, and selenium isotopes makes the identification of these species unequivocal. In addition to monomer bands, a number of other features were observed at lower frequencies in the sulphide and selenide systems, and these were assigned as dimers and higher polymers. Although it was not possible to resolve isotope fine structure on these bands, the absorptions associated with dimers Ge_2S_2 and Ge_2Se_2 could be identified on the basis of intensity versus matrix concentration studies. The vibration frequencies of GeX and Ge_2X_2 are listed in Table 5, together with their assignments. In the case of the dimers, band assignments were made on the basis of the four-membered ring model, and force constant analyses similar to that for Ge_2O_2 yield principal force constants k_{Ge-S} and k_{Ge-Se}, which are approximately half those observed for the monomers, and bond angles at germanium of ~86° and 85° for Ge_2S_2 and Ge_2Se_2, respectively.

Table 5. Infrared Frequency and Force Constant Data for Group IV Metal Chalcogenides

Monomers MX	ν_{M-X}	k_{M-X}	Ref.	Dimers M_2X_2	$\nu_{B_{2u}}$	$\nu_{B_{3u}}$	Angle X-M-X	k_{M-X}	Ref.
^{74}GeO	973.4	7.34	41	Ge_2O_2	667.0	599.0	83°	2.98	41
^{120}SnO	816.1	5.54	43	Sn_2O_2	612.2	523.2	80°	2.59	43
PbO	718.4	4.51	44	Pb_2O_2	557.4	467.7	79°	2.23	44
^{74}GeS	566.6	4.22	42	Ge_2S_2	386.7	361.3	86°	1.84	42
^{120}SnS	480.5	3.44	42	Sn_2S_2	336.3	333.9	90°	1.67	42
PbS	423.1	2.93	42	Pb_2S_2	304.7	300.2	89°	1.50	42
$^{74}Ge^{80}Se$	397.9	3.58	42	Ge_2Se_2	278.4	256.7	85°	1.60	42
$^{120}Sn^{80}Se$	325.2	2.99	42	Sn_2Se_2	233.3	219.3	86°	1.44	42
$Pb^{80}Se$	275.1	2.57	42						
$^{74}GeTe$	317.6	2.80	42						

Trimers	Frequencies	Tetramers	Frequencies	Ref.
Ge_3O_3 (N_2)	824, 440	Ge_4O_4 (N_2)	553.3, 490.0, 476.0, 457.5	41
Sn_3O_3 (Ar)	762	Sn_4O_4 (Ar)	516.6, 417.0	43
Pb_3O_3(?) (Ar)	690.4, 292.7	Pb_4O_4 (N_2)	474.4, 374.4	44

Dioxides MO_2	ν_3	Angle O-M-O	k_{M-O}	Ref.
$^{74}GeO_2$ (N_2)	1061.6	180°	7.32	50
$^{120}SnO_2$ (Kr)	863.1	180°	5.36	2

Tin

The ir spectra obtained when tin oxide vapors are isolated in low-temperature matrices have been described by Ogden and Ricks.[43] Four distinct species could be observed, and these were identified as the known vapor-phase molecules, SnO, Sn_2O_2, Sn_3O_3, and Sn_4O_4. SnO and Sn_2O_2 were characterized by $^{18}O/^{16}O$ isotope patterns, and a planar D_{2h} ring structure was proposed for Sn_2O_2 by analogy with earlier work on Si_2O_2 and Ge_2O_2. Bands due to these species were the only prominent new features from the vaporization of solid SnO_2, but when oxygen gas was passed over heated tin, or when a mixture of solid SnO_2 and elemental Sn was used, additional bands were produced, were assigned to a D_{3h} trimer and the tetramer, Sn_4O_4. The ir spectrum of Sn_4O_4 was considerably simpler than that observed for Ge_4O_4, and it was suggested that Sn_4O_4 has cubic (T_d) symmetry with a structure similar to the metal-oxygen framework in the ion $Pb_4(OH)_4^{4+}$ (Figure 6). Matrix-dilution experiments and controlled-diffusion studies indicated that

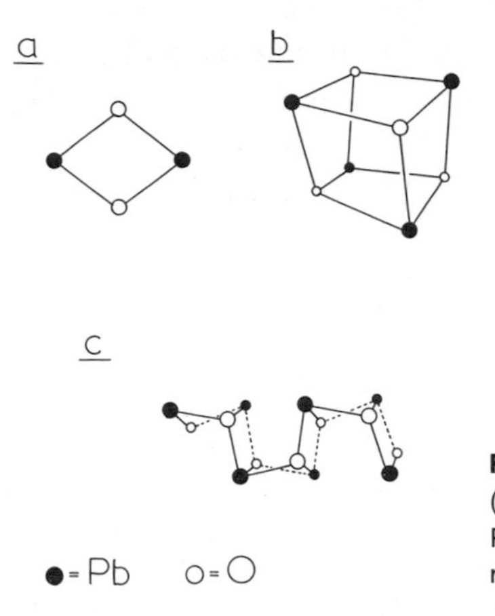

Figure 6. (*a*) Structure proposed for molecular Pb_2O_2, (*b*) Pb—O framework proposed for Pb_4O_4 and $Pb_4(OH)_4^{4+}$, (*c*) part of the layer structure in orthorhombic lead monoxide.

the trimer was relatively unimportant in the polymerization reactions of SnO, and that in contrast to the germanium oxide system, the principal reaction sequence appears to be $SnO \rightarrow Sn_2O_2 \rightarrow Sn_4O_4$.

Corresponding studies on tin sulfides and selenides have been reported by Marino et al.[42] Their spectra are interpreted in terms of the isolation of monomers and D_{2h} ring dimers on the basis of S and Se isotope data and concentration studies. The molecular parameters for all the species identified in the Sn/O, Sn/S, and Sn/Se systems are included in Table 5.

Lead

The vapor above heated lead oxide is surprisingly complex, and mass spectrometric studies[40] have shown the presence of PbO, Pb_2O_2, Pb_3O_3, Pb_4O_4, Pb_5O_5, and Pb_6O_6. Two of these, PbO and Pb_2O_2, have been positively identified in low-temperature matrices using ^{18}O enrichment,[44] and a third species, characterized by an intense pair of bands at 474.4 and 374.4 cm^{-1} (N_2 matrix), is identified as Pb_4O_4 on the basis of controlled-diffusion studies and by analogy with the ir spectrum[45] of $Pb_4(OH)_4^{4+}$. This ion is present in aqueous solutions of Pb(II) salts, and its structure has been determined by X-ray difffraction (Figure 6).[46] Two ir-active Pb—O stretching vibrations are expected for this symmetry (T_d) and the ir spectrum of $Pb_4(OH)_4^{4+}{\cdot}4ClO_4^-$ shows two strong bands at 346 and 505 cm^{-1}, which compare favorably with the nitrogen matrix spectrum of Pb_4O_4 (Table 5). Matrix spectra of the related sulfide and selenide systems have been obtained by Marino et al.,[42] and this data is also included in Table 5.

If one compares the matrix-isolation results for all these systems with the quantitative mass spectrometric data available, it is apparent that the *proportions* of species present in the vapor phase are only of marginal assistance in the interpretation of relative ir band intensities in the matrix, and at least one species has been detected in the matrix that is unknown in the vapor (e.g., Ge_4O_4).

This would suggest that extensive polymerization of monomers must be taking place here during deposition, and the fact that further reaction can be observed in controlled-diffusion studies indicates that the activation energies for these processes must be quite low.

It is also interesting to note that the four-membered ring dimer, M_2X_2, is common to all these systems, and that this structural unit is found in a wide range of Group IV–VI compounds. It occurs extensively in organometallic systems, where species such as $R_2Sn{=}O$ are more stable as dimers and cross-linked polymers than as simple monomers. In a few cases, these units persist in aqueous solution,[46] and in the lead system, we have the complex cation, $Pb_4(OH)_4^{4+}$, which may be regarded simply as protonated Pb_4O_4. Finally, the solids, MX, are frequently found to have either a face-centered-cubic structure (e.g., PbS, PbSe) or, alternatively, an orthorhombic structure that is closely related to the sodium chloride structure and in which the almost square M_2X_2 ring persists.[47] The close correspondence between Pb_2O_2, Pb_4O_4, $Pb_4(OH)_4^{4+}$, and the structure of orthorhombic PbO is shown in Figure 6, and this kind of relationship may be traced for other chalcogenides.

4.3 Cocondensation Reactions of Germanium, Tin, and Lead

Relatively little matrix work has been published on the chemistry of Group IV metal vapors, but the few studies that have been carried out provide some rather interesting chemistry.

Reactions with CO

Preliminary experiments on the cocondensation of germanium and tin vapors with either pure carbon monoxide or with CO diluted with Kr have been described by Bos,[48] and it is clear that binary molecular carbonyls of these elements may be produced in this way. For the tin system, a large number of characteristically sharp ir bands were observed in the frequency range, 1700–2100 cm^{-1}, and their assignment as carbonyl vibrations was confirmed by $C^{18}O$ enrichment. Weak absorption was also reported in the region, 900–950 cm^{-1}. The number of bands observed in the carbonyl region appears to be considerably larger than one would predict from the presence of simple binary species, $Sn(CO)_y$, and it seems likely that some of the species observed contain more than one atom of tin. The simplest spectrum was obtained with a very low proportion of CO in krypton ($\sim 0.1\%$), and the single new band observed at 1921 cm^{-1} under these conditions was assigned to SnCO.

The cocondensation of germanium with 1% CO in Krypton produced a relatively simple spectrum consisting of three peaks at 1873, 1878, and 1908 cm^{-1}, and it was suggested that the 1908 cm^{-1} band is a simple monocarbonyl GeCO. This assignment has been supported by additional experiments carried out on the system by Simpson,[49] who showed that a characteristic doublet spectrum was observed when $C^{16}O + C^{18}O$ mixtures were employed.

Reactions with O_2

Considerably more information is available on the $Ge + O_2$ and $Sn + O_2$ reactions in low-temperature matrices. In their first paper on Group-IV-atom–O_2 reactions, Bos and Ogden[2] describe a series of experiments on the cocondensation of tin vapor with nitrogen and krypton matrices containing between 1 and 20 mole% O_2. Molecular SnO_2 and Sn_2O_2 were identified as the major reaction products, and small amounts of O_3, SnO, and Sn_3O_3 were also produced. The relative amounts of these products were found to be dependent both upon the matrix deposition rate and upon the nature of the matrix gas (N_2 or Kr). Molecular SnO_2 was identified on the basis of the tin and oxygen isotope patterns observed for the antisymmetric stretching mode, and the most satisfactory agreement between observed and calculated isotopic frequencies was obtained by assuming a linear $D_{\infty h}$ structure (see Figure 1). The antisymmetric stretching vibration for $^{120}Sn^{16}O_2$ was observed at 877.8 cm^{-1} in a nitrogen matrix, and a value of 5.57 mdyn/Å was derived for the principal Sn—O force constant. This value is very close to that observed in diatomic SnO, and it is suggested that considerable multiple bonding is present in SnO_2.

The use of unscrambled oxygen isotope mixtures in this work provides a clue to the principal mechanism of formation of SnO_2 and Sn_2O_2. When tin atoms were condensed with a nitrogen matrix containing principally $^{16}O_2 + {}^{18}O_2$, only small amounts of $Sn^{16}O^{18}O$ and $Sn_2{}^{16}O^{18}O$ were observed, and it was, therefore, argued that the most important reactions occurring here were

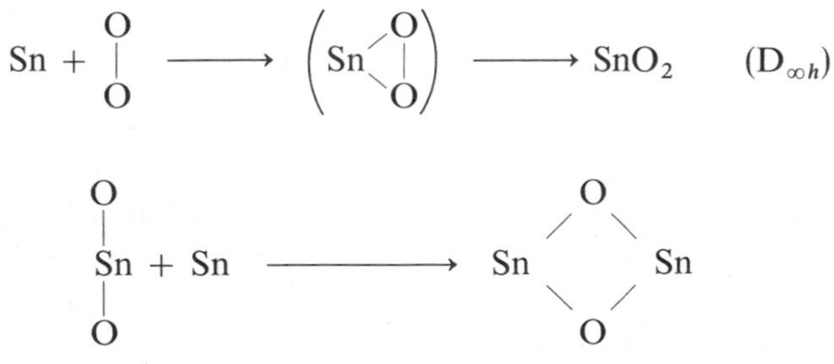

and

The additional formation of O_3 and SnO suggested that a second reaction sequence

$$Sn + O_2 \rightarrow SnO + O$$
$$O + O_2 \rightarrow O_3$$

was also taking place. A number of weaker features were observed in both matrices, which were tentatively assigned as reaction products of Sn_2 molecules.

A subsequent paper[50] on the reaction of germanium atoms with O_2 in low-temperature matrices demonstrated that molecular GeO_2 was formed by a similar insertion mechanism. Infrared bands assigned to the species were considerably more prominent in nitrogen than in krypton, and the stoichiometry was again established by characteristic isotope patterns. A normal coordinate analysis showed that GeO_2 is linear $D_{\infty h}$, and the molecule is, therefore, isostructural with CO_2. Diatomic GeO and ozone were also observed in this system, and in addition to providing mechanistic information on the Ge + O_2 reaction, experiments with unscrambled oxygen isotope mixtures showed that the ozone was produced essentially by end-on attack

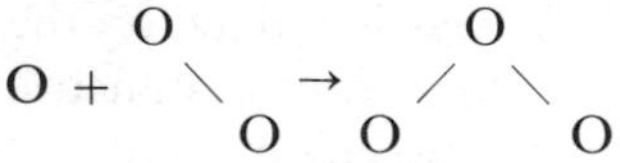

```
     O            O
O +   \    →     / \
       O  O     O   O
```

rather than via a D_{3h} cyclic intermediate.

The positive identification of the new species, GeO_2 and SnO_2, in these experiments compares very favorably with related gas-phase studies on the Pb atom + O_2 reaction,[51] where kinetic evidence is presented for the existence of a molecular intermediate PbO_2. Matrix-isolation studies on the Pb + O_2 system are difficult to carry out owing to the reluctance of lead atoms to react in low-temperature matrices, but preliminary studies indicate that PbO_2 and Pb_2O_2 are the principal reaction products.[52]

5. CONCLUSIONS

It is evident from this survey that matrix isolation has made a considerable impact on the chemistry of the main group metals and that very much more information is potentially available. In particular, the study of metal atom cocondensation reactions and the characterization of ternary oxide and halide systems are two fields where little experimental data is available at the present time, and the fact that many of these metals have two or more naturally occurring isotopes makes the characterization of new species by vibrational spectroscopy very attractive.

At the same time, it is also apparent from the conflicting reports that have appeared concerning simple triatomics such as MgF_2 or Al_2O, that even with the aid of isotope labeling, spectral interpretation can be ambiguous if unsuspected polymer formation takes place. However, the facility for carrying out controlled diffusion experiments is of considerable assistance, and, quite apart from aiding spectral interpretation, it also provides a means of studying mechanisms of polymer formation including, for example, the clustering of metal atoms.

Throughout this chapter, the main emphasis is on vibrational spectroscopy, with only passing reference being made to other spectroscopic techniques. The principal reason for this is that the majority of matrix work that has appeared so far in the literature concerning these systems has relied almost totally on ir and Raman spectra as the principal method of characterization. However, an increasing number of papers are now reporting studies on matrix-isolated species in which vibrational spectroscopy is coupled with, for example, Mossbauer or epr spectroscopy, and where the main emphasis is no longer simply the identification of the trapped species. In experiments of this kind, vibrational spectroscopy is used principally to monitor the molecule of interest, and the more detailed studies that have been carried out using these combined techniques are described in Chapters 8 and 10.

REFERENCES

1. A. J. Downs and S. C. Peake, *Molecular Spectroscopy*, Specialist Periodical Report of The Chemical Society, London, 1973, Vol I, p. 523. B. M. Chadwick, *Molecular Spectroscopy*, Specialist Periodical Report of The Chemical Society, London, 1975, Vol. III, p. 281.
2. A. Bos and J. S. Ogden, *J. Phys. Chem.*, **77**, 1513 (1973).
3. M. Allavena, R. Rysnik, D. White, G. V. Calder, and D. E. Mann, *J. Chem. Phys.*, **50**, 3399 (1969).
4. J. S. Anderson and J. S. Ogden, *J. Chem. Phys.*, **51**, 4189 (1969).
5. See, for example, J. L. Margrave, *The Characterization of High Temperature Vapors*, Wiley, New York, 1967.
6. C. P. Marino and D. White, *J. Phys. Chem.*, **77**, 2929 (1973).
7. See, for example, J. W. Hastie, R. H. Hauge, and J. L. Margrave, *Annu. Rev. Phys. Chem.*, **21**, 475 (1970).
8. A. Loewenschuss, A. Ron, and O. Schnepp, *J. Chem. Phys.*, **49**, 272 (1968); ibid., **50**, 2502 (1969).
9. J. W. Hastie, R. H. Hauge, and J. L. Margrave, *High Temp. Sci.*, **1**, 76 (1969).
10. D. White, G. V. Calder, S. Hemple, and D. E. Mann, *J. Chem. Phys.*, **59**, 6645 (1973).
11. G. V. Calder, D. E. Mann, K. S. Seshadri, M. Allavena, and D. White, *J. Chem. Phys.*, **51**, 2093 (1969).
12. R. H. Hauge, J. L. Margrave, and A. S. Kanaan, *J. Chem. Soc., Faraday II*, **71**, 1082 (1975).
13. D. L. Cocke, C. A. Chang, and K. A. Gingerich, *Appl. Spectrosc.*, **27**, 260 (1973).
14. L. B. Knight, Jr., W. C. Easley, W. Weltner, Jr., and M. Wilson, *J. Chem. Phys.*, **54**, 322 (1971).
15. L. B. Knight Jr., and W. Weltner, Jr., *J. Chem. Phys.*, **54**, 3875 (1971).

16. L. B. Knight, Jr., and W. Weltner, Jr., *J. Chem. Phys.*, **55**, 2061 (1971).
17. S. Abramowitz and N. Acquista, *J. Res. Nat. Bur. Stand., A*, **75**, 23 (1971).
18. L. Brewer and J. L. -F. Wang, *J. Chem. Phys.*, **56**, 4305 (1972).
19. M. Spoliti, G. Marini, S. N. Cesaro, and G. DeMaria, *J. Mol. Struct.*, **19**, 563 (1973).
20. D. M. Thomas and L. Andrews, *J. Mol. Spectrosc.*, **50**, 220 (1974).
21. See, for example, S. M. Tolmachev and N. G. Rambidi, *High Temp. Sci.*, **5**, 385 (1973).
22. D. A. Lynch, Jr., M. J. Zehe, and K. D. Carlson, *J. Phys. Chem.*, **78**, 236 (1974).
23. A. J. Hinchcliffe and J. S. Ogden, *J. Phys. Chem.*, **77**, 2537 (1973).
24. A. Buchler, J. L. Stauffer, W. Klemperer, and L. Wharton, *J. Chem. Phys.*, **39**, 2299 (1963).
25. J. M. Brom, Jr., T. Devore, and H. F. Franzen, *J. Chem. Phys.*, **54**, 2742 (1971).
26. J. M. Brom and H. F. Franzen, *J. Chem. Phys.*, **54**, 2874 (1971).
27. M. L. Lesiecki and J. W. Nibler, *J. Chem. Phys.*, **63**, 3452 (1975).
28. M. L. Lesiecki and J. S. Shirk, *J. Chem. Phys.*, **56**, 4171 (1972). I. R. Beattie, H. E. Blayden, and J. S. Ogden, *J. Chem. Phys.*, **64**, 909 (1976). J. S. Shirk and A. E. Shirk, *J. Chem. Phys.*, **64**, 910 (1976).
29. A. Snelson, *J. Phys. Chem.*, **71**, 3202 (1967).
30. J. W. Hastie, R. H. Hauge, and J. L. Margrave, *J. Fluorine Chem.*, **3**, 285 (1974).
31. S. J. Cyvin, B. N. Cyvin, and A. Snelson, *J. Phys. Chem.*, **75**, 2609 (1971).
32. A. Snelson, *High Temp. Sci.*, **5**, 77 (1973).
33. A. J. Hinchcliffe, D. D. Oswald, and J. S. Ogden, *Chem. Commun.*, 338 (1972).
34. A. J. Hinchcliffe and D. D. Oswald, unpublished observations.
35. G. A. Ozin and A. Vander Voet, *J. Chem. Phys.*, **56**, 4768 (1972).
36. J. W. Hastie, R. H. Hauge, and J. L. Margrave, in *Spectroscopy in Inorganic Chemistry*, C. N. R. Rao and J. R. Ferraro, Eds., Vol I, p. 57, Academic, New York, 1970.
37. G. A. Ozin, *The Spex Speaker*, **16**, Nos. 4 and 7 (1971).
38. J. W. Hastie, R. H. Hauge, and J. L. Margrave, *J. Phys. Chem.*, **72**, 4492 (1968).
39. H. Takeo, R. F. Curl, Jr., and P. W. Wilson, *J. Mol. Spectrosc.*, **38**, 464 (1971).
40. See, for example, J. Drowart and P. Goldfinger, *Angew. Chem.* (*Int. Edit.*), **6**, 581 (1967).
41. J. S. Ogden and M. J. Ricks, *J. Chem. Phys.*, **52**, 352 (1970).
42. C. P. Marino, J. D. Guerin, and E. R. Nixon, *J. Mol. Spectrosc.*, **51**, 160 (1974).
43. J. S. Ogden and M. J. Ricks, *J. Chem. Phys.*, **53**, 896 (1970).
44. J. S. Ogden and M. J. Ricks, *J. Chem. Phys.*, **56**, 1658 (1972).
45. V. A. Maroni and T. G. Spiro, *Inorg. Chem.*, **7**, 188 (1968).
46. O. E. Esval, Ph.D. thesis, University of North Carolina, 1962.
47. See, for example, B. Dickens, *J. Inorg. Nucl. Chem.*, **27**, 1495 (1965).
48. A. Bos, *Chem. Commun.*, 26 (1972).
49. M. J. Simpson, unpublished observations.
50. A. Bos, J. S. Ogden, and L. Orgee, *J. Phys. Chem.*, **78**, 1763 (1974).
51. P. R. Ryason and E. A. Smith, *J. Phys. Chem.*, **75**, 2259 (1971).
52. A. Bos and J. S. Ogden, to be published.
53. A. Snelson, *J. Phys. Chem.*, **70**, 3208 (1966).
54. A. Snelson, *J. Phys. Chem.*, **72**, 250 (1968).

55. J. W. Hastie, R. H. Hauge, and J. L. Margrave, *High Temp. Sci.*, **3**, 56 (1971).
56. R. H. Hauge, J. W. Hastie, and J. L. Margrave, *J. Mol. Spectrosc.*, **45**, 420 (1973).
57. L. Andrews and D. L. Frederick, *J. Amer. Chem. Soc.*, **92**, 775 (1970).
58. I. R. Beattie and R. O Perry, *J. Chem. Soc., A*, 2429 (1970).
59. D. M. Makowiecki, D. A. Lynch, Jr., and K. D. Carlson, *J. Phys. Chem.*, **75**, 1963 (1971).
60. M. J. Linevsky, D. White, and D. E. Mann, *J. Chem. Phys.*, **41**, 542 (1964).
61. A. J. Hinchcliffe and J. S. Ogden, *Chem. Commun.*, 1053 (1969).
62. C. A. Coulson, *Israel J. Chem.*, **11**, 683 (1973).
63. G. E. Kimball, *J. Chem. Phys.*, **8**, 188 (1940).
64. A. Snelson, *High Temp. Sci.*, **4**, 141 (1972).
65. A. Snelson, *High Temp. Sci.*, **4**, 318 (1972).
66. S. C. Peake and A. J. Downs, *J. Chem. Soc., Dalton*, 859 (1974).
67. R. G. S. Pong, R. A. Stachnik, A. E. Shirk, and J. S. Shirk, *J. Chem. Phys.*, **63**, 1525 (1975).
68. J. W. Hastie, R. H. Hauge, and J. L. Margrave, *J. Less Common Metals*, **39**, 309 (1975).
69. I. R. Beattie, H. E. Blayden, S. M. Hall, S. N. Jenny, and J. S. Ogden, *J. Chem. Soc. Dalton*, 666 (1976).

Matrix Cryochemistry Using Transition Metal Atoms

8

M. Moskovits and G. A. Ozin

1. INTRODUCTION

The technique whereby a normally unstable chemical species is stabilized by immobilizing it in an inert host lattice at temperatures sufficiently low so as to prevent diffusion is referred to as matrix isolation. In this chapter

Table 1. Mononuclear Transition Metal Complexes Synthesized by Metal Atom Cocondensation Reactions

Complex	n^a	Ref.	Complex	n^a	Ref.
Carbonyl complexes			Dinitrogen complexes		
$V(CO)_n$	1–6	9, 46	$V(N_2)_n$	6	9
$Ta(CO)_n$	1–6	3, 46	$Ta(N_2)_n$	6	46
$Cr(CO)_n$	1–6	21	$Cr(N_2)_n$	6	46
$Mo(CO)_n$	1–6	21	$Co(N_2)_n$	1	109
$W(CO)_n$	1–6	21	$Rh(N_2)_n$	1–4	26
$Mn(CO)_n$	5	51	$Ni(N_2)_n$	1–4	11
$Re(CO)_n$	5	8	$Pd(N_2)_n$	1–3	11
$Fe(CO)_n$	1–5	46	$Pt(N_2)_n$	1–3	25
$Co(CO)_n$	1–4	18, 47			
$Rh(CO)_n$	1–4	46	Dioxygen complexes		
$Ir(CO)_n$	1–4	46	$Cr(O_2)_n$	2	110
$Ni(CO)_n$	1–4	3	$Rh(O_2)_n$	1–3	46
$Pd(CO)_n$	1–4	2	$Ni(O_2)_n$	1–2	4
$Pt(CO)_n$	1–4	11	$Pd(O_2)_n$	1–2	4
$Cu(CO)_n$	1–3	44	$Pt(O_2)_n$	1–2	4
$Ag(CO)_n$	1–3	6	$Cu(O_2)_n$	2	5, 110
$Au(CO)_n$	1–2	5	$Ag(O_2)_n$	2	5
			$Au(O_2)_n$	1	5
Mixed complexes					
$Ni(N_2)_n(CO)_{4-n}$	1–3	12	Miscellaneous[b]		
$Ni(N_2)_n(O_2)$	1–2	28	$Ni(CS)_n$	1–4	111
$Pd(N_2)_n(O_2)$	1–2	28	$Ni(PN)_n$	4	112
$Pt(N_2)_n(O_2)$	1–2	28	$Ag(PN)_n$	1–2	112
			$Ni(C_2H_4)_n$	1–3	115, 118a
			$Pd(C_2H_4)_n$	1–3	115, 118b
			$Co(C_2H_4)_n$	1–3	115, 118b
			$Cu(C_2H_4)_n$	1–3	119
			$Ag(C_2H_4)_n$	1	120
			$Au(C_2H_4)_n$	1	121
			$Fe(COD)_n$	2	115
			$Co(COD)_n$	2	115
			$Pd(C_7H_{10})_n$	3	115
			$Fe(C_6H_6)_n$	2	115
			$Cr(C_6H_6)_n$	2	115

[a] Refers only to those complexes that have been reasonably well characterized, where n is the coordination number of the complex.
[b] COD = cycloocta-1,5-diene, C_7H_{10} = norbornene.

Table 2. Binuclear Transition Metal Complexes Synthesized by Metal Atom Cocondensation Reactions

Complex	Remarks	Ref.
Carbonyl complexes		
$V_2(CO)_{12}$	Two forms: (a) bridge-bonded $(CO)_5V(\mu\text{—}CO)_2V(CO)_5$, (b) metal-metal bonded	9
$Cr_2(CO)_{10}$	Weakly interacting metal-metal bonded dimer	113
$Mn_2(CO)_{10}$	Normal metal-metal bonded forms	51
$Re_2(CO)_{10}$	Normal metal-metal bonded forms	113
$Fe_2(CO)_8$	Bridge-bonded form, $(CO)_3Fe(\mu\text{—}CO)_2Fe(CO)_3$	46, 117
$Fe_2(CO)_9$	Bridge-bonded form	46
$Co_2(CO)_8$	Bridge-bonded and metal-metal bonded forms	18
$Rh_2(CO)_8$	Bridge-bonded form analogous to $Co_2(CO)_8$	7
$Ir_2(CO)_8$	Bridge-bonded form analogous to $Co_2(CO)_8$	7
$Ni_2(CO)_7$	Bridge-bonded structure, $(CO)_2Ni(\mu\text{—}CO)_3Ni(CO)_2$	114
$Cu_2(CO)_6$	Metal-metal bonded form	44
$Ag_2(CO)_6$	Metal-metal bonded form	6
$Mn_2(CO_b)$	b = Bridge-bonded CO; t = Terminally-bonded CO	51
$Mn_2(CO_b)_2$		51
$Mn_2(CO_t)$		51
$Ni_2(CO_b)$		114
$Ni_2(CO_t)$		114
$Cu_2(CO_b)$		114
$Cu_2(CO_t)$		114
Dinitrogen complexes		
$Rh_2(N_2)_8$	Two forms probably analogous to $Rh_2(CO)_8$	26
$V_2(N_2)_{12}$	Two forms probably analogous to $V_2(CO)_{12}$	9

we are concerned with matrix synthesis using transition metal atoms, particularly the types of experiments available to the cryogenic chemist who, having synthesized a new compound in a metal-atom–molecule cocondensation reaction, wishes to characterize as completely as possible the structure and properties of the trapped product(s). X-ray crystallographic methods cannot be applied to matrix-isolated species that are, by the very nature of the technique, in a noncrystalline form. Moreover, the diffuse rings that would be obtained in such an experiment would be characteristic of the polycrystalline matrix rather than the molecule of interest. Consequently, as complete a characterization as possible must be attempted by other means, by far the most useful of which are infrared, laser-Raman, uv-visible and esr spectroscopy. A clever application of these techniques can yield information about the identity of the product, stoichiometry with respect to ligand and metal, mode of ligand attachment, molecular and electronic structure, reaction mechanisms, thermodynamic and bonding properties. Each of these categories are illustrated by reference to cocondensation reactions involving transition metals and a variety of small gaseous molecules.

The use of metal vapors as starting materials in chemical reactions is a fairly recent development. Metal atoms formed at high temperatures under high vacuum conditions can be made to combine in a controlled manner with a reacting gas by bringing them together simultaneously onto a surface cooled to cryogenic temperatures (4.2–20°K). Low-temperature matrix cocondensation techniques of this type take full advantage of the extreme reactivity of metal atoms and are currently providing inorganic and organometallic chemists with new chemical pathways for the synthesis and stabilization of chemical species that would have been difficult if not impossible to prepare and study by conventional chemical techniques.

A partial list of the compounds synthesized and characterized using metal-atom-matrix techniques, up to the time of writing, are listed in Tables 1 and 2.

2. PRODUCT IDENTIFICATION

In general, when metal atoms are cocondensed with gaseous molecules, several products are formed simultaneously. For example, when Ni vapor is cocondensed with N_2 in Ar under appropriate conditions of temperature and composition, all four species, $Ni(N_2)$, $Ni(N_2)_2$, $Ni(N_2)_3$, and $Ni(N_2)_4$, are formed.[1] The number of different products formed in a matrix cocondensation reaction is often established by noting the effect of warm-up and varying concentration on the matrix spectra.

Warm-Up Experiments

In experiments of this type, the matrix containing the reaction products formed on deposition is allowed to warm up to various predetermined temperatures and held constant for a period of time governed by the reactivity and mobility of the trapped species. Warming the matrix causes its rigidity to decrease, thereby permitting diffusion of some or all of the species within to take place. Various encounters between the species in the matrix can then be anticipated, and, as a result, stepwise association, dissociation, replacement, and so forth, reactions can be expected to occur. As a result of these matrix reactions, the concentrations of the various reaction products formed in the matrix in the initial deposition change

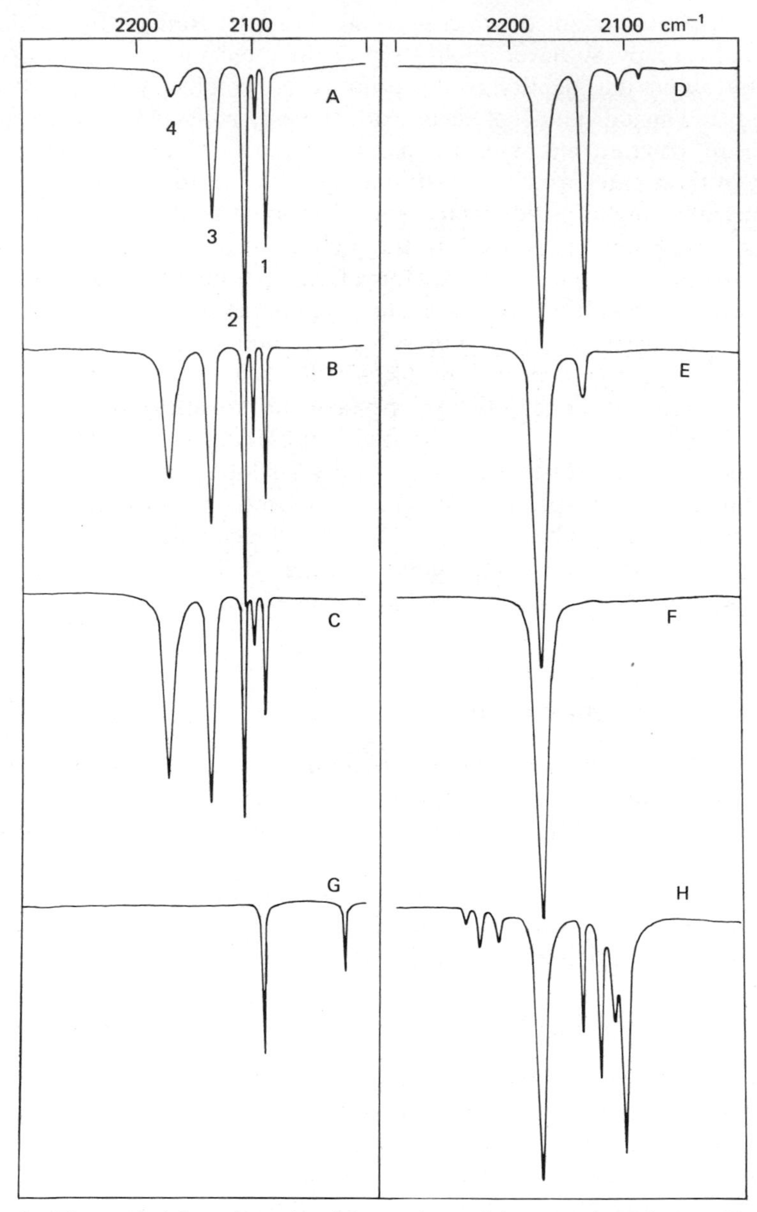

Figure 1. The matrix infrared spectra of the products of the cocondensation reaction of Ni atoms with (*a*) a $^{14}N_2$: Ar = 1 : 10 mixture at 10°K, (*b*)–(*f*) after successive warm-up experiments at 15, 20, 25, 30, and 35°K and then recooled to 10°K, (*g*) a $^{14}N_2$: $^{15}N_2$: Ar = 1 : 1 : 2000 mixture at 10°K, and (*h*) a $^{14}N_2$: $^{15}N_2$: Ar = 1 : 1 : 20 mixture deposited at 10°K, allowed to diffuse at 35°K, and recooled to 10°K.

until either the reaction is quenched by recooling the matrix or only the thermodynamically stable or kinetically inert species at that particular temperature remain.

The matrix reactions occurring during warm-up experiments often reveal which spectral lines belong to an individual species, since those either grow or decay at a constant rate relative to each other, but at a different rate from those belonging to other species. Moreover, warm-up experiments sometimes establish the order of stability of the various species within the matrix. Spectral changes observed in matrix warm-up studies are best represented by graphical plots of band intensity as a function of warm-up temperature or time. This may be illustrated by reference to the matrix infrared spectra of the cocondensation reaction between nickel atoms and N_2:Ar = 1:10 mixtures.[1] *Four* lines are observed in the NN stretching region on the initial deposition (Figure 1*a*). The spectacular warm-up behavior is illustrated in Figure 1*b*–*f*. As the Ni/Ar dilution was very high in these experiments (arranged so that the probability that a nickel atom had another nickel atom as nearest neighbor in the fcc lattice of Ar was less than 1 in 10^3), the warm-up results indicate the formation of *four mononuclear* nickel-dinitrogen complexes of the form, $Ni(N_2)_n$.

Having determined the number of different species present within the matrix, one still has the task of identifying them. This must be done by combining spectral data with chemical intuition and in one way or another forms a large part of the remainder of the chapter. What we do here is to illustrate how warm-up data were used to assign lines to particular chemical species.

When Pd atoms are cocondensed with a CO:Ar = 1:1000 mixture, four CO stretching modes are observed in the infrared spectrum[2] with band intensities at the end of the 10°K deposition in the ratio, $I_{2050}:I_{2044}:I_{2060}:I_{2070} =$ 1:0.125:0.039:0.009 (Figure 2*a*).

Thus a prima facie assignment of the four bands at 2070, 2060, 2050, and 2044 cm^{-1}, respectively, to the four species, $Pd(CO)_4$, $Pd(CO)_3$, Pd(CO), and $Pd(CO)_2$, on the basis of their relative intensities would appear at first sight to be justified, as the relative probabilities of forming $Pd(CO)_n$ decrease monotonically with increasing *n*. The extent to which statistical arguments of this type can be trusted for making vibrational assignments can be investigated by calculating the probability, P_n, of forming a $Pd(CO)_n$ molecule in the fcc lattice of solid Ar.[2] Let us assume, first of all, that the *metal* atoms are isolated from one another and that they occupy substitutional sites in the Ar lattice. We assume also that the CO molecules are in substitutional sites and react with the metal only if it comes within *one* nearest neighbor distance of it. If there is no diffusion of either species before the matrix is warmed up and if the stabilities of the $Pd(CO)_n$ are in the order, $Pd(CO)_4 \gg Pd(CO)_3 \gg$

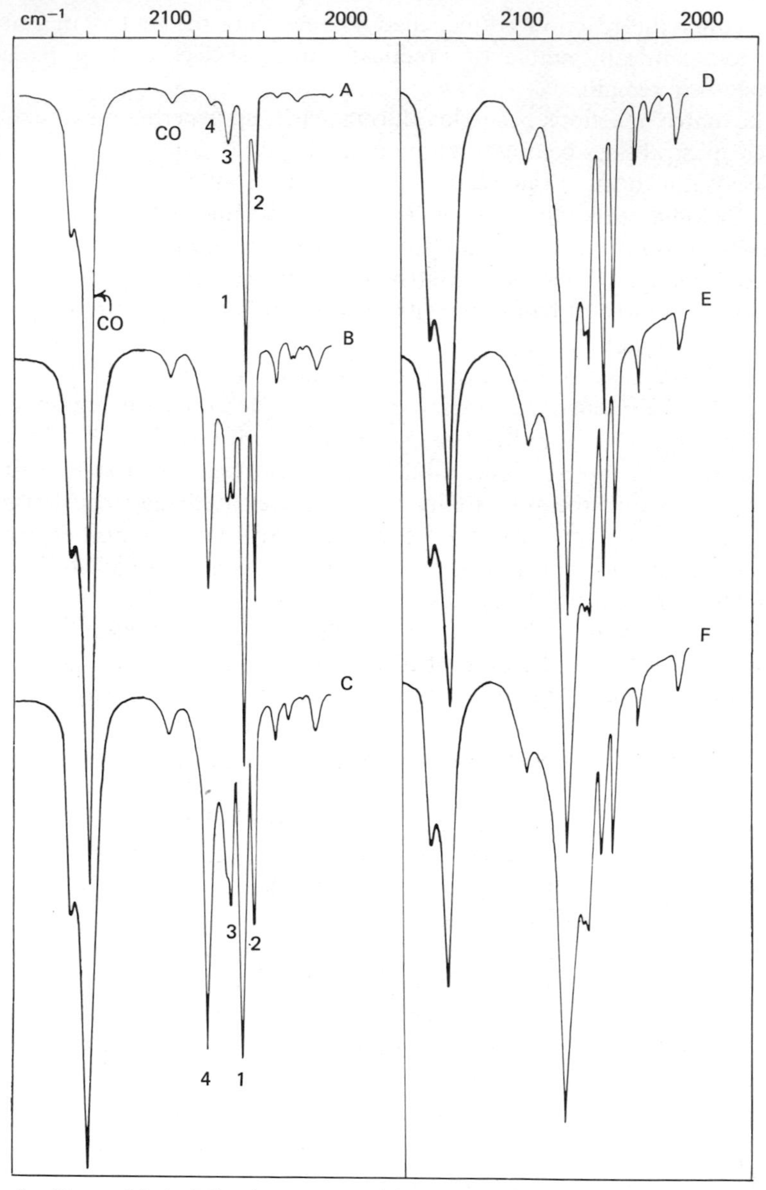

Figure 2. The matrix isolation ir spectrum of the products of the cocondensation reaction between atomic palladium and CO in Ar (1:1000), (*a*) on deposition at 10°K; (*b*)–(*f*) after successive diffusion warm-up experiments at 20–30°K.

$Pd(CO)_2 \gg Pd(CO)$, then P_n is given by

$$P_n = [{}^{12}_{n}C]\,\alpha^n(1 - \alpha)^{12-n}$$

for $n = 1, 2, 3$, and

$$P_4 = \sum_{n=4}^{12} [{}^{12}_{n}C]\,\alpha^n(1 - \alpha)^{12-n}$$

where $[{}^{12}_{n}C]$ is the binomial coefficient and α is the CO:Ar ratio. For $\alpha = 10^{-3}$, the above formulas predict a product ratio, $PdCO:Pd(CO)_2:Pd(CO)_3:Pd(CO)_4$ of $1:5.5 \times 10^{-3}:1.8 \times 10^{-5}:4.1 \times 10^{-8}$, significantly different from that observed. Based on this calculation alone, therefore, one concludes that during deposition there is *considerable* surface diffusion and reaction before the cocondensate is quenched and that assignments of $Pd(CO)_n$ to the four bands on the basis of their intensities is not unambiguous.

An analysis of the diffusion-controlled kinetics during warm-up (Figure 2*b*–*f*) (see Section 12.3 for details) indicated, however, that the tentative assignment was correct. Briefly, it was shown that the initial slope of the rate of change of concentration of the four species during warm-up was only consistent with the initial assignments. Thus, although assignments based on band intensities can be misleading, they are often useful in making tentative assignments.

3. DETERMINATION OF STOICHIOMETRY

A knowledge of product stoichiometry in a metal-atom–molecule cocondensation reaction involves the determination of n and m in an equation of the type

$$n\text{M} + m\text{X} \rightarrow \text{M}_n\text{X}_m$$

where M is the metal atom and X is the substrate molecule.

In this section we are concerned only with methods for defining *ligand* stoichiometry. The approaches for defining stoichiometry with respect to the metal are discussed in Chapter 9.

Generally speaking, there are three main approaches for defining the ligand stoichiometry of the reaction products. These are ligand-concentration dependence, ligand-isotope substitution, and mixed-ligand studies. Depending on the nature of the system under investigation, some or all of these techniques may be employed and, in principle, are capable of yielding vibrational data that can be used to define unambiguously the ligand stoichiometry of the product(s).

3.1 Ligand-Concentration Dependence

Having established the *number* of different products formed in a cocondensation reaction from their spectral behavior during warm-up, the identity of the lowest and highest coordination complexes with respect to the ligand can, in principle, be determined by ligand-concentration-dependence studies.

Lowest Stoichiometry Complexes

As described earlier for $Pd(CO)_n$, statistical factors favor the lowest coordination compounds in dilute matrices, where at $M:A \leq 1:250$ the monoligand species usually predominate. The following are some examples of *mononuclear* monoligand complexes formed under dilute matrix conditions.

$$Ni + N_2 \rightarrow Ni(N_2) \tag{1}$$

$$Ni + CO \rightarrow Ni(CO) \tag{2}$$

$$Ni + O_2 \rightarrow Ni(O_2) \tag{3}$$

In all of the above cases, the vibrational spectra of the monoligand complexes display a *single* line in the ligand-stretching region. Often these lines may be split as a result of matrix site effects that are considered in Section 8. Although the ligand-stretching modes of these complexes are infrared and Raman active for both homonuclear and heteronuclear diatomic ligands, the infrared absorption intensity associated with (1) and (3) (particularly, when sideways bonded to a metal) might be very small, or even nonobservable. The corresponding line in the Raman spectrum, however, may be quite intense.

Highest Stoichiometry Complexes

The spectrum of the highest coordination complex can often be associated with the product remaining after warm-up, providing the ligand dilution is such that reaction to completion is possible. Supporting evidence can usually be obtained by cocondensing the metal with the pure ligand, under which circumstances the ligand is acting both as reactant and matrix (referred to as a reactive matrix). Alternatively, the metal can be cocondensed with concentrated matrices ($M:A \geq 1:10$), but at annealing temperatures (usually 20–30°K), considerable surface and bulk diffusion can be expected to take place and the highest stoichiometry complex usually forms. Some metal-carbon monoxide cocondensation reactions that yield the product of highest stoichiometry in pure or concentrated matrices are shown below:

$$Au + CO \rightarrow Au(CO)_2 \tag{4}$$

$$Ag + CO \rightarrow Ag(CO)_3 \tag{5}$$

$$Ir + CO \rightarrow Ir(CO)_4 \tag{6}$$

$$Re + CO \rightarrow Re(CO)_5 \tag{7}$$

$$V + CO \rightarrow V(CO)_6 \tag{8}$$

In the warm-deposition method it is of central importance to have a knowledge of the diffusing species in the matrix reaction. This, of course, depends on a number of factors relating to the nature of the matrix material (melting point, size of sites, latent heat of condensation, etc.), diffusion properties of the metal atom and ligand, as well as the temperature of the cold surface and rate of deposition of the matrix gas (see Chapter 9).

3.2 Mixed-Ligand Isotopic Substitution

Probably one of the most informative and reliable methods for defining the stoichiometry and structure of a matrix-isolated species involves the utilization of mixed-isotopic substitution experiments. Two distinct situations can be considered. On the one hand, one can cocondense an isotopically *pure* metal atom with a *mixture* of isotopic ligands such as ${}^{n}X_2/{}^{m}X_2$; alternatively one can cocondense a *mixture* of metal atom isotopes ${}^{p}M/{}^{q}M$ with a *pure* ligand. Barring an isotope effect in the kinetics of formation, the product distribution can be expected to be statistical, and an analysis of the number, frequencies, and infrared and Raman intensities of the isotopic components can, in principle, determine uniquely the product stoichiometry with respect to both ligand and metal, the molecular structure of the product, the vibrational assignments, and force fields.

Studies of this type are fundamental to almost any matrix isolation experiment, and what follows is a selection of examples that serve to define ligand stoichiometry in mononuclear complexes.

For example, the Pt/CO/Ar cocondensation reaction has been studied both in the infrared and the Raman[10] at 20–25°K, and the proposed product was the then unknown $Pt(CO)_4$ molecule (Figure 3). Although the data were consistent with the product being a regular tetrahedral tetracarbonyl, one could not rule out the possibility of its being, for example, a tricarbonyl $Pt(CO)_3$ or a dicarbonyl species $Pt(CO)_2$, both of which could have spectra similar to those observed. However, when Pt atoms were cocondensed with ${}^{12}C^{16}O:{}^{13}C^{16}O:Ar = 1:1:20$ mixtures[11] at 20–25°K (Figure 4), the matrix infrared spectrum showed *five* lines at 2053.3, 2033.0, 2021.9, 2013.9, and 2006.9 cm^{-1}, as well as the two very intense absorptions at 2138 and

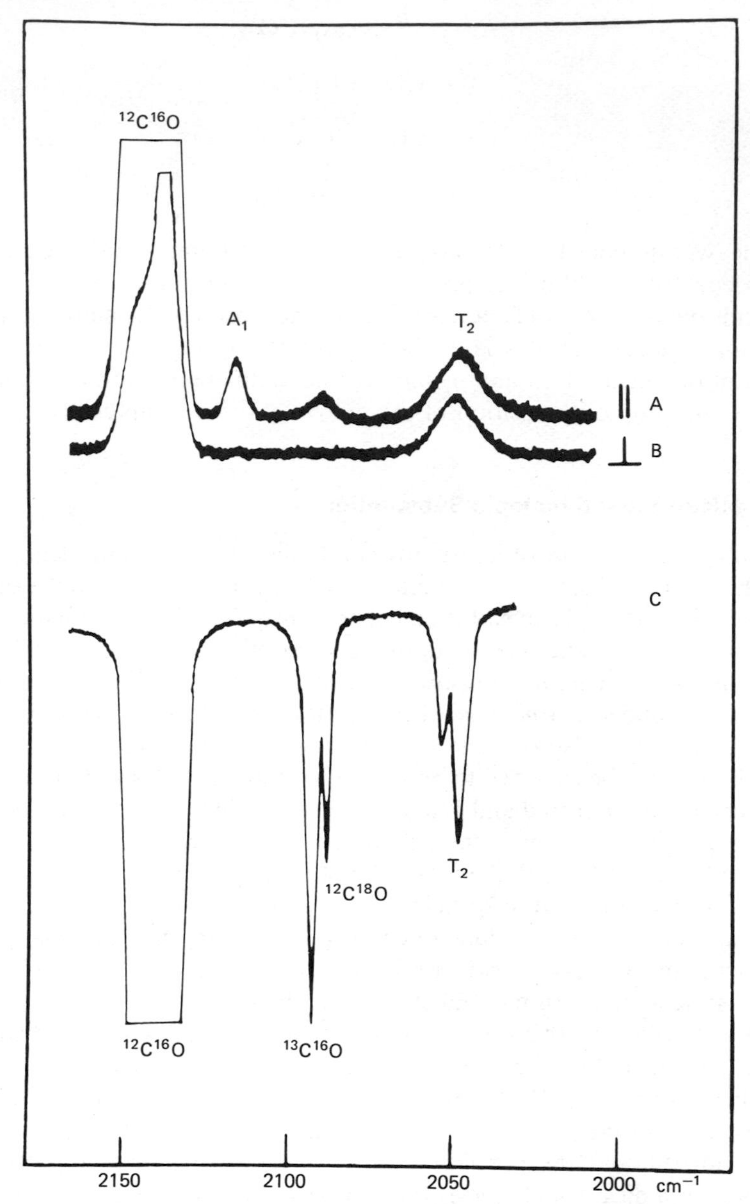

Figure 3. The matrix Raman and infrared spectra of the products of the cocondensation reaction of Pt atoms with $^{12}C^{16}O$ at 6–10°K, (*a*) Raman parallel, (*b*) Raman crossed polarization, and (*c*) infrared, showing $Pt(^{12}C^{16}O)_4$.

2093 cm^{-1} (due to unreacted $^{12}C^{16}O$ and $^{13}C^{16}O$). A very weak line was also observed at approximately 2108 cm^{-1} (see Figure 4 and Table 3).

It is easily seen that the maximum number of distinct infrared active CO stretching modes expected for the five mixed-isotopic "tetracarbonyl" molecules, $M(^{12}C^{16}O)_n(^{13}C^{16}O)_{4-n}$ ($n = 0$–4), shown below is eight:

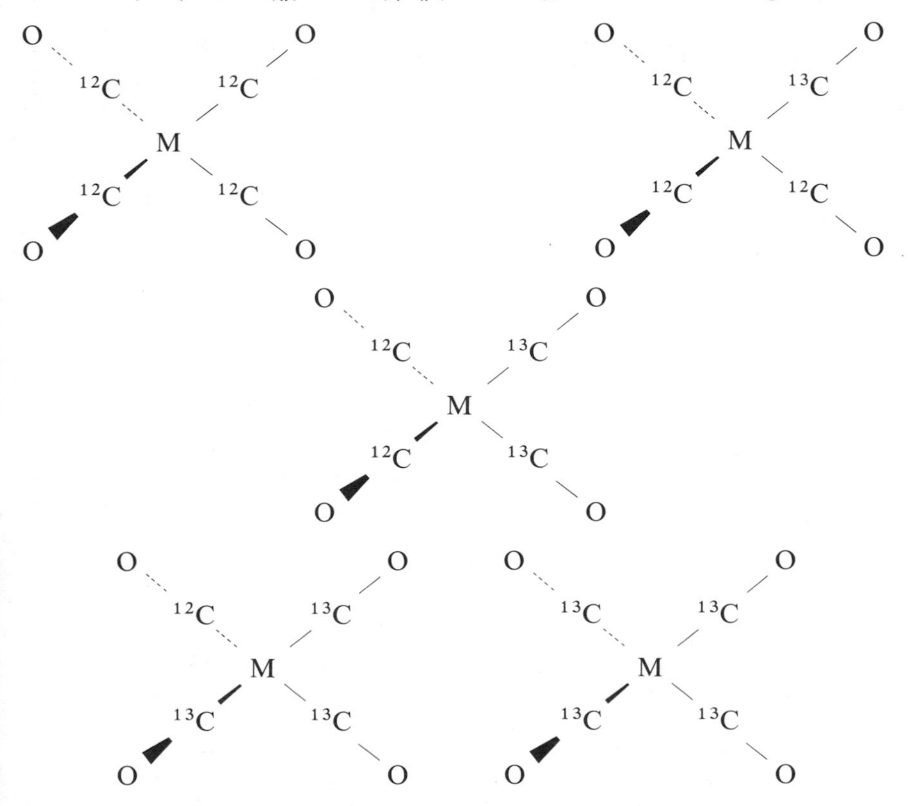

The *six* observed lines listed above were tentatively assigned to these species. The two missing lines were shown by frequency calculations (Sections 6 and 7) to overlap with the unreacted $^{13}C^{16}O$. The close agreement between the calculated and observed frequency and intensities when tetrahedral geometry was assumed provided simultaneous support for the vibrational assignments as well as the assumed structure. Therefore, it was concluded that $Pt(CO)_4$ is a regular tetrahedral molecule.

A similar set of infrared data was obtained for the Ni/N_2 reaction using $^{14}N_2$:$^{15}N_2$:Ar = 1:1:20 mixtures. In this case, overlap of free ligand and bonded ligand modes did not occur since the NN stretching mode of unreacted N_2 is infrared inactive. The full *eight*-line isotope pattern was observed in the NN stretching region (Figure 1*h*, Table 4), proving that the

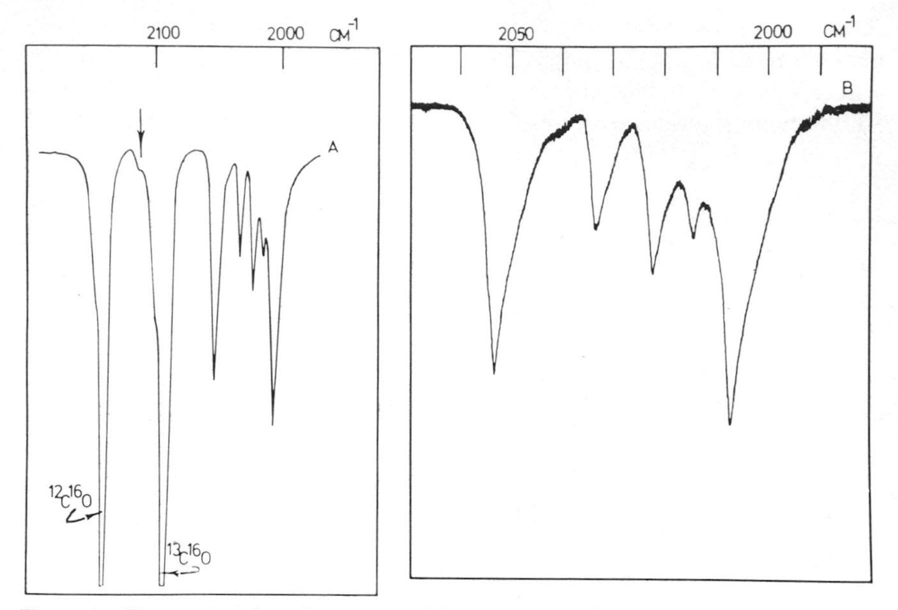

Figure 4. The matrix infrared spectrum of the products of the cocondensation reaction of Pt atoms with $^{12}C^{16}O:^{13}C^{16}O:Ar = 1:1:20$ at 20–25°K, showing the formation of the species, $Pt(^{12}C^{16}O)_n(^{13}C^{16}O)_{4-n}$ (where $n = 0$–4), (*a*) complete spectrum, normal scan, (*b*) high-resolution scan of the 2060–2000 cm^{-1} region.

Table 3. Isotopic Frequency and Infrared Absorption Intensity Calculations for the CO-Stretching Modes of $Pt(^{12}C^{16}O)_n(^{13}C^{16}O)_{4-n}$ (where $n = 0$–4) in Solid Argon and Best-Fit Cotton–Kraihanzel Force Constants[a]

Observed Frequency	Calculated Frequency	Observed Intensity[b]	Calculated Intensity	Assignment
2006.9	2007.2	10.000	10.000	$III(B_2) + IV(E) + V(T_2)$
2013.9	2014.0	1.615	1.303	$II(A_1)$
2021.9	2022.2	3.281	2.461	$III(A_1)$
2033.0	2032.7	1.856	2.420	$IV(A_1)$
2053.3	2052.8	6.258	6.561	$I(T_2) + II(E) + III(B_1)$
—	2088.1	—	0.015	$IV(A_1)$
—	2099.0	—	0.317	$III(A_1)$
~2108[c]	2107.4	—	0.074	$II(A_1)$

Table 3. (continued)

Notation	Isotopic Molecule	Point Symmetry
I	$Pt(^{12}C^{16}O)_4$	T_d
II	$Pt(^{12}C^{16}O)_3(^{13}C^{16}O)$	C_{3v}
III	$Pt(^{12}C^{16}O)_2(^{13}C^{16}O)_2$	C_{2v}
IV	$Pt(^{12}C^{16}O)(^{13}C^{16}O)_3$	C_{3v}
V	$Pt(^{13}C^{16}O)_4$	T_d

[a] The best-fit Cotton–Kraihanzel CO force constants were $k_{CO} = 17.28$ and $k_{CO \cdot CO} = 0.26$ mdyn/Å.
[b] The observed intensities were measured by assuming a triangular contour for the infrared band and using the expression

$$I = \Delta v_{1/2}\left(1 + \frac{\gamma \log_e \gamma}{1 - \gamma}\right)$$

where $\Delta v_{1/2}$ = width of the infrared band at half height, γ = fractional transmittance, and $I = -\int \log_e \gamma \, dv$.
[c] Very weak line indicated by arrow in Figure 7.

Table 4. Isotopic Frequency and Infrared Absorption Intensity Calculations for the NN Stretching Modes of $Ni(^{14}N_2)(^{15}N_2)_{4-n}$ in Solid Argon[a]

Observed Frequency	Calculated Frequency[b]	Observed Intensity[c]	Calculated Intensity	Assignment (see text for notation)
2102.0	2101.6	8.15	9.3	T_2(V), E(IV), B_2(III)
2112.0	2111.2	2.42	2.1	A_1(II)
2123.0	2122.7	2.87	2.7	A_1(III)
2138.0	2137.8	1.20	1.7	A_1(IV)
2174.0	2175.3	10.00	10.00	T_2(I), E(II), B_1(III)
2212.0	2212.2	0.48	0.57	A_1(IV)
2228.0	2227.9	0.63	0.69	A_1(III)
2240.0	2240.1	0.14	0.14	A_1(II)

[a] Note that this is the regular tetrahedral form of $Ni(N_2)_4$ (see text).
[b] The best fit NN force constants were $f_r = 19.84$, $f_{rr} = 0.34$ mdyn/Å.
[c] The observed intensities were measured by the method described for $Pt(CO)_4$.

product was $Ni(N_2)_4$, having regular tetrahedral symmetry and thereby implying "end-on linear" bonding for the dinitrogen ligand. Aside from nickel tetradinitrogen being isoelectronic and isostructural with nickel tetracarbonyl and the first example of a dinitrogen complex containing more than two dinitrogen ligands, it proved to be particularly interesting spectroscopically in that the molecule could be shown to be regular tetrahedron in argon but a distorted tetrahedron in pure nitrogen.[1] This aspect of the work is discussed in Section 8.2.

3.3 Mixed-Ligand Studies

In Chapters 6 and 9 it is demonstrated that mixed-metal studies can be used to define the metal content of the product(s) of a cocondensation reaction and hence are useful in verifying stoichiometry and molecular structure, as well as aiding vibrational assignments. The success of this technique depends to a large extent on the ability to be able to select metals whose chemical properties are very similar and thereby obviate any tendency for the species to undergo stoichiometric and structural changes.

Any change of the ligand, however, in a metal-ligand cocondensation reaction would be expected to be more dramatic and would inevitably affect the nature of the chemistry involved. Data of this kind would of course usually be of little direct value in defining the stoichiometry of the original "nonmixed" product.

However, the mixed-ligand complexes are, in themselves, of considerable chemical interest, particularly when assessing the bonding characteristics of different ligands coordinated to the same metal atom and when directly competing with each other for bonding electrons. This aspect of mixed-ligand studies is discussed in detail in Section 11 on molecular properties.

For the purposes of the present discussion we restrict ourselves to the use of mixed-ligand isotopic substitution for defining the individual ligand stoichiometries in some mixed-ligand complexes. For example, in the cocondensation reaction of nickel atoms with dilute CO/Ar matrices, De Kock[3] was able to assign the four carbonyl stretching modes associated with the compound $Ni(CO)_n$ (where $n = 1$–4).

Recently, the cocondensation reaction of nickel atoms with dilute N_2/Ar matrices[1] was shown to yield $Ni(N_2)_n$ (where $n = 1$–4). Warm-up of both of these systems yielded the most stable species, namely $Ni(CO)_4$ and $Ni(N_2)_4$. It is immediately obvious from these data that cocondensation of nickel atoms with N_2/CO/Ar mixtures should provide a simple and direct route to the synthesis of binary mixed-carbonyl–dinitrogen complexes of nickel.[12] Two different approaches were used to achieve the desired "tetracoordinate" complexes $Ni(CO)_n(N_2)_{4-n}$ ($n = 0$–4): (a) cocondensation of Ni atoms with $CO{:}N_2{:}Ar$ mixtures (total $CO{:}N_2{:}Ar = 1{:}10$) at 6–10°K, which yields all of the lower coordination number, mixed species, $Ni(CO)_x(N_2)_y$ (where $x + y \leq 4$), followed by matrix annealing at 30–35°K to yield the tetracoordinate species, or (b) cocondensation of Ni atoms with the same $CO/N_2/Ar$ mixtures, but at approximately 30°K. Method (b) apparently yields a thermodynamically equilibrated mixture, since little change could be induced by further annealing, while method (a) required prolonged annealing to achieve the same results. The infrared and Raman spectra of the products using mixtures of $^{12}C^{16}O$ and $^{14}N_2$ with various amounts of argon were consistent with the formation of the *five* possible tetracoordinate molecules, $Ni(CO)_n(N_2)_{4-n}$ ($n = 0$–4) (Figures 5 and 6 and Table 5). Isotope-substitution

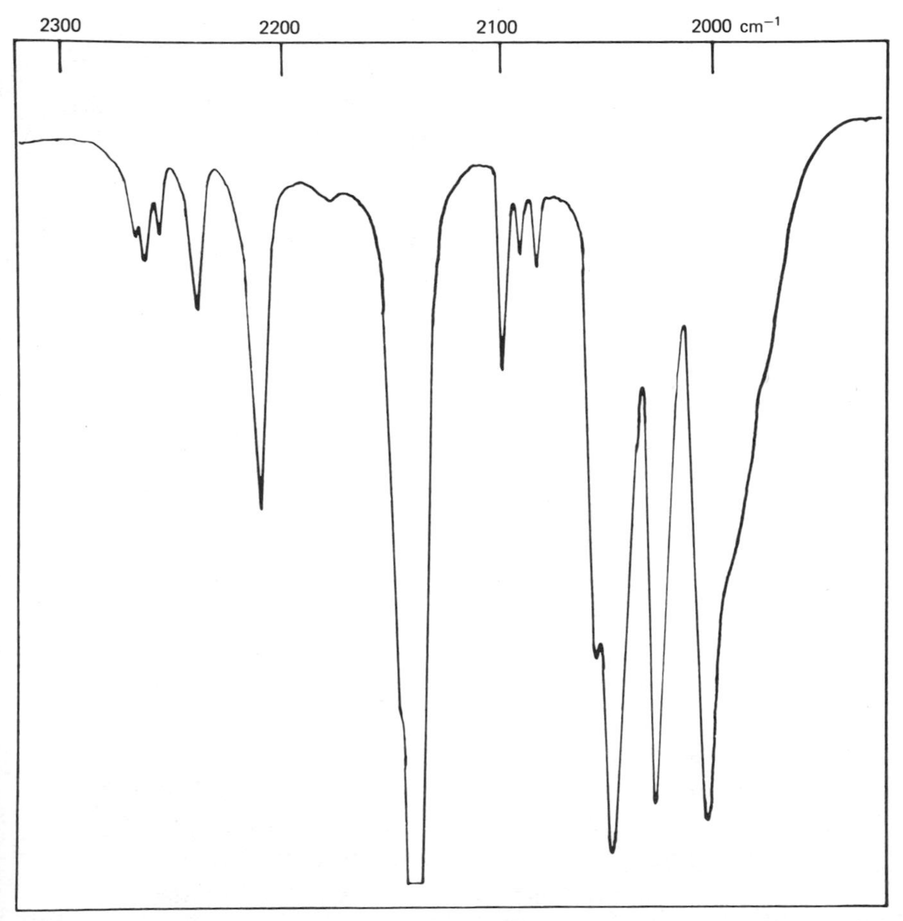

Figure 5. The matrix infrared spectrum of the products of the cocondensation reaction of Ni atoms with a $^{12}C^{16}O$: $^{14}N_2$: Ar = 1:1:20 mixture at 10°K, showing the formation of $Ni(CO)_n(N_2)_{4-n}$ (where n = 0–4).

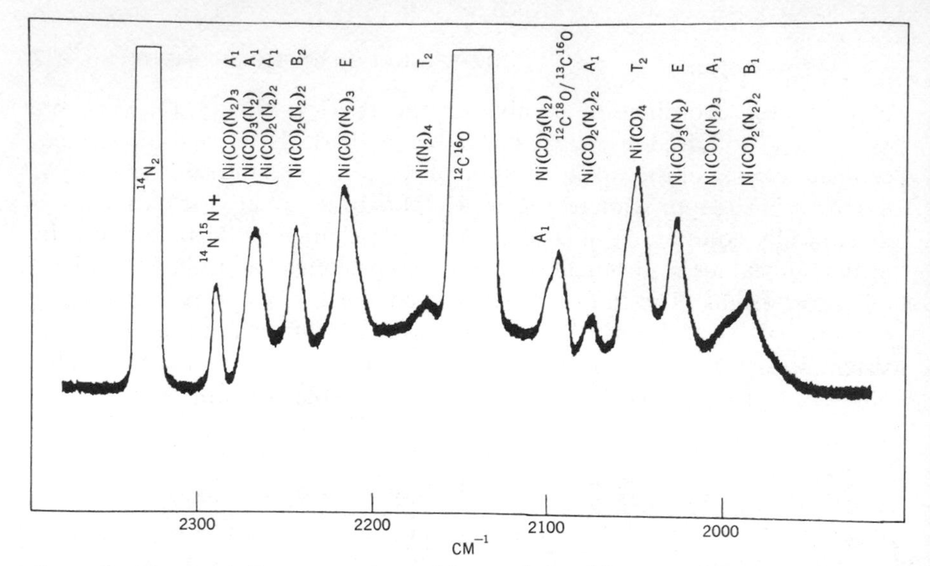

Figure 6. The matrix Raman spectrum of the products of the cocondensation reaction of Ni atoms with $^{12}C^{16}O : ^{14}N_2 = 1:3$ at 25–30°K (spectrum recorded at 6°K), showing all of the species, $Ni(CO)_n(N_2)_{4-n}$ (where $n = 0$–4).

Table 5. Observed and Calculated Frequencies and Observed Integrated Absorbances due to $Ni(CO)_m(N_2)_{4-m}$ (where $m = 1, 2, 3$) Obtained by Cocondensing Ni Atoms with $^{12}C^{16}O$ and $^{14}N_2$ in Argon Matrices

ν_{obsd}	ν_{calc}	A[a]	Mode	Assignment
2269.6	2268.0	0.393	$\nu NN(A_1)$	$Ni(CO)_2(N_2)_2$
2264.4	2264.3	0.622	$\nu NN(A_1)$	$Ni(CO)_3N_2$
2257.6	2257.7	0.227	$\nu NN(A_1)$	$NiCO(N_2)_3$
2240.0	2240.0	0.672	$\nu NN(B_2)$	$Ni(CO)_2(N_2)_2$
2210.3	2210.4	2.240	$\nu NN(E)$	$NiCO(N_2)_3$
2100.6	2100.6	0.629	$\nu CO(A_1)$	$Ni(CO)_3N_2$
2084.3	2086.2	0.230	$\nu CO(A_1)$	$Ni(CO)_2(N_2)_2$
2049.7 } 2058.6 }	—	20.950	$\nu CO(T_2)$	$Ni(CO)_4$
2029.4	2030.6	8.950	$\nu CO(E)$	$Ni(CO)_3N_2$
2005.0	2005.9	11.280	$\nu CO(A_1)$	$NiCO(N_2)_3$
1991.3	1992.2	10.820	$\nu CO(B_1)$	$Ni(CO)_2(N_2)_2$

[a] Integrated absorbance (units cm^{-1}) $= \int \log \frac{I_0}{I} d\nu$. Where lines overlap, as, for example, those at 2269.6, 2264.4, and 2257.6, A was obtained by fitting Lorentz line shapes to the bands and integrating numerically.

experiments were performed using $^{12}C^{18}O$ with $^{14}N_2/^{15}N_2/Ar$ mixtures of various ratios. The isotope multiplet patterns served to characterize the three mixed species, $Ni(CO)_3(N_2)$, $Ni(CO)_2(N_2)_2$, and $Ni(CO)(N_2)_3$, and placed the ligand assignments on a firm basis.

The interpretation of the data can best be illustrated by reference to $Ni(N_2)_3(CO)$. Tentative assignments from $^{12}C^{16}O$:$^{14}N_2$:Ar = 1:9:20 experiments are shown below:

$Ni(N_2)_3(CO)$	$\nu_{A_1}^{CO}$	2005.0 cm^{-1}
(C_{3v})	$\nu_{B_2}^{NN}$	2210.3 cm^{-1}
	$\nu_{A_1}^{NN}$	2257.6 cm^{-1}

The observed $^{12}C^{18}O$ isotopic shift for $Ni(N_2)_3(CO)$ from 2005 cm^{-1} to 1959 cm^{-1} is in excellent agreement with the calculated value using an isotopic mixture, $^{12}C^{18}O$:$^{14}N_2$:$^{15}N_2$:Ar = 1:1:1:30. The multiplet structure of the NN stretching mode at 2210 cm^{-1} was carefully examined under conditions of high resolution (see, for example, Figure 7).

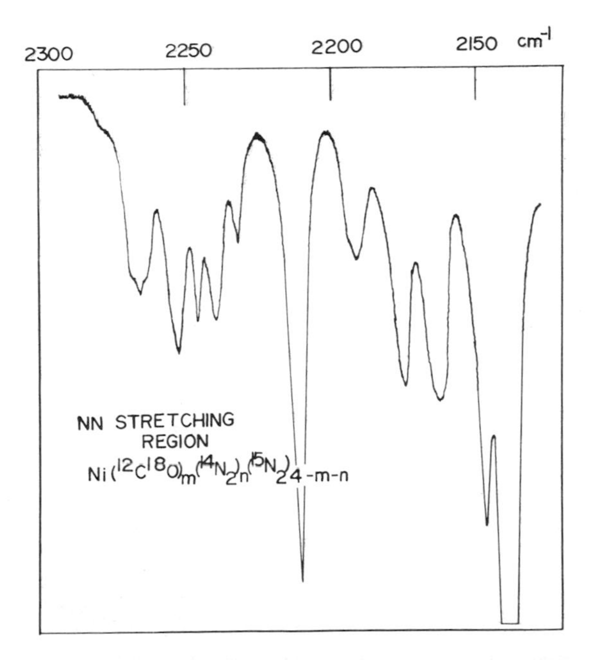

Figure 7. The matrix infrared spectrum of the products of the cocondensation reaction of Ni atoms with $^{12}C^{18}O$: $^{14}N_2$: $^{15}N_2$: Ar = 1 : 1 : 1 : 30, showing the ^{14}N and ^{15}N isotope multiplet structure for the series of complexes, $Ni(^{12}C^{18}O)_m(^{14}N_2)_n(^{15}N_2)_{4-n-m}$ (where m = 1–3 and n = 0–3).

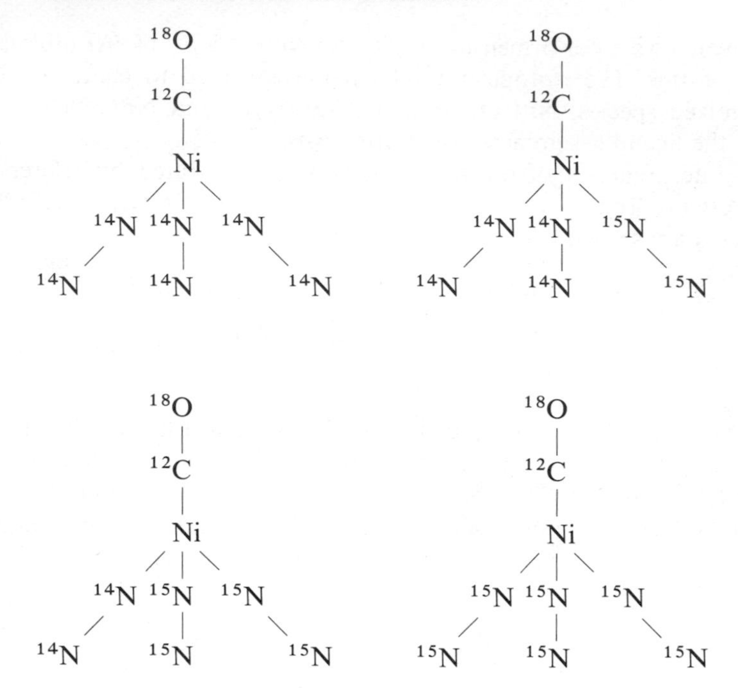

Four isotopic molecules for $Ni(N_2)_3(CO)$ are expected using $^{12}C^{18}O/^{14}N_2/^{15}N_2/Ar$ isotopic mixtures as shown above. Because the nitrogen-15 isotopic shift for the E type infrared NN stretching mode of $Ni(N_2)_3(CO)$ is expected to be about 75 cm^{-1}, it was necessary to use $^{12}C^{18}O$ isotopic substitution to shift the interfering strong $^{12}C^{16}O$ absorption (2138–2148 cm^{-1}) to lower wavenumbers.

The expected $^{14}N_2/^{15}N_2$ multiplets for a tris-dinitrogen molecule assuming C_{3v} symmetry is then a *six*-line pattern (the two A_1 νNN modes of $Ni(^{14}N_2)_3(^{12}C^{18}O)$ and $Ni(^{15}N_2)_3(^{12}C^{18}O)$ are expected to be very weak), that is, a triplet in the $^{14}N_2$ region and a triplet in the $^{15}N_2$ region. The $^{14}N_2$ triplet was well resolved, although the lowest component of the $^{15}N_2$ triplet expected at 2135 cm^{-1} was obscured by the 7% residual $^{12}C^{16}O$ in the isotopic $^{12}C^{18}O$ and the highest component overlapped with the B_2 $^{15}N_2$ stretching mode of $Ni(^{12}C^{18}O)_2(^{15}N_2)_2$ (Table 6). These data served to characterize $Ni(N_2)_3(CO)$, and from analogous experiments $Ni(CO)_2(N_2)_2$ and $Ni(CO)_3(N_2)$ were similarly characterized.

4. DETERMINATION OF MOLECULAR STRUCTURE

In the pre-matrix Raman period it was often difficult if not impossible to make reliable structural and vibrational assignments for new chemical species

Table 6. Observed and Calculated Frequencies due to $Ni(CO)_m(N_2)_{4-m}$ (where $m = 1, 2, 3$) Obtained by Cocondensing Ni Atoms with $^{12}C^{18}O$ and $^{14}N_2/^{15}N_2$ Mixtures in Argon Matrices

ν_{obsd}	ν_{calc}	Assignment
2265 (br)[a]	2268.0	(A_1) $Ni(C^{18}O)_2(^{14}N_2)_2$
	2264.3	(A_1) $Ni(C^{18}O)_3{}^{14}N_2$
	2257.7	(A_1) $Ni(C^{18}O)(^{14}N_2)_3$
2253	2255.6	(A') $Ni(C^{18}O)_2{}^{14}N_2{}^{15}N_2$
2246	2245.9	(A') $Ni(C^{18}O)(^{14}N_2)_2{}^{15}N_2$
2240	2239.1	(B_2) $Ni(C^{18}O)_2(^{14}N_2)_2$
2232	2232.1	(A') $Ni(C^{18}O)^{14}N_2(^{15}N_2)_2$
2211	2210.4	(E) $Ni(C^{18}O)(^{14}N_2)_3$, (A'') $Ni(C^{18}O)(^{14}N_2)_2{}^{15}N_2$
2191	2191.7	(A_1) $Ni(C^{18}O)_2(^{15}N_2)_2$, (A_1) $Ni(C^{18}O)_3{}^{15}N_2$
2174	2175.0	(A') $Ni(C^{18}O)_2{}^{14}N_2{}^{15}N_2$
2161	2163.2	(B_2) $Ni(C^{18}O)_2(^{15}N_2)_2$, (A') $Ni(C^{18}O)(^{14}N_2)(^{15}N_2)_2$
2147	2146.6	(A') $Ni(C^{18}O)(^{14}N_2)_2{}^{15}N_2$
2138	2135.4	(A'') $Ni(C^{18}O)^{14}N_2(^{15}N_2)_2$, (E) $NiC^{18}O(^{15}N_2)_3$ and $^{12}C^{16}O$[b]
2053		(T_2) $Ni(^{12}C^{16}O)_4$[b]
2046	2046.4	(A_1) $Ni(C^{18}O)_3(^{14}N_2)$, (A_1) $Ni(C^{18}O)_3(^{15}N_2)$
2031	2034.0	(A_1) $Ni(C^{18}O)_2(^{14}N_2)_2$, (A') $Ni(C^{18}O)_2(^{14}N_2)(^{15}N_2)$
		(A_1) $Ni(C^{18}O)_2(^{15}N_2)_2$
1983	1981.7	(E) $Ni(C^{18}O)_3(^{14}N_2)$, (E) $Ni(C^{18}O)_3(^{15}N_2)$
1959	1958.0	(A') $Ni(C^{18}O)(^{14}N_2)_2(^{15}N_2)$
	1957.7	(A_1) $Ni(C^{18}O)(^{15}N_2)_3$, (A_1) $Ni(C^{18}O)(^{14}N_2)_3$
	1957.4	(A') $Ni(C^{18}O)(^{14}N_2)(^{15}N_2)_2$
1945	1944.2	(B_1) $Ni(C^{18}O)_2(^{14}N_2)_2$
		(A'') $Ni(C^{18}O)_2(^{14}N_2)(^{15}N_2)$
		(B_1) $Ni(C^{18}O)_2(^{15}N_2)_2$

[a] Broad band.
[b] 7% $^{12}C^{16}O$ impurity in commercially available $^{12}C^{18}O$ (93%).

with the matrix-infrared data alone. The need for reliable matrix Raman data was obvious, and early studies by Andrews,[13] Claassen,[14] Nibler,[15] and Ozin[16] unequivocally demonstrated the usefulness of the matrix-Raman experiment for obtaining the data complementary to any infrared work. Their results demonstrated that the spectral sensitivities and resolution that can be achieved were often comparable with the complementary infrared data. Nibler[15] and Ozin[17] also demonstrated the viability of matrix-Raman-depolarization measurements, which added a valuable new tool for matrix-isolation studies. The success of the depolarization measurement can be easily understood by applying the cold-gas model to matrix-Raman spectra, where the guest species are assumed to be a perfectly random collection of non-interacting particles in a weakly interacting host. In this treatment of Raman-depolarization measurements, the isotropic optical behavior of a transparent medium (for example, the fcc lattice of the solid inert gases) facilitates the observations. Provided the matrix support is transparent, depolarization of the incident laser beam and of the scattered Raman light should be minimal, and experimental depolarization ratios should be meaningful for matrix-isolated species. Although perfectly transparent matrices are difficult to prepare in practice and nonideal depolarization ratios inevitably result from translucent matrices, the data do permit symmetry assignments. Using the classical expressions for randomly oriented molecules in a matrix (or fluid), it is found that for ideal Raman scattering from matrices (with 90° laser illumination collection) the same equations apply as for fluids, $0 \leq \rho_p \leq \frac{3}{4}$, and a band may be assigned to a totally symmetrical mode if its depolarization ratio is less than $\frac{3}{4}$.

Recently matrix infrared and Raman data have become available for the products of a number of transition metal atom cocondensation reactions, and in this section some of the examples that prove the usefulness of the combined techniques are discussed.

4.1 Platinum Tetracarbonyl, $Pt(CO)_4$

When Pt atoms were cocondensed with pure $^{12}C^{16}O$ at 10°K, a strong absorption at 2047.8 cm^{-1} with a weaker one at 2055.0 cm^{-1} was observed in the matrix infrared spectrum (Figure 3). The corresponding matrix Raman experiment showed an unresolved broad line centered at about 2049 cm^{-1} and a weaker line at 2119 cm^{-1}, where the former was depolarized and the latter was "totally" polarized. The matrix infrared and Raman activities and depolarization measurements are thus consistent with the product being a regular tetrahedral tetracarbonyl. These data, together with the excellent agreement between the observed and calculated infrared isotopic frequencies and absorption intensities when tetrahedral geometry was assumed for $Pt(CO)_4$, provide convincing simultaneous support for both the vibrational assignments as well as the regular tetrahedral structure.[11]

4.2 Binary Dinitrogen Complexes of Palladium, $Pd(N_2)_n$ ($n = 1$–3)

When Pd atoms were cocondensed with pure N_2, the matrix infrared and Raman spectra shown in Figure 8 were obtained. The corresponding

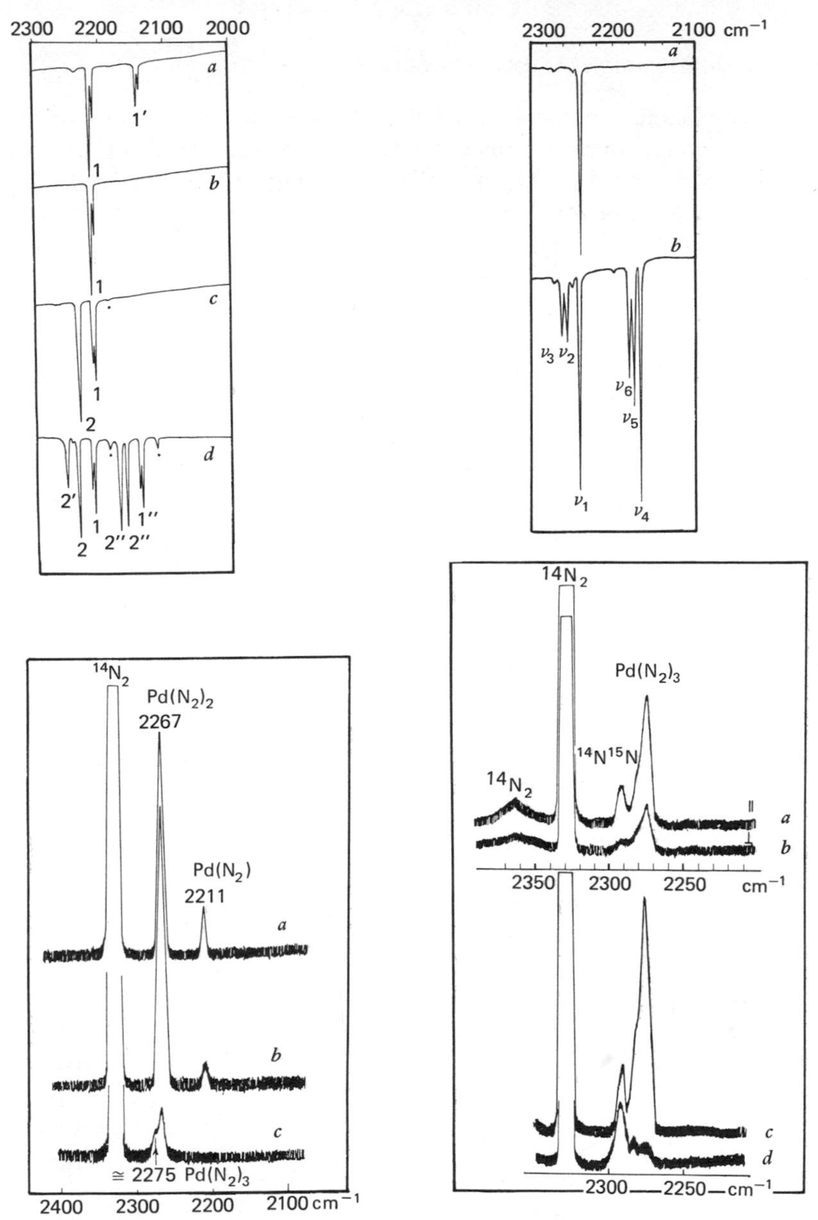

Figure 8. Top left: The matrix infrared spectra of the cocondensation reaction of Pd atoms with (*a*) a $^{14}N_2:^{15}N_2:Ar = 1:1:2000$ mixture at 10°K, (*b*) a $^{14}N_2:Ar = 1:1000$ mixture; (*c*) a $^{14}N_2:Ar = 1:10$ mixture; (*d*) a $^{14}N_2:^{15}N_2:Ar = 1:1:20$ mixture. Top right: The matrix infrared spectrum of the cocondensation reaction of Pd atoms with (*a*) pure $^{14}N_2$ and (*b*) an equimolar mixture of $^{14}N_2:^{15}N_2$ at 10°K. Bottom left: The matrix Raman spectrum of the cocondensation reaction of Pd atoms with a $^{14}N_2:Ar = 1:10$ mixture, (*a*) at 6°K; (*b*)–(*c*) after successive diffusion-controlled warm-up experiments at 20°K and 34°K and recooled to 6°K. Bottom right: The matrix Raman spectrum of the cocondensation reaction of Pd atoms with pure $^{14}N_2$ at 6°K, (*a*) parallel polarization, (*b*) crossed polarization, (*c*) after diffusion-controlled warm-up at 30°K and recooled to 6°K (see Ref. 11).

$^{14}N_2/^{15}N_2$ infrared experiment yielded an isotope multiplet pattern characteristic of a trisdinitrogen complex, $Pd(N_2)_3$, as the species with the highest stoichiometry (cf. $Ni(N_2)_4$ in pure N_2). The slight infrared activity of the lines at 2251 and 2274 cm^{-1} suggested that in the α-N_2 lattice the formally D_{3h} original planar $Pd(N_2)_3$ molecule does not have a threehold axis of symmetry. Inspection of the α-N_2 lattice indicates a C_2 site symmetry perturbation due to the inclination of N_2 molecules to the $Pd(N_2)_3$ molecular plane. Hence, the infrared active E′ νNN stretching mode is split into an A and a B mode (2251 and 2241 cm^{-1}) under C_2 symmetry, with the $A_1'(D_{3h}) \rightarrow A(C_2)$ Raman active mode (2274 cm^{-1}) becoming weakly infrared active. Assuming the Cotton–Kraihanzel approximation and C_2 symmetry for the $Pd(N_2)_3$ molecule, the force constants were calculated by applying a least-squares analysis to the observed isotopic frequencies.[1] The agreement between the observed and calculated frequencies and infrared absorption intensities (Table 7) is excellent and provides convincing evidence in favor of the C_2 substitutional site symmetry for $Pd(N_2)_3$ in solid α-N_2.

When Pd atoms were cocondensed with N_2:Ar = 1:10 matrices, the infrared and Raman spectra shown in Figure 8 were observed. The warm-up

Table 7. Isotope Frequency and Integrated Infrared Absorption Intensities for $Pd(^{14}N_2)_m(^{15}N_2)_{3-m}$ in Solid Nitrogen

Frequency Notation	Calculated Frequency	Observed Frequency	Calculated Intensity	Observed Intensity	Assignment[b]
Infrared					
ν_1	2242.8	2241.4 }	10.0	10.0	A(I)
ν_2	2250.5	2250.8 }			B(I) B(II)
ν_3	2258.2	2256.3	2.9	1.8	A(V)
ν_4	2262.8	2263.4	1.3	1.8	A(II)
ν_5	2166.7	2166.9 }	9.3	8.3	A(VI)
ν_6	2174.2	2175.5 }			B(VI) B(V)
ν_7	2176.4	2175.6	4.6	4.7	A(II)
ν_8	2180.8	2182.5	3.0	3.9	A(V)
Raman					
ν_9	2272.9	2273[a]	—	—	A(I)

[a] This frequency and its $^{15}N_2$ counterpart are formally infrared active in the system C_2. The intensity calculations predict that its intensity is negligibly small, as observed experimentally.

[b] See reference 1 for notation.

characteristics showed the presence of two species, and the $^{14}N_2/^{15}N_2/Ar$ mixed isotope experiments established them to be $Pd(N_2)_2$ and $Pd(N_2)$. The infrared and Raman activities for $Pd(N_2)_2$ show that mutual exclusion is operative and end-on bonded dinitrogen is implied in a linear $D_{\infty h}$ structure. The doublet splitting for $Pd(N_2)$ at 2215, 2211 cm^{-1} is shown in a later section to be a matrix site effect, where each line of the doublet belongs to a $Pd(N_2)$ molecule containing end-on bonded dinitrogen, but subject to slightly different matrix perturbations. The combined matrix infrared and Raman data for the Pd/N_2 cocondensation reaction are thus able to provide a great deal of information on the identification and structural assignments of the binary dinitrogen complexes of palladium, $Pd(N_2)_n$ (where $n = 1$–3), all of which were confirmed by frequency and intensity calculations.[1]

4.3 Cobalt Tetracarbonyl, $Co(CO)_4$

By analogy with Ni, Pd, or Pt, the cocondensation of Co atoms with pure CO would be expected to produce the d^9 Jahn–Teller $Co(CO)_4$ molecule, which can be considered to be the monomer residue of the well-known dicobalt octacarbonyl, $Co_2(CO)_8$. The matrix infrared and Raman spectra together with matrix Raman depolarization measurements[18] are shown in Figure 9. The existence of at least two infrared active CO stretching modes at 2029 and 2011 cm^{-1} rules out an "undistorted" tetrahedral geometry for $Co(CO)_4$ but is compatible with $D_{2d}[A_1(R) + B_2(IR/R) + E(IR/R)]$ or $C_{3v}[2A_1(IR/R) + E(IR/R)]$ selection rules. If one examines the alternative vibrational assignments, then one is forced to base a C_{3v} conclusion on (a) the observation of the infrared active, high-frequency A_1 νCO stretching mode of the C_{3v} molecule, which is predicted to be weak and could be obscured by the $^{12}C^{18}O/^{13}C^{16}O$ (2092–2088 cm^{-1}) natural abundance isotope bands, and (b) the observation of polarization on the 2029 cm^{-1} totally symmetric A_1 νCO stretching mode of the C_{3v} molecule, which is, however, expected to be slight because of "out-of-phase" vibrational coupling with the high-frequency (2107 cm^{-1}) A_1 νCO mode.

Clearly, to make a structural decision on the basis of the above infrared and Raman data alone would be extremely unwise, and one is forced to resort to other matrices, mixed-isotopic substitution, and other forms of spectroscopy to help resolve the problem.

As it turns out, the isotope-frequency calculation for

$$Co(^{12}C^{16}O)_n(^{13}C^{16}O)_{4-n} \qquad (\text{where } n = 0\text{–}4)$$

in CO matrices favors the C_{3v} distortion (Table 8). Moreover, the esr spectra of $Co(^{12}C^{16}O)_4$ and $Co(^{13}C^{16}O)_4$ in $^{12}C^{16}O$ and $^{13}C^{16}O$ matrices, respectively (see Section 10.5 for details) support the contention that the molecule

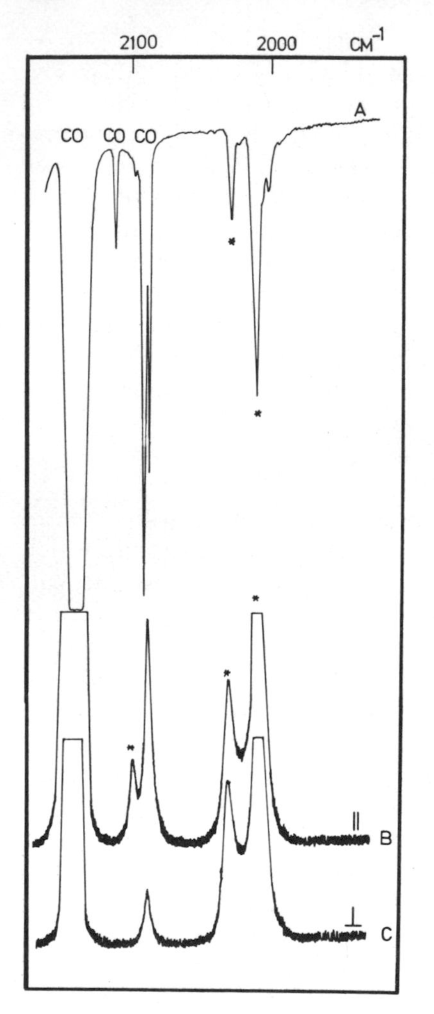

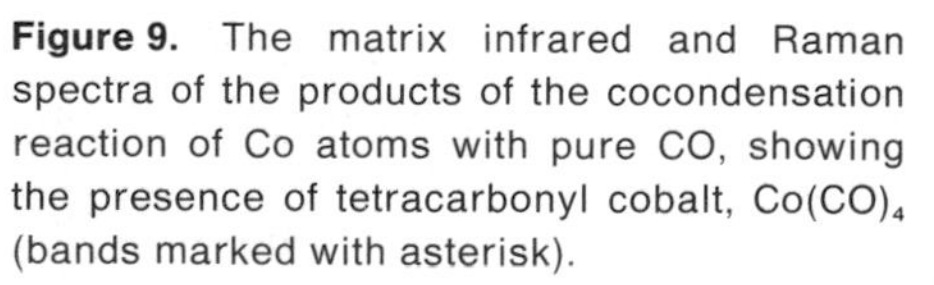
Figure 9. The matrix infrared and Raman spectra of the products of the cocondensation reaction of Co atoms with pure CO, showing the presence of tetracarbonyl cobalt, $Co(CO)_4$ (bands marked with asterisk).

has axial C_{3v} symmetry, the superhyperfine coupling to the "unique" CO ligand being easily resolved (ca. 30 G).[18] Additional support for this assignment is obtained from the uv-visible spectrum, which displays ligand-field and charge-transfer transitions characteristic of a distorted tetrahedral d^9 tetracarbonyl (Figure 10). That the observed distortion is electronic in origin rather than a matrix-site perturbation is convincingly demonstrated by studying the matrix-induced vibrational frequency shifts[18] in Ne, Ar, Kr, and Xe matrices (see Section 8 on matrix effects for details).

Table 8. Observed and Calculated Frequencies for $Co(^{12}C^{16}O)_n(^{13}C^{16}O)_{4-n}$ (where n = 0–4): C_{3v} Model

	Observed (cm^{-1})	Calculated (cm^{-1})[a]	Symmetry Assignment
Molecule (I), $Co(^{12}C^{16}O)_4$, C_{3v}	2010.7	2010.5	E
	2028.8	2028.5	A_1
	2107.0	2107.0	A_1
Molecule (II), $Co(^{12}C^{16}O)_3(^{13}C^{16}O)$, C_{3v}	1991.3	1990.9	A_1
	2010.7	2010.5	E
	—	2099.0	A_1
Molecule (III), $Co(^{12}C^{16}O)_3(^{13}C^{16}O)$, C_s	1974.6	1975.0	A′
	2026.4	2027.1	A′
	—	2098.6	A′
	2010.7	2010.5	A″
Molecule (IV), $Co(^{12}C^{16}O)_2(^{13}C^{16}O)_2$, C_s	1986.8	1986.7	A′
	2024.6	2024.3	A′
	—	2089.2	A′
	1965.7	1965.8	A″
Molecule (V), $Co(^{12}C^{16}O)_2(^{13}C^{16}O)_2$, C_s	1973.3	1973.2	A′
	1992.7	1992.9	A′
	—	2089.1	A′
	2010.7	2010.5	A″
Molecule (VI), $Co(^{12}C^{16}O)(^{13}C^{16}O)_3$, C_s	1979.6	1979.6	A′
	1997.9	1997.9	A′
	1965.7	1965.8	A″
	—	2077.1	A′
Molecule (VII), $Co(^{12}C^{16}O)(^{13}C^{16}O)_3$, C_{3v}	1965.7	1965.8	E
	2010.7	2010.7	A_1
	—	2078.4	A_1
Molecule (VIII), $Co(^{13}C^{16}O)_4$, C_{3v}	1965.7	1965.8	E
	—[b]	1983.4	A_1
	—	2060.2	A_1

[a] The best fit force constants for the C_{3v} model were: $f_r = 16.952$, $f_{r'} = 16.755$, $f_{rr'} = 0.328$, and $f_{r'r'} = 0.425$ mdyn/Å. After the final iteration the standard deviation was 0.611×10^{-5}, which statistically represents a good fit, almost two orders of magnitude better than the D_{2d} model.

[b] This frequency was observed by Rest et al. as a shoulder at 1983.6 cm^{-1}, in close agreement with the calculated value.

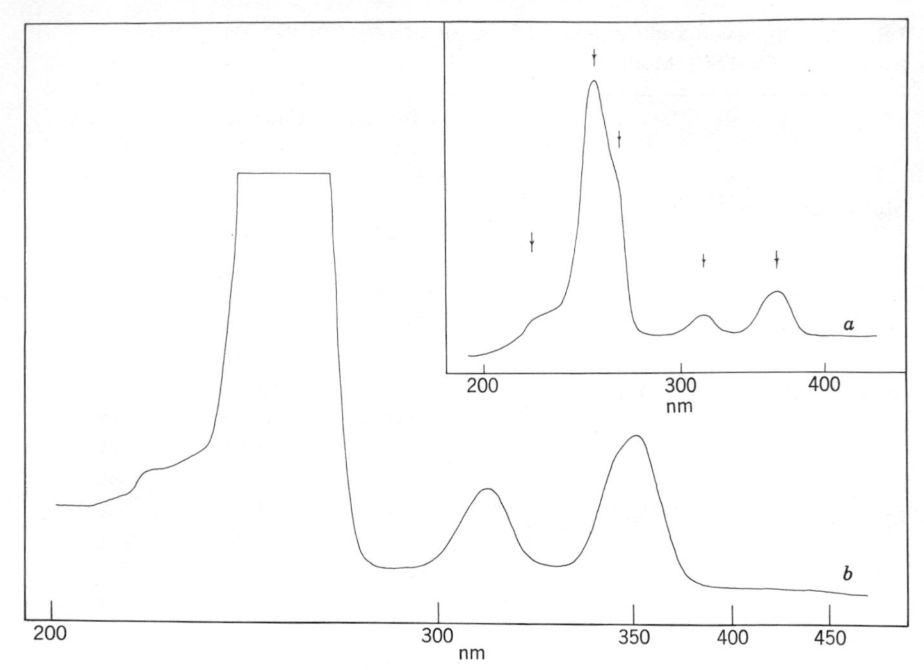

Figure 10. The uv-visible spectrum of $Co(CO)_4$ in solid CO at 6°K, (*a*) low sensitivity and (*b*) high sensitivity scans.

4.4 Rhenium Pentacarbonyl, $Re(CO)_5$

The $Re(CO)_5$ radical was first detected by Junk and Svec[19] in a mass spectroscopic study of $Re_2(CO)_{10}$ vapor. These data yielded useful information concerning the Re—Re and Re—C bond dissociation energies. More recently, Wrighton and Bredesen[20] have investigated the photoreaction of $Re_2(CO)_{10}$ with CCl_4 and have provided evidence for an efficient homolytic fission mechanism and a $Re(CO)_5$ radical intermediate.

However, prior to metal-atom–synthetic techniques it had not proven possible to obtain pure samples of $Re(CO)_5$ for vibrational or uv-visible studies, both of which are capable of yielding pertinent information concerning the nature of the bonding and the molecular and electronic properties of the radical.

Experiences with reactive matrices suggested that the cocondensation of Re atoms and pure CO should provide a direct route to $Re(CO)_5$. Indeed this proved to be the case in practice.[8] The matrix infrared spectrum obtained in $^{12}C^{16}O$ matrices (Figure 11*a*) shows two lines in the CO-stretching region at 1995*s* and 1977*w* cm^{-1} with absorption intensities approximately 4:1, respectively. Warm-up experiments confirmed their assignment to a single

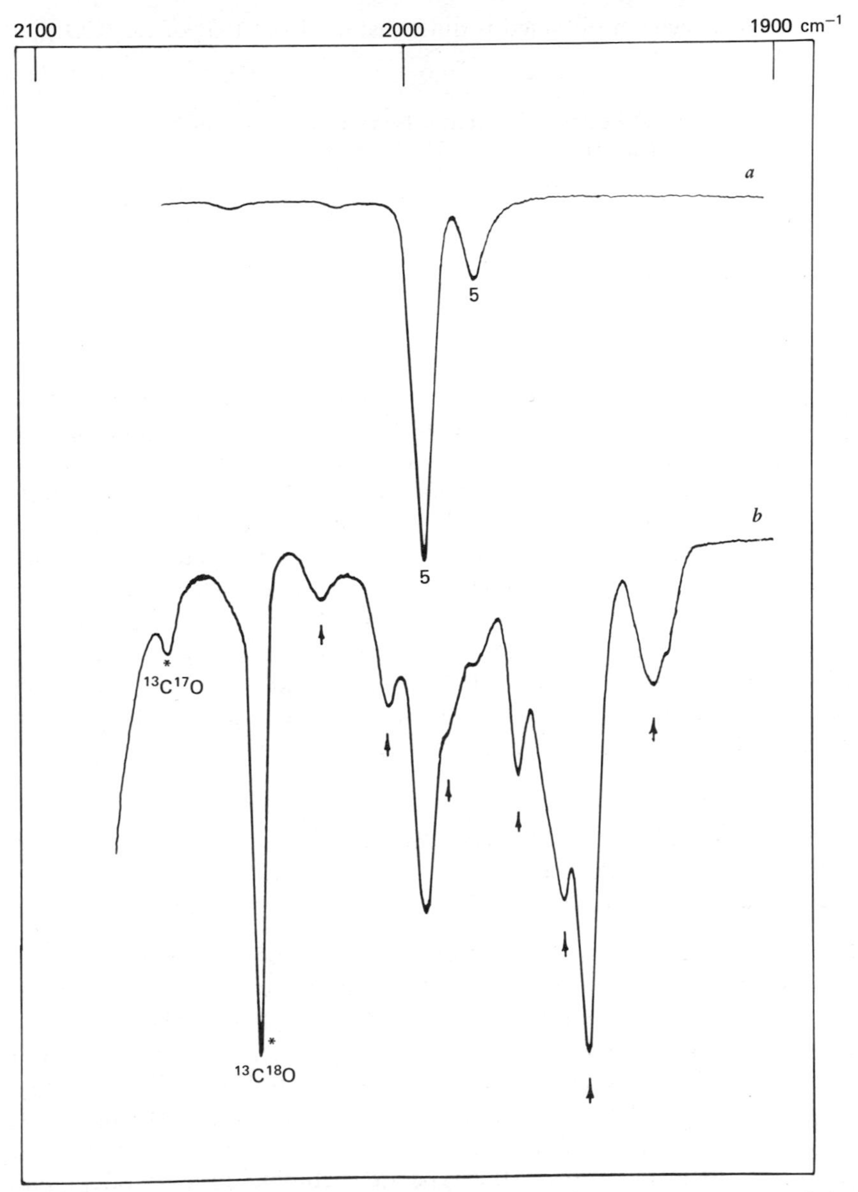

Figure 11. Matrix infrared spectrum of the products of the cocondensation reaction of Re atoms, (*a*) with $^{12}C^{16}O:Ar \simeq 1:10$ and (*b*) $^{12}C^{16}O:{}^{13}C^{16}O:Ar \simeq 1:1:20$ at 10°K. The asterisk indicates traces of $^{13}C^{18}O$ and $^{13}C^{17}O$, which are always present in the commercially available $^{12}C^{16}O/^{13}C^{16}O$ mixtures, and the arrows indicate new isotope lines.

species. The spectrum obtained is quite distinct from that of $Re_2(CO)_{10}$ as seen below.

Re/CO Reaction (cm^{-1})	Matrix-Isolated $Re_2(CO)_{10}$ in Ar (cm^{-1})
—	2070*w*
—	2018*vs*
1995.5*s*	—
1977*w*	—
—	1976*mw*

The general pattern of CO-stretching modes compared to square pyramidal $Cr(CO)_5$[21] and $Cr(CO)_5^-$,[22] the general shift to lower frequencies compared to $Re_2(CO)_{10}$, and the approximately 4:1 relative intensity pattern are all consistent with the product being square pyramidal $Re(CO)_5$ rather than the trigonal bipyramidal form (the latter being expected to display a 2:3 intensity pattern). The nonobservation of the A_1 equatorial CO-stretching mode implies that the C_{apical}—Re—C_{eq} angle is very close to 90°. When the cocondensation experiments were repeated with a $^{12}C^{16}O/^{13}C^{16}O$ isotopic mixture, the pattern of lines so obtained (Figure 11*b*) characterized a pentacarbonyl containing a unique CO ligand (apical) and a group of four equivalent (equatorial) CO ligands, consistent with the pattern predicted for square pyramidal $Re(^{12}C^{16}O)_n(^{13}C^{16}O)_{5-n}$ (where $n = 0$–5).[8]

5. MODE OF BONDING AND NATURE OF THE INTERACTION BETWEEN THE METAL AND THE LIGAND(S)

A number of matrix vibrational spectroscopic methods are available for determining product stoichiometry. When the mode of attachment of the ligand(s) to the metal atom is the subject of investigation, this information can sometimes be deduced from simple point-symmetry considerations. For example, $Ni(N_2)_4$ was shown to have regular tetrahedral T_d symmetry, implying "end-on" bonded dinitrogen.[1] However, situations often arise where symmetry criteria alone cannot define the mode of bonding of the ligand(s) to the metal. Cases of this type are becoming quite prevalent with the increasing number of reports of stable and matrix-isolated complexes containing dioxygen and dinitrogen ligands. Here again, vibrational spectroscopic investigations using mixed, isotopically substituted ligands (such as $^{16}O^{18}O$ and $^{14}N^{15}N$) are able to distinguish between the end-on and side-on bonding types. This is due to the nonequivalence of the vibrational frequencies of $M^{n}X^{m}X$ and $M^{m}X^{n}X$ in both the M—X and X—X stretching regions and the expected splitting of these modes. Using statistically scrambled mixtures of X_2 isotopes, ${}^{n}X_2:{}^{n}X^{m}X:{}^{m}X_2 = 1:2:1$, the expected patterns of XX stretching modes for end-on and side-on bonded X_2 in MX_2 are shown below:

a. Side-on bonded X_2: three equally spaced lines with absorption intensities in the ratio 1:2:1.

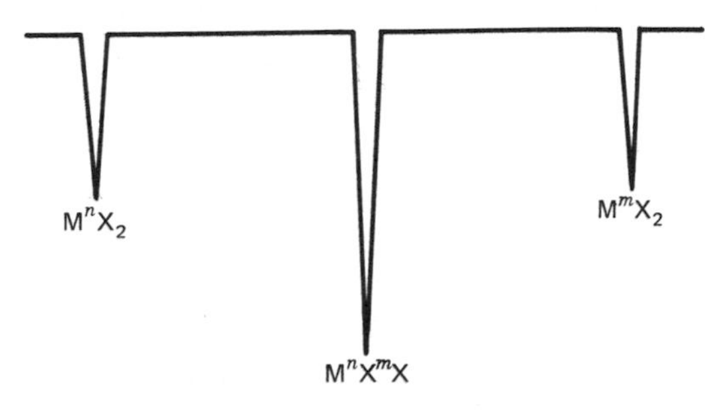

b. End-on bonded X_2: four lines with a closely spaced central doublet where the absorption intensities of the four lines are 1:1:1:1.

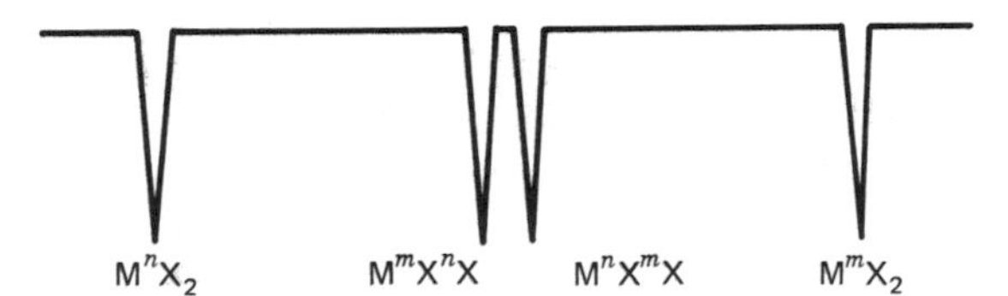

The central doublet for end-on bonded X_2 arises from the nonequivalence of the mixed isotopic species M—${}^{n}X{\equiv}{}^{m}X$ and M—${}^{m}X{\equiv}{}^{n}X$. In the cases of $Ni(N_2)$,[23] $Pd(N_2)$,[24] $Pt(N_2)$,[25] and $Rh(N_2)$,[26] the four-line pattern was observed, proving end-on bonded N_2 where the magnitude of the splitting of the central doublet was 3.8, 4.0, 4.0 and 1.7 cm^{-1}, respectively. Assuming these end-on bonded monodinitrogen complexes are linear triatomic molecules, one can, in principle, obtain from the set of four lines associated with each isotopic molecule, $M^{14}N_2$, $M^{14}N^{15}N$, $M^{15}N^{14}N$, and $M^{15}N_2$, plus the M—N stretching modes, the three stretching force constants, k_{NN}, k_{MN}, and $k_{MN \cdot NN}$, and hence an insight into the nature of the interaction of the N_2 ligand with the metal atom, M (see Section 6 for details of these calculations). On some occasions the interpretation of the scrambled isotopic results has been complicated slightly by a matrix splitting on the NN absorption associated with

MN_2. An example of this is found in PtN_2,[25] where the NN stretching mode occurs as a doublet at 2170/2166 cm^{-1}. The type of bonding of the N_2 to the Pt, therefore, does not immediately follow from the $^{14}N_2/^{14}N^{15}N/^{15}N_2/Ar$ experiments and necessitates the following analysis.

The two components of the doublet at 2170/2166 cm^{-1} may in turn arise from two end-on bonded PtN_2 species, two side-on bonded PtN_2 species, or one side and one end-on bonded PtN_2 species. The "difference" between the two PtN_2 species is likely a multiple-trapping-site effect. Assuming these four possible causes for the 2170/2166 cm^{-1} doublet labeled below for convenience

a.	2170 s	A	c.	2170 s	A
	2166 s′	B		2166 e′	B
b.	2170 e	A	d.	2170 e	A
	2166 s′	B		2166 e′	B

(where e, e′ = end-on bonded dinitrogen, and s, s′ = side-on bonded dinitrogen), the patterns that one would expect to observe in the $Pt/^{14}N^{15}N$ region (where the splitting of the $Pt^{14}N^{15}N/Pt^{15}N^{14}N$ is of the same order as the matrix splitting, that is, approximately 4 cm^{-1}) are shown in Figure 12 for each of the four situations. Only the triplet pattern corresponding to case (d) matches that observed experimentally; hence *both* components 2170 and 2166 cm^{-1} are associated with end-on bonded dinitrogen in PtN_2.

When the $M^{14}N^{15}N/M^{15}N^{14}N$ splitting is *less* than the matrix splitting as found in PdN_2, the $Pd^{14}N^{15}N/Pd^{15}N^{14}N$ region shows a doublet of doublets pattern (Figure 13), again characteristic of end-on bonded N_2,[24] rather than the triplet pattern observed for end-on bonded PtN_2.[25]

Dioxygen appears to be a more versatile ligand than dinitrogen in its bonding characteristics. X-ray crystallography has shown that dioxygen in a number of stable dioxygen complexes prefers to bond to transition metals in a side-on fashion, although a number of cobalt complexes are known with the O_2 bonded in a end-on, nonlinear fashion as well as a bridging group between the metals.[27]

Using similar experiments to those previously described for MN_2, Ni, Pd, and Pt atoms have been cocondensed with $^{16}O_2:^{16}O^{18}O:^{18}O_2:Ar = 1:2:1:800$ mixtures.[4] In all cases, matrix-infrared spectra were obtained characteristic of complexes containing a single molecule of dioxygen having oxygen atoms in equivalent environments, that is, $M{<}\begin{smallmatrix}O\\|\\O\end{smallmatrix}$. As found for $M(O_2)$ and $(O_2)M(O_2)$ and proposed for many synthetic dioxygen carriers, the side-on

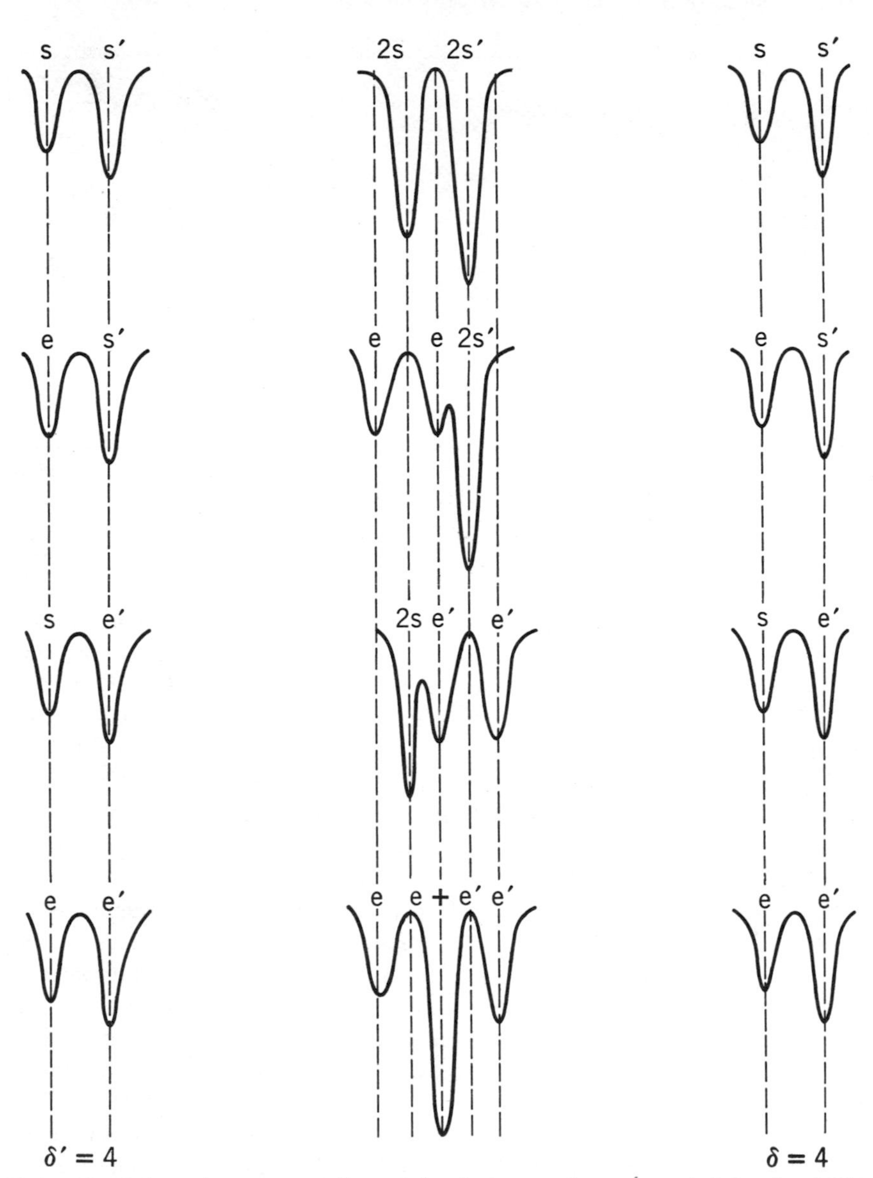

Figure 12. Schematic representation of the isotope patterns expected for the $Pt(N_2)$ molecule when using statistical mixtures of nitrogen isotopes $^{14}N_2 : ^{14}N^{15}N : ^{15}N_2 = 1:2:1$ in argon (see text for notation).

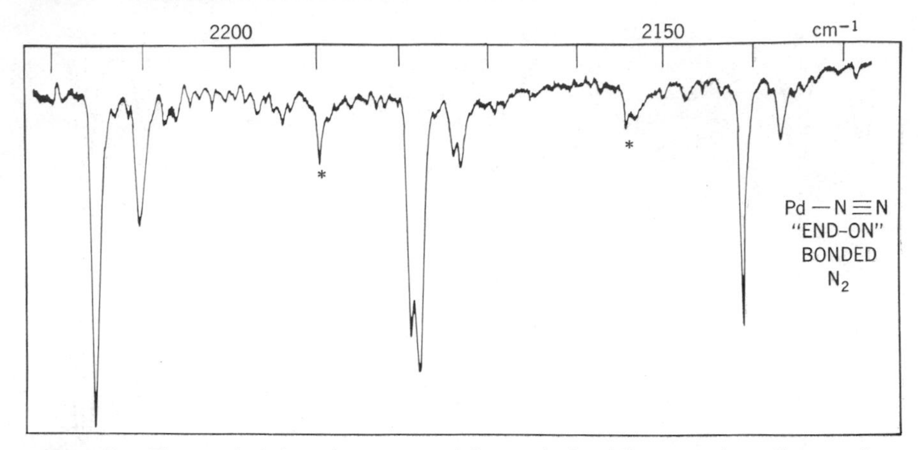

Figure 13. The matrix infrared spectrum of the products of the cocondensation reaction of Pd atoms with $^{14}N_2:^{14}N^{15}N:^{15}N_2:Ar = 1:2:1:2000$, showing the presence of $Pd(N_2)$ with end-on bonded dinitrogen.

mode of bonding is the more common form, a not unexpected result for transition metals in relatively low oxidation states. The O—O stretching frequencies corresponding to the three isotopic molecules, $M(^{16}O_2)$, $M(^{16}O^{18}O)$, and $M(^{18}O_2)$, are shown in Table 9. When Ni, Pd, or Pt atoms are cocondensed with a dilute $^{16}O_2/^{18}O_2/Ar$ matrix at 6–10°K, in addition to the *doublet* previously assigned to the O—O stretching modes of $M(^{16}O_2)$ and $M(^{18}O_2)$, three other O—O stretching modes are observed, indicating the presence of dioxygen complexes containing *two* equivalent dioxygen molecules, and the lines are easily assigned to the isotopic species, $(^{16}O_2)M(^{16}O_2)$, $(^{16}O_2)M(^{18}O_2)$, and $(^{18}O_2)M(^{18}O_2)$. In principle, the mode of attachment of the dioxygen molecules to the metal in these bisdioxygen complexes can be determined from the $^{16}O_2/^{16}O^{18}O/^{18}O_2/Ar$ experiments. For example, the original absorption at 1111.5 cm^{-1} for $(^{16}O_2)Pd(^{16}O_2)$ split into a *six*-line spectrum (Figure 14), which could be assigned to the possible isotopic combinations of bisdioxygen palladium shown in Table 13. The "sextet pattern" indicates a nondegenerate vibration of two equivalent dioxygen molecules, each containing equivalent oxygen atoms. The second and fifth components of the multiplet indicate the equivalence of the oxygen atoms in each O_2 unit, since further splitting or line broadening due to rearrangement of the $^{16}O^{18}O$ component would be expected if the molecules were attached in an end-on fashion. This broadening is not observed, all six components of the sextet being equally sharp. The absence of lines assigned to the species, $Pd(^{16}O^{18}O)_2$, in $^{16}O_2/^{18}O_2/Ar$ matrices and the nonequivalence of the $(^{16}O_2)Pd(^{18}O_2)$ and $(^{16}O^{18}O)Pd(^{16}O^{18}O)$ species, clearly indicated in the spectrum, show

Table 9. Vibrational Assignments[a] of the Oxygen Stretching Modes in Binary Dioxygen Complexes, $M(O_2)$ (where M = Ni, Pd, or Pt)

	M		
Molecule	Ni	Pd	Pt
$M(^{16}O_2)$	966.2	1024.0	926.6
$M^{16}O^{18}O)$	940.1	995.5	901.4
$M(^{18}O_2)$	913.6	967.0	875.2

[a] Frequencies in cm^{-1}.

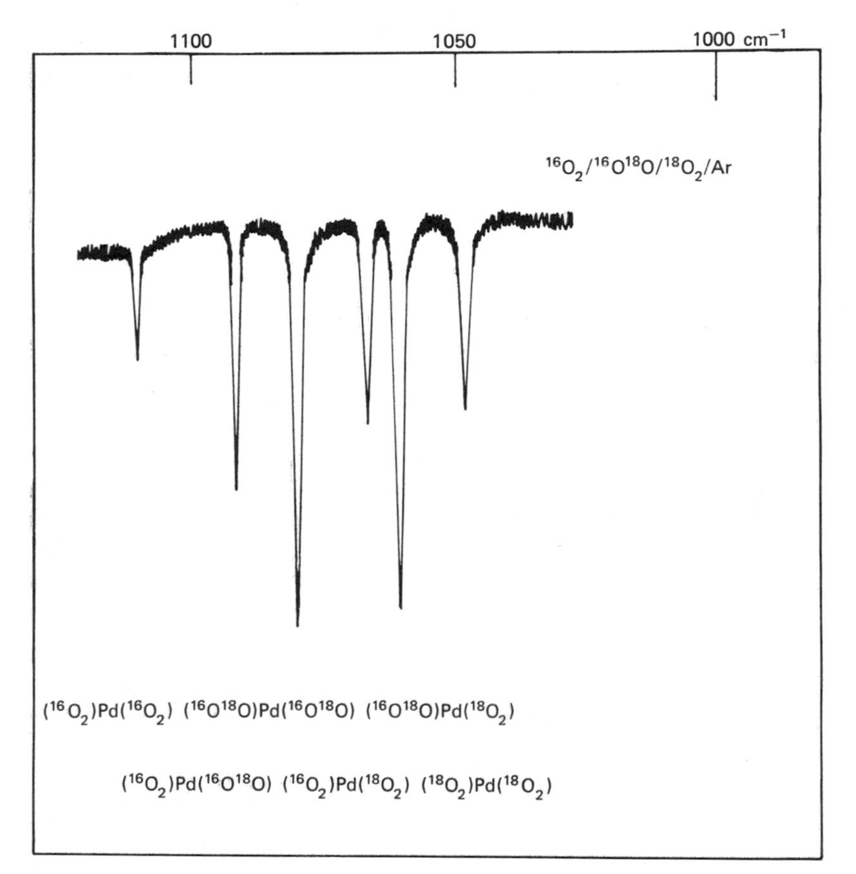

Figure 14. The matrix infrared spectrum of the products of the cocondensation reaction of Pd atoms, $^{16}O_2 : ^{16}O^{18}O : ^{18}O_2 : Ar \simeq 1:2:1:800$ at 6°K, showing the characteristic isotope pattern of $(O_2)Pd(O_2)$.

that the original oxygen molecules remain intact upon bonding to the Pd atom. Since a planar D_{2h} structure would have *cis* and *trans* isomers for the $(^{16}O^{18}O)Pd(^{16}O^{18}O)$ molecule, and since no splitting is observed at the frequency assigned to the species (1808.5 cm^{-1}), the D_{2d} "unique spiro" structure is proposed (Figure 15). This symmetry model is also favored from Dewar–Chatt maximum π-bond overlap considerations. Very similar results were obtained for both Ni and Pt atoms and established the existence of bis-

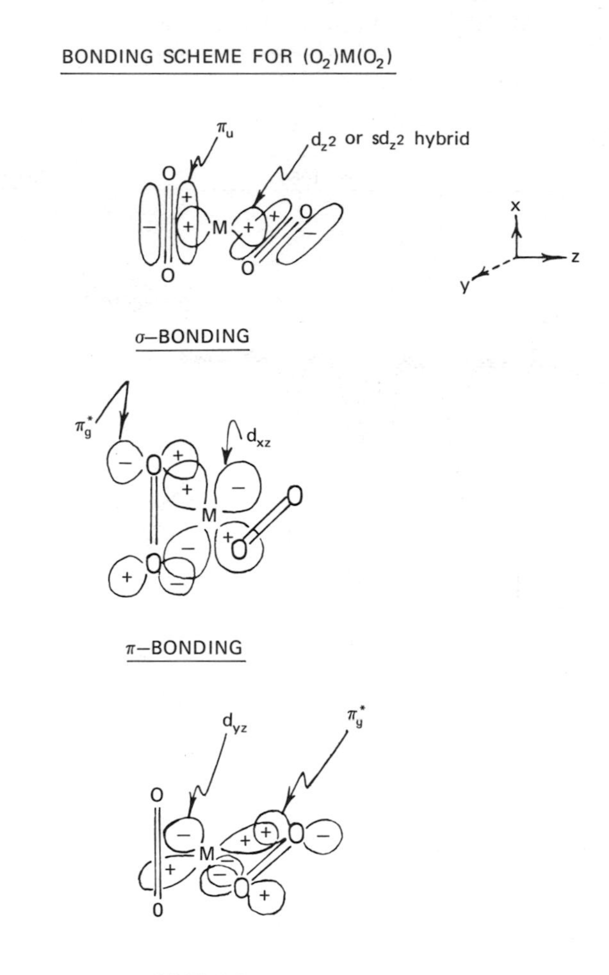

Figure 15. Schematic molecular orbital σ- and π-bonding model for a bisdioxygen complex, $(O_2)M(O_2)$, containing two equivalent dioxygen molecules (where all oxygen atoms are equivalent) bonding in a sideways fashion in a D_{2d} configuration to the central metal atom.

dioxygen complexes with side-on bonded dioxygen for all three metals. Further confirmation of these vibrational and structural assignments based on frequency and intensity calculations is described in Sections 6 and 7.

In the light of the $Ni(N_2)_4$ and $Ni(O_2)_2$ results, it was thought that the cocondensation of nickel atoms with dilute $O_2/N_2/Ar$ mixtures might prove to be a simple and direct route to the first examples of transition metal complexes containing *both* molecular dinitrogen and molecular dioxygen coordinated to the same metal atom. Indeed this turned out to be the case.[28] From the number and frequencies of the NN and OO stretching modes, their warm-up behavior and $^{16}O_2/^{14}N_2/Ar$, $^{16}O_2/^{18}O_2/^{14}N_2/Ar$, $^{16}O_2/^{16}O^{18}O/^{18}O_2/^{14}N_2/Ar$, $^{16}O_2/^{14}N_2/^{15}N_2/Ar$, $^{16}O_2/^{14}N_2/^{14}N^{15}N/^{15}N_2/Ar$ isotope multiplet patterns (see, for example, Tables 10 and 11), it was established that the new complexes were monodinitrogen-monodioxygen nickel,

Table 10. Observed and Calculated Frequencies of Isotopically Substituted $(^pO^qO)Ni(^nN^mN)$ (where $n, m = 14$ or 15 and $p, q = 16$ or 18)

Molecule	Observed (cm^{-1})	Calculated (cm^{-1})	Mode
$^{16}O_2Ni^{14}N_2$	977	978.1	νO—O
	478	478.4	νM—O symm.
	2243	2244.2	νN—N
	368	367.4	νM—N
$^{16}O_2Ni^{15}N_2$	977	978.1	νO—O
	478	477.7	νM—O symm.
	2169	2168.1	νN—N
	360	358.5	νM—N
$^{16}O_2Ni^{14}N^{15}N$	977	978.1	νO—O
	—	478.1	νM—O symm.
	2207	2207.0	νN—N
	—	362.7	νM—N
$^{16}O_2Ni^{15}N^{14}N$	977	978.0	νO−O
	—	478.1	νM—O symm.
	2206.2	2206.0	νN—N
	—	362.9	νM—N
$^{18}O_2Ni^{14}N_2$	925	923.6	νO—O
	468	464.7	νM—O symm.
	2243	2244.2	νN—N
	—	362.0	νM—N
$^{16}O^{18}ONi^{14}N_2$	951	951.3	νO—O
	—	471.7	νM—O symm.
	2244	2244.2	νN—N
	—	366.5	νM—N

Table 11. Observed and Calculated[a] Frequencies of Isotopically Substituted $(^{p}O^{q}O)Ni(^{n}N^{m}N)_2$ (where n, m = 14 or 15 and p, q = 16 or 18)

Molecule	Observed (cm^{-1})	Calculated (cm^{-1})	Mode
$^{16}O_2Ni(^{14}N_2)_2$	972	972.7	νO—O
	—	488.2	νM—O symm.
	2282	2283.3	νN—N symm.
	2260	2260.7	νN—N asym.
	345	342.6	νM—N symm.
$^{16}O_2Ni(^{15}N_2)_2$	972	972.6	νO—O
	—	486.9	νM—O symm.
	2207	2205.9	νN—N symm.
	2183	2184.1	νN—N asym.
	—	334.0	νM—N symm.
$^{18}O_2Ni(^{14}N_2)_2$	920	918.3	νO—O
	—	475.2	νM—O symm.
	2283	2283.2	νN—N symm.
	2260	2260.7	νN—N asym.
	—	337.5	νM—N symm.
$^{16}O_2Ni^{14}N_2{}^{15}N_2$	972	972.6	νO—O
	—	487.5	νM—O symm.
	2273	2273.6	ν^{14}N—^{14}N
	2195	2193.4	ν^{15}N—^{15}N
	—	348.9	νM—^{14}N
	—	337.8	νM—^{15}N
$^{16}O^{18}ONi(^{14}N_2)_2$	945	945.9	νO—O
	2283	2283.3	νN—N symm.
	2260	2260.7	νN—N asym.
$^{16}O_2Ni(^{14}N^{15}N)_2$	(2246)	2246.0	νN—N symm.
	(2226)	2223.4	νN—N asym.
	—	338.1	νM—N symm.
$^{16}O_2Ni(^{15}N^{14}N)_2$	(2246)	2243.9	νN—N symm.
	(2226)	2222.0	νN—N asym.
	—	338.4	νM—N symm.
$^{16}O_2Ni(^{14}N_2)(^{14}N^{15}N)$	(2276)	2275.2	ν^{14}N—^{14}N
	(2234)	2231.6	ν^{14}N—^{15}N
$^{16}O_2Ni(^{14}N_2)(^{15}N^{14}N)$	(2276)	2274.9	ν^{14}N—^{14}N
	(2234)	2230.0	ν^{15}N—^{14}N
$^{16}O_2Ni(^{14}N^{15}N)(^{15}N^{14}N)$	(2246)	2245.0	ν^{14}N—^{15}N
	(2226)	2222.7	ν^{15}N—^{14}N
$^{16}O_2Ni(^{14}N^{15}N)(^{15}N_2)$	2236	2237.6	ν^{14}N—^{15}N
	(2195)	2192.1	ν^{15}N—^{15}N
$^{16}O_2Ni(^{15}N^{14}N)(^{15}N_2)$	(2236)	2235.9	ν^{15}N—^{14}N
	(2195)	2192.1	ν^{15}N—^{15}N

[a] Numbers in parentheses represent frequencies assigned after calculating their values. These frequencies were usually part of a complex overlapping band structure.

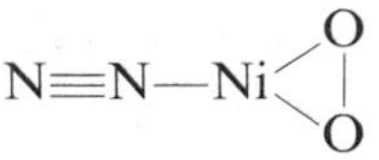

$N{\equiv}N{-}Ni\langle {}^{O}_{O}$ and bisdinitrogen-monodioxygen nickel, $(N{\equiv}N)_2Ni\langle {}^{O}_{O}$,

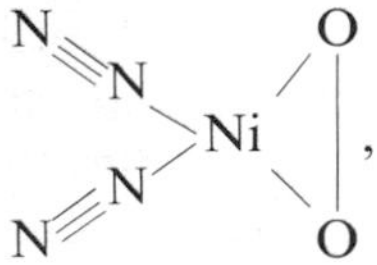

containing end-on bonded dinitrogen and side-on bonded dioxygen. The discovery and characterization of these complexes proved to be of considerable importance in assessing the bonding characteristics of O_2 and N_2 ligands when coordinated to the same metal atom and when competing with each other for bonding electrons (see Section 11.3 on molecular properties).

6. FREQUENCY AND FORCE-FIELD CALCULATIONS FOR ISOTOPICALLY SUBSTITUTED SPECIES

Since isotopic molecules have essentially the same electronic structure, the potential function describing their vibrations should be approximately the same. That is, the force constants are virtually insensitive to isotopic substitution. In this section and Section 11.3, we show how the vibrational frequencies of isotopic species can be used to provide sufficient independent pieces of information to calculate the force field of some matrix-isolated carbonyl, dinitrogen, and dioxygen complexes. Most of the calculations to be described employ a least-squares calculation of the force constants from the vibrational frequencies of the isotopic species.

For isotopically substituted carbonyl, dinitrogen, and dioxygen complexes the force-constant calculation can be greatly simplified if one assumes the Cotton–Kraihanzel force-field approximation, which essentially decouples the high-frequency ligand-stretching modes from all other vibrations of the molecule. Let us consider as an example $Ni(N_2)_4$, the data for which in solid argon are consistent with a regular tetrahedral molecule. In the mixed $^{14}N_2/^{15}N_2/Ar$ experiment, one has to consider the formation of the five possible complexes listed in Table 4, together with the symmetry species of their NN stretching modes and their respective Cotton–Kraihanzel secular equations.

A least-squares analysis of the data was performed by adjusting the values of the two parameters, k_{NN} and $k_{NN \cdot NN}$, to yield the set of frequencies shown in Table 4. The agreement between the observed and calculated frequencies is excellent for all NN stretching modes and provides convincing evidence that the T_d $Ni(N_2)_4$ structure based on the infrared and Raman activities and depolarization measurements is correct.

Although the Cotton–Kraihanzel approximation is a useful tool in discussing trends in bond strengths in closely related molecules, one tries, when possible, to do a more complete normal coordinate analysis, thereby determining a more complete and meaningful set of force constants. Although it is tempting to believe that the results of a calculation involving both ligand-stretching and metal-ligand-stretching frequencies are more trustworthy than from one involving only ligand-stretching frequencies (Cotton–Kraihanzel approximation), this will not always be the case unless, of course, enough data is present to do a complete force constant calculation. Examples of situations where more complete force-field calculations have proved useful are described in Section 11.

For the remainder of this section we briefly discuss the monodinitrogen complexes, MN_2 (where M = Ni, Pd, or Pt), the low-frequency metal-nitrogen stretching modes of which have recently been observed at 466, 378, and 394 cm^{-1}, respectively.[24] With these frequencies and the respective NN stretching modes of the four isotopic molecules, $M^{14}N_2$, $M^{14}N^{15}N$, $M^{15}N^{14}N$, and $M^{15}N_2$ (listed in Table 12), sufficient independent data are available to solve exactly for the three force constants, k_{NN}, k_{MN}, and $k_{MN \cdot NN}$. A least-squares analysis was performed on the observed spectra where the parameters, k_{NN}, k_{MN}, and $k_{MN \cdot NN}$, were adjusted to give the best fit for an assumed linear MN_2 molecule. The calculated frequencies listed in Table 12 are in excellent agreement with the observed for all lines and all molecules. The best fit force constants are also listed in Table 12.

Table 12. Observed and Calculated Frequencies[a] and Force Constants[b] for Linear M^nN^mN (where M = Ni, Pd, or Pt; *m*, *n* = 14, 15)

NiN_2		PdN_2		PtN_2		
Calcd.	Obsd.	Calcd.	Obsd.	Calcd.	Obsd.	Assignment
2019.8	2020.6	2139.4	2138.7	2099.5	2100.5	$M^{15}N_2$
2053.5	2053.6	2176.9	2178.0	2134.7	2134.5	$M^{15}N^{14}N$
2057.6	2057.4	2177.7	2178.8	2138.7	2138.5	$M^{14}N^{15}N$
2090.7	2089.0	2214.5	2213.0	2173.2	2172.8	$M^{14}N_2$
466.0	466.0	377.8	378.0	393.9	394	νMN
2.48	—	1.86	—	2.29	—	k_{MN}
17.62	—	20.46	—	19.00	—	k_{NN}
0.25	—	0.72	—	0.12	—	$k_{MN \cdot NN}$

[a] Frequencies in cm^{-1}.
[b] Force constants in mdyn/Å.

Some interesting trends are apparent on examining the force-constant data for the Ni, Pd, and Pt monodinitrogen complexes. Probably of greatest significance are the observations that

$$k_{MN}^{NiN_2} > k_{MN}^{PtN_2} > k_{MN}^{PdN_2}$$

which, as expected, parallel in an inverse fashion the Cotton–Kraihanzel bond stretching force constants

$$k_{MN}^{PdN_2} > k_{MN}^{PtN_2} > k_{MN}^{NiN_2}$$

This result is reminiscent of those observed for the tetracarbonyls of the corresponding metals, demonstrating the similarity in the bonding properties of N_2 and CO to transition metals.

7. INTENSITY CALCULATIONS FOR ISOTOPICALLY SUBSTITUTED SPECIES

The integrated intensity of a fundamental infrared band corresponding to normal coordinate, Q_k, is proportional to the expression

$$I_k = \left[\frac{\partial \mu_x}{\partial Q_k}\right]_0^2 + \left[\frac{\partial \mu_y}{\partial Q_k}\right]_0^2 + \left[\frac{\partial \mu_z}{\partial Q_k}\right]_0^2 = \left[\frac{\partial \vec{\mu}}{\partial Q_k}\right]_0 \cdot \left[\frac{\partial \vec{\mu}}{\partial Q_k}\right]_0$$

in which μ_x, μ_y, and μ_z are components of the dipole moment along axes attached to the molecule and Q_k is the normal coordinate of the fundamental mode under discussion.[29] It is more convenient, however, to describe the vibrational modes in terms of their symmetry coordinates. This can be done with the aid of the transformation connecting the normal coordinates with real internal coordinates:[29]

$$S_{k'} = \sum L_{k'k} Q_k$$

The expression for I_k then becomes

$$I_k = \left[\frac{\partial \vec{\mu}}{\partial Q_k}\right]_0 \cdot \left[\frac{\partial \vec{\mu}}{\partial Q_k}\right]_0 = \sum_{k'k''} \frac{\partial \vec{\mu}}{\partial S_{k'}} \cdot \frac{\partial \mu}{\partial S_{k''}} L_{k'k} L_{k''k} \tag{1}$$

The transformation coefficients, $L_{kk'}$ are related to the G matrix by

$$\sum_k L_{k'k} L_{k''k} = G_{k'k''}. \tag{2}$$

An expression related to the F matrix and λ_k can also be generated

$$\sum_k L_{k'k} \lambda_k^{-1} L_{k''k} = F_{k'k''}^{-1} \tag{3}$$

where $F_{k'k''}^{-1}$ is an element of the matrix inverse to the force-constant matrix. Thus two intensity sum rules can be formed over all coordinates of a given symmetry species. These can be used to evaluate the intensities of two modes belonging to the same symmetry type, that couple. Their application to isotopic molecules is apparent in (4) and (5).

$$\sum_k I_k = \sum_{k'k''} \frac{\partial \vec{\mu}}{\partial S_{k'}} \cdot \frac{\partial \vec{\mu}}{\partial S_{k''}} G_{k'k''} \tag{4}$$

$$\sum_k \frac{I_k}{\lambda_k} = \sum_{k'k''} \frac{\partial \vec{\mu}}{\partial S_{k'}} \cdot \frac{\partial \vec{\mu}}{\partial S_{k''}} F_{k'k''}^{-1} \tag{5}$$

Bos,[30] Haas and Sheline,[31] and more recently Ogden[32] and Moskovits and Ozin[1] found that these intensity sum rules could be used in conjunction with the Cotton–Kraihanzel force-field approximation[33] to predict relative absorption intensities for NN and CO infrared-active stretching modes of metal carbonyl and dinitrogen complexes, respectively. Their application to these systems simply involves the determination of the components of the transition dipole-moment vectors with respect to each symmetry coordinate, S_k, where in the Cotton–Kraihanzel approximation $G_{k'k''}$ simply equals the reduced mass of the ligand and $F^{-1}_{k'k''}$ is an element of the matrix inverse to the force constant matrix, $F_{k'k''}$.

Despite the approximations, close agreement between the observed and calculated intensities was obtained. As an example, we outline part of the calculation for $Ni(N_2)_4$.

Intensity Calculations for the NN Stretching Modes of $Ni(^{14}N_2)_n(^{15}N_2)_{4-n}$ (where $n = 0–4$).

For example, in $Ni(^{14}N_2)_3(^{15}N_2)$, which has C_{3v} symmetry, one can construct the symmetry coordinates of the NN stretching modes using the internal coordinates listed above as follows.

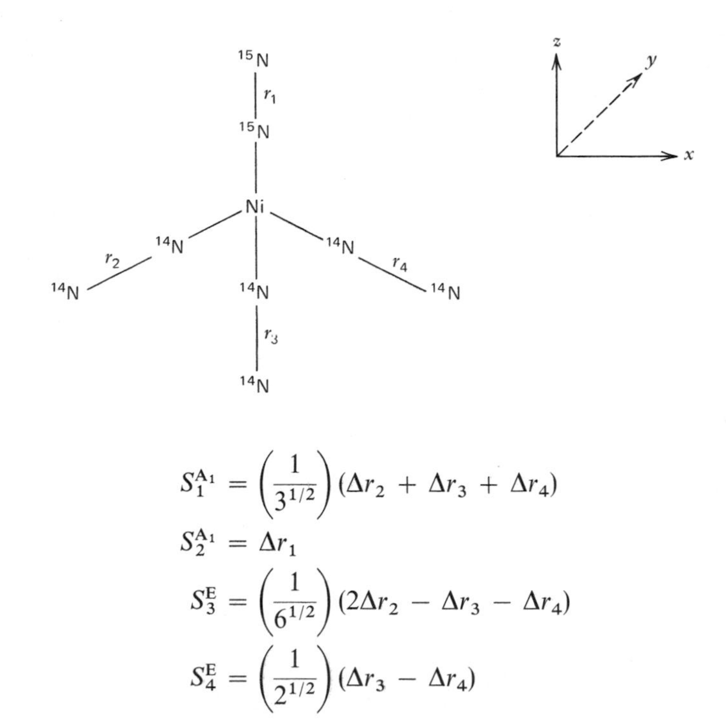

$$S_1^{A_1} = \left(\frac{1}{3^{1/2}}\right)(\Delta r_2 + \Delta r_3 + \Delta r_4)$$

$$S_2^{A_1} = \Delta r_1$$

$$S_3^{E} = \left(\frac{1}{6^{1/2}}\right)(2\Delta r_2 - \Delta r_3 - \Delta r_4)$$

$$S_4^{E} = \left(\frac{1}{2^{1/2}}\right)(\Delta r_3 - \Delta r_4)$$

The components along S_k of the transition dipole moment can then be tabulated together with the G matrix elements.

k	$\partial\mu_x/\partial S_k$	$\partial\mu_y/\partial S_k$	$\partial\mu_z/\partial S_k$	$G_{k'k''}$
1	0	0	−0.58	μ_{14}
2	0	0	1	μ_{15}
3	1.15	0	0	μ_{14}
4	0	1.15	0	μ_{14}

The inverse $F_{k'k''}^{-1}$ matrices are calculated from the $F_{k'k''}$ matrices of the different symmetry species to yield (6) and (7).

$$F^{-1}(\mathrm{A}) = \frac{1}{(f_r + 3f_{rr})(f_r - f_{rr})} \begin{vmatrix} f_r & -3^{1/2}f_{rr} \\ -3^{1/2}f_{rr} & f_r + 2f_{rr} \end{vmatrix} \tag{6}$$

$$F^{-1}(\mathrm{E}) = \frac{1}{(f_r - f_{rr})} \tag{7}$$

The intensities of the two coupled A_1 modes can then be calculated from the two simultaneous linear equations (4) and (5), while that of the E can be obtained from (4) directly. Applying this method individually to each of the molecules in the isotopic series, $Ni(^{14}N_2)_n(^{15}N_2)_{4-n}$, and multiplying the intensity of each by its expected stastical abundance, a set of expected relative intensities may be obtained. Table 4 summarizes the results of the calculation.

The agreement between measured integrated absorption and calculated intensities is remarkably close and provides further evidence for the formulation of $Ni(N_2)_4$ as a regular tetrahedral complex with end-on bonded dinitrogen.

The use of these sum rules has so far been applied only to complexes containing CO and N_2 ligands bonded in an end-on fashion to the metal atom. We conclude this section by investigating the extension of these sum rules to include complexes containing ligands bonded in a side-on manner to the metal atom, such as those found in bis-dioxygen complexes of Ni, Pd, and Pt.

Intensity Calculations for the O—O Stretching Modes of Isotopically Substituted $(^nO^mO)Pd(^nO^mO)$ (where n, m = 16 or 18).

Since homonuclear diatomic molecules do not undergo a net dipole change during bond stretching, the resultant mode is not infrared active. However, when the diatomic molecule is coordinated to a metal atom in an end-on fashion, ligand stretching is accompanied by a net dipole change as shown below,

$$\mathrm{M{-}N\overset{\rightleftarrows}{\equiv}N} \quad \vec{\mu} \neq 0 \qquad \vec{\mu}' \neq 0$$

resulting in infrared activity of the mode. If one considers an oxygen molecule coordinated to a metal atom in a side-on fashion as postulated for $M(O_2)$ and $(O_2)M(O_2)$, it is apparent that the oxygen-oxygen stretching mode does not involve a dipole change and must gain its infrared activity from a transition dipole moment induced by coupling with the metal-oxygen-stretching mode of the same symmetry. Therefore it should be possible to define a transition dipole-moment vector, $\vec{\mu}'$, for the O—O stretching mode that is directionally dependent on the transition dipole-moment vectors for the M—O stretching modes as shown below:

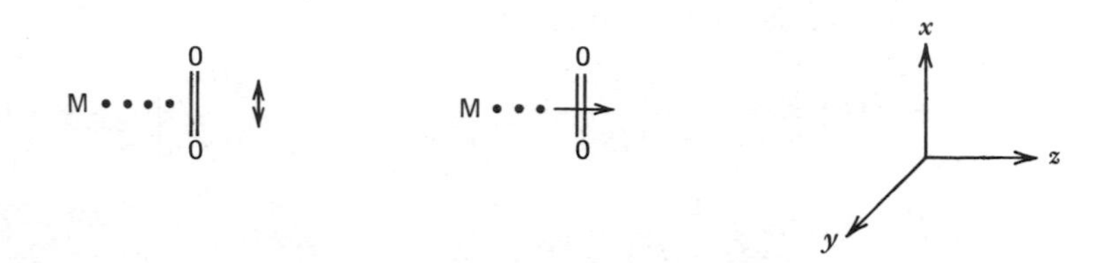

It was felt that within the limits of the Cotton–Kraihanzel approximation, the nonzero components of the "decoupled" transition dipole moment vectors ($\vec{\mu}'_z$) could be used to predict relative absorption intensities for the O—O stretching modes of the isotopic bis-dioxygen molecules. As an example, let us consider the schematic model for $(^{16}O_2)M(^{16}O_2)$ shown below:

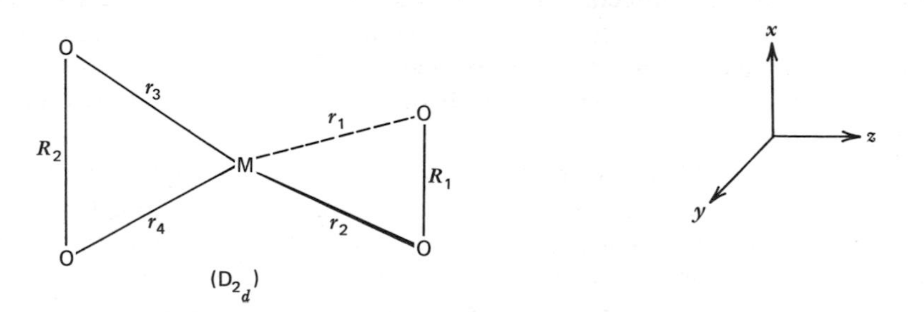

It is possible to depict diagramatically the two possible O—O stretching modes of symmetry type, A_1 and B_2, with the corresponding M—O modes:

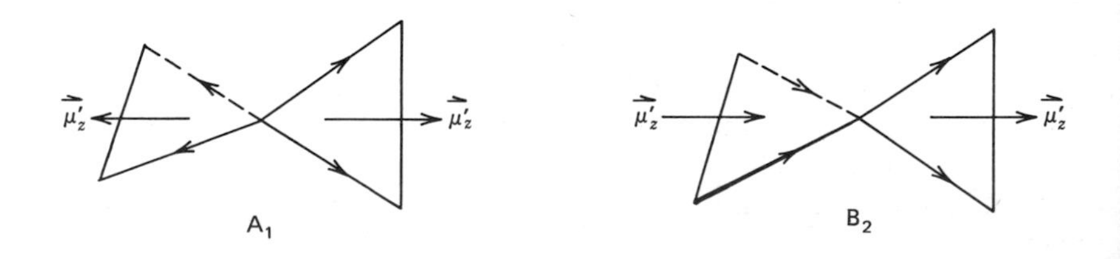

$$S_1 = \frac{1}{2^{1/2}}(\Delta R_1 + \Delta R_2) \qquad S_1 = \frac{1}{2^{1/2}}(\Delta R_1 - \Delta R_2)$$

$$S_2 = \frac{1}{2}(\Delta r_1 + \Delta r_2 + \Delta r_3 + \Delta r_4) \qquad S_2 = \frac{1}{2}(\Delta r_1 + \Delta r_2 - \Delta r_3 - \Delta r_4)$$

The vector sum of the transition dipole moment vectors for the A_1 mode is, of course, equal to zero, making this mode infrared inactive. However, the B_2 mode is expected to be infrared active, and the intensity sum rules may be applied.

Applying this method individually to each of the six isotopic molecules $(^nO^mO)M(^nO^mO)$ (where n, m = 16 or 18) and multiplying the intensity of each by its expected statistical abundance (and correcting for the true concentration ratio of $^{16}O_2$:$^{16}O^{18}O$:$^{18}O_2$), a set of expected relative intensities may be obtained as summarized in Table 13. Comparison of these calculated absorption intensities with those observed for $(O_2)Pd(O_2)$ shows remarkably close agreement and supports the validity of the model and the frequency assignments used in the computations.

Table 13. Observed and Calculated Frequencies[a] and Relative Absorption Intensities[b] for the Isotopic Bisdioxygen Palladium Complexes $(^nO^mO)Pd(^nO^mO)$ (where n, m = 16 or 18) Based on the Cotton–Kraihanzel Force-Field Approximation

	Frequencies		Intensities	
Molecule	Obsd.	Calcd.	Calcd.	Obsd.
$(^{16}O_2)Pd(^{16}O_2)$	1111.5	1111.5	1.80	1.68
$(^{16}O_2)Pd(^{16}O^{18}O)$	1092.2	1092.5	7.81	6.34
$(^{16}O^{18}O)Pd(^{16}O^{18}O)$	1080.5	1080.1	9.29	9.19
$(^{16}O_2)Pd(^{18}O_2)$	1067.2	1067.3	3.97	3.83
$(^{16}O^{18}O)Pd(^{18}O_2)$	1060.5	1060.4	10.00	10.00
$(^{18}O_2)Pd(^{18}O_2)$	1048.5	1047.9	2.99	3.69
f_R[c]	6.236		—	—
f_{RR}	0.413		—	—

[a] Frequencies in cm^{-1}.

[b] The observed intensities were measured as outlined in the footnote of Table 3.

[c] Force constant units, mdyn/Å.

8. MATRIX EFFECTS

We divide this section into three main subsections, each concerned with a different aspect of the general term, "matrix effect," namely matrix-induced frequency shifts, matrix site effects, and molecular distortions.

8.1 Matrix-Induced Vibrational Frequency Shifts

The theoretical treatment of matrix environmental effects, specifically frequency shifts of spectral bands of trapped species, is a complicated problem even for atomic and diatomic molecules. As the complete subject has recently been reviewed in some detail by Barnes,[34] we only superficially examine the topic in the context of metal atom cocondensation reactions.

In order to evaluate a vibrational shift from the gas phase to a matrix environment, defined as $\Delta\nu = \nu_{gas} - \nu_{matrix}$, one is confronted with an intermolecular interaction problem that usually involves the determination of inductive, dispersive, and repulsive contributions to $\Delta\nu$. Even in the simplest of systems this has proven to be a complex problem and generally speaking has made quantitative evaluations of matrix shifts unpracticable. One would intuitively think, however, that the early theories proposed to account for nonspecific solvent shifts,[35] especially for nonpolar solvent, should be applicable to the related problem of solutes in matrices, the main difference being a flexible solvent cage for solutions, but a rigid cavity for matrices.

The $Co(CO)_n$ molecules synthesized from Co atoms and CO have been examined in Ne, Ar, Kr, and Xe,[18] and display a series of interesting matrix frequency shifts. Let us for the purposes of the present discussion assume a similar model for solvent and matrix shifts and examine the data in terms of the Buckingham equation[35d]

$$\frac{\Delta\nu}{\nu} = C_1 + \tfrac{1}{2}(C_2 + C_3)\left[\frac{\varepsilon' - 1}{2\varepsilon' + 1}\right]$$

where C_1, C_2, and C_3 are constants, ε' is the dielectric constant of the solvent (or matrix, in our case), and ν is the mean of the gas phase and matrix frequencies, $\Delta\nu$ having been defined earlier. Buckingham originally put forward this equation to explain nonspecific solvent shifts for a solute molecule in a flexible solvent cage. Let us try to apply this equation to $Co(CO)_n$ (where (where $n = 1$–4) in nonpolar noble gas matrices.

Without exception, the CO stretching frequencies of each $Co(CO)_n$ complex show a red shift (a loose cage environment, according to Pimentel and Charles[36]) on passing from Ar through to Xe. The corresponding frequencies, as might have been expected, show a roughly linear correlation with the matrix polarizability (see, for example, $Co(CO)_4$ in Figure 16).

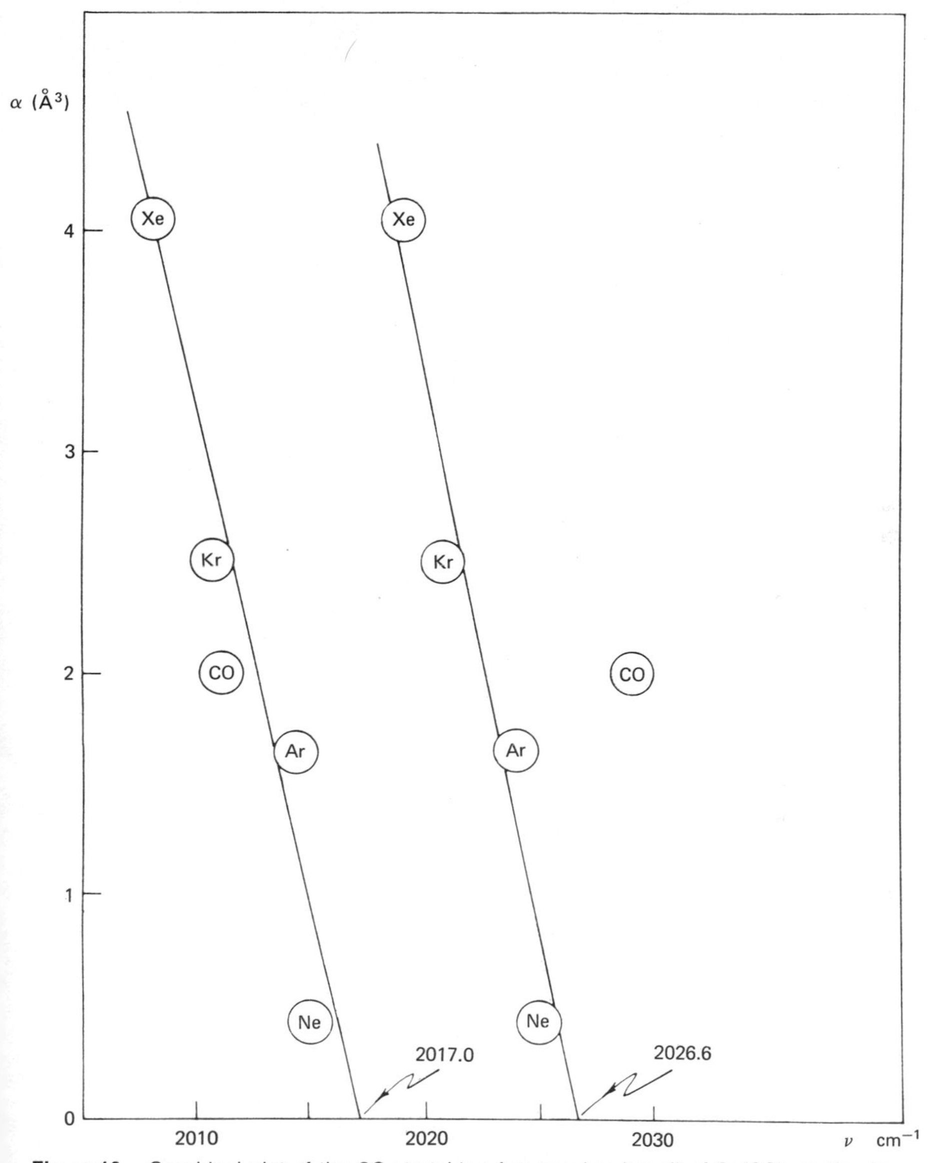

Figure 16. Graphical plot of the CO-stretching frequencies (cm^{-1}) of $Co(CO)_4$ in Ne, Ar, Kr, Xe, and CO matrices as a function of the matrix polarizability, $\alpha\,(A^3)$.

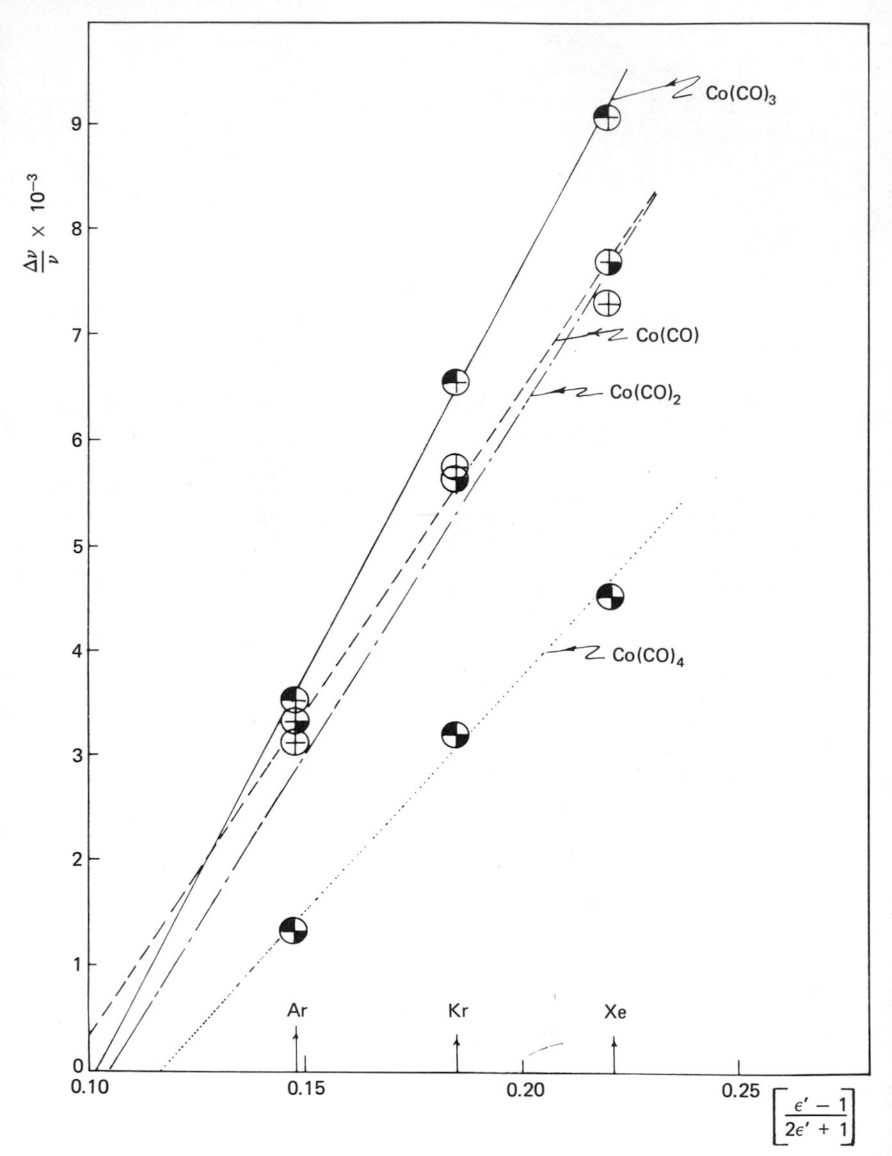

Figure 17. Graphical plot of $\Delta\nu/\nu$ versus $(\varepsilon' - 1)/(2\varepsilon' + 1)$ for $Co(CO)_4$ in Ar, Kr, and Xe matrices (see text for notation).

Let us postulate that an extrapolation of these plots to zero polarizability yields what we shall term "estimated gas phase frequencies" for $Co(CO)_n$. Using these estimated values of ν_{gas}, one can determine the frequency shift ratio, $\Delta\nu/\nu$. These are plotted as a function of $(\varepsilon' - 1)/(2\varepsilon' + 1)$ for Ar, Kr, and Xe matrices in Figure 17.

The predicted linear dependence of $\Delta\nu/\nu$ versus $(\varepsilon' - 1)/(2\varepsilon' + 1)$ on the basis of Buckingham's theory appears to hold well for all $Co(CO)_n$ species (where $n = 1$–4), although the data are limited by the scarcity of ε' (20°K) values for inert matrices.

Similar data have now been obtained for the products of a number of other metal atom cocondensation reactions (see, for example, $Ag(CO)_3$ and $Ag(CO)$,[6] $V(CO)_6$ and $V(CO)_5$[9]). It would appear, therefore, that linear Buckingham plots may prove useful for differentiating matrix-induced vibrational frequency shifts originating from nonspecific, solute-matrix interactions from situations in which "weak chemical bonds" (specific solute-matrix interactions) actually exist.

A final point to which we draw particular attention is the "parallel" behavior exhibited in the polarizability-frequency plots of the *two* components of the CO stretching mode of $Co(CO)_4$ in Ne, Ar, Kr, and Xe matrices (Figure 16). Extrapolation of these data to zero polarizability yields a "gas-phase splitting" of approximately 9.4 cm^{-1}. In the context of this section, this is a convincing demonstration of how a distortion originating in a genuine electronic effect (Jahn–Teller or spin-orbit coupling semantics) can be nicely distinguished from a matrix perturbation. Moreover, note the anomalous position for CO in Figure 16, which is entirely consistent with the proposal that the *larger* splitting observed for $Co(CO)_4$ in solid CO arises from a combination of a "matrix-site splitting" (ca. 6–7 cm^{-1}) superimposed on a small but genuine "distortion splitting" (ca. 10 cm^{-1}).[18]

8.2 Matrix-Site Effects and Molecular Distortions

Molecular distortions from regularity may arise from external influences such as lattice-site effects or internally through special electronic requirements. These distortions can often be recognized in the vibrational spectrum as a splitting of certain lines whose degeneracy has been wholly or partially removed. However, the observation of multiplet structure on vibrational lines does not necessarily imply a measurable distortion, but may arise from slightly different intermolecular interactions sensed by the isolated species when trapped in different sites of the host lattice (referred to as multiple trapping-site effects).

Lattice- or matrix-site effects are in fact quite common and can usually be distinguished from genuine molecular distortions (such as those expected for molecules subject to a Jahn–Teller distortion) by annealing or dilution of the matrix as well as by experimenting with different matrix materials. Whereas matrix-site splittings would normally be sensitive to a changing matrix environment, splittings originating from a true molecular distortion would be expected to be invariant.

Molecular Distortions Imposed by the Lattice: Nickel Tetradinitrogen in Solid Nitrogen

$Ni(N_2)_4$ in argon matrices is a regular tetrahedral molecule showing a single infrared-active T_2 νNN stretching mode at 2174 cm^{-1}. However, the spectrum in nitrogen matrices, rather than showing a single line, shows a well-resolved triplet at 2179.7, 2169.0, and 2161.5 cm^{-1}, which is insensitive to matrix annealing and indicates that the $Ni(N_2)_4$ molecule is somehow perturbed from its regular tetrahedral symmetry.[1]

In order to clarify the structure of $Ni(N_2)_4$ in solid N_2, one intuitively assumes the perturbation to be caused by the coordinated dinitrogen molecules attempting to adopt, as much as possible, orientations similar to those they would have had in the pure unsubstituted nitrogen lattice. As described in detail in Ref. 1, the nickel atom would probably choose to reside in the more spacious substitutional site of the α-N_2 (Pa 3) lattice. If the preferred interaction of the nickel were with four pseudotetrahedrally disposed nearest neighbor N_2 molecules, then the "distortion" of the end-on bonded N_2 molecules would be toward the substitutional site symmetry of N_2 in the unsubstituted lattice, that is, C_2. In order to assume C_2 symmetry, the Ni—N≡N angles must be distorted away from linearity. In addition, the pseudotetrahedral arrangement of the N_2s about the Ni atom implies a deformation of the skeletal N—Ni—N angles as well. The Cotton–Kraihanzel force field cannot in general differentiate between skeletal or

Table 14. Calculated and Observed Frequencies for $Ni(N_2)_4$ in a C_2 Substitutional Site in Solid N_2

$Ni(^{14}N_2)_4$ (T_d in Ar)		$Ni(^{14}N_2)_4$ (C_2 in $^{14}N_2$)		
	Observed Frequency	Calculated Frequency	Observed Frequency	Observed Intensity
	B	2162.3	2161.5	1.60
T_2(IR/R)	2174—B	2170.0	2169.0	10.00
	A	2180.3	2179.7	4.50
A_1(R)	2148—A	2249.0	2248.0	0.00

linear angle deformations. All one can say is that the lattice distortion in α-N_2 is large enough to remove completely the degeneracy of the T_2 mode, giving the A + 2B modes corresponding to C_2 site symmetry. The intensities of the modes resulting from a lattice distortion can often be used to calculate its magnitude (Table 14). Although this calculation cannot be carried out for the case of the distortion of $Ni(N_2)_4$ to C_2 symmetry since there are six direction cosines associated with the dinitrogens needed in general, such a calculation can be performed if we assume the distortions to be C_{2v}. In that case, the distortion could be *either* skeletal or linear angle (Ni—N≡N) bending or both. These two cases are illustrated in Figure 18. In either case the symmetry coordinates for the stretching modes are:

$$S_1^A = \frac{1}{2^{1/2}}(\Delta r_1 + \Delta r_2)$$

$$S_2^A = \frac{1}{2^{1/2}}(\Delta r_3 + \Delta r_4)$$

$$S_3^B = \frac{1}{2^{1/2}}(\Delta r_1 - \Delta r_2)$$

$$S_4^B = \frac{1}{2^{1/2}}(\Delta r_3 - \Delta r_4)$$

Associating f_{rr} with θ_2, $f_{rr'}$ with θ_3, and $f_{rr''}$ with θ_4, the secular equations for the A and B symmetry species are:

$$\begin{vmatrix} f_r + f_{rr} - \dfrac{\lambda}{\mu_{14}} & 2f_{rr''} \\ 2f_{rr''} & f_r + f_{rr'} - \dfrac{\lambda}{\mu_{14}} \end{vmatrix} = 0$$

$$\begin{vmatrix} f_r - f_{rr} - \dfrac{\lambda}{\mu_{14}} & 0 \\ 0 & f_r - f_{rr'} - \dfrac{\lambda}{\mu_{14}} \end{vmatrix} = 0$$

The analysis† yielded the values, $f_r = 19.79$, $f_{rr} = 0.48$, $f_{rr'} = 0.37$, and $f_{rr''} = 0.32$ mdyn/Å with the calculated frequencies shown in Table 14.

Recalling that the Cotton–Kraihanzel force constant approximation does not distinguish between skeletal and Ni—N≡N angular distortions, one

† It should be pointed out that the force constant analysis is rigorously correct (assuming the Cotton–Kraihanzel approximation) for C_2 symmetry as well as C_{2v}.

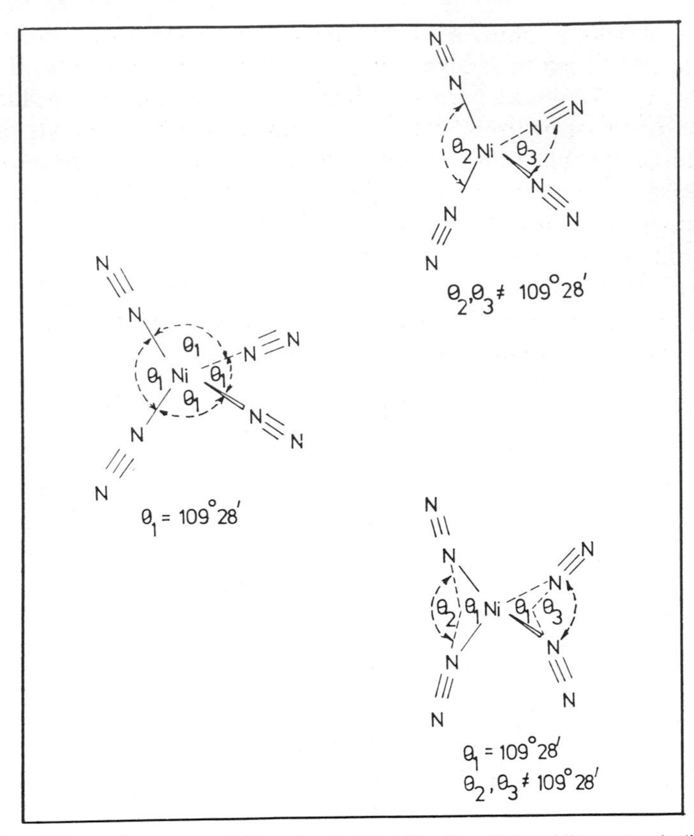

Figure 18. Schematic representation of a skeletal (N—Ni—N) I and linear angle (Ni—N≡N) bending II distortion for $Ni(N_2)_4$ in solid $\alpha - N_2$.

can show the nonzero components of the transition dipole moments with respect to the symmetry coordinates to be:

	$\partial\mu_x/\partial S_k$	$\partial\mu_y/\partial S_k$	$\partial\mu_z/\partial S_k$
S_1^A	0	0	$2^{1/2}\cos(\theta_3/2)$
S_2^A	0	0	$-2^{1/2}\cos(\theta_2/2)$
S_3^B	0	$2^{1/2}\sin(\theta_3/2)$	0
S_4^B	$2^{1/2}\sin(\theta_2/2)$	0	0

Using the measured infrared intensities of the three lines belonging to distorted $Ni(N_2)_4$, one can calculate the values of θ_2 and θ_3. These are $\theta_2 = 53°$ and $\theta_3 = 180°$.

Matrix-Site Splitting Imposed by the Lattice: Bis-Dioxygen Palladium, $Pd(O_2)_2$.

The Pd/O_2 reaction demonstrates an interesting case of matrix-site splitting.[4] When Pd atoms are deposited into a pure $^{16}O_2$ matrix at 10°K, the infrared spectrum [assigned earlier to $(O_2)Pd(O_2)$] initially shows a *single* OO absorption at 1112.3 cm^{-1} (Figure 19). However, on warming the matrix to 15–30°K, this single absorption splits into a closely spaced doublet at 1114.7 and 1110.0 cm^{-1} (Figure 19) whose intensity ratio varies as a function of annealing temperature. As described in detail in the original paper,[4] the data indicate that when first deposited at 10°K the oxygen lattice can be considered to be in a disordered glassy phase (evidence in favor of this stems from Pimentel's infrared data of O_2 when deposited at 4.2°K, which shows significant lattice imperfections in the solid so prepared[37]), and the $(O_2)Pd(O_2)$ molecule experiences an averaged local field and thereby gives rise to a single oxygen absorption (1112.3 cm^{-1}). When the oxygen matrix is warmed, it anneals (presumably in a somewhat nonuniform manner), thereby increasing the regularity of its α-O_2 ($C2/m$; C_{2h}^3) and β-O_2 ($R\bar{3}m$; D_{3d}^5) crystalline lattice portions. The $(O_2)Pd(O_2)$ molecule, formed by preferential interaction of the Pd atom in a substitutional site with two nearest neighbor O_2 molecules, now feels the perturbing effect of the two different substitutional sites (C_{2h} and D_{3d} in α-O_2 and β-O_2 lattices, respectively) and gives rise to the observed closely spaced doublet structure or matrix splitting.

The Gold Dicarbonyl Problem

The matrix reactions of Au atoms and carbon monoxide are in many ways more straightforward than the analogous Cu and Ag reactions, particularly as Au only forms two products, namely, $Au(CO)_2$ and $Au(CO)$.[5] Despite the superficial simplicity of the Au/CO system, one observes some intriguing effects, particularly for $Au(CO)_2$, which deserve special attention in the context of matrix distortions imposed by the lattice as well as matrix site effects. In brief, when Au atoms are cocondensed with a $^{12}C^{16}O:Ne \simeq 1:10$ mixture, a strong absorption is observed at 1935.8 cm^{-1}, which, in $^{12}C^{16}O$:$^{13}C^{16}O:Ne \simeq 1:1:20$ mixtures, is replaced by an approximately 1:2:1 triplet at 1935.8, 1909.5, and 1890.6 cm^{-1}, establishing the product to be $Au(CO)_2$ (Figure 20), most probably with a linear, symmetrical geometry. The observed and calculated frequencies and intensities for $Au(^{12}C^{16}O)_n(^{13}C^{16}O)_{2-n}$ (where $n = 0$–2) shown in Table 15 are in excellent agreement for the three observed modes and confirm that the missing $^{12}C^{16}O$ stretching mode predicted to occur at 2046.2 cm^{-1} is probably not

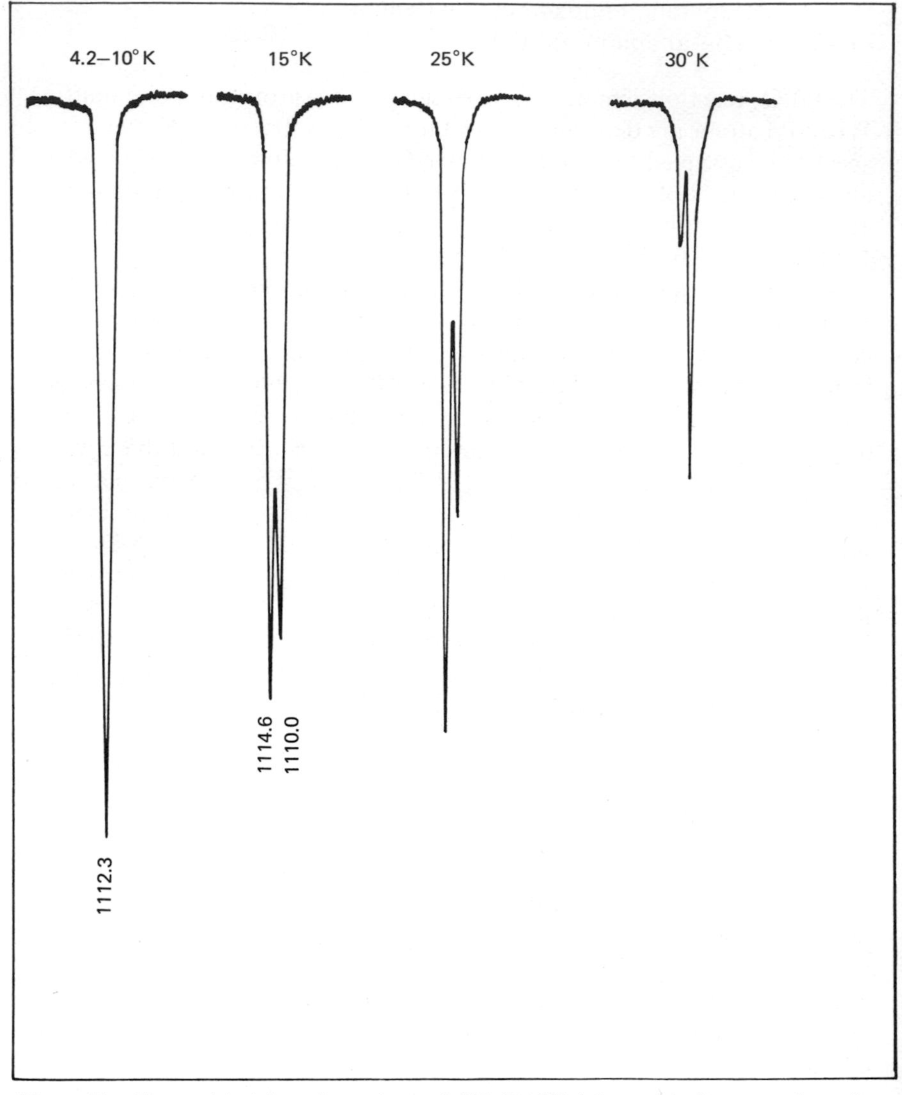

Figure 19. The matrix infrared spectrum of $(^{16}O_2)Pd(^{16}O_2)$ formed in the cocondensation reaction of Pd atoms with pure $^{16}O_2$ on deposition at 6–10°K, and after successive warm-up experiments at 15, 25, and 30°K each for 10 min.

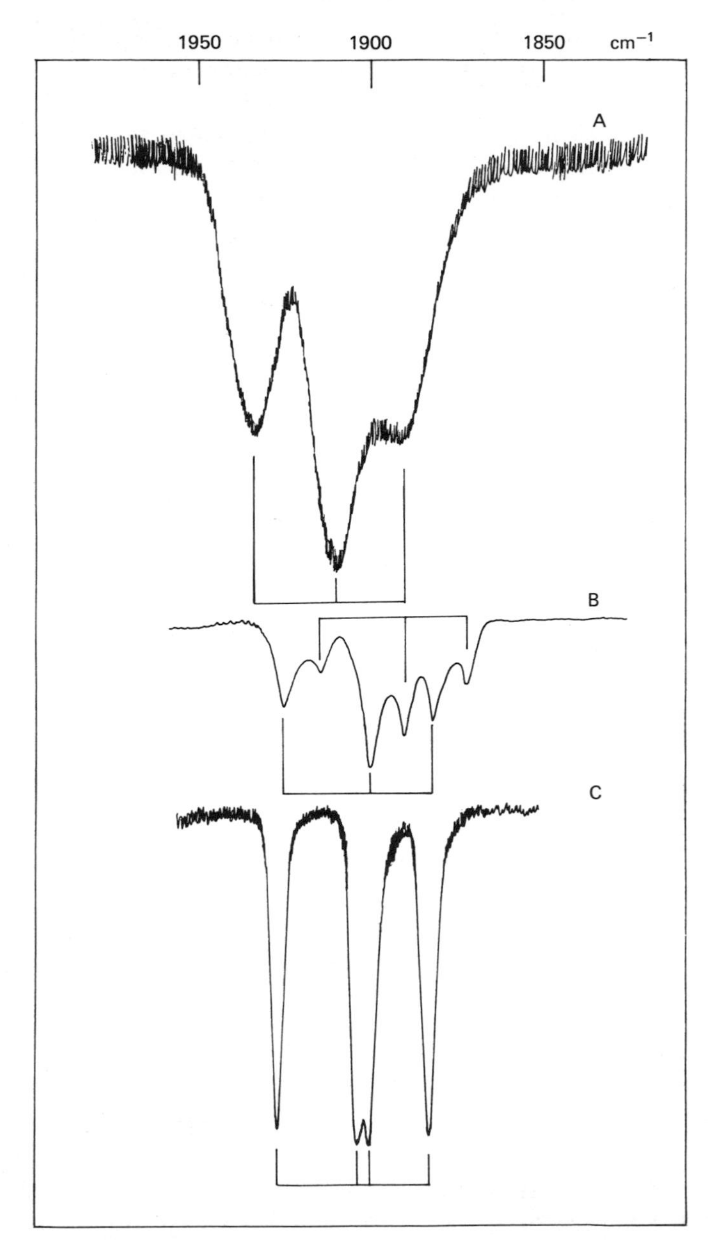

Figure 20. The infrared spectrum of $Au(^{12}C^{16}O)_n(^{13}C^{16}O)_{2-n}$ (where $n = 0$–2) in Ne, Ar, and $^{12}C^{16}O/^{13}C^{16}O$ matrices at 6°K.

observed because of its inherently low absorbance. When the gold-carbon monoxide reaction in $^{12}C^{16}O:Ar \simeq 1:10$ matrices is examined, the original single absorption observed in Ne appears as a well-resolved doublet at 1926.6/1914.8 cm^{-1} in Ar. That this is a case of a matrix multiple-trapping-site effect stems from the corresponding $^{12}C^{16}O:^{13}C^{16}O:Ar \simeq 1:1:20$ experiment that shows a doublet of triplets isotope pattern (Figure 20) consistent with the presence of two slightly different forms of $Au(CO)_2$. Geometrically, both forms contain equivalent CO groups, but differ presumably in the site perturbation that each senses in the Ar lattice.

Table 15. Observed and Calculated Isotopic Spectra of Gold Dicarbonyl in CO and Ne Matrices

Observed Frequency	Calculated Frequency	Observed Intensity	Calculated Intensity	Assignment
CO Matrices[a, b, c]				
1927.8	1926.9	—[d]	10.00	$(^{16}O^{12}C)Au(^{12}C^{16}O)^*$; Σ^+
1903.0	1903.0	—[d]	9.42	$(^{16}O^{12}C)Au(^{13}C^{16}O)^*$; Σ^+
1899.0	1898.7	—[d]	9.06	$(^{16}O^{13}C)Au(^{12}C^{16}O)^*$; Σ^+
1882.9	1884.1	—[d]	9.35	$(^{16}O^{13}C)Au(^{13}C^{16}O)^*$; Σ^+
Ne Matrices[e]				
—	2046.2	—	0.60	$(^{16}O^{12}C)Au(^{13}C^{16}O)$; Σ^+
1935.8	1934.7	10.00	10.00	$(^{16}O^{12}C)Au(^{12}C^{16}O)$; Σ_u^+
1909.5	1909.5	21.23	18.96	$(^{16}O^{12}C)Au(^{13}C^{16}O)$; Σ^+
1890.6	1891.7	8.54	9.56	$(^{16}O^{13}C)Au(^{13}C^{16}O)$; Σ_u^+

[a] Best-fit Cotton–Kraihanzel force constants are $f_r = 15.71$, $f_{r'} = 15.88$, and $f_{rr} = 0.79$ mdyn/Å.

[b] The corresponding in-phase modes were calculated at 1981, 2006, 2010, and 2026 cm^{-1} with intensities between 1 and 6% of the components of the observed quartet.

[c] It should be pointed out that O≡C—Au—O≡C is also consistent with the observed data.

[d] The four observed lines were approximately equal in intensity.

[e] Best-fit Cotton–Kraihanzel force constants are $f_r = 16.18$ and $f_{rr} = 0.95$ mdyn/Å.

On the other hand, the data obtained for $Au(^{12}C^{16}O)_n(^{13}C^{16}O)_{2-n}$ (where $n = 0$–2) in $^{12}C^{16}O/^{13}C^{16}O$ matrices (Figure 20) are quite distinct from those in Ne and Ar matrices and serve to illustrate a matrix effect of a kind that has not previously been experienced. Instead of the characteristic 1:2:1 triplet pattern of lines observed for the dicarbonyl in Ne and Ar, one observes a well-resolved quartet of lines (Figure 20) having approximately equal intensities. There is no doubt that the molecule obtained is gold dicarbonyl, but one is forced to the inescapable conclusion that in a CO matrix the molecule contains *two* kinds of CO ligand, physically different as evidenced by the closely spaced central doublet (1903.0/1899.0 cm^{-1}) of the observed quartet pattern. This type of situation, schematically illustrated as

$$O{\equiv}C{-}Au{-}\overset{*}{C}{\equiv}O$$

can be numerically simulated by computing the frequencies and intensities of the four expected isotopic molecules shown in Table 15.

The calculated isotopic spectrum for $Au(CO)_2$ in CO is in close agreement with that observed and lends credence to the idea of a lattice distortion effect resulting in physical inequivalence of the two CO groups.†

8.3 Molecular Distortions Arising from Electronic Effects: The Structures of Binary Transition Metal Carbonyl Fragments M(CO) (where n = 3–5)

The object of this discussion is to try to provide a unified view of the geometries of $M(CO)_n$ fragments (where n = 3–5) in the light of recent experimental results and theoretical calculations. We restrict our discussions to metal atoms of the first transition series.

Calculations upon which geometric conclusions have been based were generally of the extended Hückel type with minimum energy criteria.[38,39]. The third model proposed[40] is based on overlap between ligand σ pairs and the holes in a non-spherically symmetrical charge distribution. This last scheme can be regarded as an extension of the well-known Gillespie–Nyholm valence shell electron pair repulsion (VSEPR) method.[41] With some exceptions, all these methods—Burdett's orbital overlap and Jahn–Teller arguments,[38] Hoffmann's molecular orbital calculations,[39] as well as Burdett's hole-pair overlap theory[40]—lead to the same general conclusions.

It should be pointed out that a knowledge of exact level ordering is a prerequisite for application of Jahn–Teller considerations and Burdett's hole-pair theory; whereas the direct orbital approach of Hoffmann carries information not only on level ordering but also on geometrical trends and reactivities of the various molecules. As such, this method seems better suited to predict molecular geometries, whereas the former theories, giving more readily available results, are mainly useful in the interpretation of the observed shape of the molecules.

† Recent experiments have shown that an isocarbonyl–carbonyl formulation (CO)Au(CO) is most likely for the product of the Au/CO cocondensation reaction (5).

Table 16. Predicted and Observed Geometries for $M(CO)_3$ and $M(N_2)_3$ Species

	$V(CO)_3$	$Cr(CO)_3$	$Mn(CO)_3$	$Fe(CO)_3$	$Co(CO)_3$	$Ni(CO)_3$	$Cu(CO)_3$
Hoffmann's molecular orbital calculation	$C_{3v}{}^{a}$	$C_{3v}{}^{a}$	$C_{3v}{}^{a}$	$C_{2v}{}^{a}$	C_{3v}	C_{3v}	—
Burdett's minimum internal energy calculation	$C_{3v}{}^{a}$ $C_{2v}{}^{b}$ $D_{3h}{}^{c}$	$C_{3v}{}^{a}$ $C_{2v}{}^{b}$ $D_{3h}{}^{c}$	$C_{2v}{}^{a}$ $D_{3h}{}^{c}$	$C_{2v}{}^{a}$ $C_{3v}{}^{b}$	C_{2v}	D_{3h}	D_{3h}
Burdett's hole-pair overlap	$C_{3v}{}^{a}$	$C_{3v}{}^{a}$ $C_{2v}{}^{b}$	$C_{2v}{}^{a}$ $D_{3h}{}^{c}$	$C_{2v}{}^{a}$	C_{2v}	D_{3h}	D_{3h}
Jahn–Teller considerations	$C_{3v}{}^{a}$ $C_{2v}{}^{b}$	$C_{3v}{}^{a}$ $C_{2v}{}^{b}$ $C_{2v}{}^{c}$	$C_{2v}{}^{a}$	$C_{2v}{}^{a}$	C_{2v}	—	—
Observed geometry	—[d]	C_{3v} (Ref. 42)	—[d]	C_{3v} (Ref. 43)	C_{3v} (Ref. 18)	D_{3h} (Ref. 3) $Ni(N_2)_3$ D_{3h} (Ref. 1)	D_{3h} (Ref. 44) $Ni(CO)_3^-$ D_{3h} (Ref. 45)

[a] Low spin.
[b] Intermediate spin.
[c] High spin.
[d] Geometry not yet established.

Table 17. Predicted and Observed Geometries for $M(CO)_4$ Species

	$V(CO)_4$	$Cr(CO)_4$	$Mn(CO)_4$	$Fe(CO)_4$	$Co(CO)_4$	$Ni(CO)_4$
Hoffmann's molecular orbital calculation	$C_{2v}{}^{a}$	$C_{2v}{}^{a}$	$C_{2v}{}^{a}$	$D_{2d}{}^{a}$	D_{2d}	T_d
Burdett's minimum internal energy calculation	$C_{2v}{}^{a}$ $D_{4h}{}^{b}$ $T_d{}^{c}$	$C_{2v}{}^{a}$ $D_{4h}{}^{b}$ $T_d{}^{c}$	$D_{4h}{}^{a}$ $T_d{}^{c}$	$D_{4h}{}^{a}$ $C_{2v}{}^{c}$	D_{2d}	T_d
Burdett's hole-pair overlap	$C_{2v}{}^{a}$	$C_{2v}{}^{a}$ $D_{4h}{}^{b}$	$D_{4h}{}^{a}$	$D_{4h}{}^{c}$	D_{4h} or $D_{2d}{}^{a}$	T_d
Jahn–Teller considerations	$C_{2v}{}^{a}$	—	$D_{4h}{}^{a}$ $C_{2v}{}^{c}$	$D_{4h}{}^{a}$ $C_{2v}{}^{c}$	D_{2d}	—
Observed geometry	—[d]	C_{2v} (Ref. 42)	—[d]	C_{2v} (Ref. 48)	C_{3v} (Refs. 18, 47)	T_d (Ref. 3) $Ni(N_2)_4$ T_d (Ref. 1)

[a] Low spin.
[b] Intermediate spin.
[c] High spin

The $M(CO)_n$ (where $n = 3-5$) complexes for which structural information is available at this time are listed in Tables 16–18 and are compared with the predicted geometries.

Table 18. Predicted and Observed Geometries for $M(CO)_5$ Species

	$V(CO)_5$	$Cr(CO)_5$	$Mn(CO)_5$	$Fe(CO)_5$
Hoffman's molecular orbital calculations	$C_{4v}{}^a$	$C_{4v}{}^a$ $D_{3h}{}^b$	$C_{4v}{}^a$	$D_{3h}{}^a$
Burdett's minimum internal energy calculation	$D_{3h}{}^a$	$C_{4v}{}^a$ $D_{3h}{}^b$	$C_{4v}{}^a$	$D_{3h}{}^a$
Burdett's hole-pair overlap	$C_{4v}{}^a$	$C_{4v}{}^a$ $D_{3h}{}^b$	$C_{4v}{}^a$	$C_{4v}{}^a$
Jahn–Teller considerations	$C_{2v}{}^a$	$C_{4v}{}^a$	$C_{2v}{}^a$	C_{2v}
Observed geometry	D_{3h} (Ref. 46)	C_{4v} (Ref. 21)	C_{4v} (Ref. 51)	D_{3h} (Ref. 53)
		D_{3h} C_{4v} (Refs. 50, 51)	$Cr(CO)_5^-$	$Mn(CO)_5^-$
			C_{4v} (Ref. 22)	D_{3h} (Ref. 54)

[a] Low spin.
[b] Intermediate spin.

Tricarbonyls and Tris-Dinitrogen Complexes

The calculated optimum geometries for $M(CO)_3$ fragments shown in Table 16 reveal some minor discrepancies. Somewhat unexpectedly, Hoffmann's calculations for the d^{10} $M(CO)_3$ species favor a C_{3v} geometry (an almost flat pyramid, $\theta \simeq 98°$) that is computed to be about 1 kcal/mole lower in energy than the D_{3h} trigonal planar structure.[39] However, all known tricoordinate d^{10} $M(CO)_3$ and $M(N_2)_3$ (where M = Ni, Pd, and Pt) species appear to be trigonal planar. In this context a cautionary reminder is necessary. The planarity (or nonplanarity) of $M(CO)_3$ species (D_{3h} or C_{3v}) is usually determined by the nonobservation (or observation) of the totally symmetric CO stretching vibration in the infrared spectrum. This, however, can be fraught with danger, since for a near-planar structure, the symmetric CO stretching vibration is expected to be extremely weak compared to its asymmetric counterpart vibration and may pass undetected in the infrared spectrum, even though isotopic data and Raman data can help pinpoint the position of this mode.

This situation has been experienced with the vibrational data for $Co(CO)_3$ that favored the D_{3h} geometry for the molecule. Only with the aid of the esr spectrum was it possible to establish the C_{3v} pyramidal nature of $Co(CO)_3$.[18]

For the remaining carbonyls, d^6 $Cr(CO)_3$ and d^8 $Fe(CO)_3$, there seems to be general agreement that the low-spin C_{3v} geometry is the stable form of

$Cr(CO)_3$.[42] The observed C_{3v} geometry of $Fe(CO)_3$ would appear to suggest that $Fe(CO)_3$ is a high-spin species.[43]

The interesting $d^{10}s^1$ species, $Cu(CO)_3$, $Ag(CO)_3$, and $Ni(CO)_3^-$, have recently been synthesized and without exception appear to be planar molecules.[6,44,45]

Tetracarbonyls

A large number of possible configurations of a tetracarbonyl fragment (T_d, D_{2d}, C_{4v}, and C_{2v}) had to be examined in detail in order to make meaningful deductions concerning the geometrical preferences of various d electron configurations.[39] The predictions for d^5 $V(CO)_4$,[46] d^6 $Cr(CO)_4$,[42] and d^{10} $Ni(CO)_4$[3] and $Ni(N_2)_4$[1] agree well with the experimental data. The d^9 $Co(CO)_4$[18] molecule presents an interesting problem as both Hoffmann and Burdett give a D_{2d} structure as the most stable form. Experimentally, there now appears to be general agreement that $Co(CO)_4$ in CO matrices adopts a C_{3v} configuration.[18,47] However, in CO/Ar matrices it is quite likely that C_{3v} and D_{2d} isomers coexist, with the D_{2d} form being slightly more stable.[18] This is not totally unexpected in view of the separate local minimum of slightly higher energy calculated for the C_{3v} structure.[39]

There seems to be some disagreement regarding the stable configuration of low-spin d^8 $Fe(CO)_4$. Hoffmann prefers the D_{2d} structure, whereas Burdett favors the D_{4h} structure. As it turns out in practice, $Fe(CO)_4$ has been assigned a C_{2v} structure and is suspected to be a high-spin complex.[48]

Pentacarbonyls

The two geometries of interest for pentacarbonyl complexes are the square pyramidal structure (C_{4v}) and the trigonal bipyramidal (D_{3h}). The few discrepancies in the predictions seem to arise mainly because of the generally small calculated energy difference between the two structures for almost any electronic configuration, and the ease with which they can undergo intramolecular dynamic processes, interchanging axial and equatorial (or apical and basal) sites.[49] Spectroscopic evidence for matrix-isolated $Cr(CO)_5$ has been reported by two laboratories with somewhat conflicting structural conclusions.[50,21]

$Cr(CO)_5$ can be synthesized from both Cr atom CO/Ar matrix cocondensations[50] and $Cr(CO)_6$/Ar matrix photolysis reactions.[21] However, $^{12}C^{16}O/^{13}C^{16}O$/Ar mixed-isotope experiments have shown that the Cr atom technique yields trigonal bipyramidal $Cr(CO)_5$,[50] whereas the photochemical route[21] leads to the square pyramidal form. Particularly interesting are the warm-up experiments that have recently been performed on the trigonal bipyramidal form of $Cr(CO)_5$.[51] The results authenticate the existence of a trigonal bipyramidal form of $Cr(CO)_5$ at 10–15°K (possibly intermediate-spin, triplet $Cr(CO)_5$), which, on thermal annealing at 40–45°K, undergoes a kinetically slow skeletal rearrangement to the square pyramidal form of $Cr(CO)_5$.

Recent data[46] using vanadium-CO/Ar cocondensation techniques indicate $V(CO)_5$ to have a D_{3h} structure in accord with the structure proposed by Graham.[52] The predictions, on the basis of molecular orbital calculations, for d^7 and d^8 complexes agree well with the observed structures for $Mn(CO)_5$,[51] $Cr(CO)_5^-$,[22] and $Fe(CO)_5$,[53] $Mn(CO)_5^-$,[54] respectively.

In conclusion, it is clear that Kettle's earlier suggestion[55] that simple $M(CO)_n$ species should adopt the highest symmetry structures is no longer valid. Although there is still not complete agreement between Hoffmann's and Burdett's calculations, it is clear from the results obtained so far that the unusual and sometimes unexpected geometries adopted by binary carbonyl complexes are electronic in origin. Hopefully the few discrepancies that still prevail will be clarified, and a logical scheme will emerge.

9. DETERMINATION OF ELECTRONIC PROPERTIES

In this section we briefly examine the experiments and different types of information that can be extracted from matrix uv-visible studies of the products of some metal atom reactions. The actual experimentation itself is relatively straightforward and simply involves cocondensing the metal vapor, reaction partner, and matrix gas together onto an optical plate cooled to cryogenic temperatures. As the procedure is essentially identical to that used for matrix-infrared studies, it is advantageous to record both matrix infrared and uv-visible spectra using the same sample. For these experiments, either the cryogenics move to the spectrometer or vice versa. Both have been tried, and in our experiences the latter arrangement is the more convenient. Depending on which spectral regions are of interest, optical windows may be selected from LiF, CaF_2, NaCl, CsI, or sapphire, the latter being useful also for combined matrix esr–uv-visible studies.

In this section, we have chosen to illustrate the general principles involved and applicability of the technique by restricting ourselves to a small group of metals and their reactions with CO.

9.1 The Electronic Spectrum and Molecular Orbital Description of $Cu(CO)_3$ and $Cu_2(CO)_6$

When Cu atoms were deposited into pure CO (1:100,000), a purple-colored matrix was obtained. Ultraviolet-visible absorptions assignable to atomic copper were not observed. Instead, a molecular spectrum of $Cu(CO)_3$ was observed, characterized by very intense broad absorptions (Figure 21*a* and

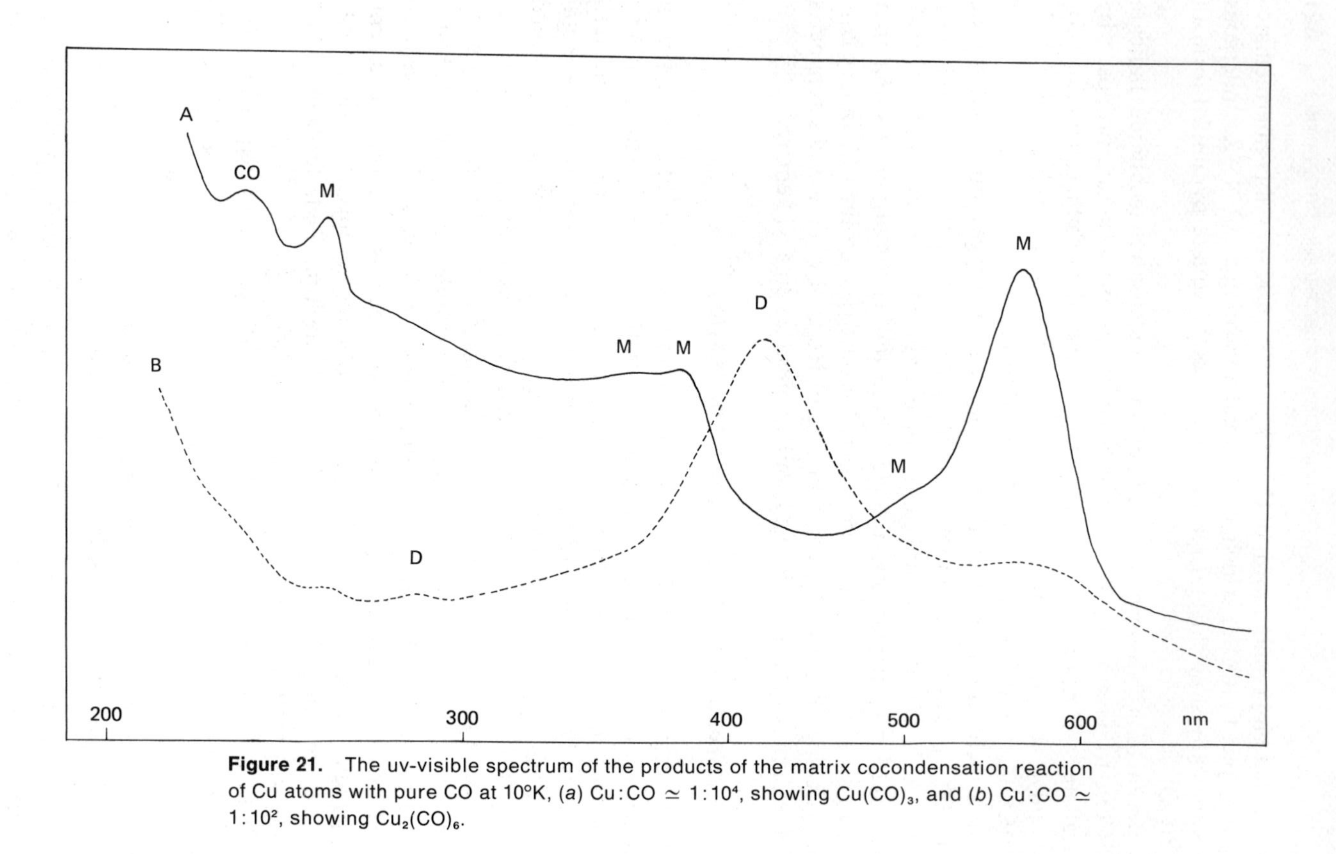

Figure 21. The uv-visible spectrum of the products of the matrix cocondensation reaction of Cu atoms with pure CO at 10°K, (*a*) $Cu:CO \simeq 1:10^4$, showing $Cu(CO)_3$, and (*b*) $Cu:CO \simeq 1:10^2$, showing $Cu_2(CO)_6$.

Table 19. Matrix Ultraviolet-Visible Spectrum of the Products of the Cu:CO Reaction Showing $Cu(CO)_3$ and $Cu_2(CO)_6$

Cu:CO $\simeq$ 1:10,000 (nm)	Cu:CO $\simeq$ 1:100 (nm)	Assignment
562 (s)		$Cu(CO)_3$
495 (wsh)		$Cu(CO)_3$
	417 (s)	$Cu_2(CO)_6$
375 (ms)		$Cu(CO)_3$
344 (wsh)		$Cu(CO)_3$
	287 (w)	$Cu_2(CO)_6$
262 (w)		$Cu(CO)_3$
	221 (s)	$Cu_2(CO)_6$

Table 19).[44] At high Cu/CO ratios (1:100), a yellow matrix was obtained. A molecular spectrum entirely different from that of $Cu(CO)_3$ was observed (Figure 21*b* and Table 19) and is assignable to $Cu_2(CO)_6$.[†] On the other hand, spectral results at low copper concentrations and in dilute CO/Ar matrices (1:100–1:500) were completely different from the above, showing intense atomic copper lines as well as a structured, low-energy absorption at 660 nm that could be assigned to either $Cu(CO)_2$ or Cu(CO). The uv-visible spectral data for $Cu(CO)_3$ and $Cu_2(CO)_6$ are discussed in the following sections.

The Molecular Orbital Description and Electronic Spectrum of $Cu(CO)_3$

In general, the observable electronic spectrum of carbonyl complexes, $M(CO)_n$, can be subdivided into two main types, namely (a) transitions between levels primarily located on the metal, and (b) transitions between levels in which there is an electron transferred from the metal to the ligand, generally called "charge transfer." Transitions located mainly on the CO ligands, for example, $\pi \rightarrow \pi^*$, are too high in energy to be observed in the easily accessible 190–700 nm region.

The electronic spectrum observed for $Cu(CO)_3$ is relatively simple, showing two main absorptions at 562 and 375 nm and two weak shoulders at 495 and 344 nm, as well as a weak feature at 262 nm (Table 19). Before concentrating on a detailed assignment of this spectrum, it is pertinent to obtain some ideas as to whether the observed transitions are of type (a) or (b), or both. Let us briefly consider the predicted and observed trends in the

† From this it is evident that the reaction $2Cu(CO)_3 \rightarrow Cu_2(CO)_6$ rather than $Cu_2 + 6CO \rightarrow Cu_2(CO)_6$ occurs as at CO:Ar $\simeq$ 1:20, only $Cu(CO)_n$ ($n = 1$–3) and $Cu_2(CO)_6$ are formed. No evidence for $Cu_2(CO)_n$ ($n = 1$–5) was found under these conditions.

metal–carbon-monoxide charge-transfer transitions in the series of complexes, $Cr(CO)_6$, $Fe(CO)_5$, $Ni(CO)_4$. In this series, the charge transfer band(s) would be predicted to move to higher energy as the effective nuclear charge on the metal increases, resulting in a stabilization of the metal orbital energies relative to the acceptor orbitals on the CO. (This trend is reflected in the valence orbital ionization potential of the metals.[56]) In addition, the charge-transfer (CT) transition(s) would be expected to move to higher energies as the number of CO ligands donating charge to the central metal atom decreases, that is, the lower CO coordination number will be expected to stabilize the metal atom orbitals by virtue of a smaller spherical term repulsion between the metal and CO electron density.

This is the trend observed in practice: $Cr(CO)_6$, 44,480, 35,780 cm^{-1} (CT); and $Ni(CO)_4$, 52,130, 47,800 cm^{-1} (CT).

Evidently the d $\rightarrow$ π^*(CO) transitions in the copper complex are expected to be at higher energies than in $Ni(CO)_4$ or even $Cr(CO)_6$. However, the lowest energy transition now involves the Cu 4*s* or 4*p* orbital. Whether this transition can be regarded as a Cu $\rightarrow$ ligand "charge transfer" in a strict sense of the expression would seem to depend on the degree to which the orbital is located on the copper atom. To obtain an insight into the spectral region where 4s $\rightarrow$ π^*(CO) charge-transfer transitions are likely to occur, one can use rough energy level schemes for $Cr(CO)_6$, $Ni(CO)_4$, and $Cu(CO)_3$. With the 4*s* orbital some 24,000 cm^{-1} above the 3*d* orbitals of Cu, the possibility of these "charge-transfer" transitions falling into the visible region must be taken into account in the discussion and assignment of the electronic spectrum of $Cu(CO)_3$.

To construct a qualitative molecular orbital energy level scheme for $Cu(CO)_3$, D_{3h} is used as the local symmetry of the copper atom. At the extreme left of Figure 22 the valence shell orbitals of Cu are shown with the symmetry labels appropriate to D_{3h} symmetry. In the center of the diagram are the energy levels of the 5σ orbitals for the three CO groups, also with symmetry designations appropriate to D_{3h}. The energy of the copper valence shell orbitals is assumed approximately equal to the appropriate valence orbital ionization potentials of atomic copper as given by Gray,[56] and the energies of the 5σ and 2π orbitals of CO were taken from the molecular orbital scheme reported by Mulliken.[57] The σ-bonding scheme is obtained by allowing copper and CO orbitals of the same symmetry type to interact where the energy level diagram is based purely on qualitative estimates of the relative energies of the molecular orbitals.

Account can be taken of the interaction of the Cu with the 2π orbitals of CO (which seem to occur, as the CO stretching frequencies of $Cu(CO)_3$ indicate substantial Cu-Cπ interactions) by finding the representations spanned by the six 2π orbitals of CO (shown with their symmetry designations

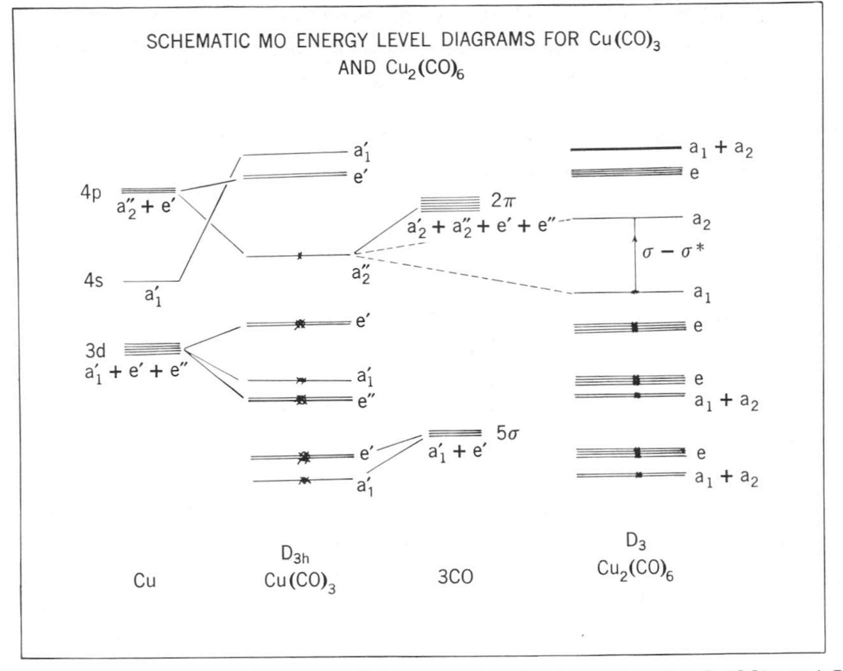

Figure 22. Qualitative molecular orbital energy level scheme for D_{3h} $Cu(CO)_3$ and D_3 $Cu_2(CO)_6$.

near the center of the diagram) and allowing them to interact with the Cu orbitals of the same symmetry type. The molecular orbital energy level scheme resulting from this analysis is shown in Figure 22, from which it can be seen that the highest filled molecular orbital in D_{3h} $Cu(CO)_3$ is of symmetry type a_2'' and contains a single electron.† The electronic ground state therefore, is, expected to be $^2A_2''$.

The electronic spectrum of $Cu(CO)_3$ recorded in a pure CO matrix is shown in Figure 21*a*. The complex has deep purple color, the spectrum being dominated by two intense absorptions at 562 and 375 nm, with two much weaker shoulders at 495 and 344 nm. The two intense bands could be assigned to the $^2A_2'' \rightarrow {}^2A_1'$ and $^2A_2'' \rightarrow {}^2E'$ transitions, respectively. The weak shoulders at approximately 495 and 344 nm most probably arise from a matrix site effect. Electronic transitions that are formally forbidden under D_{3h} could become slightly allowed in a matrix site of lower symmetry and could well account for the two shoulders at 495 and 344 nm.

† Charge iterative extended Hückel calculations have recently been performed on $Cu(CO)_3$.[63] The resulting molecular-orbital energy-level diagram supports the qualitative scheme shown in Figure 22.

The assignment of the bands in the electronic spectrum of $Cu(CO)_3$ proposed above can be tentative at best. With the present data alone, it is difficult to decide on an unequivocal assignment.

Molecular Orbital Description and Electronic Spectrum of $Cu_2(CO)_6$

$Cu_2(CO)_6$ can be viewed as a dimer derived from two $Cu(CO)_3$ residues with the Cu—Cu single bond formed by overlap of the singly occupied, highest orbital of each $Cu(CO)_3$ unit as diagrammed in Figure 22, analogous to the scheme proposed for $Mn_2(CO)_{10}$.[58] According to the infrared spectrum of $Cu_2(CO)_6$, the configuration adopted is probably that of D_3, and so the appropriate symmetry designations are used for the electronic levels of the dimer. The electronic absorption spectrum of $Cu_2(CO)_6$ is dominated by an intense transition, $\lambda_{max} = 417$ nm, which, by analogy with Gray et al.,[58] can be associated with the strongly allowed $^1A_1 \rightarrow {}^1A_2(\sigma \rightarrow \sigma^*)$ electronic transition involving orbitals arising from the copper-copper interaction. One other weaker absorption is observed for $Cu_2(CO)_6$ at 287 nm, which, by analogy with $Cu(CO)_3$, is tentatively assigned to the $^1A_1 \rightarrow {}^1E$ transition, which is expected to shift to higher energy in the dimer because of the stabilization of the $^1A_1$ ground state through the formation of the Cu—Cu bond. Poë et al.[59] have recently demonstrated that activation enthalpies for reactions of several metal-metal bonded carbonyls and bond stretching force constants for the metal-metal bond correlate well with the energies of an electronic absorption band that can be assigned to a $\sigma \rightarrow \sigma^*$ transition between orbitals of the metal-metal bond. Gray et al.[58] have made a rigorous assignment of the band at 343 nm in the electronic spectrum of $Mn_2(CO)_{10}$ to a $\sigma \rightarrow \sigma^*$ transition involving orbitals from the metal-metal interaction. Corresponding bands for $Tc_2(CO)_{10}$, $Re_2(CO)_{10}$, and $ReMn(CO)_{10}$ were also assigned by analogy:

	$\sigma \rightarrow \sigma^*$ (nm)	Ref.
$Mn_2(CO)_{10}$	343	58, 59
$Tc_2(CO)_{10}$	316	59
$Re_2(CO)_{10}$	310	59, 60
$MnRe(CO)_{10}$	324	59
$Cu_2(CO)_6$	417	44

These energies can be taken as an approximate measure of the energy required to break the metal-metal bond without drastically changing any of the other interactions in the molecule, and therefore, they should provide some measure of the relative strengths of the M—M bonds. One therefore, is, led to believe that the extremely low $\sigma \rightarrow \sigma^*$ transition energy of $Cu_2(CO)_6$

compared to the $\sigma \rightarrow \sigma^*$ energies of other M—M bonded carbonyls (that are stable at room temperature) reflects a correspondingly low Cu—Cu bond dissociation energy (or Cu—Cu bond-stretching force constant) and is consistent with the inability, so far, to synthesize the $Cu_2(CO)_6$ molecule by conventional chemical techniques.

9.2 The Electronic Properties of $Cu(CO)_3$ and $Ag(CO)_3$

In the light of the foregoing discussion it is of interest to compare the uv-visible data for $Cu(CO)_3$ with those of $Ag(CO)_3$, the latter being the product of the maxtrix reaction of Ag atoms and CO. The spectra are essentially identical (Figure 23) except that the $Ag(CO)_3$ absorptions are shifted to lower energies compared to those of $Cu(CO)_3$. In order to interpret the above data we can make use of a qualitative comparison between the molecular-orbital energy-level scheme of $Cu(CO)_3$ and $Ag(CO)_3$. As described for $Cu(CO)_3$, we assume that the energies of the silver valence-shell orbitals are approximately equal to the valence-orbital ionization potentials of atomic silver.[61]

The 5*s* and 5*p* levels of Ag are approximately 1000 cm^{-1} above the respective 4*s* and 4*p* levels of Cu. However, the 4*d* level of Ag is roughly 16,500 cm^{-1}

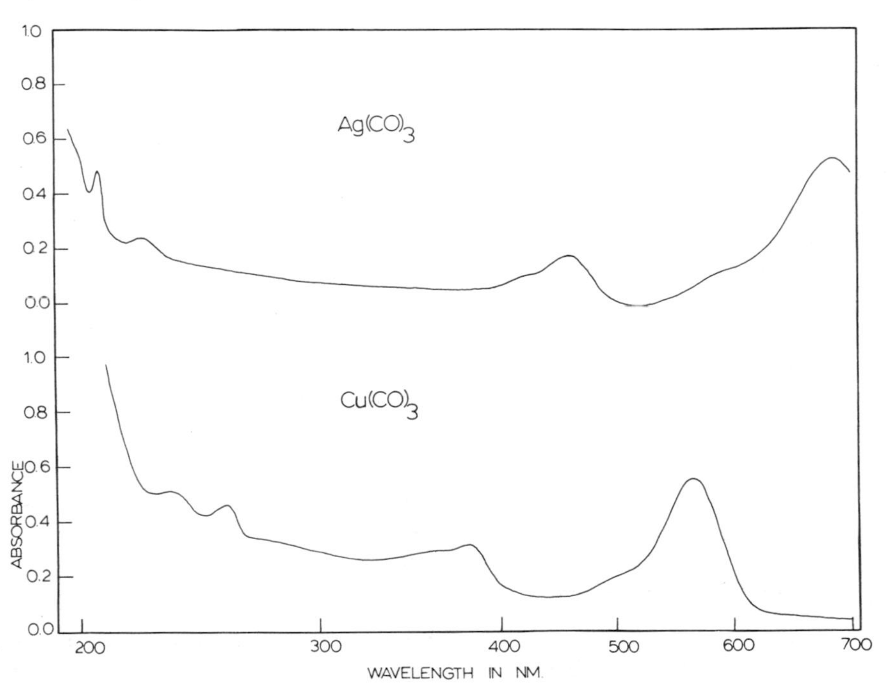

Figure 23. Ultraviolet-visible spectra of $Ag(CO)_3$ and $Cu(CO)_3$ in solid CO at 10°K.

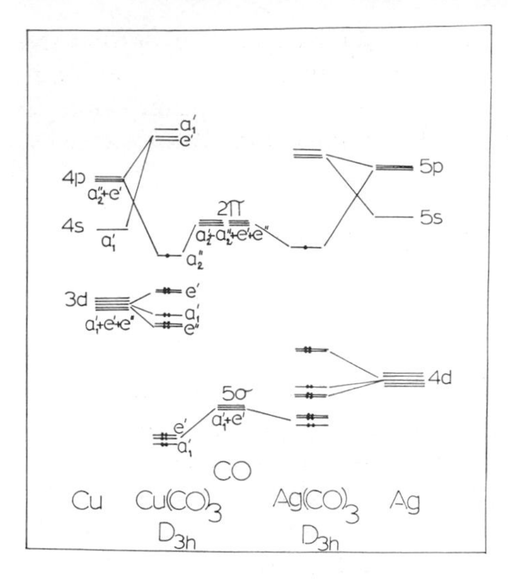

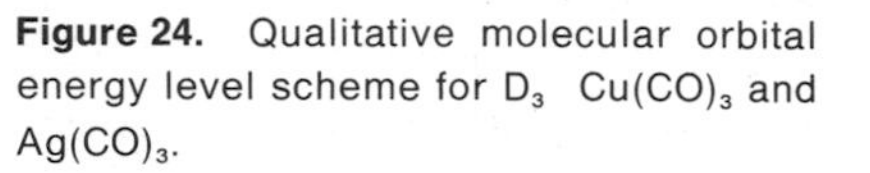

Figure 24. Qualitative molecular orbital energy level scheme for D_3 $Cu(CO)_3$ and $Ag(CO)_3$.

below that of the 3*d* level of Cu. Assuming the σ-bonding scheme can be approximated to trigonal sp^2, when account is taken of the interaction of the approximate orbitals of Cu and Ag with the 5σ and 2π orbitals of CO, one can see that both σ and π type bonding should be slightly more favorable for Cu compared to Ag (Figure 24). The outcome is a greater stabilization of the highest filled π-type molecular orbital (a_2'') and destabilization of the lowest empty σ- and π-type antibonding molecular orbitals for $Cu(CO)_3$ compared to $Ag(CO)_3$, the overall effect being a red shift of all $Ag(CO)_3$ absorptions relative to those of $Cu(CO)_3$. This prediction is in qualitative agreement with the trend observed in practice.[6]

10. MATRIX ELECTRON SPIN RESONANCE STUDIES OF METAL ATOM COCONDENSATION REACTIONS

Transition metal esr is very much the realm of the physicist, yet is of considerable interest to chemists. In this brief survey, we have kept the physics to a minimum and have emphasized the chemical information that has recently been extracted from a few selected metal atom cocondensation reactions with inert and reactive matrices.

10.1 Experimental Problems

Before discussing the esr data for some metal atom cocondensation reactions, it is worth mentioning a few of the complications that arise in this kind of matrix-isolation experiment. Because of the inherently high sensitivity of esr spectroscopy, the technique suffers from problems arising from trace amounts of paramagnetic impurities.

Experiments involving high-temperature vaporizations are particularly prone to interferences of this type, arising from pyrolysis processes occurring on, and contamination originating from, the high-temperature source.

Even with careful vacuum handling and scrupulous outgassing of the vapor source, it appears that it is extremely difficult to avoid cracking trace amounts of hydrogen, water, and hydrocarbon impurities arising either from the metal itself or from within the vacuum system. The impurities seem to appear mainly as H atoms and CH_3 radicals in inert gas matrices, although when CO is present (e.g., a M/CO/Ar reaction), they react to form the paramagnetic species, HCO and CH_3CO.

Contamination problems, particularly from species absorbing around $g = 2.0$, are a nuisance, as the reaction products often absorb in this region. There are, however, a few ways in which one can sometimes obviate the interference effects of impurities.

1. By varying the deposition time, one can often distinguish compound and impurity resonances from differences in the relative rates at which lines grow in, as the experiment progresses.
2. One can study the esr spectra at various annealing temperatures, noting changes in the relative intensities of spectral lines.
3. One can try selectively to saturate or broaden the impurity lines by examining the spectra as a function of microwave power and modulation amplitude.
4. When samples are deposited onto single-crystal sapphire rods having optically flat surfaces (a typical sapphire support might have dimensions 1.75 in. × 0.12 in. × 0.04 in.), one sometimes finds that the molecule under investigation preferentially orients itself with respect to the surface of the sapphire support. Orientation effects have been observed with flat molecules such as

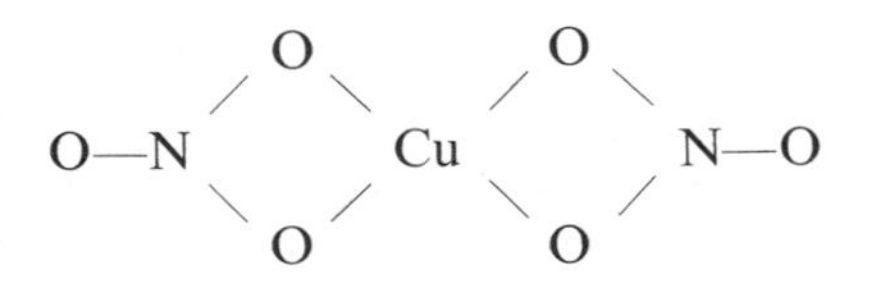

Copper nitrate[62]

O
C
Cu—C≡O
C
O

Copper tricarbonyl[63]

and linear molecules such as

F—Cu—F B=O

Cupric fluoride[62] Boron monoxide[64]

Thus, in certain fortuitous circumstances, where the impurities are randomly oriented yet the compound is oriented in the matrix, impurity lines can be distinguished from compound lines by studying the esr spectrum for various orientations of the sapphire rod with respect to the magnetic field.

10.2 Matrix Studies of Cu, Ag, and Au Atoms: Site Effects

The fate of metal atoms when deposited into inert gas matrices can be probed by optical spectroscopy and will be discussed in detail in Chapters 9 and 10. Matrix shifts, line broadening, and multiplet splittings have been observed on various occasions and have been interpreted in terms of multiple trapping site effects and/or split degeneracies by the local crystal fields. In most instances, however, a precise definition of the lattice environment of the guest atom from such data has been a fruitless task.

Electron-spin-resonance techniques, on the other hand, are potentially capable of providing specific information about matrix trapping sites, and are particularly effective when superhyperfine coupling between the matrix and the guest can be resolved. This method has recently been applied with considerable success to Cu, Ag and Au atoms in Xe matrices.[65] We illustrate the approach with reference to Cu.

To begin with, Cu atoms were studied in Ar matrices and showed the isotropic esr spectrum shown in Figure 25. The hyperfine splitting from an odd electron wholly in an s orbital is naturally very large and in such a situation its analysis requires the general Breit–Rabi formula.[66]

For a trapping site with octahedral symmetry, the resonance should be describable by the spin Hamiltonian

$$\mathrm{H} = g\beta H \cdot S + AI \cdot S \tag{8}$$

with isotropic g and A tensors. Here $S = \frac{1}{2}$ and $I = \frac{3}{2}$ for ^{63}Cu (natural abundance = 69%) and ^{65}Cu (natural abundance = 31%). The g value is expected to be very close to the free spin value, 2.0023, and the hyperfine coupling constant, A, is expected to be extremely large, manifesting the Fermi contact term, $|\psi(0)|^2$, of the half-filled s orbital.

Without going into too many details, the general Breit–Rabi solution of the Hamiltonian (8) yields the following energy expression for the states labeled $M_s = \pm\frac{1}{2}, M_I = +I, \ldots, -I$

$$W(\pm\tfrac{1}{2}, M_I) = -\tfrac{1}{4}A \pm \tfrac{1}{2}\{[g\beta H + A(M_I \pm \tfrac{1}{2})]^2 + A^2[I(I+1) - M_I(M_I \pm 1)]\}^{1/2}$$

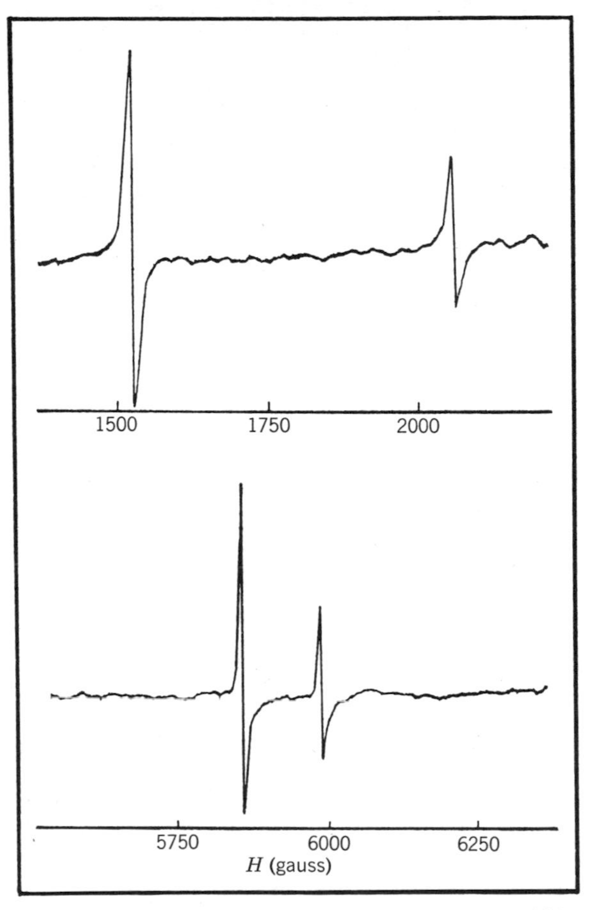

Figure 25. The esr spectrum of Cu atoms in Ar at 10°K.

Figure 26 shows how the energy levels corresponding to a copper atom would diverge. It is immediately clear that the spectrometer frequency, ν, must be larger than the zero field splitting $(I + \frac{1}{2})A = 2A$, in order to observe the "normal" esr spectrum consisting of $(2I + 1)$ hyperfine components arising from the transitions $\Delta m_S = \pm 1, \Delta m_I = 0$. The dotted lines show the possibility of observing "nmr" transitions at X-band frequencies if A is large enough.

Figure 25 shows that the esr spectrum of Cu atoms in natural abundance clearly does correspond to the latter situation. The two signals below 2000 G are assigned to the "nmr" transition, $m_S = -\frac{1}{2}$, $\Delta m_I = -\frac{3}{2} \leftrightarrow -\frac{1}{2}$, while the signals near 6000 G are assigned to the "esr" transition $\Delta m_S = \pm 1$,

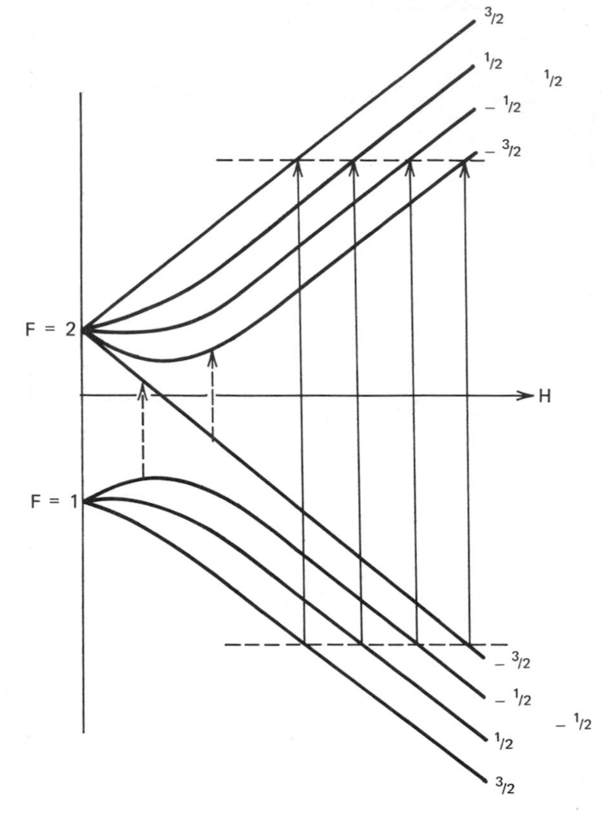

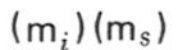

Figure 26. Splitting of energy levels of a system with $S = \frac{1}{2}$ and $I = \frac{1}{2}$ in an external magnetic field. Observable transitions are indicated for the two cases, $h\nu > 2A$ and $h\nu < 2A$.

$m_I = -\frac{3}{2}$. In each case, the stronger line arises from ^{63}Cu, the weaker from ^{65}Cu.

Several aspects are worth noting. Firstly, there is no evidence of multiple trapping sites in Ar. Secondly, each signal is extremely sharp and isotropic ($g = 1.995$), indicating little deviation, if any, of the symmetry of the trapping site from spherical symmetry. And, finally, the ratio of $A(^{65}\text{Cu}):A(^{63}\text{Cu})$ is 1.0709, in close agreement with the ratio of the nuclear magnetic moments (2.3789:2.2206 = 1.0713).

When the experiment was repeated in Xe matrices, which contain magnetic nuclei (^{129}Xe, natural abundance = 26.24%, $I = \frac{1}{2}$; and ^{131}Xe, natural abundance = 21.24%, $I = \frac{3}{2}$), superhyperfine structure was resolved on the ^{63}Cu and ^{65}Cu higher field, "esr" transitions (Figure 27) and clearly arises

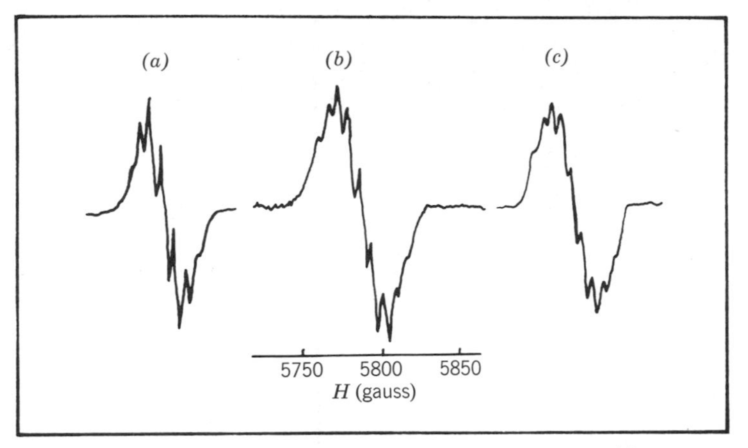

Figure 27. ^{63}Cu component of the esr transition of Cu atoms isolated in a xenon matrix, (*a*) simulated spectrum assuming six nearest neighbor xenon atoms, (*b*) observed spectrum, (*c*) simulated spectrum assuming twelve nearest neighbor xenon atoms.

from significant interaction with the nearest neighbor xenon nuclei.[65] This hyperfine interaction can be represented by an isotropic term of the form

$$\sum_{i=1}^{n} A_i(\mathrm{Xe}_i)I_i \cdot S$$

(where n is the number of nearest neighbor xenon nuclei) and, therefore, provides a unique and powerful probe into the precise environments of the copper atom.

The computer-simulated splitting patterns expected on each esr signal of the Cu atoms resulting from hyperfine interactions with the naturally abundant magnetic xenon nuclei when $n = 6$ (e.g., an octahedral interstitial site) and $n = 12$ (e.g., an octahedral substitutional site) are depicted in Figure 27. The simulated spectra are based on $A(^{129}\mathrm{Xe}) = 13.3$ G and a Lorentzian line shape.

The agreement of the simulated spectrum of Figure 27 to the observed constitutes unequivocal proof that Cu atoms isolated in a xenon matrix occupy substitutional sites with no apparent distortion in the local symmetry.

A similar situation was found to exist for Ag and Au atoms in Xe matrices with $A(^{129}\mathrm{Xe})$ assessed at 16.7 G and 15.0 G, respectively. It is worth noting that the esr spectra of Ag and Au atoms in Ar matrices show smaller hyperfine splitting than observed for copper atoms. ^{107}Ag (natural abundance = 51%, $I = \frac{1}{2}$), ^{109}Ag (natural abundance = 49%, $I = \frac{1}{2}$), and ^{197}Au (natural abundance = 100%, $I = \frac{3}{2}$) display "normal" esr patterns, completely

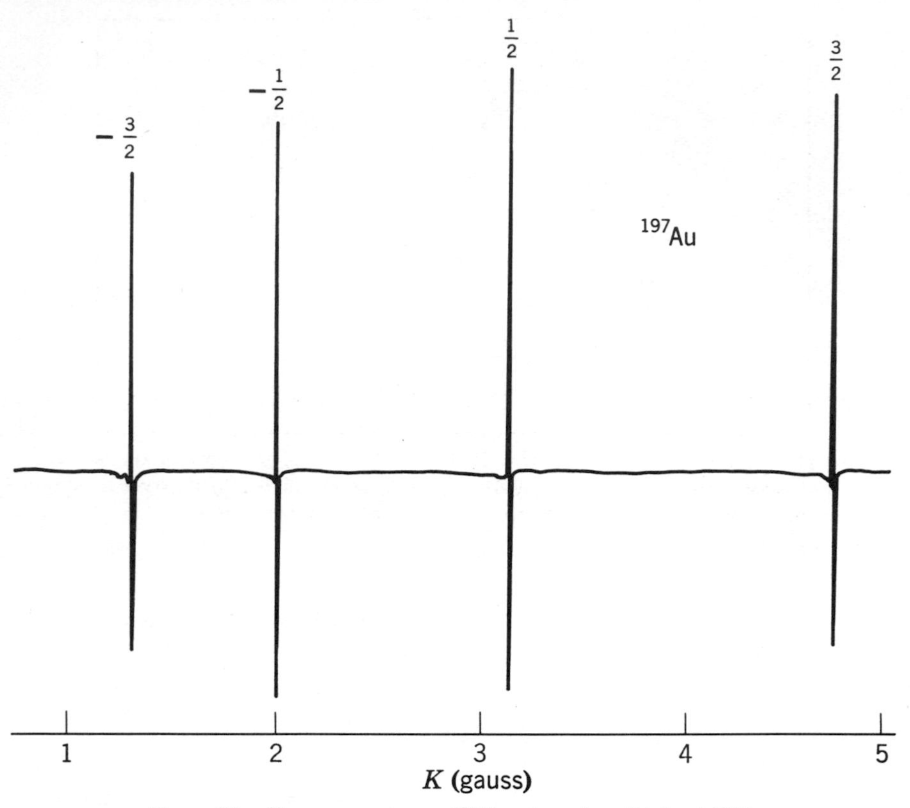

Figure 28. The esr spectrum of ^{197}Au atoms in solid Ar at 10°K.

compatible with the isotropic spin Hamiltonian (see, for example, ^{197}Au, Figure 28[65]).

An intriguing aspect of these esr studies of Cu, Ag, and Au atoms in Ar, Kr, and Xe matrices is the lack of evidence for the existence of multiple trapping sites. These results are somewhat at variance with optical matrix isolation studies of the same atoms.

Clearly, much more work remains to be done on this aspect of metal vapor cocondensations, and one would be wise, whenever possible, to combine data from optical and esr experiments when trying to quantitatively assess matrix site effects.

10.3 Matrix esr Studies of Cu Atom Diffusion in Solid Ar

While on the subject of Cu atoms isolated in Ar matrices, it is worth mentioning the potential usefulness of esr spectroscopy for studying the bulk diffusion properties of metal atoms in low-temperature solids.

One might have thought that esr would provide a more useful method for probing the metal aggregation process than the Mössbauer experiment described for Sn atoms in Section 12, as the number of potentially useful Mössbauer nucleids is far less than those accessible to esr techniques. However, the esr approach to a diffusion-controlled reaction of the type

$$\begin{aligned} M + M &\rightarrow M_2 \\ M + M_2 &\rightarrow M_3 \\ M + M_3 &\rightarrow M_4, \qquad \text{and so forth} \end{aligned}$$

suffers from the inherent weakness that not all of the M_n intermediates will be paramagnetic. Hence integration of the diffusion rate equation, at temperatures where the concentrations of dimer, trimer, and so forth, become appreciable compared to monomer, is usually not feasible, and the experiment is a pointless one. Even under conditions where $[M] \gg [M_2] \gg [M_3]$, one still requires to know the initial concentration of metal in the matrix. Assuming that this latter set of circumstances prevails, one can obtain a crude estimate of the diffusion coefficient of the metal, which could prove useful for comparative purposes.

A study of this type has recently been performed for Cu in Ar at 25, 30, and 35°K.[63] A typical diffusion run at 35°K is shown in Figure 29. Of interest is the observation that appreciable diffusion commences at 30°K.

The rate of the reaction

$$M + M \rightarrow M_2$$

is given by[67]

$$-\frac{d\nu_M}{dt} = 8\pi R D_M \nu_M^2 \left[1 + \frac{R}{(2\pi D_M t)^{1/2}}\right]$$

where ν_M is the number of metal atoms per units volume.

For a reaction radius, R, taken as 3 Å (the intersite distance in the fcc lattice of Ar) and a matrix ratio, Cu:Ar = 1:1000, a value of 4×10^{-17} cm^2 sec^{-1} is estimated for D_{Cu} at 35°K. This value should be compared with 2×10^{-16} cm^2 sec^{-1} for CO in Ar, described in Section 12.

10.4 Matrix esr Studies of Charged Metal Cations in Rare-Gas Matrices

Kasai has recently demonstrated[68] that certain metal atoms can act as electron donors to HI as an electron acceptor. HI has an extremely large cross-section for dissociative electron capture.

$$HI + e \rightarrow H\cdot + I^-$$

For Cd, Cr, and Mn, metal atoms were cocondensed with Ar containing 1% HI. The spectra in all cases demonstrated unequivocally that photo-induced

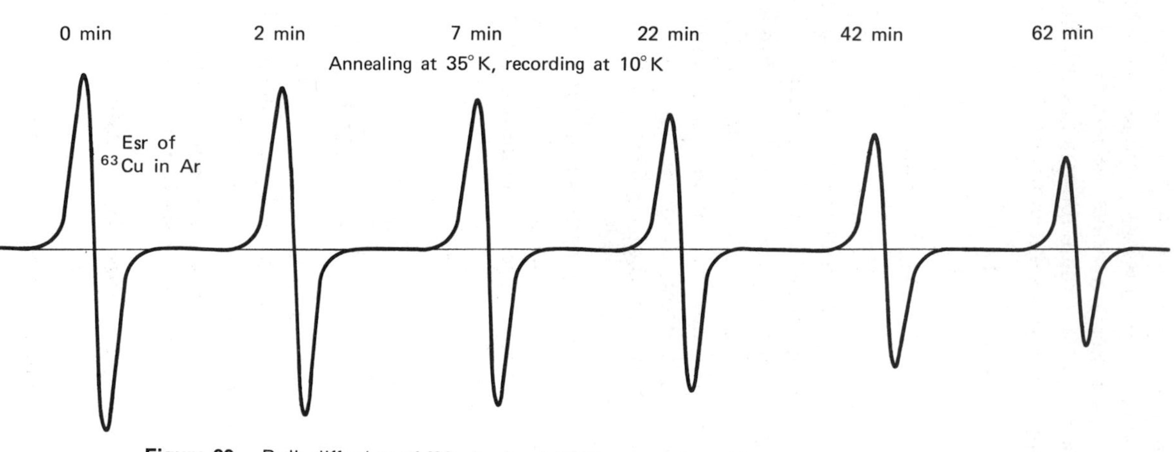

Figure 29. Bulk diffusion of ^{63}Cu in Ar at 35°K as monitored by matrix esr spectroscopy.

electron transfer is possible within a solid Ar matrix. The net reaction can be summarized as

$$M + HI \xrightarrow{h\nu} M^+ + H\cdot + I^-$$

Typical esr spectra for Cr and Cr^+ are shown in Figure 30. The esr spectra showed that both the starting metal atoms and the resulting singly ionized metal atoms can be considered isolated within the matrix.

Let us briefly discuss the highlights of the $Cr \rightarrow Cr^+$ experiment. The spectra before and after photoexcitation are compared in Figure 30. In each trace, the strong central signal is accompanied by a weak, equally spaced quartet, hyperfine structure due to ^{53}Cr (natural abundance = 9.55%, $I = \frac{3}{2}$). The ground state electronic configurations of Cr^0 and Cr^+ are $3d^5 4s^1 (^7S_3)$

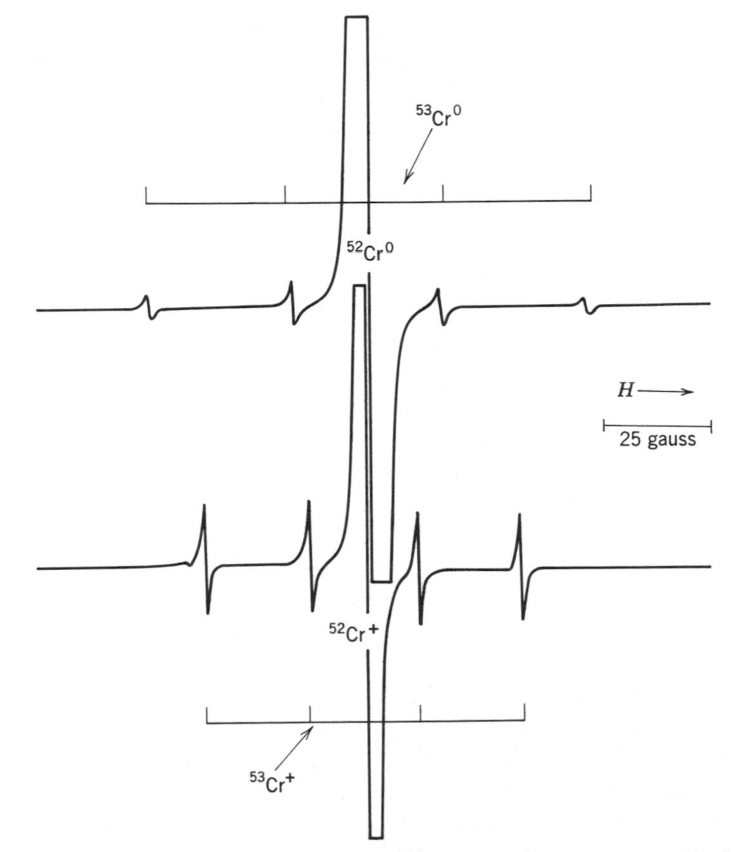

Figure 30. Electron spin resonance spectra of an argon matrix containing Cr atoms and HI molecules observed before (upper trace) and after (lower trace) the photoirradiation.

and $3d^5$ ($^6S_{5/2}$), respectively. The g values of both Cr^0 and Cr^+, therefore, are expected to be close to that of the free electron, and the spectra should be free of "fine structure," provided they are not subject to a crystal field with symmetry less than that of cubic. An important fact is that there is no evidence in the observed spectrum of any significant interaction with another magnetic nucleus such as ^1H ($I = \frac{1}{2}$) or ^{127}I ($I = \frac{5}{2}$), both 100% natural abundance. The sharpness of the observed signals indicates that there is no such magnetic nucleus within ~5 Å of the singly ionized metal atoms. A remarkable aspect of this experiment is that it sets forth direct evidence for total conversion of neutral atoms into the singly ionized form.

Paramagnetic Molecular Species

Let us now consider some recent esr studies involving molecular species formed in a few selected metal-atom–molecule cocondensation reactions. The Co/CO,[18] Cu/CO,[63] Pd/H,[69] and Li/CO[75] matrix reactions serve to illustrate some of the principles involved and information that can be extracted.

10.5 The Matrix esr Spectrum of $Co(CO)_4$

When Co atoms were cocondensed with CO at 6°K under conditions that favored the exclusive formation of $Co(CO)_4$ (see Section 4), the esr spectrum displayed in Figure 31 was obtained.[18] The close resemblance of this spectrum to the esr spectrum previously obtained for the paramagnetic species formed on subliming $Co_2(CO)_8$ onto a 77°K cold finger[70] confirm that the sublimation route does indeed yield $Co(CO)_4$ and not some other, lower $Co(CO)_n$ carbonyl complexes.

The esr spectrum shown in Figure 31 is typical of a molecule having axial symmetry, $S = \frac{1}{2}$ with hyperfine coupling to ^{59}Co ($I = \frac{7}{2}$), and, therefore, establishes general agreement between the conclusions drawn from the matrix infrared, Raman, and esr spectra of $Co(CO)_4$ in pure CO, that is, the molecule is distorted from regular T_d symmetry.

The exact nature of the distortion is revealed from the corresponding $Co/^{13}C^{16}O$ experiment (Figure 31*b*), using 93.7% enriched $^{13}C^{16}O(I(^{13}C) = \frac{1}{2})$ that shows a "doublet" $^{13}C^{16}O$ superhyperfine coupling ($A_{\parallel} = 26 \pm 2 \times 10^{-4}$; $A_{\perp} = 24 \pm 1 \times 10^{-4}$ cm^{-1}) of every hyperfine component of the original $Co(^{12}C^{16}O)_4$ molecule. Clearly, this depicts a tetracarbonyl molecule with a unique, carbonyl group and provides unequivocal evidence in favor of the axial C_{3v} distortion rather than the axial D_{2d} distortion.

A detailed analysis of the $Co(^{12}C^{16}O)_4$ and $Co(^{13}C^{16}O)_4$ esr spectra[18] was undertaken for a d^9 tetrahedral field, with a trigonal distortion (using

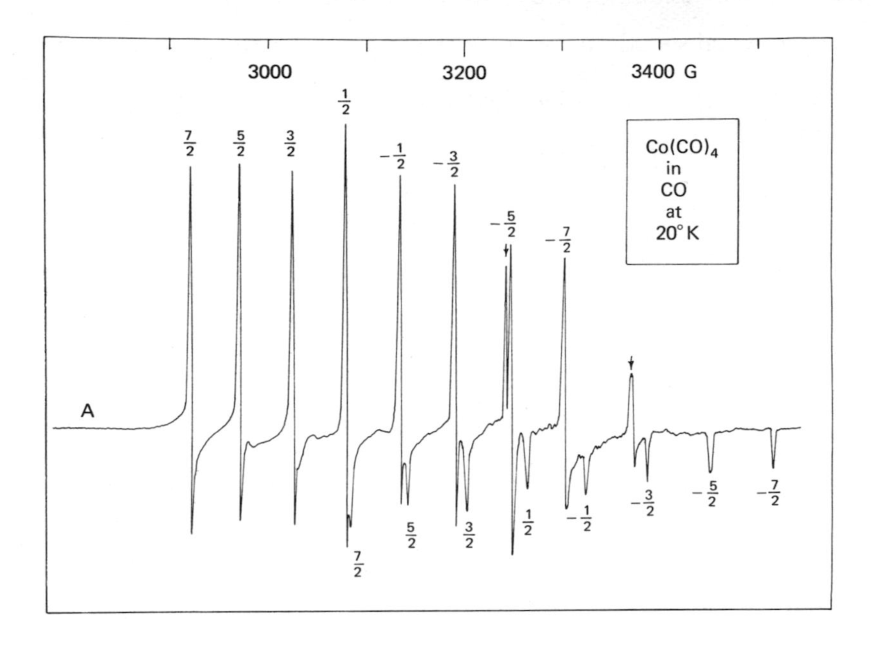

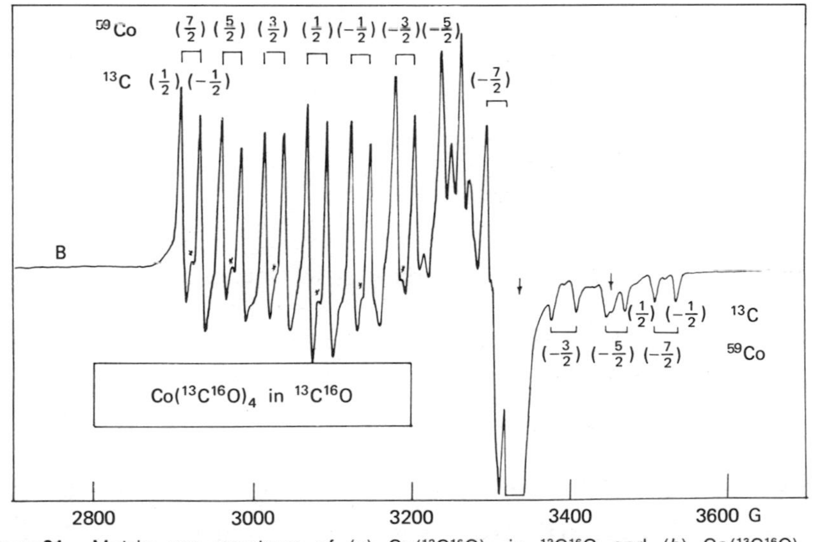

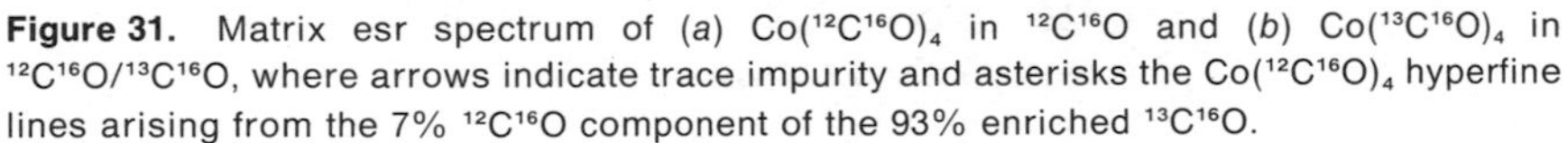

Figure 31. Matrix esr spectrum of (*a*) Co(^{12}C^{16}O)$_4$ in ^{12}C^{16}O and (*b*) Co(^{13}C^{16}O)$_4$ in ^{12}C^{16}O/^{13}C^{16}O, where arrows indicate trace impurity and asterisks the Co(^{12}C^{16}O)$_4$ hyperfine lines arising from the 7% ^{12}C^{16}O component of the 93% enriched ^{13}C^{16}O.

the hole formalism), the d functions of which are as follows:

$$\left.\begin{array}{ll} e'_{\pm} = bd_{\pm 2} \pm ad_{\mp 1} & \underline{\hspace{3cm}} \\ e_{\pm} = ad_{\pm 2} \mp bd_{\mp 1} & \underline{\hspace{3cm}} \\ a_1 = d_0 & \underline{\hspace{3cm}} \end{array}\right\} \delta \quad \Delta$$

We briefly summarize the results. For the hole formalism, the spin orbit operator is

$$\begin{aligned} \mathrm{H}_{LS} &= -\xi \vec{l} \cdot \vec{s} \\ &= -\xi[l_z \cdot s_z + \tfrac{1}{2}(l_+ s_- + l_- s_+)] \end{aligned}$$

where ξ is positive and is a one-electron spin-orbit parameter (300 cm^{-1} for $^{59}\mathrm{Co}^0$). For near tetrahedral symmetry, δ could be of the same order of magnitude as ξ, so one must solve the complete determinant for the energies of the Kramer's doublets, yielding

$$\varepsilon = \eta + \tfrac{1}{2}(2a^2 - b^2)$$
$$\varepsilon = \tfrac{1}{2}\eta - \tfrac{1}{4}(2a^2 - b^2) \pm \tfrac{1}{2}\{\eta^2 - (2a^2 - b^2)\eta + \tfrac{1}{4}(2a^2 - b^2)^2 + 6b^2\}^{1/2}$$

The wavefunctions for the Kramer's ground state doublet can be written as

$$\begin{aligned} \psi_+ &= \cos r d_0^+ + \sin r e_-^- \\ \psi_- &= \cos r d_0^- - \sin r e_+^+ \end{aligned}$$

solution of which gives

$$\tan 2r = \frac{6^{1/2} b}{\eta - \tfrac{1}{2}(2a^2 - b^2)}$$

where $\eta = \delta/\xi$ and $\varepsilon = E/\xi$. (Note that for tetrahedral symmetry, $\eta = 0$, $a^2 = \tfrac{2}{3}$, $b^2 = \tfrac{1}{3}$, and $\varepsilon = \tfrac{1}{2}, -1$. Therefore, spin-orbit interaction has removed the degeneracy of the T_2 state, and hence there is no need to invoke a Jahn–Teller distortion.)

As $g_{\parallel} = 2\langle\psi_+|l_z + 2.0023\, s_z\,|\psi_+\rangle$ and $g_{\perp} = 2\langle\psi_+|l_x + 2.0023\, s_x\,|\psi_-\rangle$,

this leads to expressions for $g_{\parallel}$ and $g_{\perp}$ as a function of the distortion, that is,

$$\begin{aligned} g_{\parallel} &= 2.0023 \cos 2r - 2(2a^2 - b^2)\sin^2 r \\ g_{\perp} &= 2.0023 \cos^2 r - 6^{1/2} b \sin 2r \end{aligned}$$

Account can also be taken of mixing with the $e'_{\pm}$ states by including a second-order perturbation. Let $\rho = \xi/\Delta$, then from first-order perturbation, ψ_+

and ψ_- are altered and lead to the following additions to g:

$$\Delta g_{\|} = 3a^2(6^{1/2}b \sin 2r - 6b^2 \sin^2 r)\rho$$

$$\Delta g_{\perp} = 3a^2\left(2\cos^2 r - \frac{6^{1/2}}{2} b \sin 2r\right)\rho$$

The experimental data were corrected to second order using expressions derived by Bleany[71] for an axial, spin $\frac{1}{2}$ system:

$$H_{\|} = \frac{h\nu_0}{g_{\|}\beta} - \frac{A_{\|}}{g_{\|}\beta} M_I - \frac{A_{\perp}^2}{2g_{\|}^2\beta^2 H}[I(I+1) - M_I^2],$$

$$H_{\perp} = \frac{h\nu_0}{g_{\perp}\beta} - \frac{A_{\perp}}{g_{\perp}\beta} M_I - \frac{(A_{\|}^2 + A_{\perp}^2)}{4g_{\perp}^2\beta^2 H}[I(I+1) - M_I^2],$$

from which one obtains

$$g_{\| } = 2.007 \pm 0.010 \qquad A_{\|} = 58 \pm 1 \times 10^{-4}\ \text{cm}^{-1}$$

$$g_{\perp} = 2.128 \pm 0.010 \qquad A_{\perp} = 55 \pm 1 \times 10^{-4}\ \text{cm}^{-1}$$

From Figure 32, which shows the dependence of $g_{\|}$ and $g_{\perp}$ on the distortion parameter, η, it can be shown that $g_{\perp} > g_{\|}$ and $g_{\|} \simeq 2$ can only occur with large values of η, a conclusion consistent with the observation of superhyperfine coupling to only one ^{13}CO ligand.

Analysis of the hyperfine coupling constant data is more complex and is not dealt with here. However, we can summarize the results by pointing out that values of $\delta \simeq 5850\ \text{cm}^{-1}$ and $\Delta \simeq 28{,}000$–$30{,}000$ can be arrived at from the magnetic data, which appear to be self-consistent with the optical

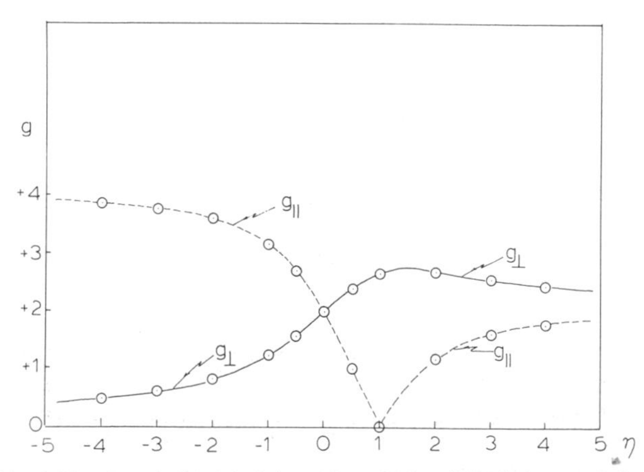

Figure 32. Dependence of $g_{\|}$ and $g_{\perp}$ on the distortion parameter, η.

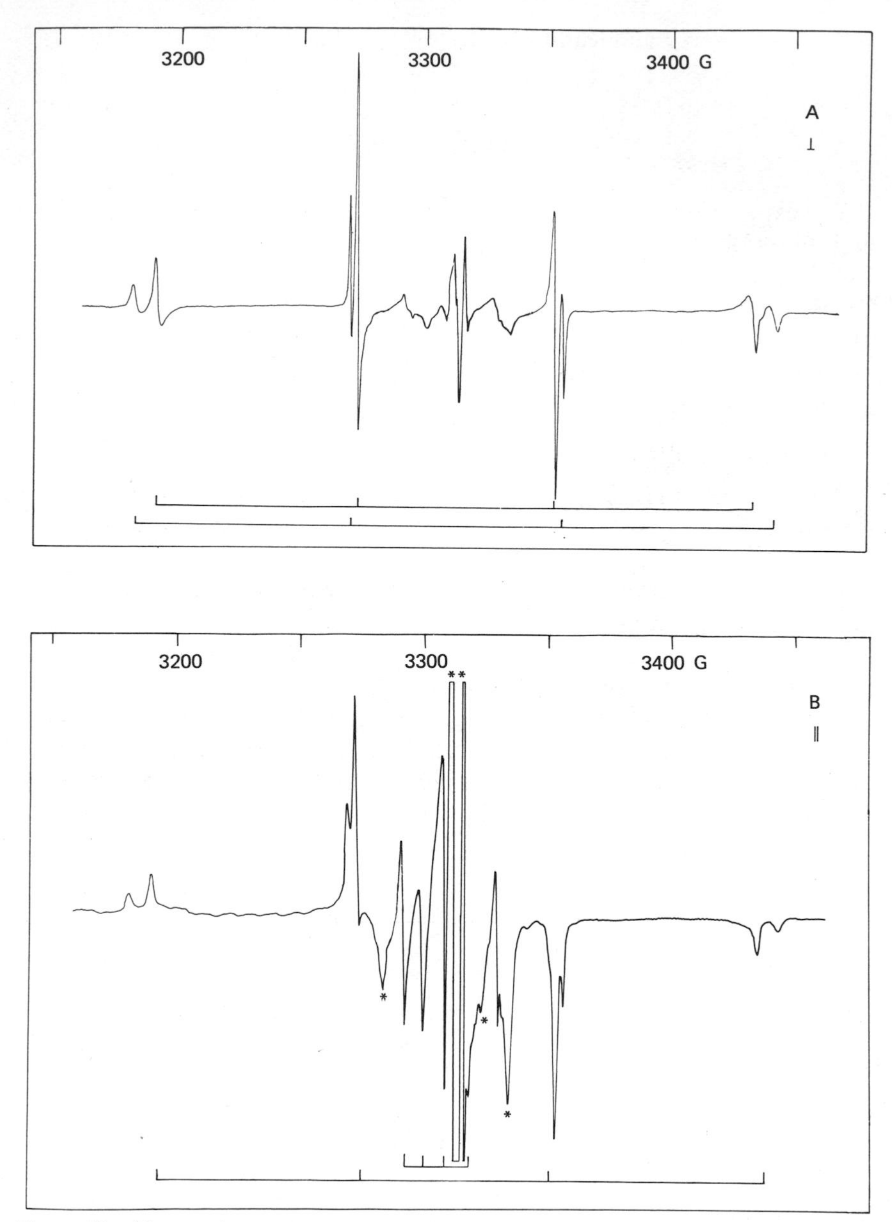

Figure 33. The matrix esr spectrum of $Cu(^{12}C^{16}O)_3$ resulting from the cocondensation of Cu atoms with a CO : Ar $\simeq$ 1 : 10 mixture at 20°K, (*a*) matrix surface perpendicular to the magnetic field, (*b*) matrix surface parallel to the magnetic field (asterisks indicate impurity lines).

data for $Co(CO)_4$, which show ligand-field transitions at 28090–31650 cm^{-1}. The hyperfine splittings indicate that the electron is on ^{13}C approximately 6% and in the ^{59}Co d orbital approximately 81%.

10.6 The Matrix esr Spectrum of $Cu(CO)_3$

When Cu atoms were cocondensed with CO:Ar = 1:10 at 20°K, the esr spectra in Figure 33 were obtained.[63] Under these conditions the only product formed is $Cu(CO)_3$. This was confirmed by infrared experiments using identical deposition conditions. The main features of the esr spectrum of $Cu(CO)_3$ in argon are:

1. The $Cu(CO)_3$ molecule is oriented in the argon matrix, its plane being parallel to the surface of the sapphire rod (Figure 33*a* and *b*).
2. The molecule has axial symmetry and $S = \frac{1}{2}$.
3. The g values are close to the free electron value ($g_{\parallel} = 2.006$; $g_{\perp} = 2.010$).
4. Hyperfine splitting is observed for the two Cu isotopes ^{63}Cu ($I = \frac{3}{2}$, natural abundance 69%) and ^{65}Cu ($I = \frac{3}{2}$, natural abundance 31%).
5. A large anisotropy in the copper hyperfine splitting is observed for the parallel and perpendicular components: $^{63}A_{\perp} = 7.8 \times 10^{-4}\ cm^{-1}$; $^{63}A_{\parallel} = 76.2 \times 10^{-4}\ cm^{-1}$.

The only way to explain the large anisotropy in the hyperfine splittings is to have the unpaired electron in either an A_2'' orbital (Cu $4p_z$, COπ^*) or in an A_1' orbital (Cu $3d_{z^2}$). For a $^2A_1'$ ground state one expects a very large hyperfine splitting with all three carbon atoms if ^{13}CO is used in the experiment, whereas for a $^2A_2''$ ground state the hyperfine splitting with ^{13}C, if observable at all, is expected to be small. Preliminary experiments using ^{13}CO/Ar gas mixtures show no splitting of the previously observed lines, although a small splitting may go undetected due to line broadening.[63]

The results so far point to a $^2A_2''$ ground state, supporting the infrared and uv-visible data discussed in Section 9. A detailed analysis of the esr data suggests that the unpaired electron is in a molecular orbital largely localized on the ligands. This would account for the g values close to the free spin value and would support the suggestion of $4p_z$(Cu) – π^*(CO) π-bonding (Ref. 6 and Section 11).

10.7 Matrix esr Studies of the Pd-Atom–H-Atom Cocondensation Reaction

Weltner has recently pioneered an important new technique whereby transition metal diatomic hydrides MH (where M = Be, Mg, Ca, Sr, Ba, Zn, Cd, and Hg) can be synthesized by simultaneously cocondensing thermally generated M and H atoms onto a cold surface at 4.2°K.[72] The furnace assembly required to perform this experiment has been described in Chapter 2.

In the context of transition metal cocondensations, we now describe Weltner's work with Pd, H, and D atoms.[69]

Among the several isotopes of palladium, only ^{105}Pd (natural abundance 22.6%) has a nuclear magnetic moment, $I = \frac{5}{2}$, that leads to hyperfine splitting (hfs). The ^{0}PdH spectrum, where ^{0}Pd indicates the absence of a nuclear moment and hfs, should then consist of two perpendicular lines and two parallel lines arising from hydrogen hfs. The esr spectrum for ^{0}PdH in Ar at 4.2°K is shown in Figure 34*a*. Two sets of perpendicular lines are seen. These have been attributed to two matrix sites, α and β, for the ^{0}PdH molecule ($g_{\perp} \sim 2.3$). The corresponding parallel lines occurred around $g \sim 2.0$ and were found to be extremely weak. Substitution of D for H yielded the ^{0}PdD spectrum shown in Figure 34*b* for the α site that shows the distinct hfs of D with $I = 1$, indicating that the magnetic molecule does indeed contain one hydrogen atom.

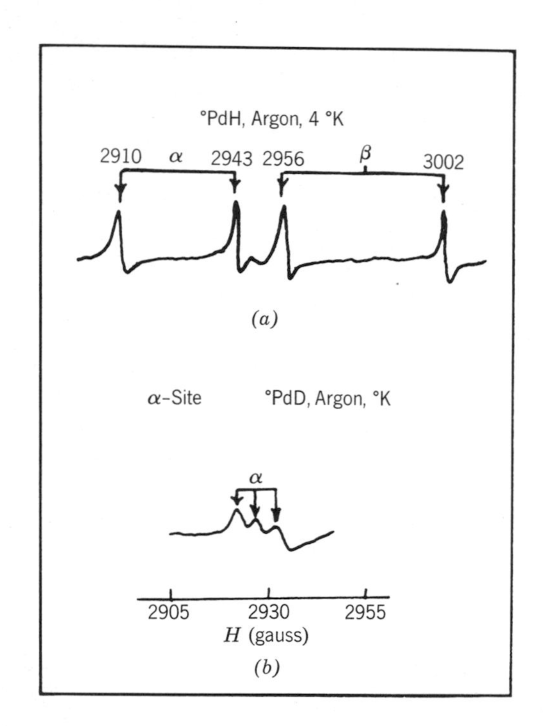

Figure 34. (*a*) Electron spin resonance spectrum of ^{0}PdH in Ar at 4.2°K formed from the matrix reaction of Pd atoms and H_2. (*b*) esr spectrum of ^{0}PdD in Ar at 4.2°K formed from the matrix reaction of Pd atoms and D_2. (α and β indicate lines attributed to two sites in the matrix; see original Ref. 69.)

Metal hfs from the 22.6% abundance of ^{105}Pd ($I = \frac{5}{2}$) was detected for both ^{105}PdH and ^{105}PdD, although the lines were very weak, and high modulation had to be used to enhance their intensity, which introduced some line broadening.

The second-order solution to the usual axial spin Hamiltonian

$$\mathscr{H} = g_{\parallel}H_zS_z + g_{\perp}(H_xS_x + H_yS_y) + A_{\parallel}I_zS_z + A_{\perp}(I_xS_x + I_yS_y)$$

was used to extract $A_{\perp}$, $A_{\parallel}$, and $g_{\perp}$ values from the spectral data.[69]

Without going into a detailed analysis of the A and g tensors, the data indicated that the odd electron in the ground $^2\Sigma^+$ state is predominantly $5d\sigma$ with perhaps 30% $5s$ character. The molecular orbital description of PdH is, therefore

$$KLMN(4d\delta)^4(4d\pi)^4(5s\sigma)^2(4d\sigma)$$

where the $4d\sigma$ and $5s\sigma$ orbitals are understood to be hybridized. This ordering of the d orbitals is also in accord with a recent crystal-field description of the molecule[73] and the known $^2\Sigma^+$ excited states[74] ($5s\sigma \rightarrow 4d\sigma$ and $4d\sigma \rightarrow 5p\sigma$). Such a model is quite reasonable for PdH, since the observed spin density on H is less than 10%, so the bond can be considered essentially ionic, that is, Pd^+H^-. Then in a crystal field of $C_{\infty v}$, the Pd^+ ion would have its degenerate shell split into δ, π, and σ levels in order of increasing energy. The nine d electrons would then fill the lowest two levels and leave one in the σ level to yield a $^2\Sigma^+$ ground state.

10.8 Matrix esr Studies of the Li–CO Cocondensation Reaction

In this concluding section we should briefly refer to some intriguing esr results recently reported by Krishnan and Margrave[75] for the Li/CO system. Using infrared spectroscopy and ^{6}Li/^{7}Li and $^{12}C^{16}O/^{13}C^{16}O$ isotopic substitution, they were able to identify four lithium carbonyls with stoichiometries Li(CO) (two isomeric forms: linear Li—CO and nonlinear Li—OC), $Li(CO)_2$ and $Li_2(CO)_2$.

Lithium monocarbonyl (C bonded) absorbs in the range, 1806–1793 cm^{-1}, and is best formulated as Li^+CO^- (note NO is isoelectronic with CO^- and absorbs at 1876 cm^{-1}). On the other hand, lithium isocarbonyl absorbs in the range, 1600–1550 cm^{-1}, and should be compared to nonlinear LiON.[76]

Of interest to the present discussion is the mononuclear dicarbonyl that absorbs at 1951/1655 cm^{-1} and contains equivalent carbon and oxygen atoms. The 1951 cm^{-1} absorption fingerprints a C≡C stretching mode, while the 1655 cm^{-1} absorption implies some double-bond contribution to the CO bonds, suggesting the formulation $Li^{\delta+}(OCCO)^{\delta-}$. The corresponding esr spectra are shown in Figure 35a, where it can be seen that the

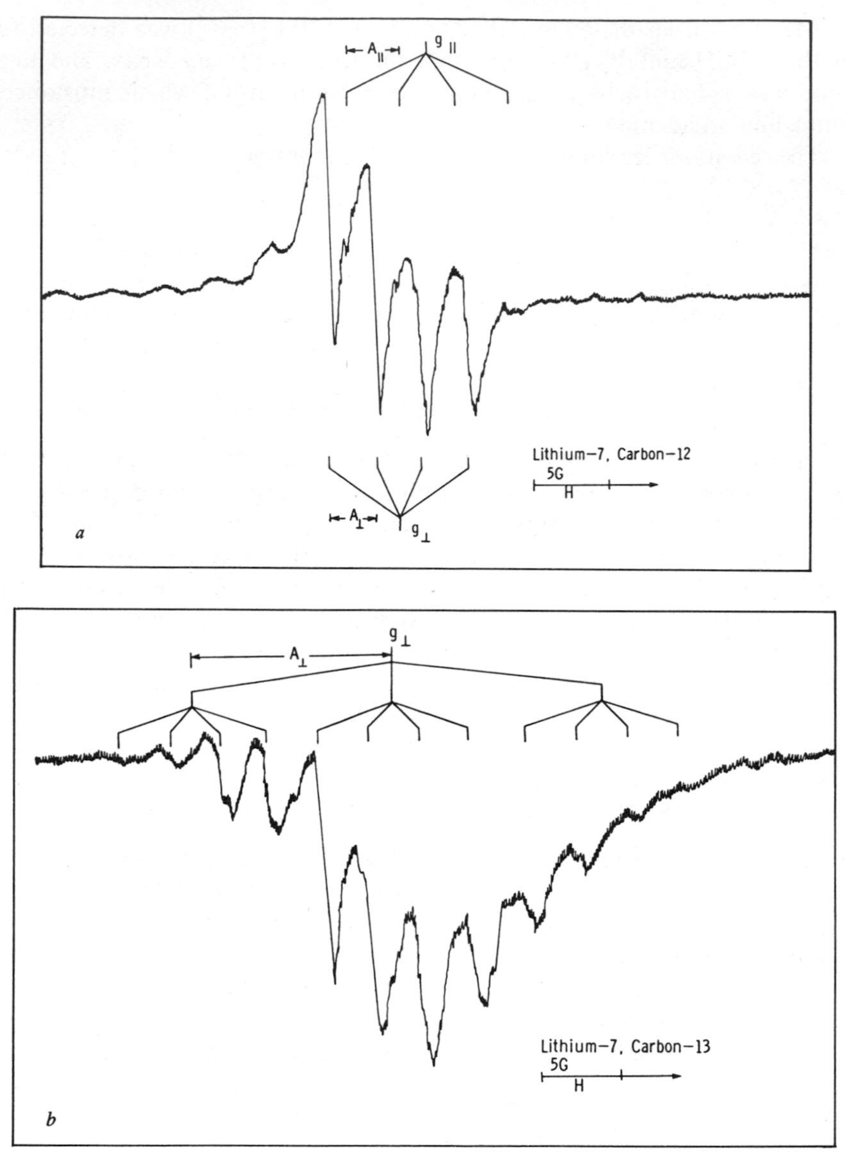

Figure 35. Electron spin resonance spectra of (*a*) Li($^{12}C^{16}O$)$_2$ and (*b*) Li($^{13}C^{16}O$)$_2$ formed from the Li/$^{12}C^{16}O$ and Li/$^{13}C^{16}O$ matrix reactions, respectively.

monolithium dicarbonyl species shows axial symmetry, $S = \frac{1}{2}$, with the unpaired electron interacting with a ^7Li nucleus having a nuclear spin of 3/2 (Figure 35*a*; $A_{\parallel}(^7Li) = 4.28$ G, $A_{\perp}(^7Li) = 3.70$ G) and two equivalent carbon atoms (Figure 35*b*; $A_{\perp}(^{13}C) = 14.9$ G). The $g_{\parallel}$ and $g_{\perp}$ values are 2.0134 and 2.0138, respectively, that is, close to the free electron g_e value. The isotropic splitting is an indication that the electron density is in an *s* orbital. Using $A_{iso} = 2/3(A_{\parallel} + 2A_{\perp})$, one obtains a value of 7.79 G for A_{iso}. The electron density, ρ, can be obtained from A_{iso} by comparing the value with A_0 (143.45 G), the isotropic hyperfine coupling for a free ^7Li atom, that is,

$$A_{iso} = \rho A_0$$

Substituting for A_{iso} and A_0 one obtains 0.055 for ρ, the spin density in the 2*s* orbital of lithium.

This suggests an almost complete transfer of the lithium 2*s* electron to the $C_2O_2^-$ unit, in agreement with the infrared measurements where the CO stretching frequency shows no lithium-isotope effect. Thus the monolithium dicarbonyl species can indeed be represented as $Li^+(OCCO)^-$.

Finally, dilithium dicarbonyl absorbs in the 1250–1300 cm^{-1} region, suggesting C—O single bonds and an acetylenediolate formulation:

$$Li^+O^-—C\equiv C—O^-Li^+$$

Interestingly, Weiss and Büchner[77] prepared similar alkali metal acetylenediolates from the reaction of liquid ammonia solutions of Na, K, Rb, or Cs with CO. An X-ray structure determination[78] showed a structure similar to that proposed for $Li_2(C_2O_2)$.

11. DETERMINATION OF MOLECULAR PROPERTIES

Infrared spectroscopy yields two kinds of data, namely, band positions and band intensities. Information regarding structure, coordination number, identity, and type of bonding may be extracted from these. The simultaneous application of matrix isolation and spectroscopy to the first three has been discussed in the previous sections. The last forms the substance of this section.

11.1 Binary Carbonyl Dinitrogen Complexes of Nickel

To illustrate the application of matrix spectroscopic data to the determination of the nature of bonding, we begin with the mixed $Ni/N_2/CO$ system.

Ever since the discovery that N_2 can form dinitrogen complexes that are isoelectronic with analogous carbonyl complexes, the similarities and differences in the nature of their bonding have been of great interest. In particular, the metal carbon bond in carbonyls is most often discussed in terms of the Dewar–Chatt model,[79] in which there is simultaneous σ donation by a lone pair on the ligand into empty *d* orbitals on the metal and π back-donation from filled metal orbitals into empty π^* orbitals on the ligand. The relative σ-donor π-acceptor properties of CO vis-à-vis those of N_2 is of some interest particularly because the decrease in bond-stretching force constant from that of the free ligand was larger in dinitrogen complexes than in the analogous carbonyls. Since it has been proposed[80] that the magnitude of the CO-stretching-force constant in metal carbonyls reflects in an inverse fashion the amount of π back-donation, the former observation led some[81] to speculate that N_2 is a better π acid than CO. It is obvious that a great deal of light would be shed on the argument if it were possible to compare the properties of the two ligands bonded to the same metal atom, thereby competing for the same electrons.

For that reason, the series of complexes $Ni(N_2)_m(CO)_{4-m}$ (where $m = 1$–3) were synthesized[12] by simultaneously cocondensing Ni vapor with matrix gases containing N_2/CO/Ar in various ratios. For a matrix ratio of 1:1:20 the ir spectrum in Figure 5 was obtained after annealing to 30°K, while for a matrix ratio of 3:1:0 the matrix Raman spectrum shown in Figure 6 was recorded. Aside from the strong infrared absorption at 2138 cm^{-1} and its isotopic counterparts, ten new absorptions, five in the NN stretching region and five in the CO stretching region, were observed. Using isotopic mixtures of $^{14}N_2$, $^{15}N_2$, and $^{12}C^{18}O$, assignments of the various lines to the species, $Ni(CO)_4$, $Ni(CO)_3N_2$, $Ni(CO)_2(N_2)_2$, and $Ni(CO)(N_2)_3$ [$Ni(N_2)_4$ was almost always absent] could be made. Intensity measurements of the bands were also made (using the linear absorbance setting on the Perkin-Elmer 180), and where serious overlaps occurred, as in the case of the 2269.6, 2264.4, and 2257.6 cm^{-1} lines, the area and position of each line were obtained by numerically fitting Lorenzians to reproduce the total measured absorbance as a function of ν.

Figure 36 shows the calculated NN and CO force constants calculated assuming the Cotton–Kraihanzel approximation for the series of molecules, $Ni(CO)_m(N_2)_{4-m}$, as a function of *m*. The curves for $Ni(N_2)_m$ and $Ni(CO)_m$ are included for comparison. Several features are at once clear. To begin with, three of the curves are monotonic with *m*, while that of the NN stretching force constant of the molecules $Ni(CO)_m(N_2)_{4-m}$ has a maximum at $m = 2$. Secondly, the CO-stretching-force constants of $Ni(CO)_m(N_2)_{4-m}$ ($m = 1$–3) lie slightly above those of the analogous $Ni(CO)_m$ and *below* that of $Ni(CO)_4$, while the NN stretching force constants of $Ni(CO)_m(N_2)_{4-m}$ are all significantly larger than those of the analogous $Ni(N_2)_m$, and all lie *above* that of $Ni(N_2)_4$.

Using a proposal made by Fenske,[82] it was assumed that the ligand-stretching-force constant was linearly related to the quantity of charge donated (σ) and accepted (π) by the ligand.

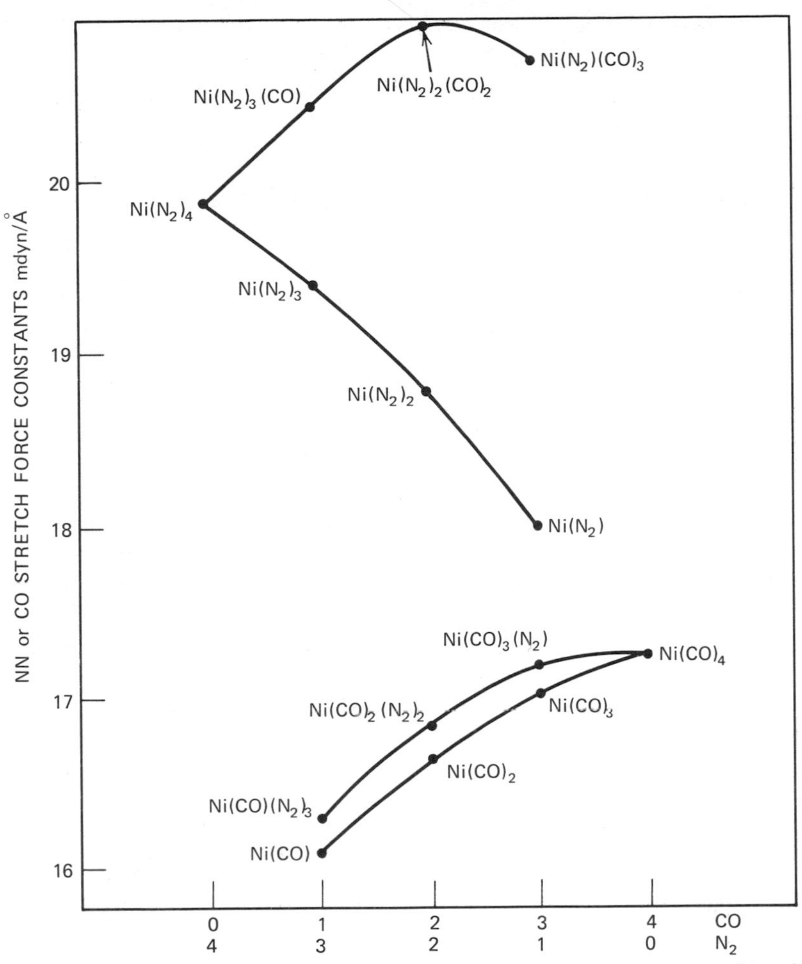

Figure 36. Graphical plot of the NN and CO Cotton–Kraihanzel bond-stretching force constants for the species, $Ni(CO)_m(N_2)_{4-m}$, $Ni(CO)_m$, and $Ni(N_2)_m$ (where $m = 1$–4) as a function of m.

Thus for $Ni(CO)_m(N_2)_{4-m}$

$$f_m^{CO} = d + |a|(\sigma_m^{CO} - \gamma\pi_m^{CO}) \tag{9}$$

and

$$f_{4-m}^{N_2} = d' - a'(\sigma_{4-m}^{N_2} + \gamma'\pi_{4-m}^{N_2}) \tag{10}$$

where f_m^{CO} and $f_{4-m}^{N_2}$ refer, respectively, to the CO and NN stretching force constants, and $|a|$, γ, and d, and a', γ', and d' are positive constants. Using the above equations it proved possible to show that the observed force constant values shown in Figure 36 were only consistent with

$$\pi^{CO} > \pi^{N_2}$$

and

$$\sigma^{CO} \geq \sigma^{N_2}$$

in all species. Moreover, equations (9) and (10) indicate that the force constants of bonded CO depend on a $(\sigma - \pi)$-like term, but those of bonded N_2 on a $(\sigma + \pi)$-like term. This explains the greater decrease in the NN-stretching force constants (from that of the free ligand) observed in N_2 complexes than the corresponding decrease in the CO-stretching force constant in the analogous carbonyls.

The greater acceptor ability of CO as compared to N_2 also explains the presence of the maximum in the $f_{4-m}^{N_2}$ versus m curve. As one proceeds from $Ni(CO)_3(N_2)$ to $Ni(CO)(N_2)_3$, $\pi_{4-m}^{N_2}$ increases because of the decrease of (electron withdrawing) COs; simultaneously and for the same reason, $\sigma_{4-m}^{N_2}$ is expected to decrease. Thus the quantity, $\sigma_{4-m}^{N_2} + \gamma'\pi_{4-m}^{N_2}$, would have a minimum, and $f_{4-m}^{N_2}$ a maximum, when the slope of $\sigma_{4-m}^{N_2}$ versus m is equal but opposite in sign to the slope of $\pi_{4-m}^{N_2}$ versus m. The maximum is observed at $f_2^{N_2}$.

The monotonic increase of f_m^{CO} with increasing m can only be explained if π_m^{CO} decreases with increasing m.

That $\pi_{4-m}^{N_2}$ varies monotonically with m despite the amonotonic behavior of $f_{4-m}^{N_2}$ is further borne out by the intensity measurements.

Because the concentrations of the absorbing species in the matrix are not known precisely, the magnitude of the individual CO and N_2 transition dipoles cannot be calculated. The ratio, $(\mu'_{CO}/\mu'_{N_2})_m$, where m refers to $Ni(CO)_m(N_2)_{4-m}$, is calculable, however, as are the quantities, $C_m(\mu'_{CO})^2$ and $C_m(\mu'_{N_2})^2$, where C_m is a constant proportional to the concentration of $Ni(CO)_m(N_2)_{4-m}$.

These quantities were obtained from the relation

$$\sum I_m = C_m \sum_{k'k''} \frac{\partial \boldsymbol{\mu}}{\partial S_{k'}} \cdot \frac{\partial \boldsymbol{\mu}}{\partial S_{k''}} G_{k'k''}$$

where $\sum I_m$ is the measured absorbance of all bands associated with $Ni(CO)_m(N_2)_{4-m}$, the other quantities having the conventional meanings. The results are shown in Table 20, which indicates that μ'_{CO}/μ'_{N_2} increases slightly as the number of COs increases. $C_m(\mu'_{CO})^2$ and $C_m(\mu'_{N_2})^2$, on the other hand, both decrease as the number of COs in the molecule increases. Although the values of C_m are unknown, one expects C_m to decrease as m decreases; thus, if anything, μ'_{CO} and μ'_{N_2} both increase even more rapidly with decreasing number of COs than do $C_m(\mu'_{N_2})^2$ and $C_m(\mu'_{CO})^2$.

Table 20. Calculated μ'_{CO}/μ'_{N_2} Values for the Molecules, $Ni(CO)_m(N_2)_{4-m}$

m	Molecule	$(\mu'_{CO}/\mu'_{N_2})_m$	$C_m(\mu'_{CO})^2$	$C_m(\mu'_{N_2})^2$
1	$NiCO(N_2)_3$	3.60	76.9	5.94
2	$Ni(CO)_2(N_2)_2$	3.97	55.7	3.53
3	$Ni(CO)_3N_2$	4.34	22.8	1.21

The monotonic decrease of both μ'_{N_2} and μ'_{CO} with increasing m in $Ni(CO)_m(N_2)_{4-m}$ is noteworthy in that it parallels the decrease of both π_m^{CO} and $\pi_{4-m}^{N_2}$ with increasing m, as previously obtained from the force constant analysis. This result is consistent with the view expressed by Darensbourg[83] and others that states that the integrated absorbances of bonded CO and N_2 increase with increased π overlap.

11.2 Tetracarbonyls of Nickel, Palladium, and Platinum

The relationship between CO-stretching force constant and σ-donated and π-accepted charge was also applied to the series $M(CO)_4$, where M = Ni, Pd, and Pt.[11]

The previously unknown species, $Pd(CO)_4$ and $Pt(CO)_4$, were synthesized by matrix isolation and characterized by isotope substitution. The observed Cotton–Kraihanzel force constants were 17.23, 17.55, and 17.28 mdyn/Å, respectively, for $Ni(CO)_4$, $Pd(CO)_4$, and $Pt(CO)_4$. (Since ν_{M-C} was also observed for these molecules, approximate k_{MC}s could be calculated assuming $M(CO)_4$ to be MX_4 where X is a mass of 28 amu. The results were k_{MC} = 1.80, 0.82, and 1.28 mdyn/Å for, respectively, $Ni(CO)_4$, $Pd(CO)_4$, and $Pt(CO)_4$, which parallels k_{CO} in an inverse fashion, as expected.) The reversal of order on going from Pd to Pt is attributed to the Lanthanide contraction.

In order to interpret the observed CO-stretching-force constants in terms of the individual σ and π contributions in the three tetracarbonyls, the assumption was made that σ_4^M, the σ-donated charge per CO in $M(CO)_4$ (where M = Ni, Pd, Pt), was proportional to the respective Allred–Rochow electronegativities of these metals. Thus

$$\sigma_4^{Pd} < \sigma_4^{Pt} < \sigma_4^{Ni} \tag{11}$$

Using the expression for K_4^M, the CO-stretching-force constant for $M(CO)_4$ given below

$$K_4^M = d + |a|(\sigma_4^M - \gamma\pi_4^M) \tag{12}$$

and inequality (11), the result

$$\pi_4^{Pd} < \pi_4^{Pt} < \pi_4^{Ni}$$

was obtained, since

$$K_4^{Pd} > K_4^{Pt} > K_4^{Ni}$$

Thus the effectiveness of CO as both a σ donor and a π acceptor toward M is in the order

$$\text{Ni} > \text{Pt} > \text{Pd}$$

Another interesting application of equation (12) is towards determining the charge on the metal. It is evident that the charge on the metal, q_M, in the three species, $Ni(CO)_4$, $Pd(CO)_4$, and $Pt(CO)_4$, is approximately given by

$$q_M \simeq -4(\sigma_4^M - \pi_4^M) \tag{13}$$

Moreover, equation (12) may be written in the form

$$\frac{K_4^M - d}{|a|} = \sigma_4^M - \gamma\pi_4^M \tag{14}$$

Equations (13) and (14) form two equations in two unknowns (σ_4^M and π_4^M) that could be solved if q_M, K_4^M, $|a|$, γ, and d were known. The last three quantities may be evaluated from Fenske's data[82] (assuming the validity of such a transfer), and K_4^M is experimentally determined. The quantity, q_M, is, in general, not available except in the case of $Ni(CO)_4$, for which SCF-MO calculations have been performed. Hillier[84] reports a value of -0.94 electrons for q_{Ni}. Using this value in equation (13) and combining with (14), one obtains $\sigma_4^{Ni} = 1.07$ and $\pi_4^{Ni} = 0.83$ electrons.

Because molecular orbital calculations have not been performed on $Pd(CO)_4$ and $Pt(CO)_4$, one cannot solve for σ_4^M and π_4^M for these complexes in the same manner as was done for $Ni(CO)_4$. Making the assumption, however, that σ_4^M is proportional to χ_M, the Allred–Rochow electronegativities of the corresponding metal, π_4^M, may be obtained from equation (14). The results are given in Table 21. These calculations indicate that Pd and Pt do not differ markedly in their σ-acceptor and π-donor properties. Ni, on the other hand, is a substantially better σ acceptor and π donor than the former two metals, resulting in the observed thermal stability of $Ni(CO)_4$.

Table 21. Calculated σ_4^M, π_4^M, and q_M Values for $Ni(CO)_4$, $Pd(CO)_4$, and $Pt(CO)_4$

M	χ_M	σ_4^M	π_4^M	q_M
Ni	1.75	1.07	0.83	−0.94
Pd	1.35	0.83	0.61	−0.88
Pt	1.44	0.88	0.67	−0.84

It is intriguing that the charge on the metal does not vary appreciably on going from $Ni(CO)_4$ to $Pd(CO)_4$ to $Pt(CO)_4$. A similar result was reported by Hillier[84] for $Ni(CO)_4$, $Cr(CO)_6$, and $Fe(CO)_5$.

11.3 Dinitrogen Dioxygen Complexes of Nickel, Palladium, and Platinum

The bonding properties of the mixed dinitrogen-dioxygen complexes are probably best described in terms of dinitrogen coordination towards a binary dioxygen complex, the latter closely approximating an ionic formulation, $M^{\delta+}(O_2^{\delta-})$, where the oxygen is somewhere between peroxide and superoxide in character. In this respect, the $(O_2)M(N_2)$ and $(O_2)M(N_2)_2$ complexes closely resemble DeKock's[85] $(N_2)MF_2$ and $(CO)MF_2$ complexes formed by cocondensing gaseous MF_2 monomers (where M = Cr, Mn, Fe, Co, Ni, and Zn) with the appropriate ligand under conditions of matrix isolation. In particular, DeKock's carbonyl complexes yielded convincing evidence of a bonding situation in which CO acts essentially as a σ donor to an ionically bound MF_2 moiety. The CO absorptions, without exception, appeared above 2138 cm^{-1}, the frequency of uncoordinated CO. In an ionic situation of the type, $(CO)M^{2+}(2F^-)$, the π donation of charge from a cationic metal center to the 2π orbitals of CO is greatly diminished compared to a formally uncharged metal center. On the other hand, the σ donor properties of the CO are considerably enhanced, the outcome being a strengthening of the CO bond (remembering the 5σ-donor orbital of CO is slightly antibonding) relative to uncoordinated CO. In fact, recent molecular orbital calculations[86] on $Ni^0(CO)$, $Ni^I(CO)$, $Ni^{II}(CO)$, and $Ni^{III}(CO)$ lend strong support to the contention of increasing σ-donor and decreasing π-acceptor participation of the CO ligand to the metal, with increasing oxidation state of the metal.

Assuming that we can carry over these ideas to N_2, it is not surprising on examining the best-fit modified valence force field force constants for $M(N_2)$, $M(N_2)_2$, $M(O_2)$, $(O_2)M(N_2)$, and $(O_2)M(N_2)_2$ (Table 22 and Figure 37) that a large increase in the k_{NN} force constant is observed on coordinating a single O_2 ligand to MN_2 to form $(O_2)M(N_2)$. In essence, the change reflects a transition from MN_2 in which N_2 bonds in the commonly encountered σ-π synergic manner to $(N_2)M^{\delta+}(O_2^{\delta-})$, in which the N_2 can be considered to be weakly coordinating in a predominantly σ-type fashion to $M^{\delta+}$ (where $2 \geq \delta^+ \geq 1$), consistent with k_{NN} force constants very close to that of uncoordinated N_2.

The coordination of a second N_2 ligand to yield $(N_2)_2M^{\delta+}(O_2^{\delta-})$ has very little effect other than causing a slight reduction in N_2 σ-bonding with a concomitant marginal increase in k_{NN} compared to that of $(N_2)M^{\delta+}(O_2^{\delta-})$ (Table 22 and Figure 37).

Table 22. Relevant Frequencies[a] and Bond-Stretching Force Constants[b] of Binary and Mixed Binary Dioxygen-Dinitrogen Complexes of Nickel, Palladium, and Platinum

	MO_2	MN_2	$M(N_2)_2$	$(O_2)M(N_2)$	$(O_2)M(N_2)_2$
Nickel					
νO—O	966	—	—	977	972
νNi—O	507	—	—	478	—
νN—N	—	—	2187	2243	2283
νN—N	—	2089	2106	—	2260
νNi—N	—	446	406	368	345
k_{OO}	3.57	—	—	3.79	3.88
k_{NiO}	1.92	—	—	1.57	1.40
k_{NN}	—	17.62	18.85	20.90	21.20
$k_{NN \cdot NN}$	—	—	0.67	—	0.16
k_{NiN}	—	2.50	1.60	1.78	1.60
$k_{NN \cdot NiN}$	—	0.25	0.25	0.59	0.33
Palladium					
νO—O	1024	—	—	1014	998
νPd—O	427	—	—	418	—
νN—N	—	—	2267	—	2310
νN—N	—	2213	2234	2288	2304
νPd—N	—	378	339	366	—
k_{OO}	4.37	—	—	4.29	4.22
k_{PdO}	1.53	—	—	1.48	1.45
k_{NN}	—	20.46	21.26	21.91	22.00
$k_{NN \cdot NN}$	—	—	0.27	—	0.02
k_{PdN}	—	1.88	1.38	1.84	1.80
$k_{NN \cdot PdN}$	—	0.72	0.72	0.77	0.50
Platinum					
νO—O	925	—	—	909	897
νPt—O	—	—	—	—	—
νN—N	—		—	2229	2267
νN—N	—	2172/2168	2197	—	2260
νPt—N	—	394	360	392	—
k_{OO}	3.53	—	—	3.37	3.30
k_{PtO}	1.20	—	—	1.20	1.10
k_{NN}	—	19.00	20.03	20.91	21.30
$k_{NN \cdot NN}$	—	—	0.43	—	0.02
k_{PtN}	—	2.26	1.87	2.26	2.00
$k_{NN \cdot PtN}$	—	0.12	0.12	0.95	0.70

[a] Units in cm^{-1}.
[b] Units in mdyn/Å.

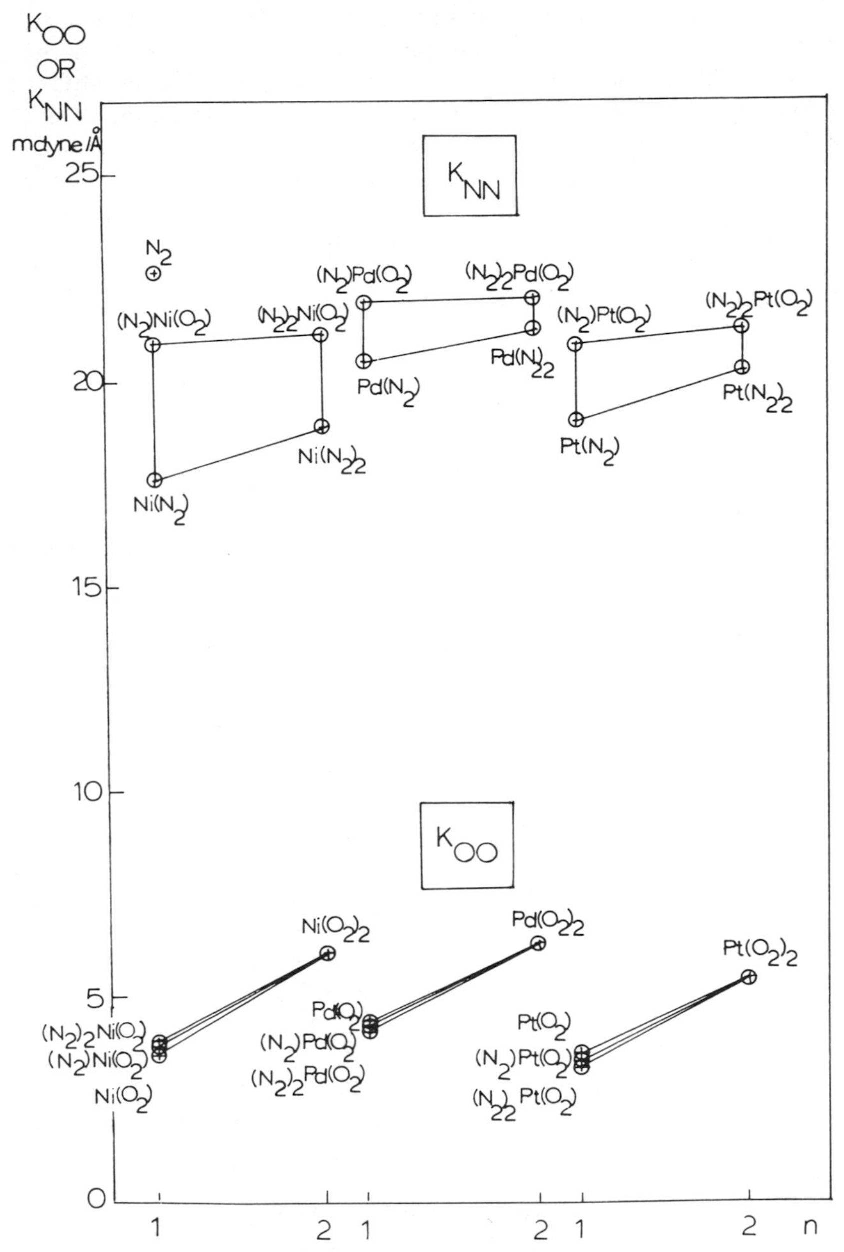

Figure 37. Graphical representation of the best-fit modified valence force field k_{NN} and k_{OO} ligand-stretching force constants for $M(N_2)$, $M(N_2)_2$, $M(O_2)_2$, $(O_2)M(N_2)$, and $(O_2)M(N_2)_2$ (where M = Ni, Pd, or Pt) as a function of the O_2 and N_2 coordination number.

Consistent with the proposal of an ionic moiety, $M^{\delta+}(O_2^{\delta-})$, to which N_2 weakly coordinates in an approximately σ-bonded fashion are the marginal changes observed in the k_{OO} force constants on passing from $M(O_2)$ to $(N_2)M(O_2)$ and $(N_2)_2M(O_2)$ (Table 22 and Figure 37). Here, the dioxygen approximately satisfies the charge requirements of the metal and barely senses the presence of weakly interacting N_2 ligands. On the other hand, a large increase in k_{OO} is observed when a second dioxygen is coordinated to $M(O_2)$ to yield $M(O_2)_2$ (Figure 37), an effect that probably reflects a transition from $M^{\delta+}(O_2^{\delta-})$, in which δ is closer to peroxide character, to $M^{2\delta+}(O_2^{\delta-})_2$, in which δ is closer to superoxide character.

11.4 The "Anomalous" Behavior of the Binary Carbonyls of Copper, Silver, and Gold

The discovery of thermally unstable carbonyls of copper, silver, and gold[5,6,44] can be considered to add a new dimension to our knowledge of binary carbonyls of the transition metals. Historically speaking, the data for copper authenticate some of the very claims[87] for unstable copper carbonyls that were postulated to exist in the vapors of copper transported by streams of heated CO. Clearly, the generally held view for the nonexistence of copper, silver, and gold carbonyls, namely, the stability of the nd^{10} valence shells, is no longer tenable in view of their apparently "normal" CO-stretching frequencies (2050–1800 cm^{-1}; cf. $Ni(CO)_4$, $\nu CO(T_2)$, 2050 cm^{-1}). Clearly, these results indicate a considerable degree of charge delocalization from the metal onto the CO ligand. Note, however, that the metal-carbon stretching frequencies (νCu—C, 375–325 cm^{-1}; νAg—C, 250 cm^{-1}) are quite low and imply a weaker M—C interaction than for $Ni(CO)_4$ ($\nu NiC(T_2)$, 438 cm^{-1}), consistent with the lower stabilities of $Cu(CO)_n$, $Ag(CO)_n$, and $Au(CO)_m$ (where $n = 1–3$, $m = 1–2$).

In this section we show how the "anomalous" trend observed in the Cotton–Kraihanzel CO force constants for $M(CO)_n$ (where M = Cu or Ag, $n = 1–3$) can in fact be fitted into a logical bonding scheme.

We have used the term "anomalous" to emphasize a striking difference between the group IB binary carbonyls and those, for example, of Ni, Pd, Pt, Co, Rh, or Ir, namely, an *amonotonic* trend in the Cotton–Kraihanzel CO-stretching force constants for the former, that is,

$$Cu(CO)_3 > Cu(CO) > Cu(CO)_2 \qquad \text{(Ref. 44)}$$

$$Ag(CO)_3 > Ag(CO) > Ag(CO)_2 \qquad \text{(Ref. 6)}$$

$$Au(CO) > Au(CO)_2 \qquad \text{(Ref. 5)}$$

yet a consistently *monotonic* trend for the latter, for example,

$$Co(CO)_4 > Co(CO)_3 > Co(CO)_2 > Co(CO) \qquad \text{(Ref. 18)}$$

$$Ni(CO)_4 > Ni(CO)_3 > Ni(CO)_2 > Ni(CO) \qquad \text{(Ref. 11)}$$

The rationalization used to explain the commonly observed trend for the binary carbonyls of the nickel and cobalt groups involves the synergic relationship between the 5σ- and 2π-orbital contributions to the bonding of the CO ligand to the metal, as described earlier.

Two assumptions are necessary to explain the amonotonic behavior of the Group IB binary carbonyls. Firstly, π-backbonding effects from the stable valence d orbitals to the 2π orbitals of CO are minimal (if any contribution at all, the effect will be greater for Cu than Ag). Secondly, the charge density on the metal remains roughly constant with coordination number, n, and so the valence d-orbital stabilities are little affected as n changes.

Let us consider each molecule in turn:

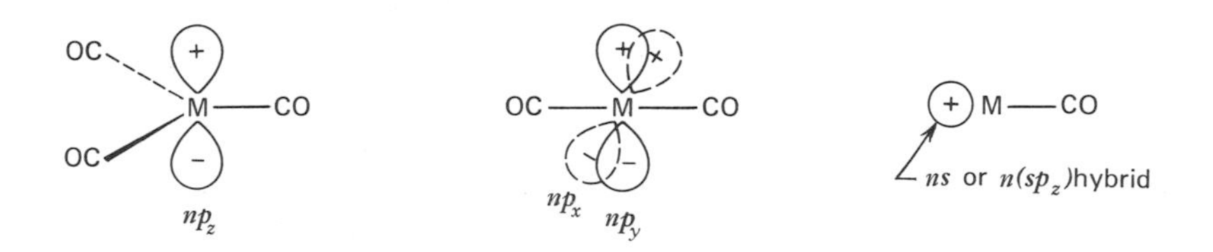

In view of the decreasing ^{63}Cu, ^{65}Cu hyperfine coupling constants on passing from Cu(CO) to $Cu(CO)_3$,[63] one can conclude that the 4*s* character of the odd electron is decreasing in the same order. In fact, the esr data for Cu(CO) and $Cu(CO)_3$ indicate that the odd electron is almost entirely localized in 4*s* and $4p_z$ orbitals, respectively.[63] The implication is that we can describe the σ-bonding scheme for the tricarbonyl as approximately sp^2 with a $^2A_2''$ electronic ground state. Therefore, the orbital most likely to π-bond to the CO is the $4p_z$, which contains the odd electron. Similarly, the dicarbonyl with approximately *sp* σ-hybrid orbitals has a $^2\Pi$ electronic ground state. The $4p_x$, $4p_y$ degenerate set containing the odd electron is most likely to π-bond to CO, and hence the tricarbonyls and dicarbonyls should display the "normal" trend in the Cotton–Kraihanzel CO-stretching-force constants, as the π- and σ-charge transfer per CO group should decrease as n increases, that is,

$$Cu(CO)_3 > Cu(CO)_2$$

and

$$Ag(CO)_3 > Ag(CO)_2$$

as observed.

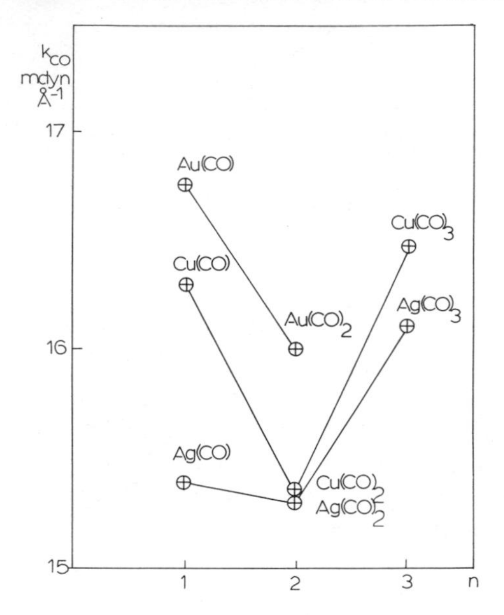

Figure 38. Graphical representation of the Cotton–Kraihanzel k_{CO} bond-stretching force constants for $M(CO)_n$ (where M = Cu or Ag, n = 1–3 and M = Au, n = 1–2) as a function of the coordination number, n.

However, the monocarbonyl is unique in one important respect, namely, the electronic ground state is $^2\Sigma^+$, having the odd electron in a σ-type orbital, in contrast to the π character of the odd electron in the tri- and dicarbonyl.

The so-called "anomaly" can now be understood, as the π contribution to the M—C bond of the monocarbonyl, instead of being the largest, turns out to be the smallest in the series. Moreover, the σ contribution gains in importance in the monocarbonyl, with the result that the Cotton–Kraihanzel CO-stretching constant is higher than anticipated, and the *amonotonic* order results. (Clearly, some $d\pi$-π^* bonding occurs, as the monocarbonyl absorbs below the frequency of free CO.) This rationale would appear to hold generally for the carbonyls of Cu, Ag, and Au (Figure 38).

11.5 The Importance of *p* Orbitals in the π-Bonding Schemes of Metal Carbonyls

In a sense, the discussions of the previous section have led us naturally into the subject of *p*-orbital π contributions to the overall bonding scheme and stability of metal carbonyl complexes. This topic is especially interesting in view of the synthesis of Ga,[88] Ge and Sn[89] carbonyls (in Chapter 7), which have one thing in common with the Group IB carbonyls, namely, a *stable, filled* valence *d* shell. The synthesis of Al carbonyls[90] adds a further dimension to this discussion, namely, how to assess the bonding scheme of a carbonyl complex whose central metal atom is devoid of valence *d* electrons.

It would seem not to be too unreasonable to extend the bonding ideas expounded for the Group IB carbonyls to the main group carbonyl complexes $Al_x(CO)_2$, $Ga_x(CO)_2$, $Ge(CO)_n$, and $Sn(CO)_m$. We, therefore, suggest that similar $p\pi$-π^* bonding with minimal $d\pi$-π^* contributions (impossible, of course, for Al) furnish sufficient stabilization to give these main group carbonyls an independent existence.

In view of the facile matrix surface diffusion and dimerization of light metal atoms, the aluminum and gallium dicarbonyl complexes are probably best formulated $Al_2(CO)_2$ and $Ga_2(CO)_2$. The aluminum complex is especially interesting. Whether it is formulated as (I) or (II) below, the main π contribution to the Al-C bonding scheme is most certainly $p\pi$-π^*.

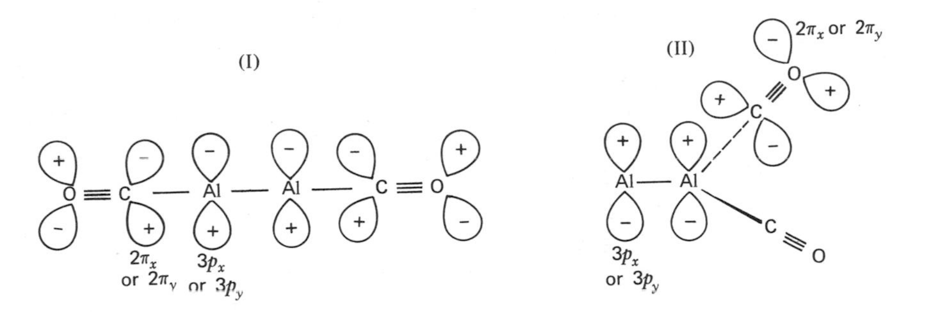

In this context, it is worth noting the results of a recent esr study of the Al/C_2H_4 matrix cocondensation reaction, the product of which is suggested to be $Al(C_2H_4)$, having axial symmetry and $S = \frac{1}{2}$.[91] The structure proposed contains side-on bonded ethylene as shown below, bonded in a Dewar–Chatt-type scheme.

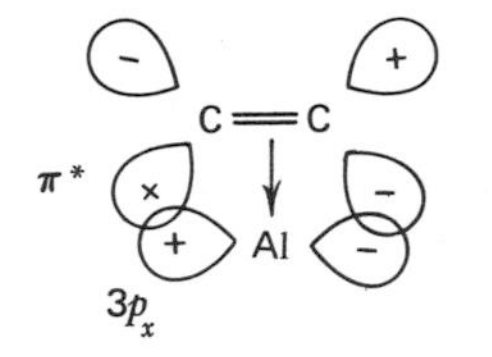

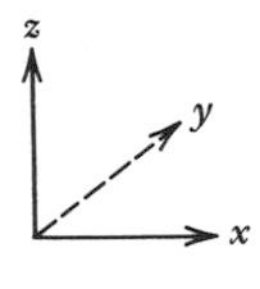

A preliminary analysis of the *g* values and hyperfine coupling constants leads one to believe that there is considerable delocalization of the odd $3p_x$ electron of the Al onto the ethylene ligand through $p\pi$-Π^* overlap, a situation analogous to that described earlier for the Group IB binary carbonyls, and $Al_2(CO)_2$.

11.6 Rare-Earth Carbonyls

One of the earliest metal-atom-matrix cocondensation reactions to be reported involved uranium vapor and carbon monoxide.[92] The infrared spectra and warm-up data were interpreted in terms of binary carbonyls of uranium, $U(CO)_n$ (where n is probably 1–6), which, incidentally, still represent the only known 5*f* carbonyl compounds. Since this report, two other papers have appeared that describe similar reactions with the lanthanides.[93,94]

Up to the time of writing, cocondensation reactions have been performed with Pr, Nd, Gd, Ho, and Eu. The infrared spectra and warm-up data paralleled those of the uranium system quite closely and have similarly been interpreted in terms of binary carbonyls, $M(CO)_n$. It should be noted here that neither the vibrational nor structural assignments were confirmed by isotopic substitution experiments, and, in this respect, the data should be treated with some caution.

Nevertheless, the data are unique, and certain interesting conclusions can be drawn from the set of vibrational data listed in Table 23. Several points are worth noting here. In all cases the high-frequency absorption

Table 23. CO Stretching Frequencies of Rare-Earth, Carbonyls, (cm^{-})

Pr	Nd	Gd	Ho	Eu	Assignment
1989	1990	1986	1982	2000	$M(CO)_6$
1965	1965	1967	1901	1974	$M(CO)_5$
1940	1940	1945	1929	1908	$M(CO)_4$
1885	1891	1901	1902	—	$M(CO)_3$
1858	1861	1864	1859	1873	$M(CO)_2$
1835	1840	1841	1830	—	$M(CO)$

persists after the final matrix warm-up (see, for example, Figure 39 for gadolinium vapor). Particularly striking is the constancy of CO stretching frequency for the various metals as a function of coordination number, n (Table 23), and also the monotonic increase in CO stretching frequency as n increases. Furthermore, the CO frequencies closely resemble those reported for $M(CO)_n$ where M = Cr, Mo, and W, $n = 1$–6). In the light of these data, the highest frequency CO absorption was ascribed to a hexacarbonyl complex.

We may summarize by stating that the regularities and trends observed in the spectra of the known lanthanide carbonyls can be explained with the classical σ-π model for transition metal carbonyls. The importance of 4*f* orbitals in the overall bonding scheme, however, cannot be assessed in any

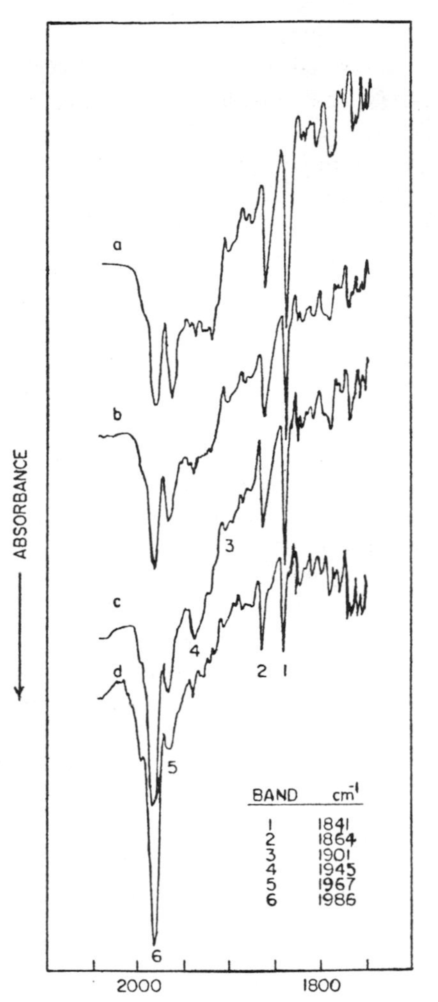

Figure 39. The infrared spectra of gadolinium carbonyl species in argon matrices during annealing, 1 mole% CO/Ar.

quantitative way from the data presently available, although it has been suggested that the lanthanide contraction of 4*f* orbitals might account for the fact that the CO stretching frequencies are almost independent of the nature of the metal for a given coordination number.[94]

12. REACTION KINETICS BY THE MATRIX-ISOLATION METHOD

12.1 Introduction

In an early paper, Pimentel[95] first outlined the unique possibilities that matrix-isolation spectroscopy offers for the study of chemical reactions with very low activation energies, such as those encountered with free radicals. Although his exploratory experiments were limited to the diffusion of NH_3 in Ar and the *cis-trans* isomerization of HNO_2 in Ar, the potential usefulness of the method was clearly demonstrated. However, since Pimentel's 1958 matrix kinetics paper, the technique had not been further explored, and until very recently its usefulness remained an unanswered question.

We begin this section with a general discussion of the various types of molecule that can result from metal-vapor cocondensation reactions and whose subsequent chemical reactions can, in principle, be monitored by matrix-isolation kinetic techniques.

The first involves a molecule that is stable to decomposition in the matrix, but which is unstable with respect to an intramolecular rearrangement within the confines of the matrix cage. Chromium pentacarbonyl, $Cr(CO)_5$, formed in the Cr/CO/Ar cocondensation reaction,[50] could possibly exemplify a case of matrix isomerization, where a kinetically stable trigonal bipyramidal form (produced on deposition) rearranges on warming to the thermodynamically stable square pyramidal form.[21] The rate of formation of the most stable isomer would be governed by first-order reaction kinetics and if the matrix cage influences intramolecular movements, it is likely that the measured activation energy would exceed that for the molecule in the gas phase.

The second type of molecule is one that is unstable to decomposition through one or more bond rupture processes. The effect of matrix isolation upon this type of molecule is to provide a sufficiently low-temperature environment to reduce the thermal energy available to the molecule, below that required to cause thermal decomposition. Bond rupture reactions of very unstable matrix-isolated species have been observed, for example,

$$Pd(N_2)_3 \rightarrow Pd(N_2)_2 \qquad \text{(Ref. 96)}$$

and would be expected to display first-order reaction kinetics. Clearly, the activation energies of bond rupture processes will depend on the dimensions of the activated complex compared to the size of the matrix cage.

Thirdly, one can envisage a situation where the product of a cocondensation reaction may be a molecule that is itself stable to decomposition, but is unstable to further reaction with another species present in the matrix. An example of this type of molecule is $Pd(CO)_n$ (where $n = 0$–3,[2] which reacts with CO upon warming according to the following scheme:

$$Pd + CO \rightarrow Pd(CO)$$
$$Pd(CO) + CO \rightarrow Pd(CO)_2$$
$$Pd(CO)_2 + CO \rightarrow Pd(CO)_3$$
$$Pd(CO)_3 + CO \rightarrow Pd(CO)_4$$

Matrix isolation stabilizes coordinatively unsaturated molecules of this type by imposing a diffusion-controlled kinetic impediment to further reaction. By following the appearance and disappearance of $Pd(CO)_n$ and the disappearance of CO, one has a means of determining the relative diffusion coefficients of the reacting species[2] (a parameter difficult to obtain by other methods).

Finally, one can devise a situation where the product is an odd electron species (atomic or molecular), unstable with respect to a dimerization or aggregation process. Reactive species of this type can be brought together within the confines of the matrix in a highly controlled manner, and measurable reaction rates can be observed even for activation energies of less than a kilocalorie, a range difficult to study by other methods. Reactions of this type, involving bond formation, are expected to be influenced to some extent by the matrix.

Atoms are, of course, potential candidates for this type of aggregation process. In principle, experiments of this type can provide a means of studying the very early stages of metal formation and can lead to a controlled synthesis of well-defined metal clusters. In this category, one can refer to the aggregation of Cu atoms in Ar matrices as monitored by esr spectroscopy (Section 10)[63] and Bos's[97] matrix Mössbauer studies of the diffusion of Sn atoms in solid N_2 to form Sn_n (where $n = 1$–4).

Examples of *molecular* species that undergo dimerization reactions during matrix warm-up experiments are the paramagnetic species, $Cu(CO)_3$[44] and $Ag(CO)_3$.[6] As the latter molecule is particularly well suited for matrix-isolation kinetic measurements, we use it as a model system to examine critically the experimental procedures, precautions, and complications involved in a matrix kinetics experiment before passing on to examine the diffusion kinetics of $Pd(CO)_n$ in Ar,[2] and the aggregation reactions of Sn atoms in α-N_2 to form Sn_n (where $n = 1$–4).[97]

Temperature Measurement and Control

An extremely critical aspect of any kinetics experiment is the determination of accurate temperatures during a run and to ensure temperature reproducibility between different runs. In matrix spectroscopy, because of the compression fitting techniques normally employed for achieving thermal contact between the cold end of the refrigeration system and the optical plate, it is most desirable that the temperature sensor be incorporated into the optical plate (with a recommended cryogenic epoxy) and as close to the matrix–optical-plate interface as possible. An arrangement of this type should ensure both accuracy and reproducibility of temperature measurement to better than $\pm 0.25°$.

One also requires in a matrix-kinetics experiment rapid attainment of a chosen temperature, stability at the chosen warm-up temperature, followed by rapid temperature quenching. In our experience, the most satisfactory experimental combination is a liquid-helium transfer system coupled to a proportional temperature controller and a digital microvoltmeter. With this arrangement temperature reproducibility and stability of $\pm 0.1°$ can be achieved with warm-up times of the order of 3–4 sec (6.0–20°K), 5–7 sec (6.0–30°K) and 8–10 sec (6.0–40°K), followed by quenching times to 6.0°K of the order of 2–3, 3–4, and 4–5 sec, respectively. Although the closed-cycle helium refrigerator competes with the helium transfer system in terms of temperature stability, it does not have the refrigeration capacity or versatility for rapid warm-up and quenching of matrix samples, especially above 30°K.

Having set up the cryogenic experiment for matrix-kinetic studies, one must be on the alert for complications relating to the particular system under study. The warm-up reaction of $Ag(CO)_3$ is particularly interesting in this respect, as two extraneous effects are operative. Let us discuss each of these effects in turn.

Photochemical Complications

During a number of recent matrix-infrared-spectroscopic investigations of transition metal carbonyls, unexpected reactions were observed that appeared to be photochemical in origin and were subsequently traced to uv-visible radiation emitted from the globar source of the infrared spectrometer. The first reported example of this was found in the matrix photochemistry of $Fe(CO)_5$,[48] the $Fe(CO)_4/CO$ products so formed undergoing "reverse" photolysis to $Fe(CO)_5$ in the beam of the infrared spectrometer. It was discovered that a germanium cutout filter inserted between the infrared source and the matrix sample eliminated unwanted matrix photochemistry.

It was discovered that $Ag(CO)_3$ was also sensitive to the radiation emitted from the globar source. Evidence for this photosensitivity originated from the observation that with a germanium filter the absorbances of the CO stretching modes of $Ag(CO)_3$ begin to decrease at a measurable rate at approximately 30°K. However, without the germanium filter, reaction begins at temperatures *below* 20°K. Although the mechanisms of the photochemical reaction are not yet fully understood, the reaction kinetics appear to be first order, indicative of a bond rupture process.

Matrix Annealing Complications

When matrices are first deposited at 4.2–15°K, the lattice can be expected to be in a disordered glassy phase. The host lattice so prepared, especially when doped with a guest molecule, can be expected to display significant imperfections; the guest molecule experiences an averaged local field in the lattice and thereby gives rise to a spectrum that may not be truly representative of the equilibrium configuration of the guest in the host lattice. However, when the matrix is warmed it anneals, thereby increasing the regularity of the crystalline portions at the particular temperature chosen. The infrared absorption spectrum of $Ag(CO)_3$ in a CO matrix was found to be quite sensitive to the deposition conditions, and reproducible kinetic data could only be obtained in matrices that had been carefully annealed just below the reaction temperature.

12.2 Kinetics of the Reaction $2Ag(CO)_3 \rightarrow Ag_2(CO)_6$ in Low-Temperature Matrices

In this section we discuss the matrix reaction

$$2Ag(CO)_3 \rightarrow Ag_2(CO)_6 \rightarrow \text{other products}$$

as performed in pure CO matrices that had been deposited at known Ag:CO ratios and temperatures, with warm-up experiments conducted in the accessible temperature range, 25–39°K. The factors governing these temperature limits are twofold. First, the lower limit is determined by the temperature at which measurable absorbance changes are observed for $Ag(CO)_3$, whereas the upper limit is critically governed by the temperature at which the matrix support is beginning to boil off at a measurable rate. The upper limit can be established by monitoring either the pressure in the sample vacuum shroud or the absorbances of the natural abundance $^{13}C^{16}O/^{12}C^{18}O$ isotope lines that serve as an "internal matrix reference" and should remain strictly invariant throughout a run. In practice, slight matrix boil-off was difficult to avoid, especially at the high-temperature limit. This effect was corrected for by normalizing all absorbances with respect to the $^{13}C^{16}O/^{12}C^{18}O$ reference lines.

Monatomic Ag vapor was cocondensed with CO onto the tip of a liquid-helium-cooled cryostat. Silver vapor flux was measured with a quartz crystal balance while gas flows were regulated with a precalibrated micrometer needle valve. Two sets of experiments were performed. In the first set four experiments were done, each with a different metal flux and a constant CO deposition rate of approximately 2 mmole/hr. The metal/CO ratios in these experiments were approximately 7.4×10^{-5}, 2.9×10^{-4}, 4.4×10^{-4}, and 5.9×10^{-4}, which correspond to initial metal concentrations of 1.3×10^{18}, 5.0×10^{18}, 7.6×10^{18}, and 10.1×10^{18} silver atoms/cm^3, respectively, assuming that solid CO has a density of 0.8 gm cm^{-3}. The temperature was fixed at 35°K by means of an electronic temperature regulator using a Au, 0.7% Fe/chromel thermocouple sensor soldered to the copper plate holder.

(The temperature of the copper block was compared during a previous set of experiments with that measured at the center of the CsI plate on which the matrix was deposited and found to be within 0.1°K for temperatures above 20°K.)

In the second set, the metal flux was kept constant so as to give a metal/CO ratio of approximately 2.9×10^{-4}, and experiments were performed at 30, 33, and 37°K, which, together with the previous set at 35°K, yielded results at four temperatures altogether. The spectral changes observed during a typical kinetic run are shown in Figure 40. In brief, the spectra show a gradual and monotonic decrease in the absorbances of the $Ag(CO)_3$ starting material, with the simultaneous growth and eventual decay of two sets of new lines at 1998, 1955 cm^{-1} and 1980, 1944 cm^{-1}. In addition to the changes described above, there was a simultaneous decrease in the absorbances of *all* species, becoming more pronounced at the high-temperature limit. This was proven not to be an artifact related to matrix boil-off, as the $^{13}C^{16}O/^{12}C^{18}O$ reference absorbances remained essentially unchanged throughout the runs.

This observed behavior can be accounted for, qualitatively, by assuming that $Ag(CO)_3$ is either decomposing unimolecularly, or dimerizing, or both. The observed metal concentration dependence of the rate of disappearance

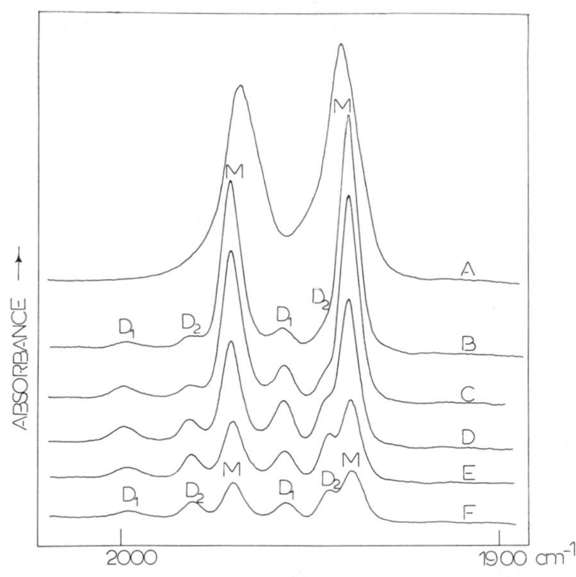

Figure 40. A portion of the spectrum of $Ag(CO)_3$ in a solid CO matrix showing the gradual decrease in reagent absorbance (M) and simultaneous growth and decay of product (D_1, D_2). A to F represent increasing time at constant temperature.

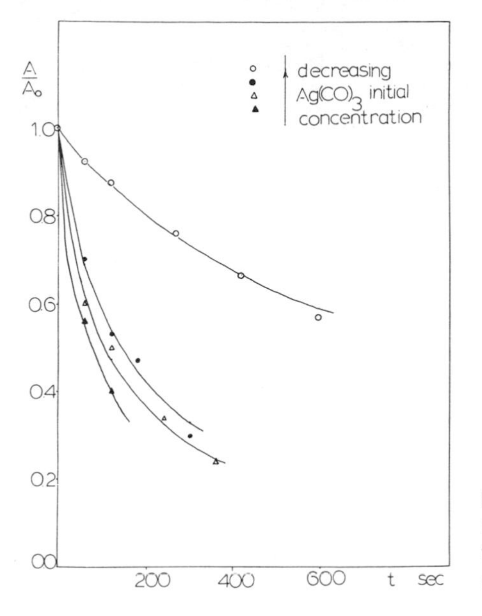

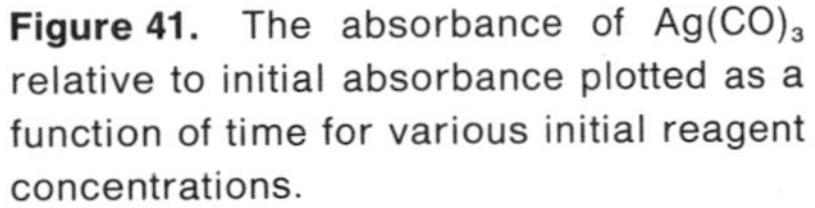

Figure 41. The absorbance of $Ag(CO)_3$ relative to initial absorbance plotted as a function of time for various initial reagent concentrations.

of the absorptions attributed to $Ag(CO)_3$ eliminates the first explanation (Figure 41). The last mechanism was also cast out on the basis of the observation that in solid Xe, which can be warmed to 80–100°K, $Ag(CO)_3$ is not found to decompose (6). We, therefore, conclude that the observed removal of $Ag(CO)_3$ results from dimerization alone (followed by other reactions). This mechanism is, therefore, considered in detail.

Diffusion-Controlled Kinetics

Reactions occurring in solids are often diffusion limited. The rate of a diffusion-limited bimolecular reaction differs from a normal one in that the apparent rate constant is time dependent.[98] Smoluchowski[99] has derived the following rate equation for such reactions:

$$-\frac{dX}{dt} = 4\pi r_0(D_X + D_Y)\left(1 + \frac{r_0}{[\pi(D_X + D_Y)t]^{1/2}}\right) XY \tag{15}$$

where X and Y are the concentrations of reagents X and Y, D_X and D_Y are their respective diffusion coefficients, and r_0 is the X—Y separation below which reaction is ensured.

For a dimerization process

$$2X \rightarrow X_2$$

equation (15) reduces to

$$-\frac{dX}{dt} = 8\pi r_0 D \left(1 + \frac{r_0}{(2\pi Dt)^{1/2}}\right) X^2 \tag{16}$$

and equation (16) integrates to give

$$\frac{X}{X_0} = \frac{1}{aX_0 t + 2abX_0 t^{1/2} + 1} \tag{17}$$

where

$$a = 8\pi r_0 D$$

and

$$b = \frac{r_0}{(2\pi D)^{1/2}} \tag{18}$$

D being the diffusion coefficient of X. For spectroscopic data, X/X_0 equals A/A_0, the absorbance ratio. Equation (17) reduces to a normal second-order rate law if b vanishes.

The kinetic data from the first set of experiments (Figure 41) were fit to equation (17) using a least-squares program. The eight adjustable parameters, aX_0 and abX_0, for each of the four runs, were reduced to five by requiring the ratio, abX_0/aX_0, to be constant. The fit so obtained was superior to that obtained by setting b identically to zero; that is, diffusion-limited kinetics describe the data better than ordinary second-order kinetics, implying that diffusion is the rate-determining step rather than dimerization. The data obtained from the fit are given in Table 24.

Using equation (18) one can show that $X_0ab^2 = 4r_0^3X_0$. This quantity can be obtained from the two parameters, aX_0 and abX_0, used in fitting the kinetic data by taking a ratio of the second parameter to the first, squaring

Table 24. Rate Constants and Diffusion Coefficients Obtained from the Dimerization of $Ag(CO)_3$ with Varying Initial Reagent Concentrations

$X_0^{\text{molecules}}$ (cm^{-3})	aX_0 (sec^{-1})	abX_0 ($sec^{-1/2}$)	r_0 (Å)[a]	D (cm^2/sec)[b]
1.3×10^{18}	1.11×10^{-3}	9.47×10^{-4}	5.4	5.2×10^{-16}
5.0×10^{18}	6.15×10^{-3}	5.23×10^{-3}	6.1	7.5×10^{-16}
7.6×10^{18}	7.83×10^{-3}	6.66×10^{-3}	5.7	6.3×10^{-16}
10.1×10^{18}	10.55×10^{-3}	8.98×10^{-3}	5.7	6.4×10^{-16}
			5.7[c]	6.8×10^{-16}[c]

[a] obtained from $ab^2 = 4r_0^3$
[b] obtained from aX_0 using equation (18) and $r = 5.7$ Å.
[c] mean values.

that quantity, and multiplying the result by aX_0. Using these quantities and the tabulated values of X_0, an average value of 5.7 Å was obtained for r_0, and with this value of r_0, D was calculated from aX_0 and equation (18). An average value of 6.8×10^{-16} cm^2/sec was obtained. This value of D is in remarkably close agreement with that obtained by Bos et al.[97] for the diffusion of tin in nitrogen and by Kündig, Moskovits, and Ozin[2] for the diffusion of CO in argon, while 5.7 Å is an entirely feasible value of r_0.

Rate of Diffusion as a Function of Temperature

The kinetic data for the dimerization of $Ag(CO)_3$ at various matrix temperatures is shown in Figure 42. These data were fit to the Smoluchowski equation as before, with the data for each temperature being treated independently.

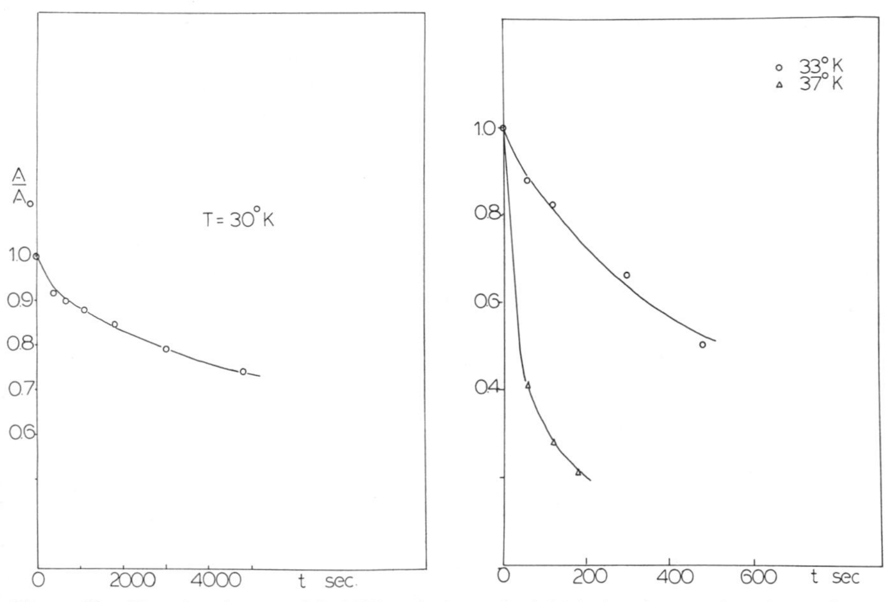

Figure 42. The absorbance of $Ag(CO)_3$ relative to its initial absorbance plotted as a function of time at 30, 33, and 37°K.

The kinetic parameters so obtained are listed in Table 25. Much greater scatter is evident in the r_0 values calculated from aX_0 and abX_0 of these experiments than in the previous set. In view of the error inherent in these measurements, however, the average value of 10 Å for r_0 is in acceptable agreement with the previously calculated value of 6 Å. The diffusion coefficient, D, was calculated from the aX_0 values, as before, for each of

Table 25. Rate Constants and Diffusion Coefficients Obtained from the Dimerization of $Ag(CO)_3$ at Various Temperatures

T (°K)	aX_0 (sec^{-1})	abX_0 (sec^{-1})	r_0 (Å)	D (cm^2/sec)[a]
30	2.32×10^{-5}	1.73×10^{-3}	18.6	1.8×10^{-18}
33	1.81×10^{-3}	7.28×10^{-4}	2.4	1.4×10^{-16}
35	6.15×10^{-3}	5.23×10^{-3}	6.1	4.7×10^{-16}
37	1.57×10^{-2}	3.15×10^{-2}	14.7	1.2×10^{-15}
			10.4[b]	

[a] obtained from aX_0 using equation (18), $r_0 = 10.4$ Å and $X_0 = 5.0 \times 10^{18}$ molecules/cm³.
[b] average.

the temperatures, and it too is listed in Table 25. An Arrhenius plot of $\ln D$ as a function of $1/T$ is shown in Figure 43, indicating acceptable Arrhenius behavior implying thereby a functional form for D of the type $D = D_0 \exp(-E_a/RT)$. A nonlinear least squares fit of the D values as a function of temperature yields a value of 1860 ± 300 cal/mol for the activation energy of the process while the preexponential factor was found to be 1.9×10^{-4} cm²/sec, with a standard deviation roughly four times this number making the diffusion coefficients uncertain by a factor of eight. Thus, as often happens, the activation energy can be calculated with considerably greater accuracy than the preexponential factor. The rather large uncertainty attests to the difficulty of performing kinetics of this sort at such low temperatures. This may also explain the paucity of data available for diffusion in solids of permanent gases at cryogenic temperatures. The calculated values of D depend critically on the value of r_0 used. By using an r_0 value of 6 Å rather than 10 Å, D would increase by the ratio of the two values. In view of the large error in the preexponential factor, however, such a correction is not justified.

Diffusion studies in low-temperature van der Waals solids are not plentiful in the literature. Self-diffusion in neither solid CO nor N_2 has been reported. Diffusion data for the noble gas solids are, however, available with Ar being the noble gas whose properties most closely approximate those of CO. Berne et al.[100] have measured the self-diffusion of Ar in the neighborhood of 79°K, where the diffusion coefficient of Ar is of the order of 10^{-10} cm² sec^{-1}, and they arrive at an expression

$$D = 45e^{-3860/RT}$$

If one were justified in using this expression at lower temperatures, which one is probably not, D for argon at 35°K would be 5×10^{-23} cm²/sec, that

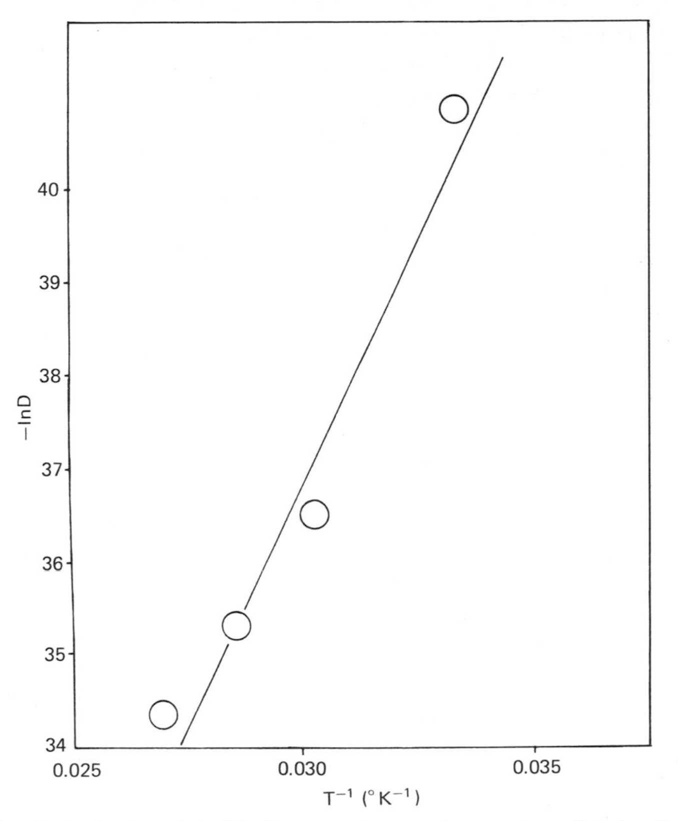

Figure 43. An Arrhenius plot of ln D versus inverse temperature, D being the diffusion coefficient of $Ag(CO)_3$ in the solid CO matrices used.

is, much lower than our measured rate of diffusion of $Ag(CO)_3$ in CO. Diffusion coefficients of the order of 10^{-16} cm^2 sec^{-1} at 30°K have been previously reported for CO in Ar,[2] and Sn in N_2.[97] However, one should keep in mind that self-diffusion studies are carried out in pure single crystals, while the system we are studying is either polycrystalline or glassy, and contains a large concentration of impurities, all of which factors tend to increase the rate of diffusion. In a glassy solid, diffusion through interstitial sites is possible as well as through vacancies, while in a polycrystalline solid the presence of grain boundaries facilitates molecular migration. The presence of impurities increases the concentration of vacancies, which would increase the preexponential factor in the expression for D but not necessarily the activation energy for vacancy migration. Accordingly, if we assume that $Ag(CO)_3$ moves through solid CO via a CO vacancy mechanism, then the activation energy for this process would not be greatly different from that of self-diffusion in CO. An estimate for the value of this activation energy may be obtained from Oshcherin's empirical relationship,[101]

$$E_a = k(4.85 \times 10^{-7})a^2\theta_D{}^2 M \text{ kcal/mole}$$

where θ_D is the Debye temperature, a is the lattice parameter, and M is the molecular weight of the solid. k is a constant depending on lattice type, with $k = 0.75$ for noble gas fcc crystals. Using Oshcherin's expression and the following data for CO, $\theta_D = 79.5°\text{K}$,[102] $a = 5.65$ Å,[103] one obtains $E_a = 2.05$ kcal mole^{-1}, in remarkably good agreement with our measured value of 1.9 kcal/mole.

12.3 Diffusion Kinetics of $Pd(CO)_n$ (where $n = 1–4$) and of CO in Ar

In the present analysis the concentrations of free CO and of the four carbonyls, $Pd(CO)_n$, are assumed to be proportional to their respective absorbances according to the Beer–Lambert relationship. Absorbances were calculated using peak height values. Thus

$$A_n = \varepsilon_n[n]l \tag{19}$$

where A_n, $[n]$, and ε_n are the absorbance, concentration, and extinction coefficient, respectively, of $Pd(CO)_n$, and l is the matrix thickness. Similarly, for the free CO, A_{CO} was set equal to $\varepsilon_{CO}l[CO]$. If one makes the assumption that

$$\varepsilon_n = n\varepsilon_1 \tag{20}$$

that is, that each of the CO ligands absorbs independently of the others, and with similar strength in all the various $Pd(CO)_n$, then combining equations (19) and (20) one obtains

$$[n] = \frac{1}{\varepsilon_1 l}\left(\frac{A_n}{n}\right) \tag{21}$$

In the absence of suitable values for ε_1 and l, we set the factor, $\varepsilon_1 l$, equal to unity and use the quantity, A_n/n, as a direct measure of $[n]$.

Analysis of the Warm-Up Kinetics

Smoluchowski[99] gives the following equation for the rate of decrease of concentrations of two species, A and B, that react by diffusion in a solid:

$$\frac{-d[A]}{dt} = \frac{-d[B]}{dt} = 4\pi r_0[D_A + D_B]\cdot\left(1 + \frac{r_0}{[\pi(D_A + D_B)t]^{1/2}}\right)[A][B] \tag{22}$$

D_A and D_B are the diffusion coefficients of species A and B, and r_0 is the radius of sphere within which A and B are assumed to react instantaneously. For typical values of diffusion coefficients in solids and for $r_0 \simeq 3$ Å, the term, $r_0/[\pi(D_A + D_B)t]^{1/2}$, is negligible for values of t in excess of 1. This, coupled with the assumption that one of the species, B, say, diffuses much more slowly than A, reduces equation (22) to the form

$$\frac{-d[\mathrm{A}]}{dt} = 4\pi r_0 D_{\mathrm{A}}[\mathrm{A}][\mathrm{B}] \tag{23}$$

which is of the form

$$\frac{d[\text{products}]}{dt} = k[\mathrm{A}][\mathrm{B}]$$

that is, that of second order kinetics. In the present case it was assumed that only CO diffuses, that trimolecular and higher encounters contribute negligibly to the formation of product, and that no decomposition of the product takes place at the temperatures of the warm-up experiments.

Accordingly, the reactions that take place during warm-up are

$$\mathrm{M} + \mathrm{CO} \rightarrow 1$$
$$1 + \mathrm{CO} \rightarrow 2$$
$$2 + \mathrm{CO} \rightarrow 3$$
$$3 + \mathrm{CO} \rightarrow 4$$

where 1, 2, 3, and 4 stand, respectively for the Pd mono-, di-, tri-, and tetracarbonyl, and M for the free metal. The kinetics, therefore, are governed by the equations

$$\frac{d[1]}{dt} = k[\mathrm{CO}]([\mathrm{M}] - [1]) \tag{24}$$

$$\frac{d[2]}{dt} = k[\mathrm{CO}]([1] - [2]) \tag{25}$$

$$\frac{d[3]}{dt} = k[\mathrm{CO}]([2] - [3]) \tag{26}$$

$$\frac{d[4]}{dt} = k[\mathrm{CO}][3] \tag{27}$$

where

$$k = 4\pi r_0 D_{\mathrm{CO}} \tag{28}$$

and [M] stands for the free metal concentration. The free CO concentration decreases according to the equation

$$\frac{-d[\mathrm{CO}]}{dt} = k[\mathrm{CO}]([\mathrm{M}] + [1] + [2] + [3])$$

which may be written

$$\frac{-d[\mathrm{CO}]}{dt} = k[\mathrm{CO}](N_0 - [4]) \tag{29}$$

where $N_0 = [\mathrm{M}] + [1] + [2] + [3] + [4]$ is a constant equal to the total metal concentration in the matrix in any form. Figure 44 shows the concentration of the various species in the matrix as a function of warm-up time.

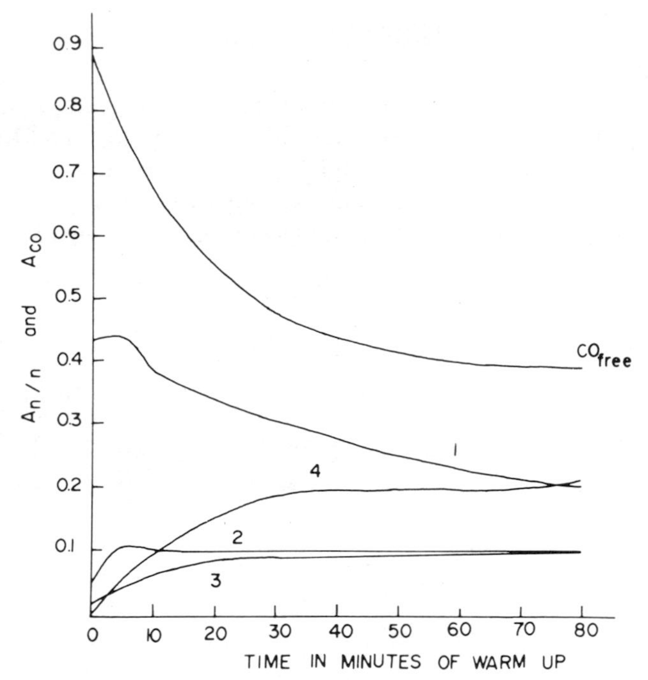

Figure 44. A_n/n and A_{CO} as a function of warm-up time, where A_n is the absorbance of $\mathrm{Pd(CO)}_n$. The numbers, 1–4, refer to the four $\mathrm{Pd(CO)}_n$ species.

It indicates that, with the exception of free CO, the concentrations of all species at first increase with time. This indicates that initially the right-hand sides of equations (24)–(27) are positive, thereby implying that

$$[\mathrm{M}]_0 > [1]_0 > [2]_0 > [3]_0 > [4]_0 \tag{30}$$

where the zero subscript indicates initial values.

This vindicates the assignments of the bands to the $\mathrm{Pd(CO)}_n$ species on the basis of their relative intensities on deposition (Figure 2). It should be pointed out that because the concentrations of the four species are proportional to

A_n/n rather than to A_n, one requires a tentative assignment of the lines to the various $Pd(CO)_n$ before the concentrations can be evaluated and compared. The initial values of the A_n for the four carbonyl species were such, however, that only the assignment stated previously is compatible with inequality (30).

Diffusion of CO in Argon

Since the free metal concentration decreases monotonically with time, the concentration of the monocarbonyl is expected according to equation (24) to come to a maximum and then decrease as indicated in Figure 44. Although the temperature control and the time measurements were not carried out with the precision required in diffusion experiments, it is instructive to see if the diffusion coefficient that one must assume for free CO in order to get the observed curves in Figure 44 is a reasonable one. To do this, one can make use of the fact that initially $[4] \ll N_0$, which yields the solution

$$[\mathrm{CO}] = [\mathrm{CO}]_0 \exp(-kN_0 t) \tag{31}$$

for equation (29).

A plot of $\log_e A_{\mathrm{CO}}$ versus t shows an initial straight-line portion having slope of 3×10^{-4} sec^{-1}. Taking N_0 to be approximately 3×10^{18} atom cm^{-3} as determined from the deposition conditions and r_0 to be 4×10^{-8} cm, one obtains approximately 2×10^{-16} cm^2/sec for D_{CO} in argon at a mean temperature of approximately 30°K.

No literature values were found for the diffusion of CO in solid argon. One can, however, calculate the self-diffusion coefficient at 30°K of Ar in Ar from the data of Berne et al.[100] to be of the order of 10^{-27} cm^2/sec. It would appear that CO diffuses through solid argon much more rapidly than does argon itself.

12.4 Aggregation of Sn Atoms in Solid N_2

In concluding this section, Bos and Howe's matrix Mössbauer studies of the diffusion behavior of Sn atoms in solid N_2 should be mentioned.[97] The experiments for Sn involve bulk diffusion during matrix warm-up after deposition, whereas those described in Chapter 9 involve the more facile process of matrix surface diffusion occurring on deposition.

The details of the matrix Mössbauer experiment can be found in the original reference.[97] The Mössbauer spectra of Sn atoms in solid N_2 were obtained over a wide range of concentration and diffusion conditions and were characterized by four basic identifiable components, one of which predominated after evaporation of the N_2 matrix (see Figure 45). On the basis of their aggregation behavior these were assigned to Sn, Sn_2, Sn_3, and higher aggregates, Sn_n, the latter showing Mössbauer parameters resembling tin aggregates.

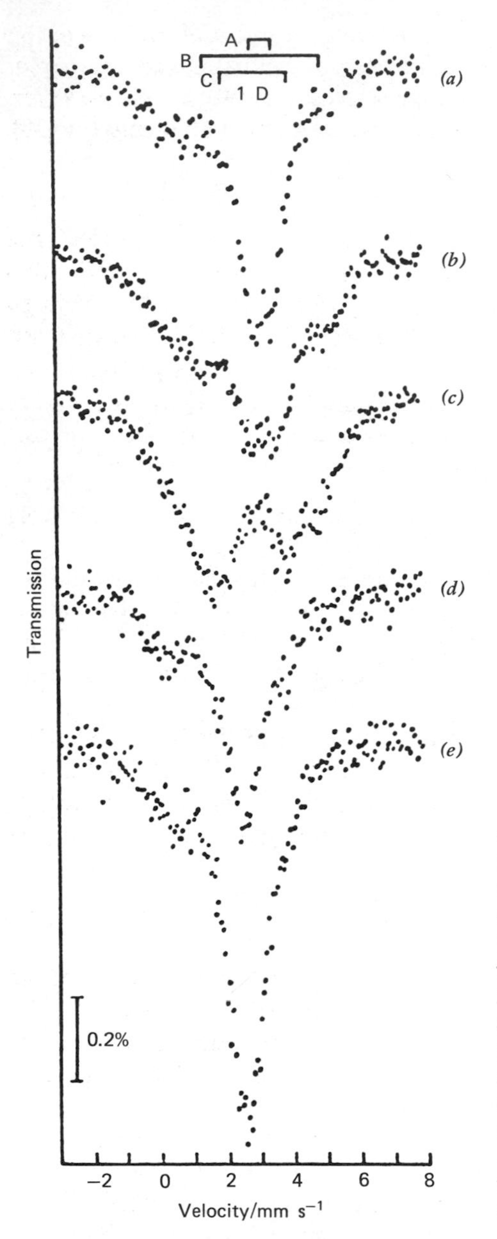

Figure 45. The changes in the Mössbauer spectrum following progressive recombination of matrix-isolated tin monomers. (*a*) shows the spectrum of the initial matrix. Annealing was performed for 20 min at 22 and 24°K, for 1 hr at 26, 28, and 30°K and for 45 min at 32, 34, 36, 38, and 41°K. The changes are best illustrated by the summed spectra after annealing at 26, 28, 30, and 32°K (*b*), and after annealing at 36, 38, and 41°K (*c*). Spectra (*d*) and (*e*) were taken after 45 min at 51 and 77°K, respectively, after the nitrogen had evaporated. A, monomer; B, dimer; C, polymer; D, aggregate.

The small quadrupole splitting observed for Sn atoms in α-N_2 was attributed to electrical asymmetry, originating in either a low substitutional lattice site (C_2) or in a weakly interacting Sn/N_2 system.

Sn_2 also showed a quadrupole splitting, indicative of an electron imbalance in the p-orbital occupancies and suggestive of a singlet Sn_2 molecule.

Diffusion kinetic measurements were performed in both α-N_2 (4.2–34°K) and β-N_2 (35.6–41°K). Typical changes in the Mössbauer spectrum following progressive recombination of matrix-isolated tin monomers are shown in Figure 45.

The diffusion equation described for the $Pd(CO)_n$ system was employed to analyze the diffusion kinetics of Sn atoms in α-N_2 up to 34°K.

$$\mathrm{Sn} + \mathrm{Sn} \rightarrow \mathrm{Sn_2}$$
$$\mathrm{Sn_2} + \mathrm{Sn} \rightarrow \mathrm{Sn_3}$$
$$\mathrm{Sn_3} + \mathrm{Sn} \rightarrow \mathrm{Sn_4}$$

It follows that

$$-\frac{dv_1}{dt} = (8\pi RDv_1^2 + 4\pi RDv_1v_2 + 4\pi RDv_1v_3)\left(1 + \frac{R}{(2\pi Dt)^{1/2}}\right) \quad (32)$$

where $v_{1,2,3}$ are the numbers of Sn, Sn_2, and Sn_3 moieties per unit volume, D is the diffusion coefficient of Sn, and R is the reaction radius (where it was assumed that under these conditions only Sn atoms are mobile). Since $v_3 < v_2 < v_1$ and v_2 and v_3 only increase at a fraction of the rate that v_1 decreases, the quantity, $(v_2 + v_3)$, was taken to be independent of time over the period of any one anneal. Equation (32) can then be solved to give

$$\ln\frac{(v_1)_0[2v_1 + v_2 + v_3]}{(v_1)[2(v_1)_0 + v_2 + v_3]} = 4\pi RDt(v_2 + v_3)\left(1 + \frac{2R}{(2\pi Dt)^{1/2}}\right) \quad (33)$$

where $(v_1)_0$ is the initial value of v_1 at any of the temperatures used. Using a reaction radius, R, of 4 Å (the intersite distance in the fcc lattice of α-N_2), D was found to be in the range, 10^{-22} m^2 sec^{-1} at 34°K.

Interestingly, the diffusion process was found to be more facile in β-N_2 (above 35.6°K) with D at least as large as 10^{-20} m^2 sec^{-1}. The greatly enhanced mobility of Sn and Sn_2 in β-N_2 was attributed to a combination of (a) the expansion of the lattice by 0.5% on passing from the α-N_2 (fcc) to the β-N_2 (hcp) phase, (b) the almost free rotation of the N_2 ellipsoids, and (c) the necessary increase in the temperature at which the measurements were made. In this context, it is worth noting the similarity in the diffusion coefficients of CO in Ar at 30°K ($D_{CO} \simeq 2 \times 10^{-20}$ m^2 sec^{-1})[2] a rather important consideration when trying to interpret matrix warmup reactions involving metal atoms and small molecules such as CO, N_2, O_2, and so forth.

13. SOME THERMODYNAMIC CONSIDERATIONS

One can enumerate three types of molecules that can be stabilized by matrix isolation. The first is a molecule or radical that is itself stable to decomposition but is unstable to further reaction in the matrix, for example, $Ni(CO)_3$, which forms $Ni(CO)_4$ upon reacting with CO. Matrix isolation stabilizes this type of molecule by imposing a diffusion-controlled kinetic impediment to further reaction. The second is a molecule that can undergo an intramolecular rearrangement. In this case, the low temperature of the matrix, hence the low thermal activation energy available to the molecule, is the important factor. The third is a molecule that is unstable to decomposition, for example, $Ni(N_2)_4$. The effect of matrix isolation upon this type of molecule aside from providing a low-temperature environment, hence reduced availability of thermal energy, deserves some discussion.

Let us assume that the matrix-isolated molecule does not interact with the matrix, the latter serving simply as a cage within which the molecule of interest is trapped. (The effects arising from the invalidity of this assumption are discussed below.) The molecule, therefore, can be considered to be quasigaseous at all temperatures.

Consider a molecule of the form, $M(L)_m$.† This molecule is stable at a (low) temperature, T, under ordinary preparative conditions (i.e., in a bottle) if ΔG of the reaction

$$M(L)_m(g) \rightarrow M(s) + mL(g) \qquad (a)$$

is positive. Note that the decomposition products of the reaction are bulk metal and gaseous ligand.

The same molecule is stable at the *same* temperature, T, under matrix isolation conditions only if ΔG of the reaction

$$M(L)_m(g) \rightarrow M(g) + mL(g) \qquad (b)$$

is positive. Note that the decomposition products are quasigaseous.

Figure 46 shows this difference graphically. The quantities shown in Figure 46 refer to temperature, T, which is presumably in the range, 4 to approximately 77°K, the usual range of temperature encountered in matrix isolation. The enthalpies, however, differ only slightly from those at 298°K. It is at once obvious that for the hypothetical example chosen, the decomposition of ML_m under ordinary conditions [reaction (a)] is thermodynamically allowed ($\Delta G < 0$), while its decomposition under matrix isolation conditions [reaction (b)] is thermodynamically forbidden ($\Delta G > 0$).

† It is assumed for simplicity that $M(L)_m$ and L are gaseous at all temperatures. The arguments are not changed substantially if account is taken of the actual state of these species under ordinary conditions.

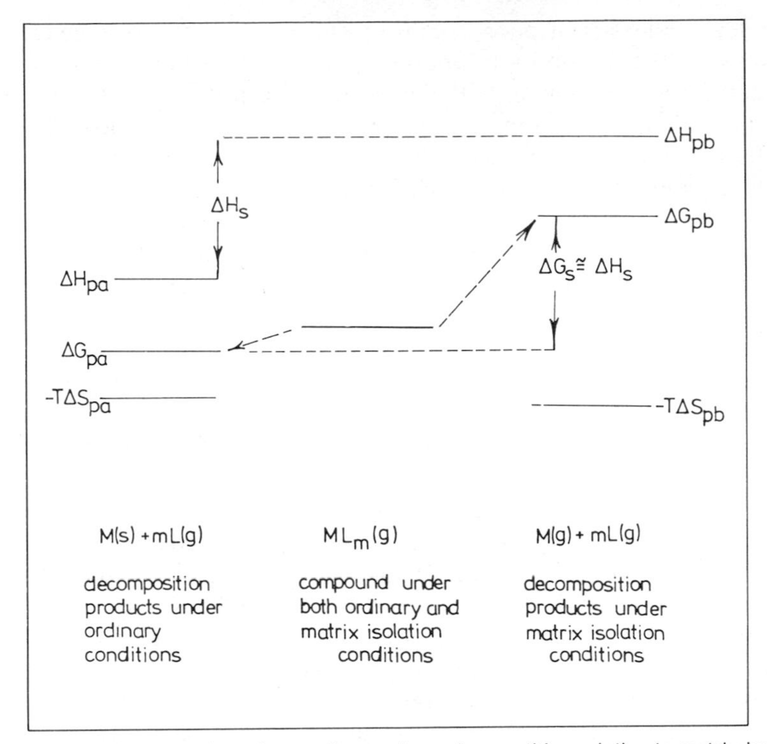

Figure 46. A schematic view of some thermodynamic quantities relating to matrix-isolated species of the form, ML_m.

The quantity ΔG_s by which ΔG_{pb} exceeds ΔG_{pa} is approximately equal to $\Delta H_s - T(\Delta H_v/T_b + \Delta H_f/T_m) \simeq \Delta H_s$, in which ΔH_s, ΔH_v, and ΔH_f are, respectively, the enthalpies of sublimation, vaporization, and fusion of the metal, and T_b and T_m are the boiling and melting temperatures of the metal.

Evidently, it is mainly the large heats of sublimation of the transition metals (e.g., $\Delta H_s = 103$, 91, and 136 kcal/mole for Ni, Pd, and Pt) that result in the stability of species such as $Ni(O_2)_2$, $Ni(O_2)(N_2)_2$, and $Ni(N_2)_4$ in the matrix.

The above analysis indicates that matrix isolation raises the decomposition temperature of a molecule of the form ML_m by a quantity approximately equal to $\Delta H_s/\Delta S$ above its ordinary decomposition temperature. It is evident, moreover, that even if the decomposition reaction (a) is thermoneutral or exothermic under ordinary conditions (i.e., even if the molecule is ordinarily unstable at any temperature), it may be possible to stabilize the species by matrix isolation. The bonds so formed could have bond energies as high as $\Delta H_s/m$ and would be in every way similar to "ordinary" covalent bonds.

Under conditions of poor isolation, where more than one metal atom lies within close proximity to another, the decomposition products would be metal clusters and ligand, with the resultant ΔG presumably lying between ΔG_{pa} and ΔG_{pb}, thereby diminishing the stabilizing effect of the matrix.

Although the stabilizing effect of the matrix has been approached in a way most convenient for those accustomed to thinking in terms of ordinary preparative conditions, it is causally inverted. In fact, the high heat of sublimation of the metal does not cause the enhanced stability in the matrix, but rather it causes the high instability of molecules such as $Ni(N_2)_4$ under ordinary conditions. Thus no bond formation takes place in the matrix or otherwise unless the process conforms to the normal thermodynamic rules. For example, in a case such as $XeCl_2$ where Xe is normally gaseous under ordinary conditions, its stability in the matrix is due mainly to the low temperature of the environment, that is, to the lack of thermal energy sufficient to break the Xe—Cl bond.

The effect of the interaction between the matrix-isolated species and the matrix would, if anything, increase the stability of isolated species yet further. The effect of the largely van der Waals interaction between the guest and host on the enthalpy of decomposition is very small. This is not so in the case of the entropy of decomposition. Upon decomposition the products could incorporate themselves into the host lattice, thereby decreasing ΔS_{pb}. This is especially so in cases where the ligand and the matrix gas are one and the same (a reactive matrix), for example, $Ni(N_2)_4$ in a N_2 matrix.

When a particular species is synthesized by matrix isolation, it is natural to consider how stable it might be under ordinary circumstances. It is obvious from the above discussion that one cannot, in general, obtain this information from warm-up experiments, because in most cases the matrix would be destroyed before the matrix decomposition temperature of the guest molecule is reached. For example, if the metal in reaction (a) were Ni, the reaction were thermoneutral, and $\Delta S_{pb} = 100$ e.u. (approximately ΔS^0_{298} of $Ni(CO)_4$), then $\Delta G_{pb} = 0$ at 1000°K. Consequently other means must be sought to gain insight into the ordinary thermal stability of matrix-isolated species.

An important technique has recently been developed[7,115] that enables one to establish not only the feasibility of a macroscale cryochemical synthesis but also the thermal stability of the reaction product(s). The method can be considered to be a marriage between matrix and macroscale cryochemistry and basically involves a matrix synthesis (usually in the temperature range, 4.2–77°K) followed by careful boil-off of the matrix. In this way, solid, unisolated reaction product(s) can be retained on the low-temperature optical window for subsequent spectroscopic examination, from cryogenic temperatures up to ambient temperature. The decomposition temperature of the reaction product(s), therefore, can be established.

Table 26. Thermal Stabilities of Metal Atom Cocondensation Products

Compound[a]	Decomposition Temperature (°C)	Ref.
$Rh_2(CO)_8$	−45	7
$Ir_2(CO)_8$	−55	7
$Ni(C_2H_4)_3$	0	115
$Pd(C_2H_4)_3$	−30	115
$Co(C_2H_4)_3$	−60	115
$Fe(COD)_2$	−20	115
$Co(COD)_2$	−10	115
$Pd(C_7H_{10})_3$	15	115
$Fe(C_6H_6)_2$	−60	115

[a] COD = cycloocta-1,5-diene, C_7H_{10} = norbornene.

Micropreparative techniques of this type can provide a vital link between matrix and macroscalc cryochemical syntheses and represents an important area for future research.

Table 26 lists some reactive compounds that have recently been synthesized by metal atom cocondensation techniques and whose thermal stabilities have been established.

The potential usefulness of the technique can be appreciated by realizing that the majority of the compounds in the above list are impossible to obtain by conventional chemical approaches.

The thermal stabilities of reaction products can also be investigated by less direct approaches, one of which is described below.

In discussing the systems, $Ni(CO)_m$, $Pd(CO)_m$, $Pt(CO)_m$ (where $m = 1–4$), Ozin, Moskovits et al.[11] have speculated that ΔF_{CO}, the decrease of the Cotton–Kraihanzel CO-stretching-force constant from that of free CO (18.46 mdyn/Å) is a measure of the bond energy (referred to the valence state of M), hence $m\Delta F_{CO} = \Delta H_C$ is a measure of the dissociation enthalpy of the species $M(CO)_m$. Since the stereochemistry and coordination number are expected to affect both the dissociation enthalpy and the CO-bond-stretching force constant, the true dependence of the bond-dissociation enthalpy upon ΔF_{CO} is expected, at least to a first approximation, to be independent of both the identity of M and the stereochemistry of $M(CO)_m$.

Cotton has introduced the concept of a M—C bond energy $\bar{D}^*$ referred to the metal in its "valence" state. $\bar{D}^*$ is obtained from $\bar{D}$, the conventional M—C bond energy, by adding the appropriate fraction of the promotion energy required to raise the metal from the ground state to its d^n valence state. In fact, when ΔF_{CO} is compared to $\bar{D}$ and $\bar{D}^*$ for the series of carbonyls,

Table 27. Bond Dissociation Energies, Metal–Valence-State Promotion Energies, and Decrease in CO-Stretching Force Constant for Various Carbonyls

	Energies in kcal/mole			
	ΔF_{CO}	$\bar{D}$[a]	E^*[b]	$\bar{D}^*$
$Cr(CO)_6$	2.00[c]	26	162	53
$Mn_2(CO)_{10}$	2.18[d]	24	172	58
$Fe(CO)_5$	2.69[e]	29	153	60
$Co_2(CO)_8$	1.46[f]	33	80	53
$Ni(CO)_4$	1.23[g]	35	42	45.5
$Mo(CO)_6$	1.96[a]	36	120	56

[a] A. Cartner, B. Robinson, and P. J. Gardner, *J. Chem. Soc., Chem. Commun.*, 295 (1973).
[b] Except for Mo all values are obtained from H. A. Skinner and F. Summer, *J. Inorg. Nucl. Chem.*, **4**, 245 (1957). E^* for Mo was calculated in the same manner as was done by Skinner for Cr using Moore's[104] atomic energy levels for Mo(I).
[c] Calculated from data reported by R. L. Amster, R. B. Hannan, and M. C. Tobin, *Spectrochim. Acta*, **19**, 1489 (1963).
[d] F. A. Cotton and R. M. Wing, *Inorg. Chem.*, **9**, 1328 (1965). A weighted average between axial and equitorial force constants is presented.
[e] W. G. Fately and E. R. Lippincott, *Spectrochim. Acta*, **10**, 8 (1957). A weighted average between axial and equitorial force constants is presented.
[f] G. Bor, *Spectrochim. Acta*, **19**, 1209 (1963). (This quantity arises from an approximation different from that of Cotton and Kraihanzel. It was, however, the only value available.)
[g] P. Kündig, M. Moskovits, and G. A. Ozin, *J. Mol. Struct.*, **14**, 137 (1972).

$Cr(CO)_6$, $Mn_2(CO)_{10}$, $Fe(CO)_5$, $Ni(CO)_4$, $Mo(CO)_6$, and $Co_2(CO)_8$ (Table 27), a poor correlation is obtained between ΔF_{CO} and $\bar{D}$, but a good one is obtained between ΔF_{CO} and $\bar{D}^*$. A more graphic comparison is obtained by plotting $\bar{D}^*$ against ΔF_{CO} (Figure 47). We see in the figure that although ΔF_{CO} is a measure of $\bar{D}^*$, it is not a linear measure, hence $m\Delta F_{CO}$ is a measure of ΔH^*, the dissociation enthalpy to the valence state of the metal, only

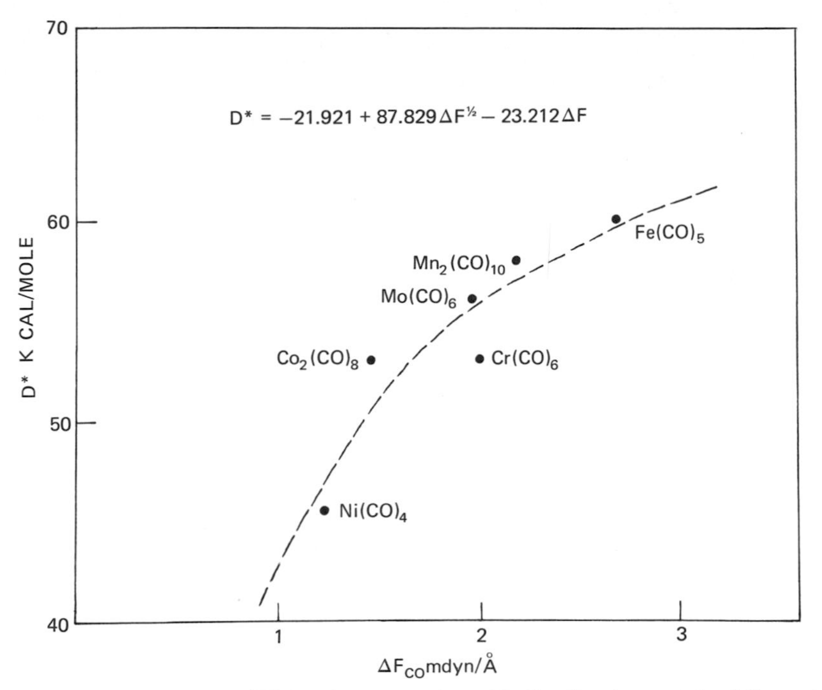

Figure 47. A plot of D^*, the MC bond energy referred to the d^n valence state of the metal, versus ΔF_{CO}, the decrease in the CO-stretching-force constant from that of free CO in various metal carbonyls.

among compounds of equal coordination number or in a series of compounds of the form, $M(CO)_m$.

In the case of metal carbonyls, the problem may be avoided by interpolating values of $\bar{D}^*$ from measured values of ΔF_{CO}. To aid in the interpolation the empirical expression

$$\bar{D}^* = -21.92 + 87.83(\Delta F_{CO})^{1/2} - 23.21\Delta F_{CO} \tag{34}$$

obtained by fitting measured $\bar{D}^*$ and ΔF_{CO} to equation (34) using a nonlinear least-squares analysis, was used. (It should be pointed out that this relationship has no physical significance whatsoever.)

In order to illustrate to what extent ΔH_c mimics $\Delta H^* = m\bar{D}^*$, these two qualities are plotted in Figure 48 for the series, $Ni(CO)_m$. $\bar{D}^*$ values were obtained from equation (34), from which ΔH^* were calculated by multiplication by the appropriate coordination number. ΔH_c values were set equal to $m\Delta F_{CO}$ and normalized so that $\Delta H_c = \Delta H^*$ at $Ni(CO)_4$. Figure 48 indicates that ΔH_c shows a more gradual increase with coordination number than does ΔH^*. Aside from that, ΔH_c indicates correctly the trend in bond energy

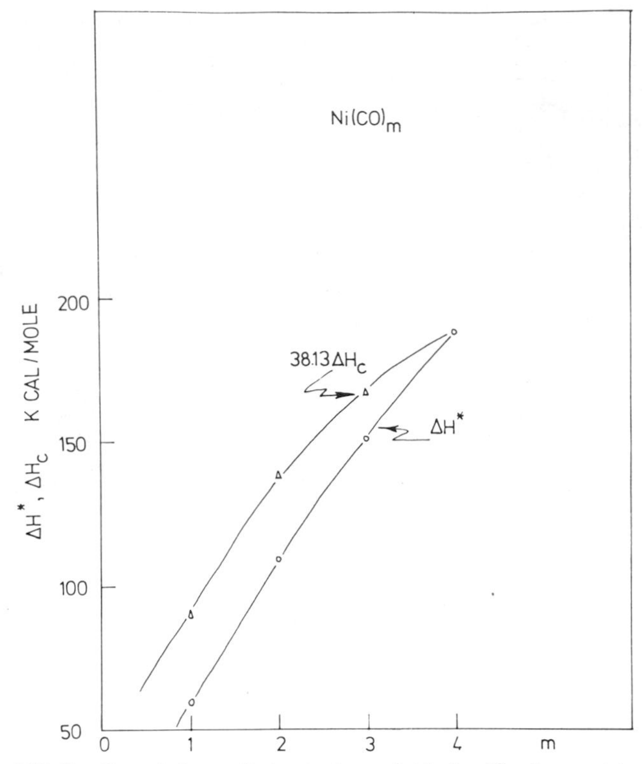

Figure 48. ΔH^*, the dissociation enthalpy to the metal in its d^{10} valence state, and ΔH_C as a function of m for $Ni(CO)_m$ ($m = 1$–4).

with coordination number (and among various carbonyls of the same coordination number). This fact is gratifying because a similar function ($m\Delta F_{N_2}$) was used to gauge the thermal stabilities of dinitrogen complexes such as $Ni(N_2)_m$, $Pd(N_2)_n$, $Pt(N_2)_n$,[25] $Rh(N_2)_m$[26] ($m = 1$–4, $n = 1$–3) for which thermodynamic data are not available, hence an empirical expression similar to equation (34) cannot be obtained.

Using equation (34), $\bar{D}^*$ values were obtained for the species, $Pd(CO)_4$ and $Pt(CO)_4$, which have, as yet, not been prepared by conventional means. Since the promotion energies to the valence (d^{10}) state are easily obtained from Moore's tables (104), one can obtain $\bar{D}$ values for these compounds from which ΔH° values may be calculated by correcting for the heat of sublimation of the metal. Details of the calculation are shown in Table 28.

As expected, $Pt(CO)_4$ is calculated to be less stable than $Ni(CO)_4$. What is intriguing is that $Pd(CO)_4$ is calculated to be *more* stable than $Ni(CO)_4$, largely as a result of the low heat of sublimation and low promotion energy of Pd metal.

Table 28. Calculation of $\Delta H^\circ_{(298)}$ for $Pd(CO)_4$ and $Pt(CO)_4$ from Values of ΔF_{CO} as Described in the Text (All Values in kcal/mole)

	ΔF_{CO}	$\bar{D}^{*a}$	E^{*b}	$\bar{D}$	$\Delta H_s^{\,c}$	ΔH°
Ni	1.23	47	42.1	36	103	43
Pd	0.92	41	0	41	91	73
Pt	1.18	46	17.5	42	136	31

[a] Obtained from equation (34).
[b] Calculated from Moore's atomic energy levels[104] assuming a d^{10} valence state.
[c] F. A. Cotton and G. Wilkinson, *Advanced Inorganic Chemistry*, 3rd Edit., Wiley-Interscience, New York, 1972.

14. METAL ATOM COCONDENSATION REACTIONS AND SURFACE CHEMISTRY

Despite the fact that the species, $Pd(CO)_4$ and $Pt(CO)_4$, had not been known before their synthesis by matrix isolation, CO chemisorbed on Pd and Pt had been observed spectroscopically. In fact, a working hypothesis was used by the authors during the initial experiments in this area that stated that any ligand that chemisorbed on a metal would form complexes with it in the matrix. To date no exceptions have been found to this rule. The spectrum of CO chemisorbed on polycrystalline metal films or suspensions usually contains more than one band; for example, CO chemisorbed on Co metal shows two bands,[105] one at 2000 cm^{-1} and one at 1880 cm^{-1}. The conventional explanation for this observation is that the higher frequency mode arises from end-on bonded CO on the metal, while the lower frequency band is due to bridge-bonded CO to two metal atoms.[106] Blyholder,[105] on the other hand, offers another explanation to account for the lower frequency bands. He performed an extended Hückel calculation considering only the π-electron system of a nine-metal atom square cluster with the CO placed at either a central or a corner position and found that there is more π-bonding between a metal in the corner position and the CO than there is between a central metal atom and the CO. He, therefore, concludes that the high-frequency band associated with chemisorbed CO is due to CO end-on chemisorbed on central sites, whereas the lower frequency bands arise from CO *end-on* chemisorbed at edge and corner metal sites.

The results of Blyholder's molecular orbital calculations are intuitively understandable. The increase in π-bonding on going from a central site to a corner site is due to the decrease in the number of metal atoms bonded to the

CO-bearing atom. The next logical step would be to have *no* other metal atoms bonded to the M—C—O group. Such a molecule is expected to have even more π overlap between metal and CO π^* orbitals, and hence a CO stretching frequency lower than the frequency due to CO end-on chemisorbed at *any* site on the same metal. Species of the form, MCO, have been produced and studied by matrix isolation.[107]

Table 29 compares the CO-stretching frequencies of CO chemisorbed on several transition metals, M, with those measured for the analogous molecules, M—CO, obtained by matrix isolation.[107] With the exception of Fe, the lower-frequency CO stretching band (band 2) associated with chemisorbed CO has a *lower* frequency than the CO stretching band associated with the corresponding triatomic MCO. This implies that band 2 cannot arise from an end-on chemisorbed CO as proposed by Blyholder,[105] but is, in all likelihood, a bridge-bonded carbonyl as was first proposed by Eischens et al.[106] This point is further strengthened by recent low-energy electron-difraction evidence[108] that CO chemisorbed on single-crystal platinum forms both bridge-bonded and end-on bonded carbonyls.

The quantity, R^{M}_{CO}, tabulated in Table 29 is defined by $R^{M}_{CO} = \Delta\nu_{MCO}/\Delta\nu_{ChCO}$, where Δ implies decrease from the frequency of free CO (2141 cm^{-1}) and the frequency of band 1, the band due to chemisorbed CO lying at the

Table 29. Comparison between the CO-Stretching Frequencies of CO Chemisorbed on Various Metals with the Analogous Frequency for the Species, MCO

Metal	ν_{ChCO} Band 1 (cm^{-1})[a]	ν_{ChCO} Band 2 (cm^{-1})	ν_{MCO} (cm^{-1})	R^{M}_{CO}[g]
V	1940[b]	1890[b]	1902[e]	1.19
Fe	1980[b]	1900[b]	1898[e]	1.51
Co	2000[b]	1880[b]	1954[e]	1.33
Ni	2075[b]	1935[b]	1996[f]	2.20
Cu	2120[b]	—	2010[h]	6.24
Pd	2080[c]	1860[c]	2050[f]	1.49
Pt	2070[d]	1800[d]	2052[f]	1.25

[a] The subscript ChCO implies chemisorbed CO.
[b] G. Blyholder and M. C. Allen, *J. Amer. Chem. Soc.*, **91**, 3158 (1969).
[c] N. N. Kavtaradze and N. P. Sokolova, *Dokl. Akad. Nauk. SSSR*, **162**, 847 (1965).
[d] N. N. Kavtaradze and N. P. Sokolova, *Dokl. Akad. Nauk. SSSR*, **168**, 140 (1966).
[e] Data of H. Huber, E. P. Kündig, M. Moskovits, and G. A. Ozin, to be published.
[f] Reference 11.
[g] $R^{M}_{CO} = \Delta\nu_{MCO}/\Delta\nu_{ChCO}$, where Δ implies decrease from $\nu_{CO} = 2141$ cm^{-1}. Band 1 of the chemisorbed CO was used in evaluating R^{M}_{CO}.
[h] Reference 44.

highest frequency, is used in calculating these values. Band 1 is assumed to arise from end-on chemisorbed CO.

It is noteworthy that all R^{M}_{CO} values are larger than unity, and with the marked exceptions of Cu and Ni all cluster around approximately the values 1.2–1.5.

An intriguing relationship exists between ΔF_{CO} of CO chemisorbed on various metals and their heats of chemisorption similar to that found for metal carbonyls. For example, plotting the quantity $\Delta H_{ads} + E^*/10$ versus ΔF_{CO} gives a monotonically increasing function almost identical to that given by $\bar{D}^*$ versus ΔF_{CO} of the metal carbonyls (ΔH_{ads} is the initial heat of adsorption of CO on the appropriate metal, while E^* is its d^n promotion energy) (see Figure 49 and Table 30). The factor of 10 by which E^* was divided was arrived at by obtaining $\bar{D}^*$ from equation (36), corresponding to the various ΔF_{CO} values for chemisorbed CO, subtracting the reported heat of adsorption from $\bar{D}^*$, and plotting the residual energy against E^*.

It should be immediately stated that the correlation between $\Delta H_{ads} + E^*/10$ and ΔF_{CO} does not imply that the surface metal atom is in a d^n valence

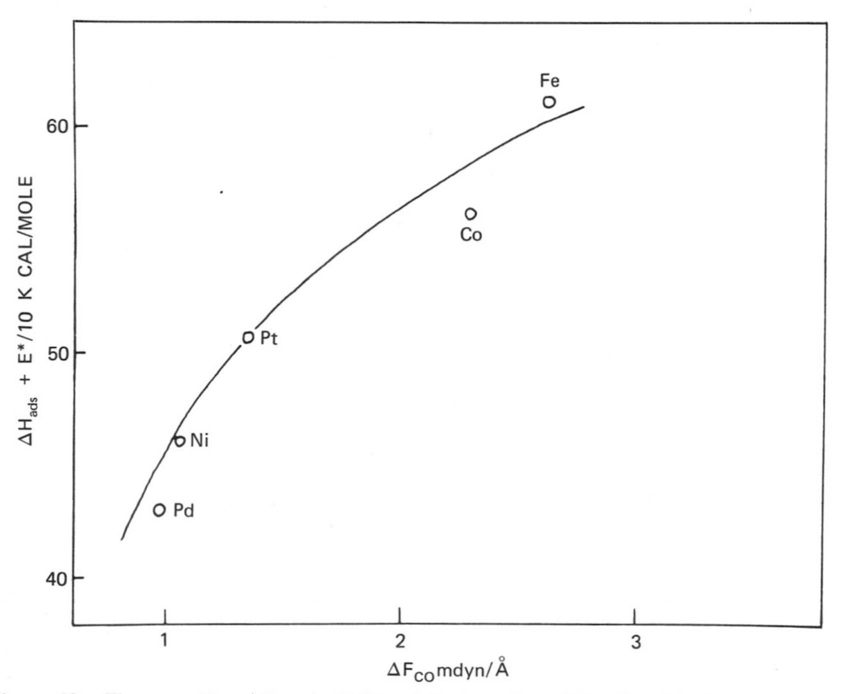

Figure 49. The quantity, $\Delta H_{ads} + E^*/10$, plotted against ΔF_{CO} for CO chemisorbed on various metals.

Table 30. Thermodynamic Properties of Chemisorbed CO

Metal	ΔF_{CO}[a]	ΔH_{ads}[b]	$\Delta H_{ads} + E^*/10$
Fe	2.63	46	61.3
Co	2.31	48	56.0
Ni	1.07	42	46.2
Pd	0.99	43	43.0
Pt	1.16	49	50.7

[a] Units are mdyn/Å; obtained from data in table using the Cotton-Kraihanzel approximation.
[b] Initial heats of adsorption in kcal/mole; D. Brennan and F. H. Hayes, *Phil. Trans. A*, **258**, 347 (1965).

state (as was assumed to be the case in metal carbonyls) with ten groups bonded to it. It is more likely that the surface metal atom is in some other valence state whose promotion energy depends monotonically on E^* as one passes from metal to metal. It is, nevertheless, fascinating that so reasonable a number of "ligands" as ten was obtained from the analysis.

The chemisorption of N_2 on various metals, although less widely studied, has produced results not unlike those discussed above. Table 31 summarizes these data. Unlike chemisorbed CO, N_2 chemisorbed on various metals

Table 31. Comparison between the NN-Stretching Frequencies of N_2 Chemisorbed on Various Metals with the Analogous Frequency for the Species, MN_2

Metal	ν_{ChN_2}	ν_{MN_2}	$R_{N_2}^{M}$
Ni	2202[a]	2090[c]	1.90
Pd	2260[a]	2213[c]	1.70
Pt	2230[a]	2168[c]	1.60
Rh	2236[b]	2156[d]	1.85

[a] R. van Hardeveld and A. van Montfoot, *Surf. Sci.*, **4**, 396 (1966).
[b] Yu. G. Borod'ko and V. S. Lyutov, *Kinet. Catal.*, **12**, 202 (1971).
[c] Reference 25.
[d] Reference 26.

gives rise to only one band, assumed to be end-on bonded N_2. This is consistent with the fact that no bridge-bonded dinitrogen complexes [with, perhaps, the exception of $Rh_2(N_2)_8$] have been reported. As shown in Table 31, $R^{M}_{N_2}$ values defined in an analogous manner to R^{M}_{CO} are, except for the case of Ni, consistently larger than the corresponding R^{M}_{CO} values. This observation can be rationalized in terms of σ and π population changes on going from ML to L chemisorbed on M, L being N_2 or CO.

No trustworthy heat of chemisorption data are available for these systems, hence an analysis similar to that made for chemisorbed CO cannot be performed. We have, however, found the assumption that $m\Delta F_{N_2}$ is a measure of the heat of decomposition of the species, $M(N_2)_m$ (to the gaseous metal atom in its valence state) useful in much the same way $m\Delta F_{CO}$ was found to be.[25]

Nature of Bonding in Chemisorbed CO and N_2

The CO-stretching force constants of $Ni(CO)_4$, $Ni(CO)_{ads}$, and NiCO are, respectively, 17.23, 17.38, and 16.09 mdyn/Å. Similarly, the NN-stretching force constants of $Ni(N_2)_4$, $Ni(N_2)_{ads}$, and NiN_2 are, respectively, 19.65, 19.99, and 17.74 mdyn/Å. (A similar trend is observed in the Fe/CO system, namely, $k_{CO}[Fe(CO)_{ads}] > k_{CO}[Fe(CO)_5] > k_{CO}[FeCO]$, but there is a reversal in this order between $Cr(CO)_6$ and $Cr(CO)_{ads}$).

In terms of equation (9) (Section 11), the above observation implies that the quantity, $\sigma - \gamma\pi$, is larger (hence π smaller) in CO chemisorbed on Ni than in Ni(CO) and, what is noteworthy, larger even than in $Ni(CO)_4$. A facile explanation is that the nickel atoms surrounding the Ni atom to which a CO molecule is chemisorbed withdraw more π charge than 3 COs. If this is the case, then the same trend should be observed in the Ni/N_2 system since N_2 is an even poorer π acceptor than CO, and indeed this is what is found.

Alternatively, one might suppose that the decreased π overlap between chemisorbed CO and a surface metal atom arises from the fact that the latter is in a different valence state from what it is in the complex.

The quantities, R^{M}_{CO} and $R^{M}_{N_2}$, measure the change in σ and π orbital population on going from, respectively, MCO and MN_2 to the chemisorbed states. Starting with equations (9) and (10) (section 11), one can derive expressions for $R^{M}_{N_2}$ and R^{M}_{CO} in terms of π and σ changes. These are

$$R^{M}_{N_2} \simeq 1 + a'\left[\frac{\gamma'\Delta\pi_{N_2} + \Delta\sigma_{N_2}}{K_{N_2} - K_{ChN_2}}\right] \tag{35}$$

$$R^{M}_{CO} \simeq 1 + |a|\left|\frac{\gamma\Delta\pi_{CO} - \Delta\sigma_{CO}}{K_{CO} - K_{ChCO}}\right| \tag{36}$$

where $\Delta\pi$ is the decrease in π charge accepted by the CO in going from MCO to chemisorbed CO, and $\Delta\sigma_{CO}$ is the decrease in σ-donated charge sustained during the same process, with an analogous definition existing for $\Delta\pi_{N_2}$ and $\Delta\sigma_{N_2}$. K_{CO}, K_{N_2}, K_{ChCO}, and K_{ChN_2} are, respectively, the ligand-stretching force constants of free CO, free N_2, CO chemisorbed on M, and N_2 chemisorbed on M.

It is at once clear that $R^{M}_{N_2} > R^{M}_{CO}$ because the second term in equation (36) is of the form, $\Delta\pi - \Delta\sigma$, while that of equation (35) is of the form, $\Delta\pi + \Delta\sigma$. [Note that since $(K_{CO} - K^{M}_{ChCO}) < (K_{N_2} - K^{M}_{ChN_2})$, the relative magnitude of $R^{M}_{N_2}$ to R^{M}_{CO} cannot be explained by the magnitude of the denominators in equations (35) and (36).]

The implication of the above result is that there are *both* σ and π decreases accompanying passage from either the monocarbonyl or monodinitrogen complex to the chemisorbed ligand on M. Moreover, $\gamma\Delta\pi_{CO}$ must be greater than $\Delta\sigma_{CO}$ in order to explain R^{M}_{CO}s being greater than unity. Since γ is approximately unity, we conclude that π-backbonding is decreased *more* than is σ-bonding when going from MCO to CO chemisorbed onto M.

From the foregoing, we conclude that the bonds between CO and N_2 chemisorbed on metal surfaces and those in ordinary metal carbonyls and metal dinitrogen complexes are very similar in both strength and character. In particular, the σ-donor–π-acceptor model adopted for these ligands in ordinary coordination complexes is capable of explaining many of the properties encountered when these molecules are chemisorbed on metals. Moreover, species such as MCO and MN_2 synthesized by matrix isolation where M is a metal atom serve as limiting cases for these ligands chemisorbed on metal particles of decreasing particle size, an important fact to know in attempting to understand the nature of surface species.

15. CONCLUSION

The wealth of information available to the chemist via spectroscopy is limited greatly by the number of systems amenable to such investigation. Matrix isolation multiplies that number by many orders of magnitude. Thus radicals such as CF_2 and molecules that can be construed to be reaction intermediates such as $Ni(CO)_3$ and Br_3 can be formed and studied at leisure rather than by fast-scan techniques. Moreover, one can often tailor-make some molecules such as NiN_2, PdO_2, and $Ni(CO)_2(N_2)_2$ that exhibit certain chemical features of interest, or form molecules that extend or complete homologous series such as $XeCl_2$, $Pd(CO)_4$, and $Pt(CO)_4$, an important undertaking to the chemist who is ultimately interested in trends. Many of the molecules so synthesized display a certain beauty or simplicity that makes their study meritorious for its own sake, for example, N_2PdO_2 in which two atmospheric molecules are bonded in different fashions to the same metal.

Finally a great deal of knowledge concerning the nature of the bonding, thermodynamic properties, and structure can be obtained by judiciously analyzing force constant and intensity data emerging from these systems.

In view of what has already been accomplished we feel confident that cryogenic techniques will provide a great deal more fascinating chemistry in the future.

REFERENCES

1. H. Huber, E. P. Kündig, M. Moskovits, and G. A. Ozin, *J. Amer. Chem. Soc.*, **95**, 332 (1973).
2. E. P. Kündig, M. Moskovits, and G. A. Ozin, *Can. J. Chem.*, **50**, 3587 (1972); J. H. Darling and J. S. Ogden, *J. Chem. Soc., Dalton*, 1079 (1973).
3. R. L. DeKock, *Inorg. Chem.*, **10**, 1205 (1971).
4. H. Huber, W. Klotzbücher, G. A. Ozin, and A. Vander Voet, *Can. J. Chem.*, **50**, 3746, (1972).
5. D. McIntosh and G. A. Ozin, *Inorg. Chem.*, 1976 (in press).
6. D. McIntosh, and G. A. Ozin, *J. Amer. Chem. Soc.*, 1976 (in press).
7. L. Hanlan and G. A. Ozin, *J. Amer. Chem. Soc.*, **96**, 6324 (1974).
8. E. P. Kündig and G. A. Ozin, *J. Amer. Chem. Soc.*, **96**, 5585 (1974).
9. H. Huber, T. A. Ford, E. P. Kündig, W. Klotzbücher, M. Moskovits, and G. A. Ozin, *J. Amer. Chem. Soc.*, 1976 (in press).
10. H. Huber, E. P. Kündig, M. Moskovits, and G. A. Ozin, *J. Mol. Struct.*, **14**, 137 (1972).
11. E. P. Kündig, D. McIntosh, M. Moskovits, and G. A. Ozin, *J. Amer. Chem. Soc.*, **95**, 7324 (1973).
12. E. P. Kündig, M. Moskovits, and G. A. Ozin, *Can. J. Chem.*, **51**, 2737 (1973).
13. L. Andrews, *J. Chem. Phys.*, **57**, 51 (1972); D. A. Hatzenbühler and L. Andrews, *J. Chem. Phys.*, **56**, 3398 (1972).
14. J. S. Shirk and H. H. Claassen, *J. Chem. Phys.*, **54**, 3237 (1971).
15. J. W. Nibler and D. A. Coe, *J. Chem. Phys.*, **55**, 5133 (1971).
16. D. Boal and G. A. Ozin, *J. Chem. Phys.*, **55**, 3598 (1971); D. Boal, G. Briggs, H. Huber, G. A. Ozin, E. A. Robinson, and A. Vander Voet, *Nature Phys. Sci.*, **237**, 174 (1971); H. Huber and G. A. Ozin, *J. Mol. Spectrosc.*, **41**, 595 (1972).
17. H. Huber, G. A. Ozin, and A. Vander Voet, *Nature Phys. Sci.*, **232**, 166 (1971); G. A. Ozin and A. Vander Voet, *J. Chem. Phys.*, **56**, 4768 (1972).
18. L. Hanlan, H. Huber, E. P. Kündig, B. McGarvey, and G. A. Ozin, *J. Amer. Chem. Soc.*, **97**, 7054 (1975).
19. G. A. Junk and H. J. Svec, *J. Chem. Soc., A*, 2102 (1970).
20. M. Wrighton and D. Bredesen, *J. Organometal. Chem.*, **50**, C35 (1973).
21. M. A. Graham, M. Poliakoff, and J. J. Turner, *J. Chem. Soc., A*, 2939 (1971).
22. P. A. Breeze and J. J. Turner, *J. Organometal. Chem.*, **7**, C44 (1972).
23. M. Moskovits and G. A. Ozin, *J. Chem. Phys.*, **58**, 1251 (1973).

24. W. Klotzbücher and G. A. Ozin, *J. Amer. Chem. Soc.*, **97**, 2672 (1975).
25. E. P. Kündig, M. Moskovits, and G. A. Ozin, *Can. J. Chem.*, **51**, 2710 (1973); D. W. Green, J. Thomas, and D. M. Gruen, *J. Chem. Phys.*, **58**, 5453 (1973).
26. G. A. Ozin and A. Vander Voet, *Can. J. Chem.*, **51**, 637 (1973).
27. L. Vaska, *Account. Chem. Res.*, 1976 (in press).
28. W. Klotzbücher and G. A. Ozin, *J. Amer. Chem. Soc.*, **95**, 3790 (1974); ibid., **97**, 3965 (1975).
29. E. B. Wilson, J. C. Decius, and P. C. Cross, *Molecular Vibrations*, McGraw-Hill, New York, 1955.
30. G. Bos, *J. Organometal. Chem.*, **10**, 343 (1967).
31. H. Haas and R. K. Sheline, *J. Chem. Phys.*, **47**, 2996 (1967).
32. J. H. Darling and J. S. Ogden, *J. Chem. Soc.*, *Dalton*, 2496 (1972).
33. F. A. Cotton and C. S. Kraihanzel, *J. Amer. Chem. Soc.*, **84**, 4432 (1962).
34. A. Barnes, "Theoretical treatment of matrix effects," in *Vibrational Spectroscopy of Trapped Species*, H. E. Hallam, Ed., Wiley, London, 1974.
35. (a) J. G. Kirkwood, *J. Chem. Phys.*, **2**, 351 (1934); (b) W. West and R. T. Edwards, *J. Chem. Phys.*, **5**, 14 (1937); (c) E. Bauer and M. Magat, *J. Phys. Rad.*, **9**, 319 (1938); and ibid., *Physica* **5**, 718 (1938); (d) A. D. Buckingham, *Proc. Roy. Soc.*, *A*, **248**, 169 (1958); ibid., **255**, 32 (1970); A. D. Buckingham, *Trans. Faraday Soc.*, **56**, 753 (1960); J. G. David and H. E. Hallam, *Spectrochim. Acta*, **23A**, 593 (1967).
36. S. W. Charles and G. C. Pimentel, *Pure Appl. Chem.*, **7**, 111 (1963).
37. B. R. Cairns and G. C. Pimentel, *J. Chem. Phys.*, **43**, 3432 (1965).
38. J. K. Burdett, *J. Chem. Soc.*, *Faraday II*, **70**, 1599 (1974).
39. M. Elian and R. Hoffman, *Inorg. Chem.*, **14**, 1058 (1975).
40. J. K. Burdett, *Inorg. Chem.*, **14**, 375 (1975).
41. R. J. Gillespie, *Molecular Geometry*, Van Nostrand-Reinhold, New York, 1972.
42. R. N. Perutz and J. J. Turner, *J. Amer. Chem. Soc.*, **97**, 4791 (1975).
43. M. Poliakoff, *J. Chem. Soc.*, *Dalton*, 210 (1974).
44. H. Huber, E. P. Kündig, M. Moskovits, and G. A. Ozin, *J. Amer. Chem. Soc.*, **97**, 2097 (1975).
45. J. K. Burdett, *J. Chem. Soc.*, *Chem. Commun.*, 763 (1973).
46. H. Huber, L. Hanlan, and G. A. Ozin, unpublished work.
47. O. Crichton, M. Poliakoff, A. J. Rest, and J. J. Turner, *J. Chem. Soc.*, *Dalton*, 1321 (1973).
48. M. Poliakoff and J. J. Turner, *J. Chem. Soc.*, *Dalton*, 1351 (1973); ibid., 2276 (1974).
49. H. Mahrike, R. J. Clark, R. Rosanke, and R. K. Sheline, *J. Chem. Phys.*, **60**, 2997 (1974).
50. E. P. Kündig and G. A. Ozin, *J. Amer. Chem. Soc.*, **96**, 3820 (1974).
51. H. Huber, E. P. Kündig, G. A. Ozin, and A. J. Poë, *J. Amer. Chem. Soc.*, **97**, 308 (1974).
52. M. A. Graham, Ph.D. thesis, University of Cambridge, 1971.
53. R. V. G. Evans and M. W. Lister, *Trans. Faraday Soc.*, **72**, 1107 (1950).
54. B. A. Frenz and J. A. Ibers, *Inorg. Chem.*, **11**, 1109 (1972).
55. S. F. A. Kettle, *J. Chem. Soc.*, *A*, 420 (1966).
56. H. Basch, A. Viste, and H. B. Gray, *J. Chem. Phys.*, **44**, 10 (1966).

57. R. S. Mulliken, *Can. J. Chem.*, **36**, 10 (1958).
58. R. A. Levenson, H. B. Gray, and G. P. Ceasar, *J. Amer. Chem. Soc.*, **92**, 3653 (1970).
59. J. P. Fawcett, A. J. Poë, and M. V. Twigg, *J. Chem. Soc., Chem. Commun.*, 267 (1973).
60. M. Wrighton and D. Bredesen, *J. Organometal. Chem.*, **50**, C35 (1973).
61. V. I. Baranovskii and A. B. Nikol'skii, *Theor. Exp. Chem.*, **3**, 309 (1967).
62. W. Weltner, P. Kasai, and E. B. Whipple, *J. Chem. Phys.*, **44**, 2581 (1966).
63. E. P. Kündig and G. A. Ozin, unpublished work.
64. W. Weltner, L. B. Knight, and W. C. Easley, *J. Chem. Phys.*, **54**, 1610 (1971).
65. P. H. Kasai and D. McCleod, Jr., *J. Chem. Phys.*, **55**, 1566 (1971).
66. G. Breit and I. I. Rabi, *Phys. Rev.*, **38**, 2082L (1931).
67. M. V. Smoluchowski, *Z. Phys. Chem.*, **92**, 192 (1917).
68. P. Kasai, *Account. Chem. Res.*, **4**, 329 (1971).
69. L. B. Knight and W. Weltner, *J. Mol. Spectrosc.*, **40**, 317 (1971).
70. H. J. Keller and H. Wawersik, *Z. Naturforsch.* **206**, 938 (1965).
71. B. Bleany, *Phil. Mag.*, **42**, 441 (1951).
72. L. B. Knight and W. Weltner, *J. Chem. Phys.*, **55**, 2061 (1971); ibid., **54**, 3875 (1971).
73. R. A. Berg and O. Sinanoglu, *J. Chem. Phys.* **32**, 1082 (1960); J. T. Hougen, G. E. Leroi, and T. C. James, ibid., **34**, 1670 (1961).
74. C. Malmberg, R. Scullman, and P. Nylén, *Ark. Fys.*, **39**, 495 (1968).
75. J. Margrave and H. Krishnan, Ph.D. thesis, Rice University, 1975.
76. L. Andrews and G. C. Pimentel, *J. Chem. Phys.*, **44**, 2361 (1966).
77. E. Weiss and W. Büchner, *Helv. Chim. Acta*, **46**, 1121 (1963); *Chem. Ber.*, **98**, 126 (1965); *Z. Anorg. Allg. Chem.*, **330**, 251 (1964).
78. W. Büchner, *Helv. Chim. Acta*, **46**, 2111 (1963).
79. J. Chatt, *J. Chem. Soc.*, 2939 (1953); M. J. S. Dewar, *Bull. Soc. Chim. Fr.*, **18**, C71 (1951).
80. J. Chatt, D. P. Melville, and R. L. Richards, *J. Chem. Soc., A*, 2841 (1969).
81. J. P. Collman, M. Kubota, F. D. Vastine, J. Y. Sun, and J. W. Kang, *J. Amer. Chem. Soc.*, **90**, 5430 (1968).
82. M. B. Hall and R. F. Fenske, *Inorg. Chem.*, **11**, 1619 (1972).
83. D. J. Darensbourg, *Inorg. Chem.*, **11**, 1436 (1972); ibid., **10**, 2399 (1971).
84. I. H. Hillier, *J. Chem. Phys.*, **52**, 1948 (1970).
85. C. W. DeKock and D. A. Van Liersburg, *J. Amer. Chem. Soc.*, **94**, 3235 (1972); ibid., *J. Phys. Chem.*, **78**, 134 (1974).
86. P. Polizer and S. D. Kasten, *Surf. Sci.*, **36**, 186 (1973).
87. H. Kunz-Krause, *Apoth. Ztg.*, **31**, 66 (1916); V. A. Plotnikov and K. N. Ivanov, *Zh. Khim. Promsti*, **7**, 1136 (1930); V. A. Plotnikov and O. K. Kudva, *Zh. Obshch. Khim.*, **1**, 1075 (1931); C. R. Bertrand, *C.R. Acad. Sci.*, **177**, 997 (1923).
88. S. Ogden, private communication.
89. A. Bos, *J. Chem. Soc., Chem. Commun.*, 26 (1972).
90. A. J. Hinchcliffe, J. S. Ogden, and D. D. Oswald, *J. Chem. Soc., Chem. Commun.*, 338 (1972).
91. P. Kasai, *J. Amer. Chem. Soc.*, **97**, 5609 (1975).

92. W. Weltner, J. L. Slater, R. K. Sheline, and K. C. Liu, *J. Chem. Phys.*, **55**, 5129 (1971).

93. J. L. Slater, T. C. DeVore, and V. Calder, *Inorg. Chem.*, **13**, 1808 (1974).

94. J. L. Slater, T. C. DeVore, and V. Calder, *Inorg. Chem.*, **12**, 1918 (1973).

95. G. C. Pimentel, *J. Amer. Chem. Soc.*, **80**, 62 (1958).

96. W. Klotzbücher and G. A. Ozin, unpublished work.

97. A. Bos and A. T. Howe, *J. Chem. Soc.*, *Faraday II*, **70**, 440, 451 (1974).

98. T. R. Waite, *Phys. Rev.*, **107**, 463 (1957).

99. M. V. Smoluchowski, *Z. Phys. Chem.*, **92**, 192 (1917).

100. A. Berne, G. Boato, and M. DePaz, *Nuovo Cimento*, **B46**, 182 (1966).

101. B. N. Oshcherin, *Phys. Status Solidi*, **B43**, K59 (1971), corrected expression.

102. E. A. Moelwyn-Hughes, *Physical Chemistry*; 2nd Edit., p. 105., Pergammon, New York, 1961.

103. R. W. G. Wyckoff, *Crystal Structures*, Vol. 1, 2nd Edit., p. 186, Wiley-Interscience, New York, 1963.

104. C. E. Moore, *Atomic Energy Levels*, Nat. Bur. Stand. circular 467.

105. G. Blyholder and M. C. Allen, *J. Amer. Chem. Soc.*, **91**, 3158 (1969).

106. R. P. Eischens, S. A. Francis, and W. A. Pliskin, *J. Phys. Chem.*, **60**, 194 (1956).

107. E. P. Kündig, M. Moskovits, and G. A. Ozin, unpublished work.

108. R. Mason, private communication.

109. G. A. Ozin and A. Vander Voet, *Can. J. Chem.*, **51**, 637 (1973).

110. J. S. Ogden, information presented at Meldola Lecture, London, 1974.

111. H. Huber, M. Moskovits, and G. A. Ozin, (unpublished work); and G. A. Ozin and A. Vander Voet, *Prog. Inorg. Chem.*, **19**, 140 (1975).

112. P. Timms, private communication; *Angew. Chem.* (*Int. Edit.*), **14**, 275 (1975).

113. E. P. Kündig, M. Moskovits, and G. A. Ozin, *Angew Chem.* (*Int. Edit.*), **14**, 292 (1975).

114. M. Moskovits and J. Hulse, unpublished work.

115. P. L. Timms, "Atomic species in chemical synthesis", A.C.S. Meeting, Chicago, 1975.

116. D. M. Gruen, J. W. Boyd, and J. M. Lavoie, *J. Chem. Phys.*, **60**, 4088 (1974).

117. J. J. Turner and M. Poliakoff, *J. Chem. Soc.*, *A*, 2403 (1971).

118(a). E. P. Kündig, M. Moskovits, and G. A. Ozin, communicated at Merck Symposium "Metal Atoms In Chemical Synthesis," Darmstadt, Germany, May 1974.

118(b). H. Huber, G. A. Ozin and W. Power, *J. Amer. Chem. Soc.*, 1976 (in press).

119. H. Huber, D. McIntosh and G. A. Ozin, *J. Organometallic Chemistry*, 1976 (in press).

120. D. McIntosh and G. A. Ozin, *J. Amer. Chem. Soc.*, 1976 (in press).

121. D. McIntosh and G. A. Ozin (in preparation).

Synthesis of Transition Metal Diatomic Molecules and Binuclear Complexes Using Metal Atom Cocondensation Techniques

9

M. Moskovits and G. A. Ozin

1. INTRODUCTION

Metal atoms react under certain matrix conditions to form diatomic molecules and binuclear complexes. A thorough understanding of these aggregation processes is crucial in any study involving metal vapors in chemical synthesis. We begin by briefly reviewing the present state of knowledge of transition metal diatomic molecules. Considerations of the conditions under which dimerization of metal atoms is favored follows, together with the various experimental techniques available for assigning spectral bands to binuclear species in the presence of mononuclears. The effect of the atomic weight of the metal atom and the influence of the matrix material on the formation of binuclear species is also discussed.

The vapors above the liquid transition metals, scandium through zinc, have been extensively studied using a combination of high-temperature Knudsen cell and mass spectrometric techniques.[1] Although small concentrations of the transition metal diatomic molecules, M_2, could be detected in the presence of the respective monatomic species, M, the latter formed more than 99% of the vapor.[1–4] It is because of this that the chemistry of transition metal diatomic molecules has not previously been exploited, although considerable information regarding their physical properties has been reported.[5]

In searching the literature for data on diatomic metal species one finds scattered reports of the diffusion and aggregation properties of monatomic metal species frozen in rare-gas solids at cryogenic temperatures. Electronic spectroscopy has proven to be of considerable value in this respect, as it is sometimes possible to observe the spectra of both the monatomic and diatomic metal species in the presence of each other.[5–13] An appropriate starting point for this discussion is, therefore, a brief account of the known physical properties of the first-row, transition metal diatomic molecules themselves.

Mainly as a result of the work of Kant et al.[1] it has proven possible to study the temperature dependence of the M/M_2 ratio using mass spectrometric techniques. These data have been analyzed and yield bond dissociation energies for M_2. The values are illustrated in graphical form in Figure 1 and show an interesting "double hump" dependence on atomic weight. Semiempirical molecular orbital calculations have been reported for the M_2 molecules,[14] and Mulliken overlap population analyses at r_e indicate that σ-bonding predominates in the metal-metal bond, mainly from the 4*s* orbital. This has recently been confirmed by an *ab initio* self-consistent field molecular-orbital calculation on Cu_2.[33] The valence-state electronic configurations for the metal atoms in M_2 are agreed to be $4s3d^{n+1}$. It is interesting to note that when the bond dissociation energy, D_0, is corrected for the promotion energy term from the electronic ground state, $4s^23d^n$, using $D_{vs} = D_0 + 2E_{vs}$, the valence state dissociation energies, D_{vs}, are all found to lie in the range, 56 ± 10 kcal/mole.[1a] The extremely low D_0 value for Mn_2 (Figure 1) is attributed to the extremely high E_{vs} for manganese, and for Zn_2, due to the presence of filled $4s^2$ and $3d^{10}$ valence shells. Aside

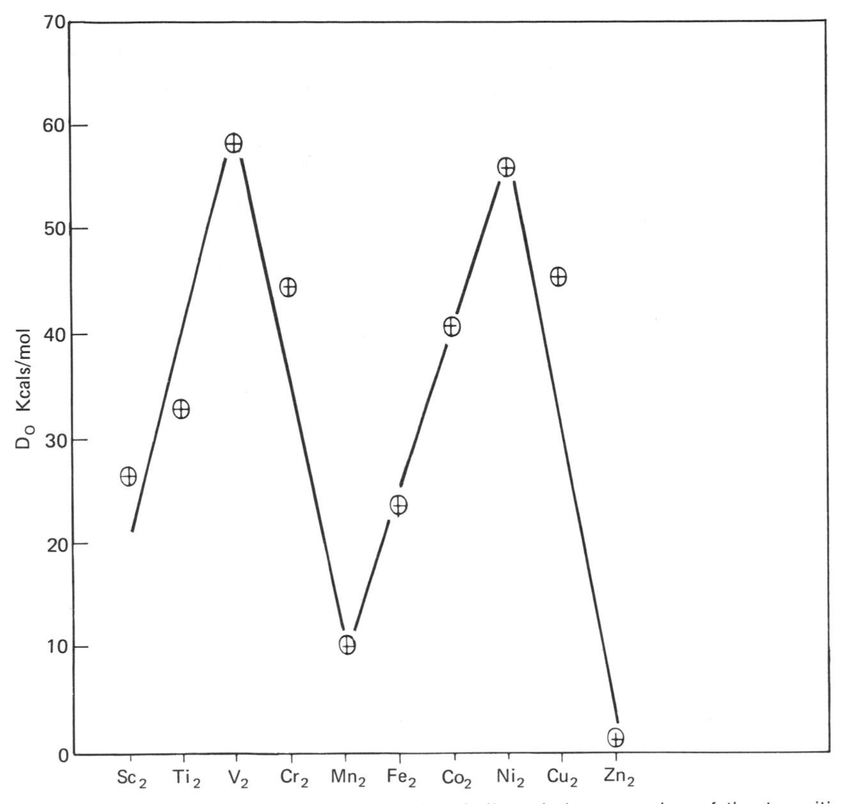

Figure 1. Graphical representation of the bond dissociation energies of the transition metal diatomic molecules, Sc_2 to Zn_2.

from Mn_2 and Zn_2, the transition metal diatomic molecules have substantial bond dissociation energies and would be expected to have an independent and interesting matrix chemistry. In Figure 2 we present a tabulation in "periodic table form" of the known diatomic molecules of the elements together with the techniques by which they have been detected.

It is generally accepted by matrix spectroscopists that diffusion of atoms or molecules occurs continually in all systems at all temperatures.[15] At low temperatures, where their mobilities are small, the rate of diffusion is experimentally insignificant. At higher temperatures diffusion does indeed occur, and matrix-annealing experiments can be a very powerful tool in the hands of a skillful experimentalist. In this way, spectral lines may be sharpened, diffusion-controlled chemical reactions monitored, *in situ* matrix photolyses studied, and, of central interest in this discussion, aggregation processes of metal atoms may be controlled.

H_2																	
abc d Li_2	c Be_2											ab B_2	abc d C_2	N_2	O_2	e F_2	a g Ne_2
ab Na_2	abc Mg_2											c Al_2	abc Si_2	ab e P_2	abc e S_2	ef Cl_2	a g Ar_2
ab K_2	abc Ca_2	a c d Sc_2	abc Ti_2	abc d V_2	abc d Cr_2	abc d Mn_2	abc i Fe_2	abc d Co_2	ab d Ni_2	abc d Cu_2	a Zn_2			ab e As_2	ab f Se_2	ef Br_2	a g Kr_2
ab Rb_2				c Nb_2	c Mo_2			c d Rh_2		abc d Ag_2			i Sn_2	a Sb_2	ab f Te_2	b ef h I_2	a g Xe_2
ab Cs_2						d Re_2		d Ir_2	c Pt_2	abc Au_2	a c Hg_2		abc Pb_2	a Bi_2			

Figure 2. Periodic table of the known homonuclear diatomic molecules. Detection by (*a*) mass spectroscopy, (*b*) UV–VIS spectroscopy in the gas phase, (*c*) UV–VIS spectroscopy in the matrix, (*d*) matrix chemistry, (*e*) Raman spectroscopy in the gas phase or in the matrix, (*f*) laser-induced resonance fluorescence in the gas phase, (*g*) microwave spectroscopy in the gas phase, (*h*) resonance-Raman effect, (*i*) matrix Mössbauer.

Even though bulk diffusion in the crystalline solid state is a well-known phenomenon, rates of diffusion in matrices are difficult to predict quantitatively owing to their nonuniform nature. Nevertheless, only low-energy mechanisms such as vacancy migration and diffusion along grain boundaries need be considered.[15] On the other hand, surface diffusion processes can occur in the liquid-like interface formed on quenching gas beams onto a cold target. These processes, although by no means well understood, turn out to be of central importance to the discussions that follow.

Several authors have reported on the electronic spectra of atoms trapped in matrices that generally show multiplets, the centers of which are usually shifted to energies higher than the energy of the gaseous atom. A number of explanations have been offered to explain the origin of the multiplet structures of the spectral lines, such as removal of orbital degeneracy by low site symmetries in the host lattice, Jahn–Teller effects, interactions between pairs of trapped atoms on non-nearest neighbor sites in the host lattice, and so forth. The generally accepted rationale of the matrix shifts is in terms of a van der Waals interaction between the trapped atom and its matrix host (see Chapter 10 for details).

Other spectral features, usually an order of magnitude less intense than the atomic lines, that escaped reasonable assignments to atomic lines, have sometimes been attributed to dimers or higher metal aggregates. Recently, some detailed investigations[10,12] of the experimental parameters that contribute to the formation of these "aggregates" have begun to appear in the literature.

2. DIFFUSION OF METAL ATOMS IN MATRICES

The isolation of metal atoms in inert-gas matrices is a well-known technique in atomic spectroscopy. A number of metals have been isolated as atoms in inert matrices in order to gain an insight into the effect of the matrix upon the spectra of trapped species and in order to study the formation of dimers, higher polymers, and the diffusion mechanism of metal atoms in low-temperature matrices. Extremely light atoms such as Li[12] and Be[16] have been observed to diffuse readily in rare gas solids below 40°K, whereas heavier atoms such as Mg,[9] Ca,[17] Sn,[18] Pb,[10] Cu,[11] Ag,[7,8,11] Au,[11] and so forth, have much less tendency to diffuse and aggregate after the deposition has been completed. Matrix annealing with heavier atoms often has little effect other than causing the sharpening and/or shifting of spectral lines usually attributable to the increasing regularity and/or phase changes in the host matrix. The low diffusion coefficients of most metal atoms in rare gas solids can be attributed, either or both, to their large masses and/or to van der Waals interaction of M with the matrix. At matrix temperatures the van der Waals energy between large atoms can be several times larger than kT. This essentially increases the size of the diffusing species, it now being a van der Waals molecule that diffuses.

3. ORIGIN OF M_2 MOLECULES

The origin of M_2 molecules is of central concern in investigations of the diffusion properties of metal atoms in inert gas matrices. A detailed knowledge of the aggregation process is also vital to our understanding of the early stages of metal particle formation and the production of binuclear complexes in metal-atom–ligand matrix cocondensation reactions. Experiments that have been designed to give a detailed insight into both the origin of M_2 molecules and their respective M_2L_n complexes are described in the following sections.

4. MATRIX SYNTHESIS AND CHARACTERIZATION OF M_2 MOLECULES

It has recently been discovered[19] that metal dimers can be formed readily in large quantities under matrix conditions and that the bands associated with these entities can be readily identified using a novel technique that makes use of the fact that the metal atoms being deposited are capable of diffusion either on the matrix surface or within a narrow region (the reaction zone) near the surface before its kinetic energy is dissipated sufficiently to immobilize it. This process is amenable to analysis in terms of a kinetic model assuming that at constant deposition rate of matrix material a steady state exists in the concentrations of the various species within the moving reaction zone. Metal dimers and higher aggregates form within the reaction zone to a much higher extent than one would calculate from a statistical analysis based upon the matrix ratio. The dependence of the concentration of the various species upon the rate of deposition of metal can then be calculated and compared with the experimentally observed dependence of the absorbance of bands on metal deposition rate.

Using this technique, the dimers of V,[20] Cr,[21] Mo,[22] Cu,[23] and Ag[24] have been formed and identified from their uv-visible spectra. In this section, we use chromium to illustrate this technique.[21]

The crucial aspect of this experiment is the deposition of metal at an accurately known and constant rate. A quartz crystal microbalance is used for this purpose. Argon is deposited at approximately 2 mmole/hr, and chromium is deposited from a Knudsen cell at various rates ranging from a few parts per thousand to a few percent of the argon rate.

Chromium Atoms, Cr

Figure 3 shows a portion of the uv-visible spectrum of chromium in argon. As the rate of chromium metal deposition is increased, the features at 260 and 455 nm grow in (Figure 3*a*–*c*). Figure 3*a* was obtained by depositing Cr into argon at such a slow rate that mainly chromium atoms are found in the

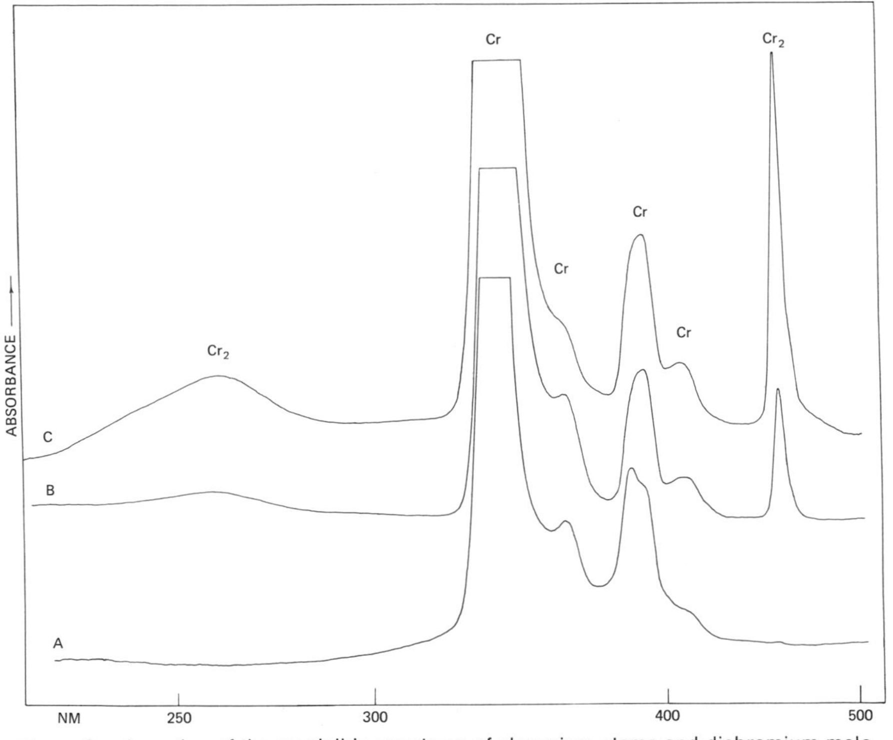

Figure 3. A portion of the uv-visible spectrum of chromium atoms and dichromium molecules in argon. (*a*) low metal flow; (*b* and *c*) progressively higher metal flow.

matrix. Table 1 compares the frequencies of the lines measured for Cr in argon with those of the gaseous atom.[25] A blue matrix shift of approximately 2000 cm^{-1} is observed for all the lines.

Diatomic Chromium, Cr_2

Unambiguous assignment of the broad band at 260 nm has not been made. However, it is known that at least the band at 455 nm belongs to Cr_2 as a result of the kinetic analysis summarized below.

Table 1. Frequencies and Assignments for Matrix-Isolated Cr and Cr_2 in Argon

$\nu_{\text{Ar matrix}}$	ν_{gas}[a]	$\Delta\nu = \nu_g - \nu_{\text{Ar}}$	Assignment	Species
29,400 (vs)	27,935	~ -1600	$a^7S \rightarrow y^7P^0$	Cr
	27,820			
	27,728			
26,900 (w)	24,971	~ -1700	$a^7S \rightarrow z^7F^0$	Cr
	25,010			
	25,089			
	25,206			
	25,359			
	25,548			
	25,771			
25,800 (s),	23,499	~ -2200	$a^7S \rightarrow z^7P^0$	Cr
25,500 (s)	23,386			
	23,305			
24,300 (vw)	—	—	?	Cr
38,500	—	—	—	Cr_2
22,000	—	—	—	Cr_2

[a] Reference 25.

Consider the reaction

$$M \underset{k_1}{\overset{M}{\rightarrow}} M_2 \underset{k_1}{\overset{M}{\rightarrow}} M_3$$

which is assumed takes place in the reaction zone. At constant metal deposition rate, R_0, a steady state exists in the concentrations of M, M_2, and M_3 given by the equations

$$\frac{dM}{dt} = R_0 - k_1M^2 - k_2MM_2 - rM = 0 \tag{1}$$

$$\frac{dM_2}{dt} = \tfrac{1}{2}k_1M^2 - k_2MM_2 - rM_2 = 0 \tag{2}$$

$$\frac{dM_3}{dt} = k_2MM_2 - rM_3 = 0 \tag{3}$$

where r is related to the rate at which these species are frozen out and is proportional to the argon deposition rate. At low metal deposition rates one can show that the ratio of dimer concentration to atom concentration is proportional to R_0, while the ratio of trimer concentration to atom concentration is proportional to R_0^2.

Figure 4 shows a plot of log A_{455}/A_{390} versus log R_0 together with a line of unit slope, where A_{455} and A_{390} are, respectively, the absorbances of the unknown 455 nm line and the 388–392 nm Cr atom doublet. The excellent correlation between the unit slope line and the experimental points suggests that the 455 nm line belongs to Cr_2 and *not* to a higher aggregate.

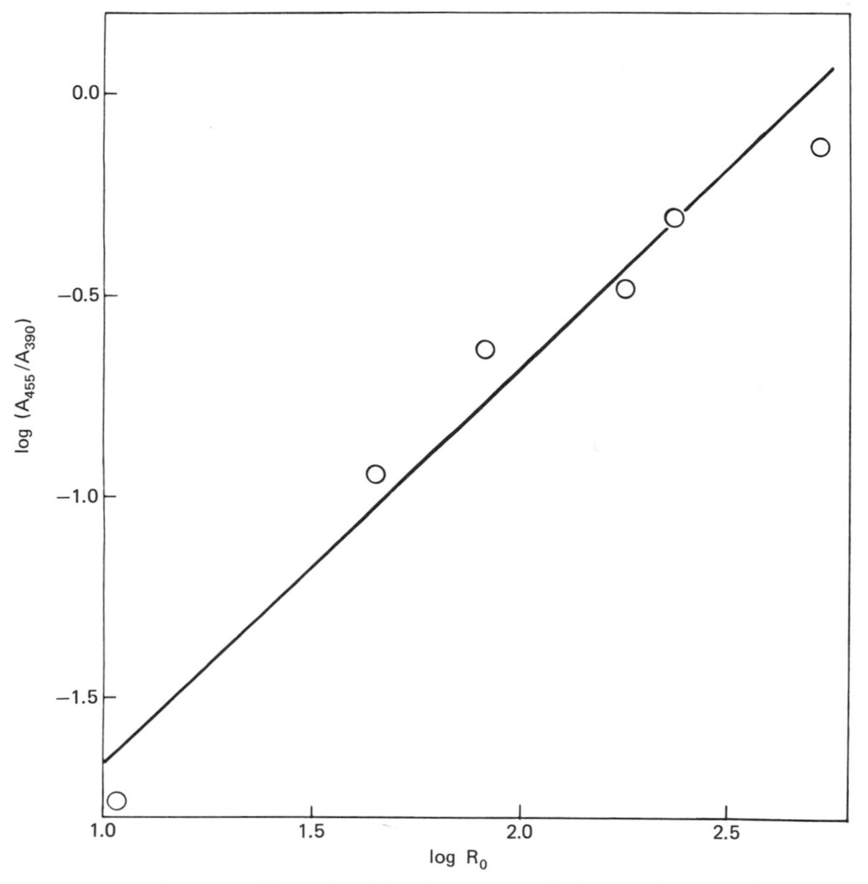

Figure 4. A log-log plot of the ratio of the absorbance of a line attributed to Cr_2 to that of a Cr resonance absorption as a function of the chromium metal deposition rate at constant argon deposition rate.

The apparent absence of Cr_3 could be a result of a low extinction coefficient, or a low k_2 value, or both. The latter reason may be a result of the activation energy required in the reaction

$$Cr_2 + Cr \rightarrow Cr_3$$

in which the Cr_2 bond may have to be weakened in order for the reaction to proceed. The kinetic result, incidentally, also suggests that Cr_2 is formed in or on the matrix rather than in the gas phase. This last fact is also borne out by the intriguing observation that for a given metal deposition rate the dimer/monomer ratio decreases as one increases the molecular weight of the noble gas used to isolate them, with argon giving the most dimers and xenon the least (see later).

5. EXTINCTION COEFFICIENT MEASUREMENTS FOR M AND M_2 SPECIES

In order to establish spectroscopically the relative proportions of metal atom to metal dimer in a matrix cocondensation reaction, it is necessary to know the extinction coefficients of metal atom M and the corresponding M_2 molecule. This can be achieved from a careful study of the electronic spectra of the metal atom deposited in a solid inert matrix as a function of the metal concentration in the matrix.

Since the M_2 molecules generally account for less than 1% of the metal vapor during deposition, at very low metal concentrations only metal atoms should be isolated. We, therefore, can write

$$A_M = \varepsilon_M n_M \tag{4}$$

where n_M is the number of M atoms per unit area, ε_M is the extinction coefficient of M, and A_M is the absorbance of M.

Experiments are then conducted at higher metal concentrations where isolation of M atoms and M_2 molecules is achieved. Under these conditions we can write

$$A'_M = \varepsilon_M n'_M \tag{5}$$

and

$$A'_{M_2} = \varepsilon_{M_2} n'_{M_2} \tag{6}$$

where n'_M and n'_{M_2} are the number of M atoms and M_2 molecules per unit area, ε_{M_2} is the extinction coefficient of M_2, and A'_M and A'_{M_2} are the absorbances of M and M_2, respectively. Both sets of experiments are designed so that the total metal concentration is retained constant. Thus

$$n'_M + 2n'_{M_2} = n_M \tag{7}$$

From equations (4)–(7) we can solve for the ratio of the extinction coefficients, yielding:

$$\frac{\varepsilon_M}{\varepsilon_{M_2}} = \frac{1}{2}\left(1 - \frac{A'_M}{A_M}\right)\frac{A_M}{A'_{M_2}} \tag{8}$$

Although in principle two experiments are sufficient to evaluate $\varepsilon_M/\varepsilon_{M_2}$, it is found in practice that due to the error sources in the experiment (for example, small fluctuations in tip temperature, gas and metal deposition rates, etc.), as well as in the absorption measurements (for example, weak bands, rising base lines due to scattering, etc.), a series of experiments yields a more realistic average $\varepsilon_M/\varepsilon_{M_2}$ value.†

For example, experiments of this type were performed for the two most intense and well-resolved bands of V_2 (see, for example, Figure 5), absorbing at 587/494 nm in Ar.[20] The extinction coefficient data yield an average value for $\varepsilon_V/\varepsilon_{V_2}$ of 0.465/0.510 for the 587/494 nm lines in Ar with respect to the 471 nm resonance absorption of atomic V. Although these ratios seem quite small, one must remember that the 471 nm resonance absorption of atomic V is one of the weakest lines in the optical spectrum and was employed only to improve the accuracy of the absorbance measurements. More realistically, $\varepsilon_V/\varepsilon_{V_2}$ should be quoted with respect to the most prominent V atomic line (314 nm in Ar), which yields a $\varepsilon_V/\varepsilon_{V_2}$ value in the range of 2–4.

It is found that the matrix dependence of the $\varepsilon_V/\varepsilon_{V_2}$ ratio does not vary by more than a factor of about two on passing from Ar to Kr, a not unreasonable observation in view of the comparable but weak guest-host van der Waals interactions for atoms entrapped in the noble gas solids.[20]

Finally, it is worth noting that Andrews and Pimentel[26] have reported such data for lithium in inert gas matrices from which they calculated a lower limit of about 300 for $\varepsilon_{Li}/\varepsilon_{Li_2}$. The trend so far is to somewhat lower values for the transition metals compared to lithium, for example, for selected lines of medium absorbance,

$$\frac{\varepsilon_{Cr}}{\varepsilon_{Cr_2}} \simeq 4.0 \qquad \frac{\varepsilon_{Mo}}{\varepsilon_{Mo_2}} \simeq 15.0$$

$$\frac{\varepsilon_{Cu}}{\varepsilon_{Cu_2}} \simeq 13.0 \qquad \frac{\varepsilon_{Ag}}{\varepsilon_{Ag_2}} \simeq 2.0$$

$$\frac{\varepsilon_{V}}{\varepsilon_{V_2}} \simeq 3.0 \qquad \frac{\varepsilon_{Rh}}{\varepsilon_{Rh_2}} \simeq 5.0$$

Nevertheless, the results strongly imply that the electronic spectrum of a metal atom is considerably more intense than the spectrum of the respective

† It is noteworthy that this technique can in principle be extended to any number of metal clusters. For example, if one can obtain M_3 in the presence of M and M_2, then three different metal concentration experiments allow one to solve for $\varepsilon_M/\varepsilon_{M_3}$ (same notation as above) using the expression:

$$\frac{\varepsilon_M}{\varepsilon_{M_3}} = \frac{1}{3}\frac{A_M}{A''_{M_3}}\left[1 - \frac{A''_M}{A_M} - 2\cdot\frac{A'_M}{A_M}\cdot\frac{A''_{M_2}}{A'_{M_2}}\right]$$

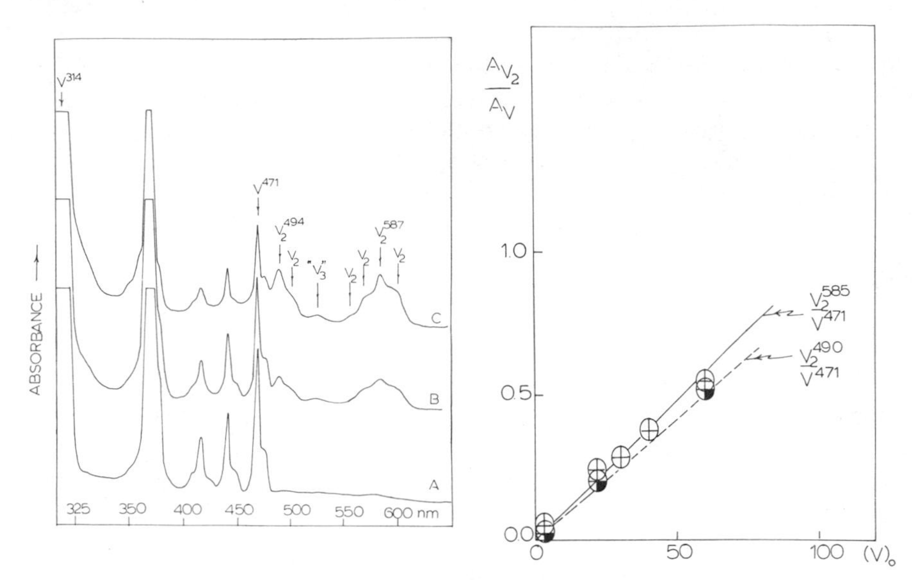

Figure 5. The uv-visible spectrum of V in Ar matrices at 6–10°K, (*a*) at low V concentrations showing isolated V atoms; (*b* and *c*) at progressively higher V concentrations, showing both V atoms and the growth of V_2 molecules; and a plot of the absorbance ratio of typical lines attributed to V_2 to that of a V resonance absorption as a function of the V metal deposition rate at constant Ar deposition rate.

M_2 molecule under comparable conditions. Therefore, weak broad spectral features, which are often observed in the presence of extremely intense atomic absorptions and have generally been regarded as parasitic and their origin not pursued, should not be overlooked as they could well reflect the presence of substantial concentrations of M_2 molecules.

6. THE EFFECT OF THE ATOMIC WEIGHT OF M ON THE DIMERIZATION PROCESS

The data collected for the first-row transition metal atoms indicate quite clearly that dimerization can occur readily during matrix condensation at metal-matrix ratios greater than 1/1000 and can be attributed to a surface diffusion process occurring during deposition. One might have anticipated that on proceeding to the heavier transition metals, dimerization would be less facile (under comparable conditions) on at least two accounts. First, the rate of diffusion of a metal atom under matrix conditions has a mass dependence of between $M^{-1/2}$ and M^{-1}, depending on whether the diffusion layer is more closely approximated by a gas-like or a liquid-like region, which would reduce the tendency of the heavier transition metal atoms to dimerize compared to the first-row metals. Second, as the heavier metals are more polarizable, van der Waals type interaction with the matrix material will increase, thereby tending to stabilize atoms rather than M_2 molecules. In the absence of "special electronic effects," it appears that dimerization is less pronounced for the heavier transition metals, as borne out by recent experiments with Cu/Ag,[24] Mn/Re,[19] and Co/Rh/Ir.[27] For example, the electronic spectra of Cu and Ag in argon as a function of metal concentration were investigated initially to establish the existence of Cu_2 and Ag_2 molecules.[24] Techniques similar to those described for V_2 and Cr_2 were employed, and the results are displayed in Figures 6 and 7. Having identified the uv-visible spectral absorptions of Cu_2 and Ag_2, the experiments were repeated under identical conditions of total metal concentration. Using conditions that favored large amounts of Cu_2, it was found that the only observable reaction product in the case of silver was atomic silver. Reasonable quantities of Ag_2 could only be induced to form by increasing the Ag concentration by a factor of at least 10.

7. THE EFFECT OF THE MATRIX MATERIAL ON THE DIMERIZATION PROCESS

The parameters that control the degree of isolation of unstable molecules, free radicals, and molecular ions are well documented. Less well understood are the properties of the matrix that can affect the isolation of metal atoms and diatomic molecules.

To gain some insight into this problem, the uv-visible spectra of Cr[21] and Cu[23] in various inert gas matrices were investigated. The results of a typical series of experiments for V/V_2 are shown in Figure 8. Similar data for Cu/Cu_2 in Ne and Ar are shown in Figure 9, from which it can be seen that the efficiency of isolation of atoms follows the order, Xe $>$ Kr $>$ Ar $>$ Ne (a result observed for other metals and by other authors[10,16,19,28]). This is understood to be a consequence of the formation of van der Waals complexes between the metal and the noble gas atom, with the most polarizable noble gas forming the strongest complexes. A metal atom complexed in this fashion can either be considered to be quenched or at least its diffusion slowed down considerably.

Few other matrix supports have been investigated from this point of view, although preliminary studies with SF_6[10] show that it is ideal for isolating Pb_2 molecules, implying that the polarizability of the fluorine atoms in SF_6 is very small. SF_6 has also been used as a matrix for studies of the electronic spectra of Cu, Ag, and Au atoms,[11] where the degree of isolation of atoms appears to be greatest for Au, which is not unexpected in the light of our earlier discussions of metal atom diffusion as a function of its atomic weight and polarizability.

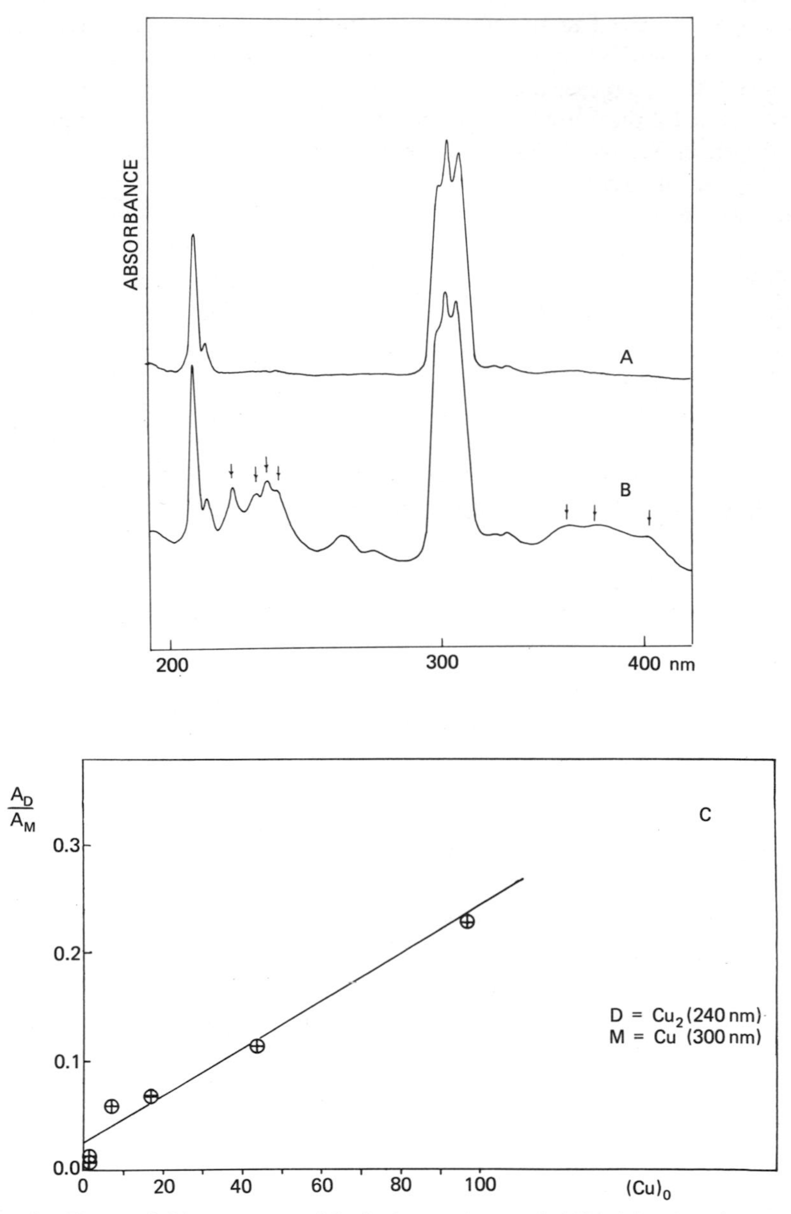

Figure 6. The uv-visible spectrum of Cu in Ar matrices at 6–10°K, (*a*) at low Cu concentrations showing isolated Cu atoms; (*b*) at higher Cu concentrations showing, in addition to atomic Cu, new spectral features that can be assigned to Cu_2 (marked with arrows); and (*c*) a plot of the absorbance ratio of typical lines attributed to Cu_2 to that of a Cu resonance absorption as a function of the Cu metal deposition rate at constant Ar deposition rate.

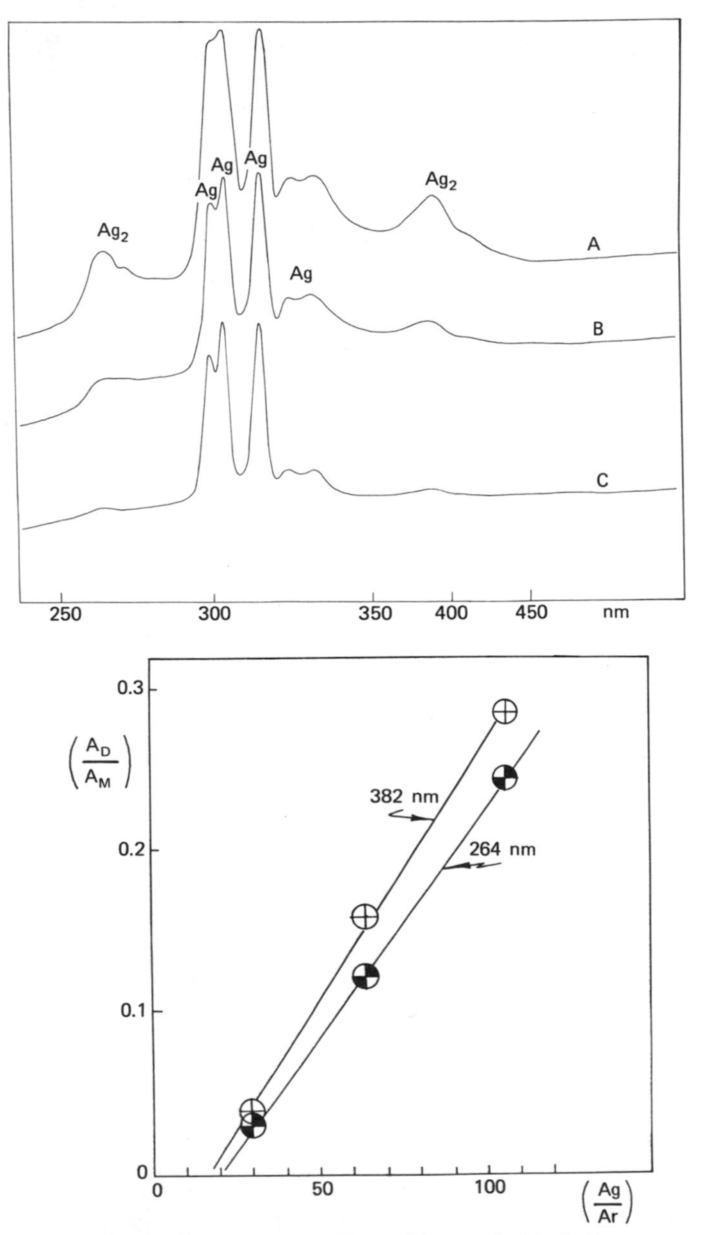

Figure 7. The same as Figure 6 but applied to Ag/Ag_2.

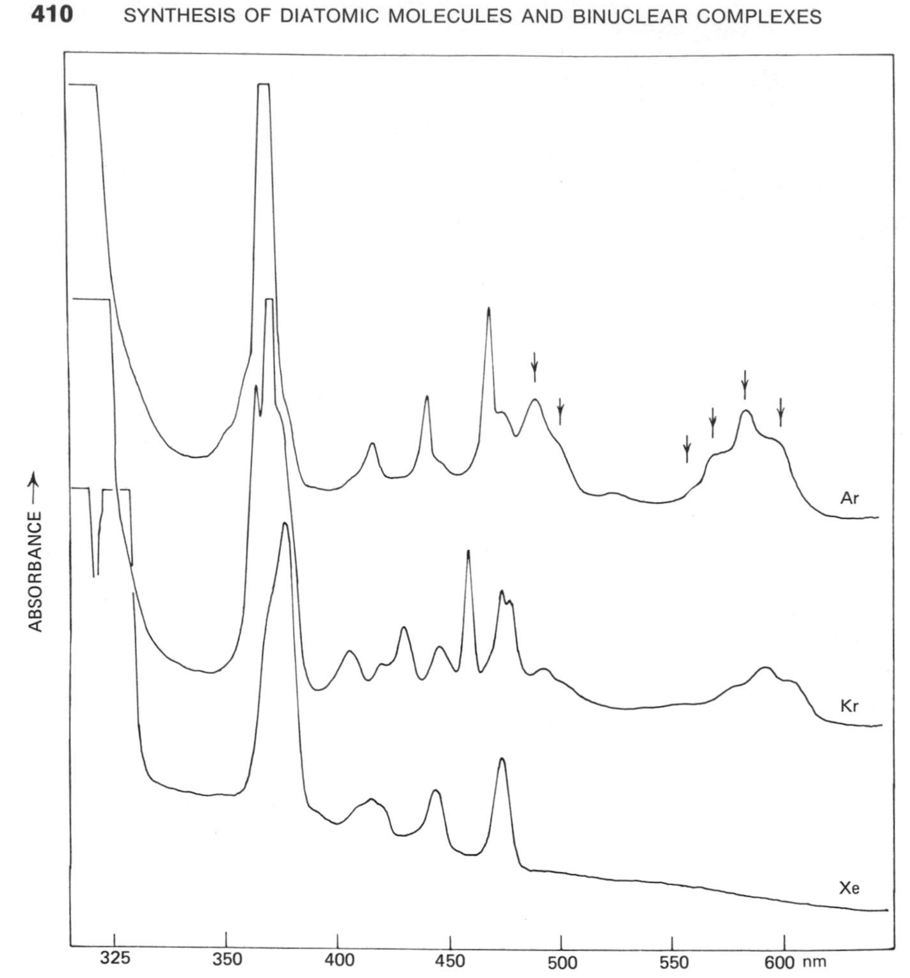

Figure 8. The uv-visible spectrum of V atoms deposited into Ar and Kr and Xe matrices under identical conditions of temperature, V metal deposition rate, and inert gas deposition rate.

8. ELECTRONIC PROPERTIES OF SOME TRANSITION METAL DIATOMIC MOLECULES

Quantum-mechanical calculations of the electronic properties of metal clusters as a function of the cluster size are proving to be particularly important for understanding, for example, the early stages of metal nucleation and the reactivity of metal particles in heterogeneous catalytic processes.

Considerable progress has been made since the early calculation of Hoffman[29] on Cu_2. Hoffman first demonstrated the usefulness of extended

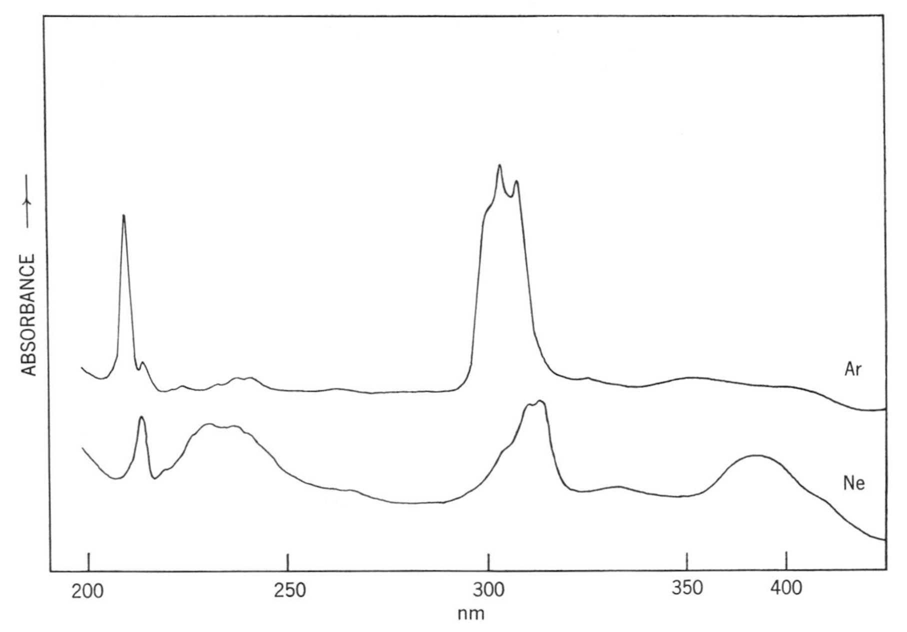

Figure 9. The uv-visible spectrum of Cu atoms deposited into Ne and Ar matrices under identical conditions of temperature, Cu deposition rate, and inert gas deposition rate.

Hückel techniques for providing a description of the bonding and potential energy surface of a transition metal diatomic molecule. For example, Cooper et al.[30] carried out extended Hückel calculations on the diatomic molecules of the first-row transition metals, and Baetzold[31] has extended these studies still further to include extended Hückel and complete neglect of differential overlap (CNDO) calculations on a number of second- and third-row transition metal diatomics as well as a number of metal clusters, M_n, where $n = 2\text{–}55$.[32] Recently, too, *ab initio* self-consistent field molecular-orbital calculations have been attempted by Joyes et al.[33] for Cu_2 and generally support the conclusions drawn from extended Hückel[34] and CNDO[31] approaches, namely, that the main bonding contribution arises from 4*s*—4*s* overlap in a Cu—Cu σ-bond.

For the purposes of this brief discussion, we consider the uv-visible spectral properties of V_2, Cr_2, Cu_2, and Ag_2 in terms of the available results of semiempirical molecular orbital calculations.

Cooper et al.[30,34] have reported the results of some extended Hückel calculations for V_2, from which they deduce the level ordering to be $\sigma_{3d}^2\pi_{3d}^4\delta_{3d}^4$, having a ${}^1\Sigma_g^+$ electronic ground state. Their data imply that the 4*s* orbital contributes negligibly to the V—V bond in V_2 and that the main bonding interaction originates from $3d_z^2$ overlap. This is a surprising conclusion in view of the high bond dissociation energy of V_2 (57 kcal/mole)[35] and the small ($\sim$1 eV) energy separation between the $4s/3d_z^2$ orbitals of atomic vanadium.

The V_2 molecule has been reinvestigated[20] using extended Hückel techniques with charge iteration. These calculations arrived at a ground state configuration and bonding description for V_2 at variance with those previously reported,[30] even though the two sets of molecular and electronic parameters and approximations employed were essentially the same.

In an attempt to understand the origin of the discrepancy, a series of calculations for V_2 was performed in which the atomic $3d$ levels were retained constant at 7 eV and the $4s$ level was varied between 10 and 5 eV. The V—V internuclear distance was held at 2.30 Å. Clementi–Raimondi orbital coefficients[36] and the Cusach's approximation[37] for the calculation of the off-diagonal resonance integrals were employed with the splitting parameter, K, retained at 2 throughout the calculations. This approach has produced reasonable results for other transition metal diatomic and cluster molecules.[30–32,34]

Some of the results are summarized below and in Figure 10, which shows the situation for the two cases of the $4s$ orbital lower and higher in energy than the $3d$ orbitals.

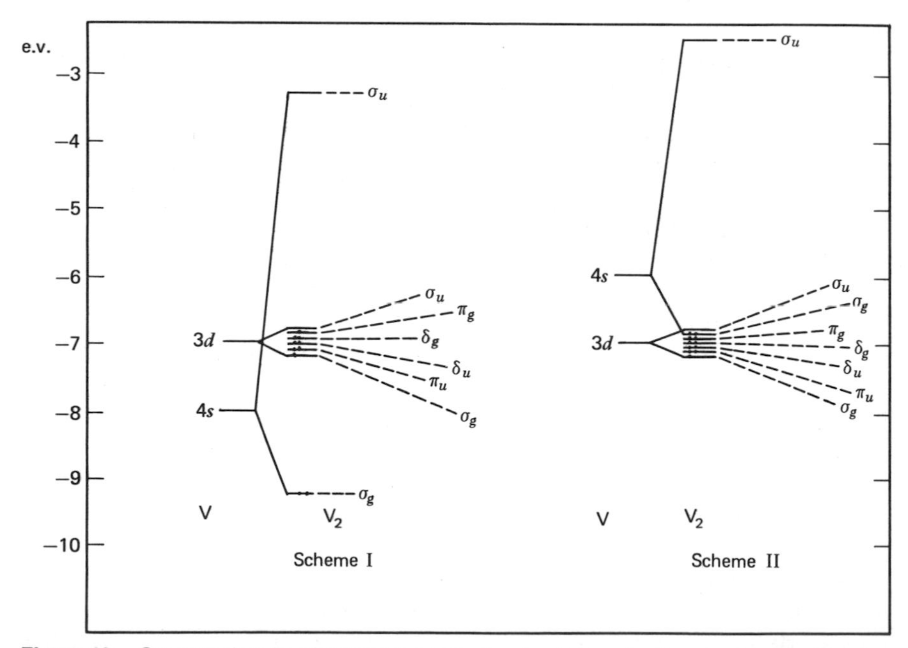

Figure 10. Computed molecular orbital energy level schemes for V_2 using charge iterative extended Hückel molecular orbital techniques.

Scheme I. When the 4*s* orbital is lower in energy than the $3d_z^2$, the following parameters are obtained:

Ground state:	High spin
Total overlap population:	0.837
Orbital overlap populations:	
$4s - 4s$	0.737
$3d_z^2 - 3d_z^2$	0.054
$3d_{xz,yz} - 3d_{xz,yz}$	0.044

Under these circumstances, the bonding is adequately described by a strong $4s - 4s$ σ interaction with negligible $4s3d_z^2$ overlap and minimal π-bonding contributions to the V—V bond. The optical spectrum anticipated for the more reasonable high spin[55] situation should contain at least two low-energy, dipole, and spin-allowed transitions:

$$\sigma_g \rightarrow \pi_u$$

$$\sigma_g \rightarrow \sigma_u$$

in the 16,000–20,000 cm^{-1} region as well as a number of other transitions to higher energies.

Scheme II. When the 4*s* orbital is higher in energy than the $3d_z^2$, the following parameters are obtained:

Ground state:	High spin
Total overlap population:	0.162
Orbital overlap populations:	
$4s - 4s$	0.049
$3d_z^2 - 3d_z^2$	0.050
$3d_{xz,yz} - 3d_{xz,yz}$	0.044
$3d_z^2 - 4s$	0.003

It would appear that this set of circumstances resembles most closely those reported by Cooper et al.,[30,34] although it is disturbing that the parameters used to generate his data are similar to those used to generate the molecular orbital picture of scheme I.[24]

However, the results of scheme II indicate a substantially weaker V—V interaction than obtained in scheme I, where the bonding originates from a mixture of $4s - 4s$ and $3d_z^2 - 3d_z^2$ σ overlaps, a comparable amount of π-bonding, and a small degree of $4s3d_z^2$ overlap. Moreover, only one spin-allowed optical transition is anticipated in scheme II, which by comparison with scheme I is likely to occur at *high* energies. This seems to be an unlikely description of the V_2, molecule and it would appear that scheme I is more suitable.

Similar calculations have been performed on Cr_2[22] and yield a molecular-orbital energy-level scheme not unlike that obtained for V_2. The bonding in Cr_2 is again adequately described by a strong $4s - 4s\ \sigma$ interaction; the optical spectrum should again display two low-energy dipole and spin-allowed transition, namely, $\sigma_g \rightarrow \pi_u$ and $\sigma_g \rightarrow \sigma_u$. At least one of these transitions is observed at 455 nm, the other could be obscured by the group of atomic Cr lines around 400 nm (Figure 3). Furthermore, the broad feature at 260 nm is probably best assigned to the predicted high-energy $\sigma_g \rightarrow \sigma_u$ transition indicated in Figure 10.

The electronic spectra obtained for Cu_2 and Ag_2 are tabulated below:

Cu_2 (nm)	Ag_2 (nm)	Assignment
~400	~380	$\Sigma_g^+ \rightarrow \Sigma_u^+$
~230	~270	$\Sigma_g^+ \rightarrow \Pi_u$

Using the results of extended Hückel and CNDO molecular-orbital calculations,[30,31,34] the energy-level schemes shown in Figure 11 can be constructed. Both sets of calculations predict that the lowest energy $\Sigma_g^+ \rightarrow \Sigma_u^+$ transition should be at slightly higher energies for Ag_2 compared to

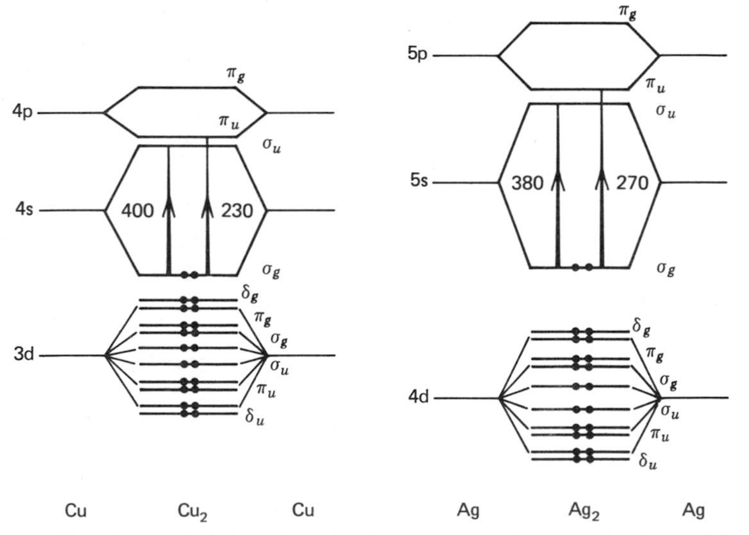

Figure 11. Schematic molecular-orbital energy-level diagrams for Cu_2 and Ag_2.

Cu_2, in agreement with the observed lines at about 380 and 400 nm, respectively. From the similarity of the $\sum_g^+ \rightarrow \sum_u^+$ transition energies, one can conclude that the $4s\sigma$ overlap in Cu_2 should not differ significantly from the $5s\sigma$ overlap in Ag_2. This proposal is roughly in line with known gas-phase bond dissociation energies, $D_0(Cu_2) = 45.5$ and $D_0(Ag_2) = 37.6$ kcal/mole.

9. CHEMICAL EVIDENCE FOR BINUCLEAR COMPLEX FORMATION

Most of the information concerning matrix reactions leading to binuclear transition metal complexes has been derived from metal-atom–carbon monoxide cocondensation reactions. Evidence for the formation of binuclear products comes from the following sources:

a. Observations of known binuclear carbonyl complexes,

b. Synthesis of new binuclear complexes,

c. Observation of bridging CO stretching modes,

d. Observation of a mononuclear carbonyl complex in the presence of its binuclear counterpart,

e. Ligand and metal concentration studies,

f. Ligand and mixed metal isotope-substitution studies,

g. Codeposition of two different metals.

The chemical evidence for the matrix formation of binuclear complexes can be formalized by discussing each of the above experiments in turn, with reference to selected examples.

9.1 Observation of Known Binuclear Complexes

When Mn or Co atoms were deposited with pure $^{12}C^{16}O$ at M/CO ratios of about 1/1000, the matrix infrared spectra showed the presence of mainly $Mn_2(CO)_{10}$[38] and $Co_2(CO)_8$.[39] The identification of these reaction products was established by comparing the spectra with those obtained by depositing $Mn_2(CO)_{10}$ and $Co_2(CO)_8$ in Ar matrices using gas-flow Knudsen cell techniques (Figures 12 and 13). The similarity of the two sets of data proves, therefore, that manganese and cobalt, under the conditions of a matrix cocondensation experiment, can be induced to yield exclusively binuclear compounds.

9.2 Synthesis of New Binuclear Complexes

The matrix reactions of transition metal atoms can, in principle, provide chemical pathways to the synthesis of new binuclear complexes that cannot

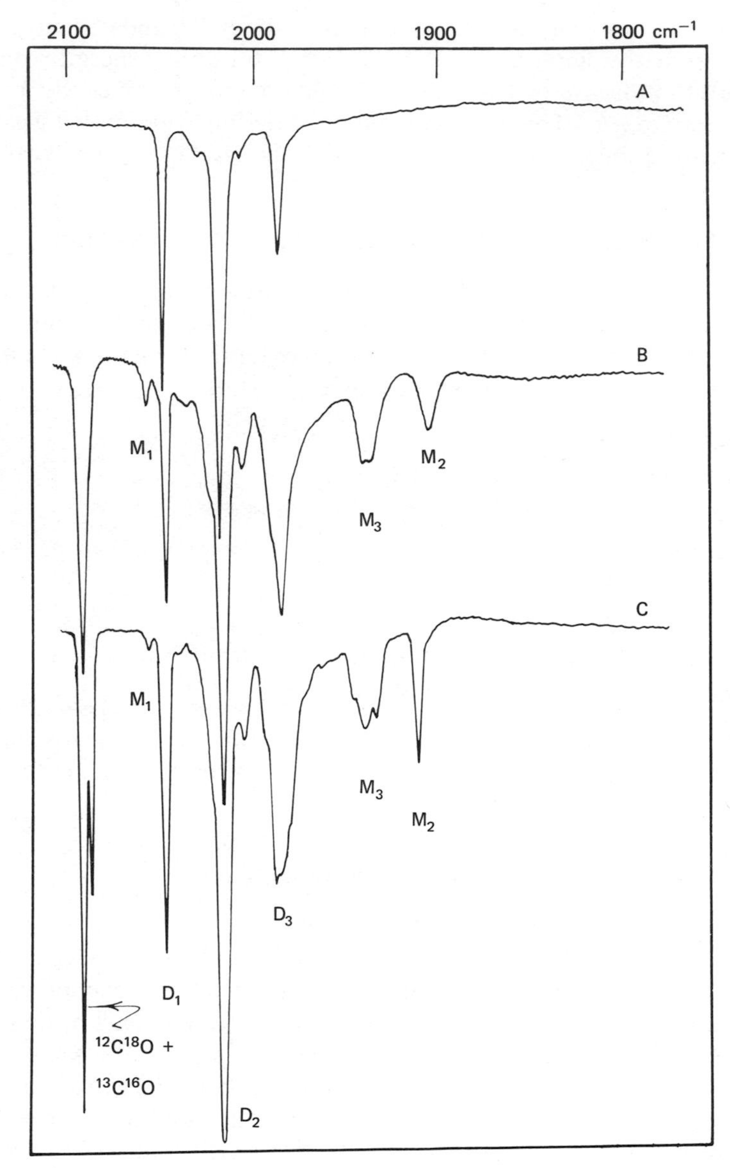

Figure 12. The infrared spectra of (*a*) $Mn_2(CO)_{10}$ isolated in Ar (approximately 1:1000), and (*b*) and (*c*) the products of the matrix cocondensation reaction of Mn atoms with CO:Ar $\simeq$ 1:5 and pure CO, respectively, showing $Mn_2(CO)_{10}$ (D) and $Mn(CO)_5$ (M) [the splitting of M_3, the E mode of $Mn(CO)_5$, was most pronounced in pure CO and is assigned to a matrix-site effect].

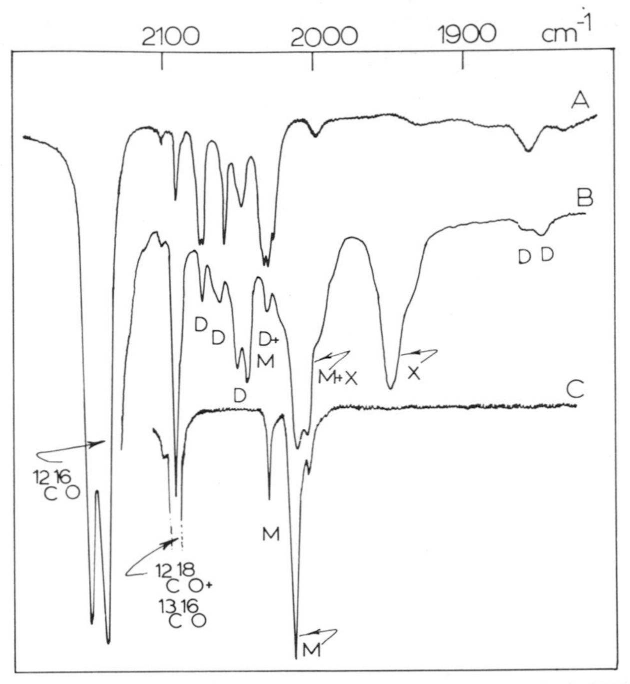

Figure 13. The infrared spectra of (*a*) $Co_2(CO)_8$ in Ar (approximately 1:1000), (*b*) and (*c*) the products of the matrix cocondensation reaction of Co atoms with CO at Co:CO $\simeq$ 1:1000 and Co:CO $\simeq$ 1:10000, showing the presence of $Co(CO)_4$ (M) and $Co_2(CO)_8$ (D), respectively [the lines marked X belong to lower cobalt carbonyls $Co(CO)_n$].

be prepared by conventional methods of chemical synthesis. The carbonyl complexes of the Co, Rh, Ir group of metals can be used to illustrate the potential usefulness of the method.

Following the synthesis of $Co(CO)_4$,[39] the analogous radicals, $Rh(CO)_4$ and $Ir(CO)_4$, were synthesized from the matrix reactions of Rh and Ir atoms with pure CO.[27] The complexes were characterized using $^{12}C^{16}O/^{13}C^{16}O$ mixed-isotope experiments. In a continuation of this work, high Rh and Ir atom concentrations that favor the matrix formation of binuclear complexes were used to synthesize the binuclear carbonyl, $Rh_2(CO)_8$, and the new iridium analogue, $Ir_2(CO)_8$.[40] On the basis of metal concentration dependence (see later) and the number, intensity, and position of the infrared bands (Table 2 and Figures 14 and 15), the bridged form was assigned to these compounds in the matrix. When $Rh_2(CO)_8$ and $Ir_2(CO)_8$ were allowed to warm up on the optical plate under conditions of dynamic vacuum, the

Table 2. Infrared Spectra of $M_2(CO)_8$ (where M = Co, Rh, or Ir)

$Co_2(CO)_8$[a]	$Co_2(CO)_8$[b]	$Rh_2(CO)_8$[a,c]	$Ir_2(CO)_8$[a,c]	$Rh_2(CO)_8$[a,d]
2071 vs	—	2060 s	2095 ms	2086 s
—	2069 vs	—	—	—
2044, 2042 } vs	—	2043 ssh, 2038 s }[e]	2068 sbr	2061 s
—	2031 ms	—	—	—
—	2022 vs	—	—	—
1866sh	—	1852 wsh	1848 wsh	1860 w
1857s	—	1830 mw	1822 m	1845 m

[a] Bridge-bonded form.
[b] Non-bridge-bonded form.
[c] This study.
[d] R. Whyman, *J. Chem. Soc., Dalton*, 1375 (1972).
[e] Appears as a doublet or a singlet (2040 cm^{-1}) depending on the deposition conditions.

infrared spectra showed that a transformation occurred at about −48°C for $Rh_2(CO)_8$ and about −58°C for $Ir_2(CO)_8$ to yield the bridge-bonded forms of $Rh_4(CO)_{12}$ and $Ir_4(CO)_{12}$. This instability of the binuclear rhodium and iridium carbonyls with respect to $Rh_4(CO)_{12}$ and $Ir_4(CO)_{12}$ may be the reason why these analogs of $Co_2(CO)_8$ have escaped synthesis so far. These data again show the potential of the matrix-synthetic method to characterize unstable complexes.

9.3 Observation of Bridging CO-Stretching Modes

Thc bonding in a bridging carbonyl ligand can be pictured as ketonic, which in a valence bond sense would imply a formal oxidation state of +1 for each bridgehead metal. Alternatively, the bonding can be discussed in a more flexible way in which a three-center molecular-orbital scheme is considered involving the 5σ orbital of the CO ligand and the symmetry-allowed σ-type orbital on each metal atom. A similar three-center scheme can be envisaged for the π-bonding description of the molecule in order to correlate the CO-stretching frequencies (or CO-bond-stretching force constants) with the extent of π- and σ-bonding. When part of the same complex, bridging carbonyl groups characteristically absorb at lower infrared frequencies than terminal carbonyl groups. Thus the observation of CO bridge stretching modes is indicative of binuclear complex formation.

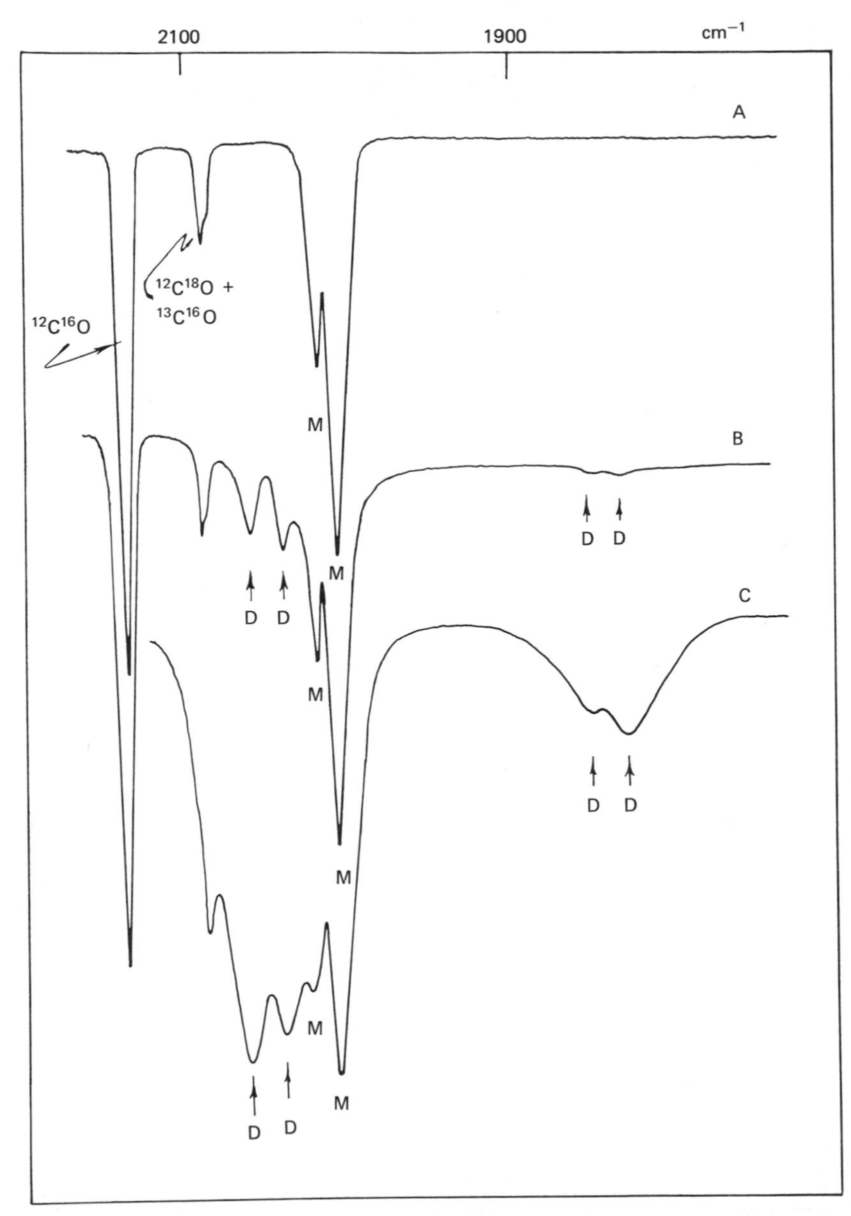

Figure 14. The matrix infrared spectra of the products of the Rh-atom–CO cocondensation reaction at 10°K. (*a*) Rh:CO $\simeq$ 1:10000, (*b*) Rh:CO $\simeq$ 1:1000, and (*c*) Rh:CO $\simeq$ 1:100, where the M indicates lines belonging to $Rh(CO)_4$ and the D indicates lines belonging to $Rh_2(CO)_8$. [On allowing the CO matrix to pump away slowly (ca. 50°K) the lines associated with $Rh_2(CO)_8$ remain up to 225°K, at which temperature $Rh_2(CO)_8$ is observed to disproportionate to $Rh_2(CO)_{12}$, and at room temperature to $Rh_6(CO)_{16}$.]

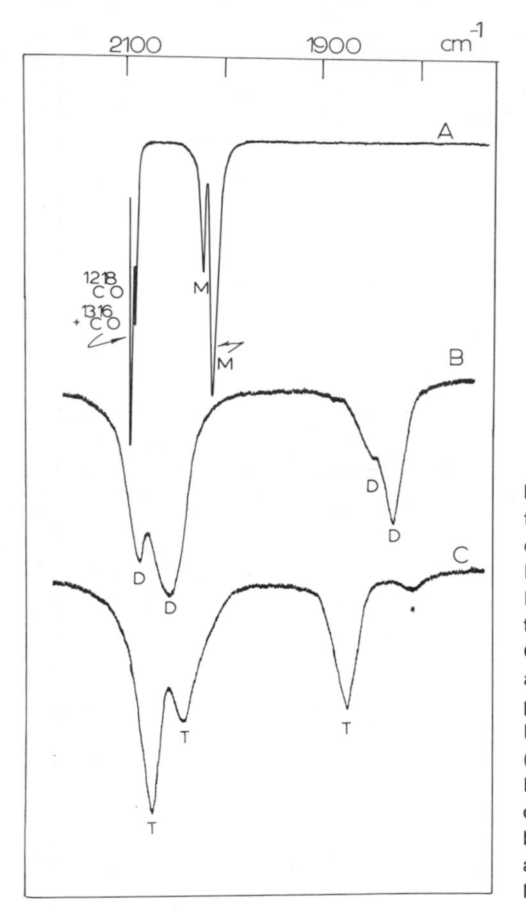

Figure 15. The matrix infrared spectra of the products of the Ir-atom–CO cocondensation reaction at 10°K. (*a*) Ir:CO $\simeq$ 1:10000 showing well-isolated $Ir(CO)_4$. (*b*) The spectrum obtained in the range 10–215°K, after allowing the CO matrix (Ir:CO $\simeq$ 1:100) to pump away slowly (ca. 50°K), showing the presence of spectroscopically pure $Ir_2(CO)_8$ in the bridge bonded form. (*c*) The spectrum obtained on warming $Ir_2(CO)_8$ above 215°K, showing the disproportionation of $Ir_2(CO)_8$ to the bridge-bonded form of $Ir_4(CO)_{12}$. The asterisk indicates a small amount of $Ir_6(CO)_{16}$ growing in.

An example of this comes from the Mn/CO/Ar matrix reaction.[38] When manganese atoms are cocondensed with $^{12}C^{16}O$:Ar = 1:250 using Mn deposition rates that favor the formation of Mn_2 (Figure 16*a*), two prominent low-frequency CO absorptions are observed at 1835.4 and 1688.2/1689.0 cm^{-1} (the latter was shown to be a matrix-site splitting), together with a high-frequency weak absorption at 1964.4 cm^{-1}. Using an isotopic mixture, $^{12}C^{16}O$:$^{13}C^{16}O$:Ar = 1:1:500, the matrix infrared spectrum shown in Figure 16*b* was obtained in which a characteristic isotope pattern of a symmetrical dicarbonyl and a low and high frequency monocarbonyl is easily discernible (Table 3).

On the basis of the frequencies of the observed carbonyl stretching modes when compared to $Mn_2(CO)_{10}$ or $Mn(CO)_5$, together with the prior knowl-

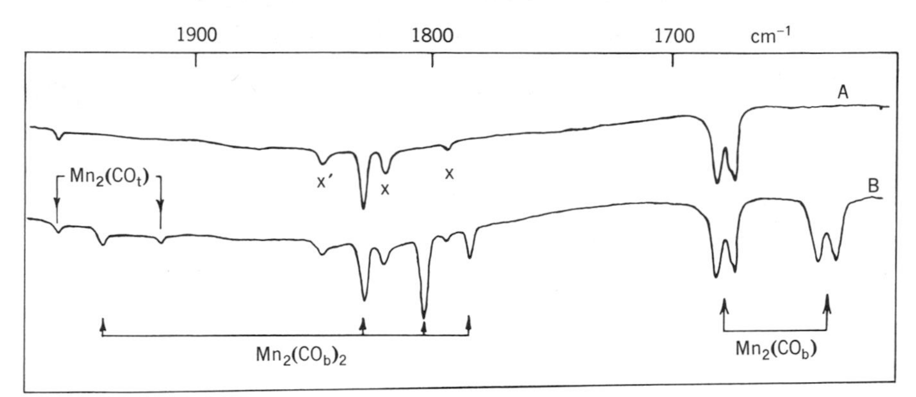

Figure 16. Infrared spectra of the products of the matrix cocondensation reaction of Mn atoms with (*a*) $^{12}C^{16}O:Ar \simeq 1:250$ and (*b*) $^{12}C^{16}O:{}^{13}C^{16}O:Ar \simeq 1:1:500$ at 6–10°K at Mn:CO $\simeq$ 1:1000, showing the formation of $Mn_2(CO_t)$, $Mn_2(CO_b)_2$, and $Mn_2(CO_b)$ [where x′ refers to $Mn(^{12}C^{16}O)$ and x to trace impurities].

Table 3. Infrared Spectra of the Products of the Matrix Reaction of Mn_2 With Dilute Co/Ar Mixtures

Observed	Calculated[a]	Assignment
1964.4	—	$Mn_2(^{12}C^{16}O_t)$
1920.6	—	$Mn_2(^{13}C^{16}O_t)$
1945.6	1945.6	$Mn_2(^{12}C^{16}O_b)(^{13}C^{16}O_b)$
1835.4	1835.3	$Mn_2(^{12}C^{16}O_b)_2$
1809.6	1809.2	$Mn_2(^{12}C^{16}O_b)(^{13}C^{16}O_b)$
1790.6	1791.1	$Mn_2(^{13}C^{16}O_b)_2$
1688.2/1680.0[b]	—	$Mn_2(^{12}C^{16}O_b)$
1646.4/1638.2[b]	—	$Mn_2(^{13}C^{16}O_b)$

[a] Calculated Cotton–Kraihanzel force constants: $Mn_2(CO_b)$, $f_r = 11.45$; $Mn_2(CO_b)_2$, $f_r = 14.61$; $f_{rr} = 1.00$; $Mn_2(CO_t)$, $f_r = 15.58$. Units in mdyn/Å, b = bridging CO group and t = terminal CO group.
[b] Matrix splitting.

edge that Mn_2 should constitute a major proportion of the reacting metal species in the matrix under the experimental conditions used, the low-frequency CO-stretching modes are best assigned to a binuclear bridged monocarbonyl species (I) and a binuclear bis-carbonyl bridged species (II). Whether or not a Mn—Mn bond should be included in a description of their bonding could be elucidated using esr experiments. The monocarbonyl absorbing at high CO-stretching frequencies is assigned to the binuclear species (III).

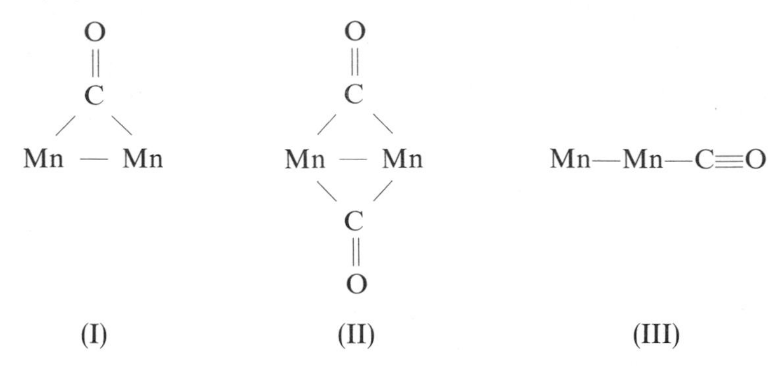

The rise to higher frequencies on passing from $Mn_2(CO_b)$ to $Mn_2(CO_b)_2$ is consistent with a decrease in the charge delocalization of the Mn_2 electrons into the π^* orbitals of the CO bridge ligand(s) concomitant with the increase in the formal oxidation state of the Mn from Mn(I) to Mn(II). The rise to higher frequencies on passing from Mn(CO) to $Mn_2(CO_t)$ can be rationalized using an inductive argument. The charge density associated with the electrons used in the Mn—Mn bond of $Mn_2(CO_t)$ is larger in the Mn electron core of Mn(CO) than of $Mn_2(CO_t)$. The outer *d* electrons of manganese are, therefore, more shielded from the nucleus in Mn(CO), allowing better π-back-bonding, which results in a lower CO-stretching frequency for Mn(CO) compared to $Mn_2(CO_t)$.

9.4 Observation of a Mononuclear Carbonyl Complex in the Presence of its Binuclear Counterpart

The Co/CO matrix reaction exemplifies the above situation. Under Co:CO $\simeq$ 1:10^3 concentration conditions, $Co_2(CO)_8$ can be identified (Figure 13). In addition, a doublet at 2029 and 2011 cm^{-1} is observed. Crichton et al.[41] have independently investigated the matrix photolysis reaction of $Co(CO)_3(NO)$ in argon matrices doped with CO:

$$Co(CO)_3(NO) \xrightarrow[Ar/CO]{h\nu} Co(CO)_4$$

The product of this reaction also shows a doublet at 2029 and 2011 cm^{-1} with an intensity ratio identical to the similar lines observed in the Co/CO reaction. Mixed $^{12}C^{16}O/^{13}C^{16}O$ isotope experiments proved that the product giving rise to the 2029/2011 cm^{-1} doublet in both the matrix photolysis[41] and the Co/CO reactions[39] is one and the same species, namely, the C_{3v} $Co(CO)_4$ radical. Further confirmation that both mononuclear and binuclear cobalt carbonyls can result from the matrix reaction of Co atoms with pure CO stems from experiments at high Co atom concentrations (Co:CO $\simeq$ 1:100) and very low Co atom concentrations (Co:CO $\simeq$ 1:10^5), the infrared of which show mainly $Co_2(CO)_8$ and mainly $Co(CO)_4$, respectively. It is interesting to note that the infrared spectrum of $Co_2(CO)_8$ in this experiment shows bands assignable to both the Co—Co bonded and CO bridge-bonded isomers. The same result is obtained when $Co_2(CO)_8$ is vaporized at 30°C and isolated in an argon matrix at 10°K (Figure 13).

9.5 Assignment of the Mononuclear Species by Inference

When the mononuclear species has not previously been detected, one has to resort to comparative methods of identification, using isostructural and isoelectronic species (or mixed-isotope experiments whenever possible).

For example, when Re atoms are cocondensed with pure CO at very low concentrations, Re:CO $\simeq$ 1:10^5; aside from the absorptions assigned to $Re_2(CO)_{10}$, two new carbonyl absorptions are observed (Figure 17).[42] These are assigned to a C_{4v} pentacarbonylrhenium radical, based on the numbers, frequencies, and relative intensities of the observed infrared active carbonyl stretching modes. The assignment is further strengthened by comparison of the infrared data with those for other known C_{4v} square pyramidal complexes, for example, $Cr(CO)_5$[43] and $Cr(CO)_5^-$.[44]

9.6 The Dependence of the Concentration of Binuclear Species on Metal/Matrix Ratios

Reactive Matrices

One can consider two ways in which binuclear carbonyl species can arise. The matrix reaction between metal atom M and CO gives rise to mononuclear carbonyls, $M(CO)_n$, where $n = 1$–3, and binuclear species, $M_2(CO)_m$, where, for example, $m = 1$–6, as in the case of the Cu/CO system. Species containing higher aggregates of metal are, in general, also possible. We assume, however, that the matrix ratio, $[M]_0$, never becomes high enough to result in the formation of appreciable quantities of these. The surface reaction and the statistical factor will be considered separately.

Statistical Concentrations of Dimers. Assuming that aggregates larger than dimers are in negligible abundance and that there are 12 nearest neighbor sites associated with every substitutional site in the CO matrix, one can write down the dimer complex and monomer complex concentrations directly:

$$[M_2(CO)_6] = 12([M]^2)(1 - [M])^{11}$$

and

$$[M(CO)_3] = [M](1 - [M])^{12}$$

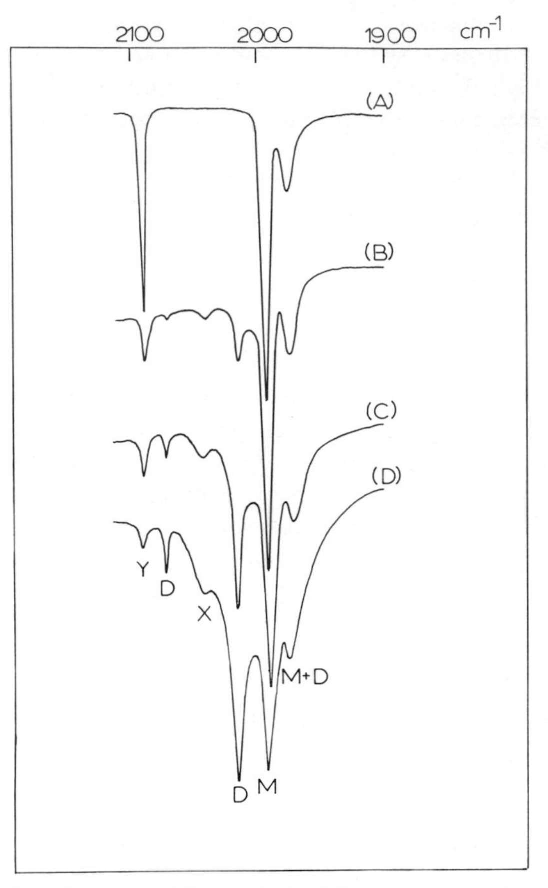

Figure 17. The infrared spectra of the products of the matrix cocondensation reaction of Re atoms with CO at 6–10°K. (*a*) Re:CO $\simeq$ 1:100000, (*b*) Re:CO $\simeq$ 1:10000, (*c*) Re:CO $\simeq$ 1:1000 and (*d*) Re:CO $\simeq$ 1:100, showing the formation of both $Re(CO)_5$ and $Re_2(CO)_{10}$ (M = $Re(CO)_5$; D = $Re_2(CO)_{10}$; X = unidentified, higher aggregate species; Y = $^{13}C^{16}O$, $^{12}C^{18}O$ in natural abundance).

therefore,

$$\frac{[M_2(CO)_6]}{[M(CO)_3]} = \frac{12[M]}{1 - [M]}$$

which reduces to approximately 12[M] for [M] not too large.

Surface Reaction. We assume that only CO and M are mobile species; accordingly, only those reactions that involve at least one mobile species are admitted. This implies that reactions of the form

$$M(CO)_n + M(CO)_{n'} \rightarrow M_2(CO)_{n+n'}$$

are inadmissable. The total reaction scheme for the formation of $M(CO)_3$ and $M_2(CO)_6$ is, therefore, as follows.

$$\begin{array}{ccccccccc}
M & \xrightarrow[CO]{k_1} & MCO & \xrightarrow[CO]{k_2} & M(CO)_2 & \xrightarrow[CO]{k_3} & M(CO)_3 & & \\
\downarrow{\scriptstyle M\,|\,k_1''} & & \downarrow{\scriptstyle M\,|\,k_2''} & & \downarrow{\scriptstyle M\,|\,k_3''} & & \downarrow{\scriptstyle M\,|\,k_4''} & & \\
M_2 & \xrightarrow[CO]{k_1'} & M_2CO & \xrightarrow[CO]{k_2'} & M_2(CO)_2 & \xrightarrow[CO]{k_3'} & M_2(CO)_3 & \xrightarrow[CO]{k_4'} & M_2(CO)_4 \\
 & & & & & & & & \downarrow{\scriptstyle k_5'\,|\,CO} \\
 & & & & & & M_2(CO)_6 & \xleftarrow[CO]{k_6'} & M_2(CO)_5
\end{array}$$

The backward steps in these reactions are assumed to be negligibly slow and are thus not considered.

Mononuclear Species. The rate equations for formation of mononuclear species can be written directly:

$$\frac{d[MCO]}{dt} = k_1[M][CO] - k_2[MCO][CO] - k_2''[MCO][M] \quad (9)$$

$$\frac{d[M(CO)_2]}{dt} = k_2[MCO][CO] - k_3[M(CO)_2][CO] - k_3''[M(CO)_2][M] \quad (10)$$

$$\frac{d[M(CO)_3]}{dt} = k_3[M(CO)_2][CO] - k_4''[M(CO)_3][M] \quad (11)$$

Since the reaction is carried out in pure CO all species of the form, $M(CO)_n$, become $M(CO)_3$ upon quenching, hence the final concentration of $M(CO)_3$ is given by

$$[M(CO)_3] = \int_0^{\tau_q} \left\{ \frac{d[MCO]}{dt} + \frac{d[M(CO)_2]}{dt} + \frac{d[M(CO)_3]}{dt} \right\} dt + [M]_{\tau_q} \quad (12)$$

where τ_q is the average quenching time, that is, the average lifetime of a species before it is immobilized within the matrix.

Substituting from equations (9), (10), and (11) into (12), one obtains

$$[M(CO)_3] = [M]_{\tau_q} + \int_0^{\tau_q} [M]\{k_1[CO] - (k_2''[MCO] + k_3''[M(CO)_2] + k_4''[M(CO)_3])\}\, dt \quad (13)$$

where $k_2'' \simeq k_3'' \simeq k_4''$ if the reactions are assumed to be diffusion limited. Integral (13) can, therefore, be written

$$[M(CO)_3] = [M]_{\tau_q} + \int_0^{\tau_q} [M]\{k_1[CO] - k_2''([M(CO)] + [M(CO)_2] + [M(CO)_3])\}\, dt$$

Under the experimental conditions considered the CO concentration far exceeds the sum of concentrations of all mononuclear species. Moreover, k_2'' is not expected to be markedly different from k_1. Accordingly, integral (13) reduces to

$$[M(CO)_3] \simeq \int_0^{\tau_q} k_1[M][CO]\, dt + [M]_{\tau_q} \tag{14}$$

The rate of change of concentration of metal can also be written down

$$\frac{-d[M]}{dt} = k_1[M][CO] + k_1''[M]^2$$

which reduces, for the same reasons stated above, to

$$\frac{-d[M]}{dt} \simeq k_1[M][CO]$$

which has the following solution

$$[M] = [M]_0 e^{-k_1[CO]t} \tag{15}$$

in which [CO] being in great excess is taken to be a constant and where $[M]_0$ is the initial metal concentration, hence proportional to the metal/CO ratio.

Substituting equation (15) into equation (14) and integrating, one obtains

$$[M(CO)_3] \simeq [M]_0$$

Binuclear Species. The kinetic equations governing the production of binuclear species may also be written down as follows:

$$\frac{d[M_2]}{dt} = \tfrac{1}{2}k_1''[M]^2 - k_1'[M_2][CO] \tag{16}$$

together with six other equations of the form

$$\frac{d[M_2(CO)_n]}{dt} = k_n'[M_2(CO)_{n-1}][CO] + k_{n+1}''[M][M(CO)_n] - k_{n+1}'[M_2(CO)_n][CO] \tag{17}$$

for $n \leq 3$ and deleting the term in k'' for $n > 3$.

Using similar arguments as those stated for production of mononuclears, the concentration of $M_2(CO)_6$ is given by

$$[M_2(CO)_6] = \tfrac{1}{2}k_1'' \int_0^{\tau_q} [M]^2\, dt + k_2'' \int_0^{\tau_q} [M] \left(\sum_{n=1}^{3} M(CO)_n \right) dt \qquad (18)$$

We have assumed throughout that the concentration of binuclear species is always smaller than that of mononuclears, consequently the quantity

$$\sum_{n=1}^{3} M(CO)_n \simeq [M]_0 - [M] \qquad (19)$$

Substituting (19) into (18) and (15) into the resulting expression and integrating, we obtain

$$[M_2(CO)_6] = K_2[M]_0^2$$

where

$$K_2 = \frac{k_1''}{4k_1[CO]}(1 - e^{-2k_1[CO]\tau_q}) + \frac{k_2''}{2k_1[CO]}[1 + e^{-2k_1[CO]\tau_q} - 2e^{-k_1[CO]\tau_q}]$$

The result of this analysis indicates, therefore, that the ratio of the concentration of binuclear species to that of mononuclear should increase proportionately to the matrix ratio $[M]_0$. A similar analysis for metal trimers indicates that the analogous ratio involving trinuclears increases as the square of the metal/CO ratio.

We see, therefore, that both statistical and surface reaction pathways lead to binuclear-mononuclear concentration ratios that are proportional to the matrix ratio, while the appropriate ratios related to species containing larger metal aggregates vary as some higher power of the matrix ratio. It is apparent, therefore, that one can distinguish binuclear species from both higher aggregates and mononuclears by observing the effect of varying the matrix ratio on the relative absorbances of the appropriate spectral bands.

Let us illustrate this by considering the results for the Cr/CO reaction. When Cr atoms are cocondensed with pure $^{12}C^{16}O$ at 10–15°K, the major product of the reactions is seen to be $Cr(CO)_6$ (Figure 18). However, two other intense carbonyl absorptions at 1962 and 1933 cm^{-1} always accompanied hexacarbonylchromium. These two new lines appeared with the same relative intensities in a number of different runs, and are ascribed to a species, X. The intense carbonyl absorptions have very similar frequencies and relative intensities (approximately 3:1) to those of Turner's pentacarbonylchromium[43] produced by photolyzing $Cr(CO)_6$ in Ar matrices. Furthermore, annealing of the CO matrix at 40–45°K had essentially no effect on the

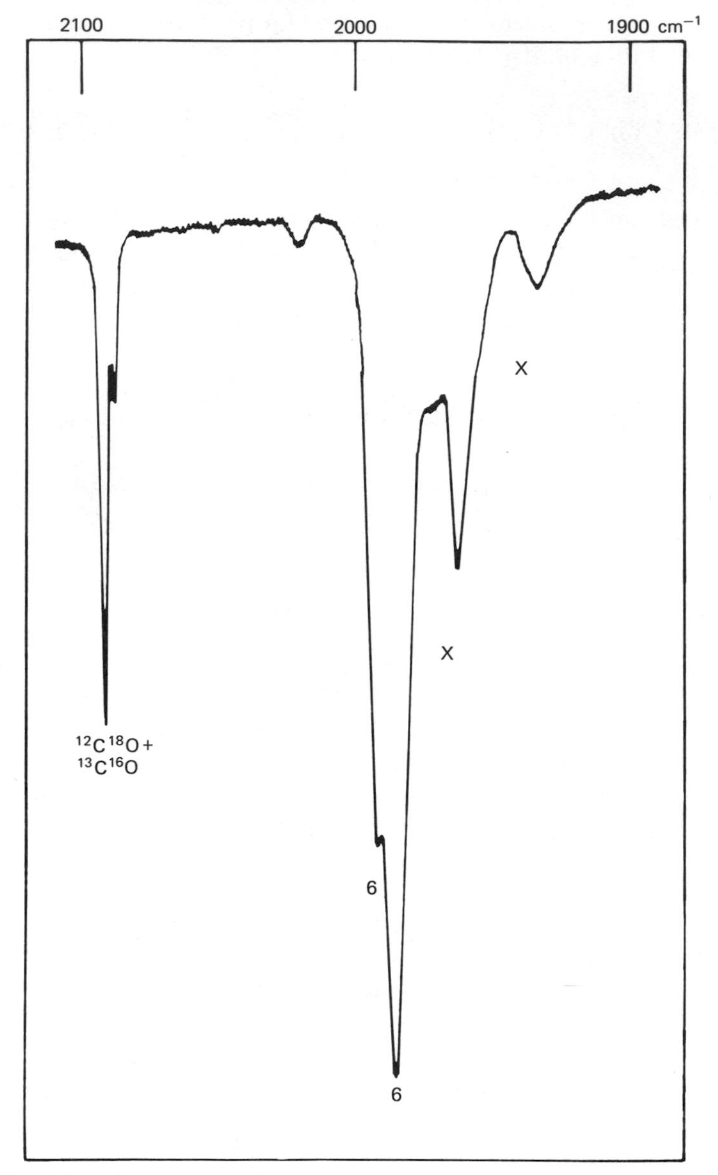

Figure 18. Infrared spectrum of the products of the matrix cocondensation reaction of Cr vapor with CO at 6–10°K with Cr:CO $\simeq$ 1:1000, showing the presence of $Cr(CO)_6$ (labeled 6) and $Cr_2(CO)_{10}$ (labeled X).

spectrum, a result paralleling the annealing behavior of Turner's pentacarbonylchromium in Ar/CO matrices.[43]

If X is a coordinatively unsaturated square-pyramidal pentacarbonylchromium, its reluctance to react with CO is most unusual, especially in a "pure" CO matrix, which leads one to believe that the vacant coordination site in the C_{4v} $Cr(CO)_5$ fragment is blocked. One explanation is that the lines belong to a weakly interacting dimer:

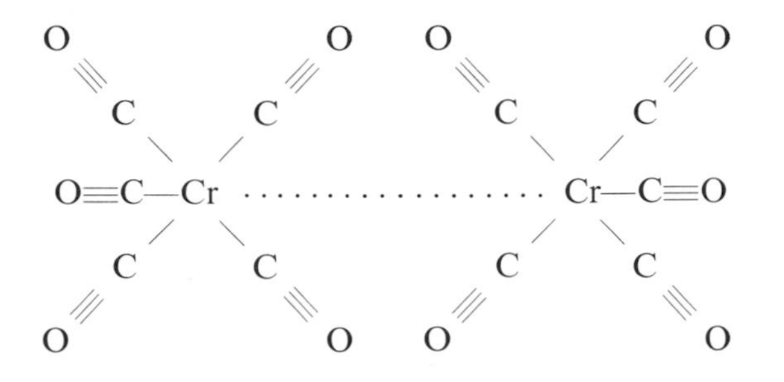

That the weakly interacting dimer, $Cr_2(CO)_{10}$, is indeed a likely formulation for this complex is in keeping with the results of a careful study of the dependence of the absorbance ratio $A_X/A_{Cr(CO)_6}$ on the Cr/CO ratio.[19] The absorbance ratio is observed to increase proportionally with increasing Cr concentration (Figure 19).

The Cu/CO cocondensation reaction in pure CO and in dilute CO/Ar matrices provides another example of the use of this method to determine

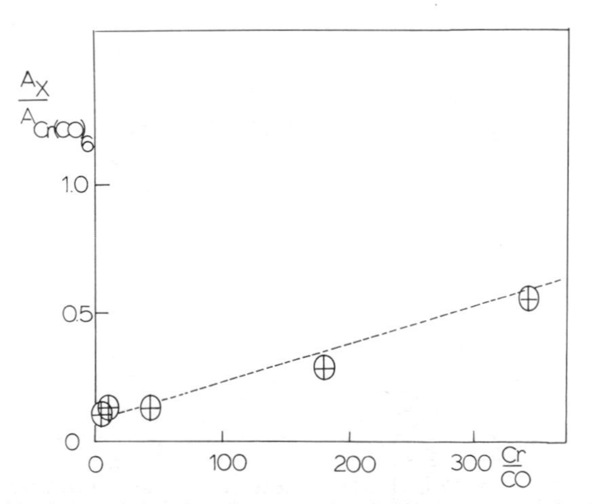

Figure 19. Graphical plot of the absorbance ratio, $A_X/A_{Cr(CO)_6}$, versus Cr/CO, proving that X is best formulated as the weakly interacting dimer, $Cr_2(CO)_{10}$.

metal stoichiometry.[23] The result of a study of the Cu/CO matrix reaction in the range, 1:10^2 to 1:10^5, is shown in Figure 20. This shows quite clearly that at low Cu/CO ratios ($\leq$1:1000), $Cu(CO)_3$ is the only observable species on deposition. However, at higher Cu/CO ratios ($\geq$1:1000), two new lines (D) begin to grow in at 2039 and 2003 cm^{-1}, coincident with the lines ascribed to $Cu_2(CO)_6$ formed in the dimerization of $Cu(CO)_3$ in Ar matrices. At

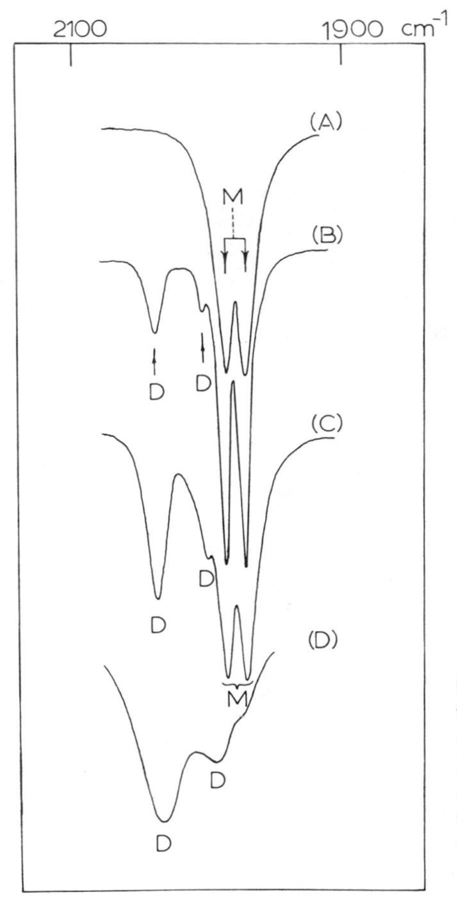

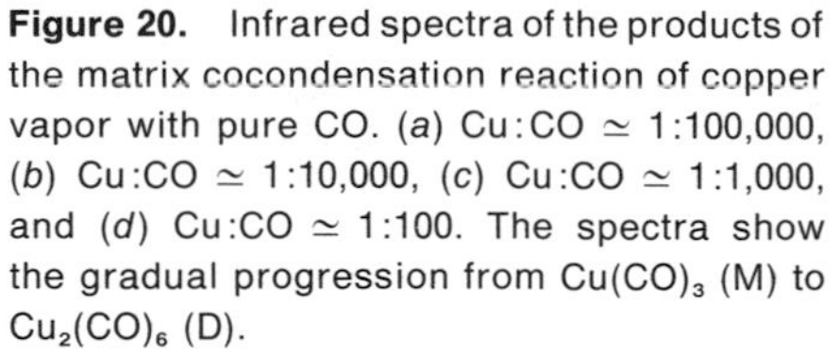

Figure 20. Infrared spectra of the products of the matrix cocondensation reaction of copper vapor with pure CO. (*a*) Cu:CO $\simeq$ 1:100,000, (*b*) Cu:CO $\simeq$ 1:10,000, (*c*) Cu:CO $\simeq$ 1:1,000, and (*d*) Cu:CO $\simeq$ 1:100. The spectra show the gradual progression from $Cu(CO)_3$ (M) to $Cu_2(CO)_6$ (D).

Cu/CO ratios approaching 1:100, the infrared absorbances of $Cu(CO)_3$ are hardly observable. Instead, the two new lines (D) dominate the spectrum. The spectra shown in Figure 20 are plotted graphically in Figure 21, where a proportional increase in the absorbance ratio, D/M, is observed with

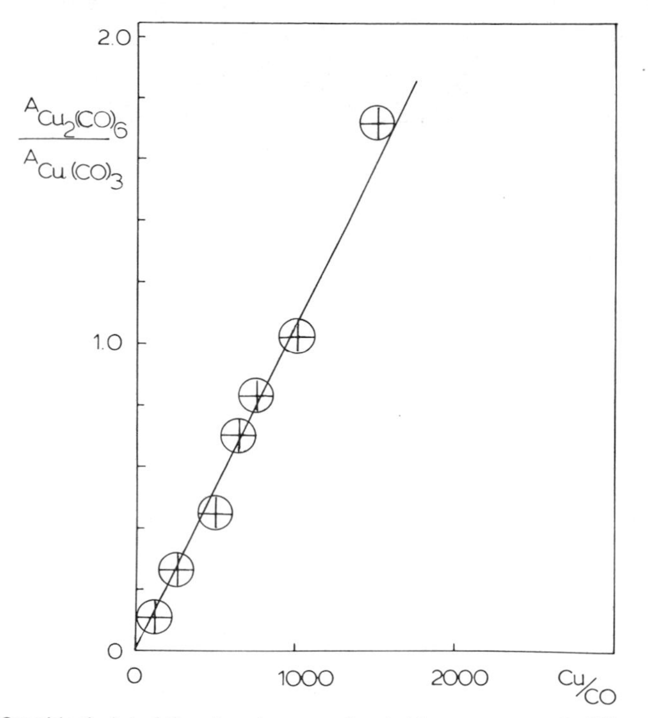

Figure 21. Graphical plot of the absorbance ratio, $A_D/A_{Cu(CO)_3}$, versus Cu/CO, proving that D is best formulated as $Cu_2(CO)_6$.

increasing Cu concentrations. These results provide convincing evidence for the binuclear formulation, $Cu_2(CO)_6$, for species D.

Using similar techniques, a study of the Rh/CO and Ir/CO matrix reactions (described earlier) as a function of the Rh/CO and Ir/CO ratios was undertaken. The results of such a series of experiments showed quite clearly that at low metal/CO ratios ($\leq$1:1000) $Rh(CO)_4$ and $Ir(CO)_4$ (M), are the only observable species formed on deposition. However, at high metal/CO ratios ($\geq$1:1000) new lines begin to grow in (Table 2), which are ascribed to $Rh_2(CO)_8$ and $Ir_2(CO)_8$ (D).

The results of the above experiments for Rh/CO and Ir/CO display a proportional increase in the absorbance ratio D/M with increasing metal concentration (see, for example, Figure 22).

In summary, the Rh and Ir concentration dependence and the formation of the new rhodium and iridium carbonyl species on allowing matrices containing $Rh(CO)_4$ and $Ir(CO)_4$, respectively, to warm up provide convincing evidence for the binuclear formulation, $Rh_2(CO)_8$ and $Ir_2(CO)_8$, for species D.

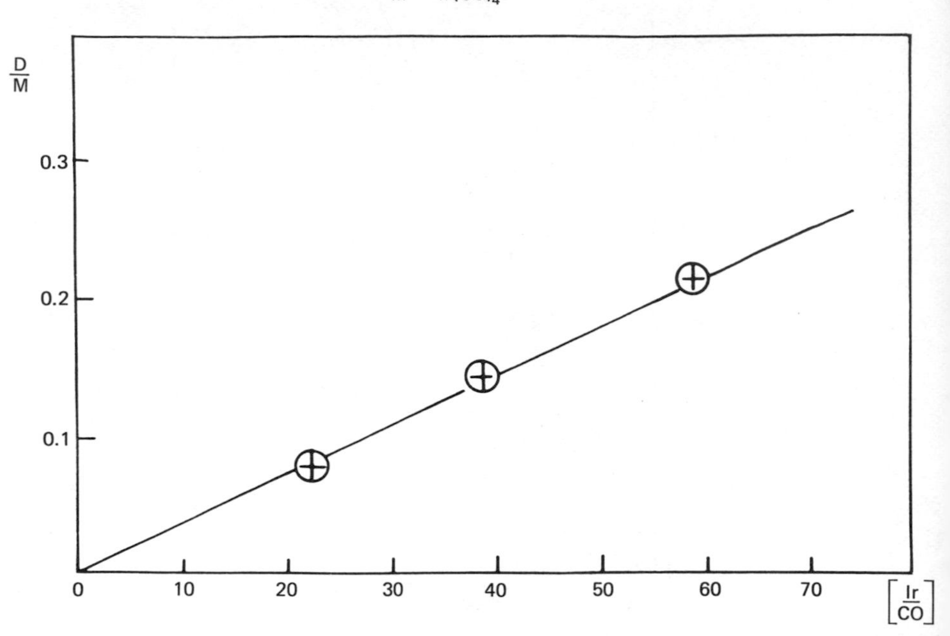

Figure 22. Graphical plot of the absorbance ratio, $A_D/A_{Ir(CO)_4}$, versus Ir/CO confirming that D is best formulated as $Ir_2(CO)_8$.

Nonreactive Matrics

If a metal atom cocondensation reaction is performed with a reactive substrate diluted in a nonreactive matrix, then metal concentration studies can again be used to establish whether or not the products are mononuclear. For example, experiments of this type proved to be particularly useful for the matrix reactions of Cu atoms with CO in Ar at *low* copper concentrations,[23] where mononuclear carbonyls, $Cu(CO)_n$ (where $n = 1–3$), were expected to predominate:

$$\mathrm{Cu} \xrightarrow[\mathrm{CO}]{k_1} \mathrm{Cu(CO)} \xrightarrow[\mathrm{CO}]{k_2} \mathrm{Cu(CO)_2} \xrightarrow[\mathrm{CO}]{k_3} \mathrm{Cu(CO)_3}$$

The following considerations served to clarify the situation. At intermediate CO/Ar ratios (1:10 to 1:100) where the ligand is arranged to be in excess of the copper and all three species coexist, let us assume that the concentration of free CO is approximately constant. Under these circumstances the rate equations for the formation of mononuclear species *only* can be solved (after some quenching time, τ_q, described earlier) to yield the following expressions:

$$[\mathrm{CuCO}] = \left\{\frac{k_1}{k_2 - k_1}\right\}[\mathrm{M}]_0\{e^{-k_1[\mathrm{CO}]_0\tau_q} - e^{-k_2[\mathrm{CO}]_0\tau_q}\}$$

$$[\mathrm{Cu(CO)_2}] = \left\{\frac{k_1k_2}{(k_2 - k_1)(k_3 - k_1)(k_3 - k_2)}\right\}[\mathrm{M}]_0\{(k_3 - k_2)e^{-k_1[\mathrm{CO}]_0\tau_q} - (k_3 - k_2)e^{-k_2[\mathrm{CO}]_0\tau_q} + (k_2 - k_1)e^{-k_3[\mathrm{CO}]_0\tau_q}\}$$

$$[\mathrm{Cu(CO)_3}] = \left\{\frac{[\mathrm{M}]_0}{(k_2 - k_1)(k_3 - k_1)(k_3 - k_2)}\right\}\{k_2k_3(k_3 - k_2)(1 - e^{-k_1[\mathrm{CO}]_0\tau_q}) - k_1k_3(k_3 - k_1)(1 - e^{-k_2[\mathrm{CO}]_0\tau_q}) + k_1k_2(k_2 - k_1)(1 - e^{-k_3[\mathrm{CO}]_0\tau_q})\}$$

The results of this analysis indicate, therefore, that the absorbance ratios, $A_{[\mathrm{CuCO}]}:A_{[\mathrm{Cu(CO)_2}]}:A_{[\mathrm{Cu(CO)_3}]}$, should be independent of the copper concentration. A similar analysis performed in the presence of binuclear species is more complex, but nevertheless indicates that the analogous ratio involving mononuclears should still be independent of the copper concentration.

It is apparent, therefore, that, having established that the product of the Cu/CO reaction at very low copper concentrations is the mononuclear tricarbonyl $\mathrm{Cu(CO)_3}$, copper concentration studies can then be usefully employed to investigate the copper stoichiometry of the mono- and dicarbonyls formed in dilute CO/Ar matrices at low copper concentrations.

Experimentally, one discovers that the absorbance ratio, $A_{[\mathrm{Cu(CO)_3}]}:A_{[\mathrm{Cu}_X\mathrm{(CO)_2}]}:A_{[\mathrm{Cu}_Y\mathrm{(CO)}]}$ remains constant over a 300-fold change in Cu concentration, proving the new complexes to be mononuclear ($X = Y = 1$) with respect to copper, namely, $\mathrm{Cu(CO)_2}$ and Cu(CO). The esr spectra of $\mathrm{Cu(^{12}CO)}$ and $\mathrm{Cu(^{13}CO)}$ have been obtained recently and confirm the mononuclear, paramagnetic nature of the monocarbonyl.[45]

9.7 Mixed Isotopic Substitution

It is well established that mixed isotopic substitution experiments provide a useful method for determining the stoichiometry and structure of a matrix-isolated species. Two distinct situations can be considered. On the one hand, one can cocondense an isotopically pure metal atom with a mixture of isotopic ligands such as $^n\mathrm{X}_2/^m\mathrm{X}_2$; alternatively, one can cocondense a mixture of metal atom isotopes $^p\mathrm{M}/^q\mathrm{M}$ with a pure ligand. Barring an isotope effect in the kinetics of formation, the product distribution can be expected to be statistical, and an analysis of the number, frequencies, and infrared and Raman intensities of the isotopic components can, in principle, determine the product stoichiometry with respect to both ligand and metal, the molecular structure of the product, the vibrational assignments, and force fields. Studies of this type are fundamental to almost any matrix-isolation experiment.

The application of this method in determining the binuclear complexes formed in matrix cocondensation reactions is nicely illustrated by the $Mn/^{12}C^{16}O/^{13}C^{16}O/Ar$ reaction described earlier in Section 9.3.

9.8 Codeposition of Two Different Metals

If the product of a matrix-cocondensation reaction is suspected to contain more than a single metal atom, then, in principle, simultaneous deposition of two different metals (chosen usually from the same group in order to maintain similar chemical properties and to avoid major structural changes in the trapped product) with the reacting gas can verify the molecular composition and aid the vibrational assignments.

This technique is particularly useful for monoisotopic metals or when isotopes of the metal are not readily accessible. Few examples using mixed metals can be referred to in the literature other than Andrews' study of alkali metal peroxides, Na_2O_2, K_2O_2, Rb_2O_2, and Cs_2O_2, formed by secondary reactions of the alkali metal atom M with the respective superoxide MO_2 in the matrix (see Ref. 46 and Chapter 6).

Brewer used metal codeposition studies (for example, Pb/Ag, Cu/Ag, Cu/Au, Ag/Au) in order to investigate the origin of the multiplet splitting in the electronic spectra of Pb, Cu, Ag, and Au atoms in inert matrices.[10,11] He discovered that even at 1 atom% of metal, the atomic spectra remained essentially unperturbed, indicating that the fine structure was not due to nearest neighbor M—M interactions, but rather site-symmetry lattice perturbations. However, detailed investigations of the electronic spectra of the diatomic species present in these mixed-metal experiments were not feasible at the time, presumably because of the low absorption intensities associated with the diatomic molecules. In the infrared the situation is quite different, as dimeric carbonyl reaction products usually have high carbonyl extinction coefficients and are easily observed, and in principle the mixed metal experiment is an ideal way by which one can ascertain the stoichiometry of the products with respect to the metal. For example, when Cu and Ag atoms were simultaneously cocondensed with pure CO at low metal concentrations, the spectrum obtained was essentially a superposition of the spectra of the respective Cu/CO and Ag/CO reaction products (Figure 23) consistent with the mononuclear formulation, namely, $Cu(CO)_3$ and $Ag(CO)_3$.[23,47]

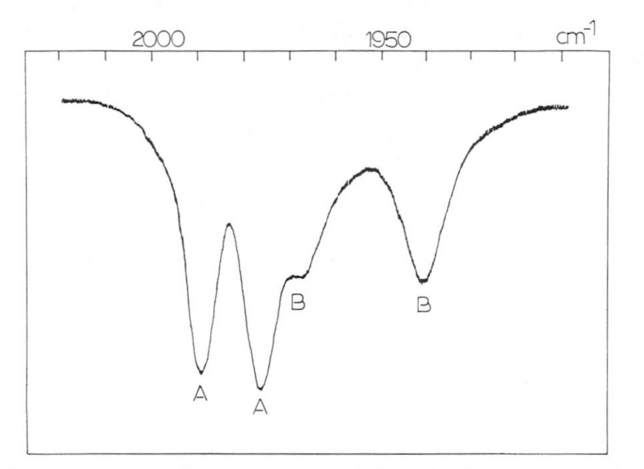

Figure 23. Infrared spectra of the products of the simultaneous matrix codeposition reaction of Cu and Ag atoms with CO at 6–10°K (Cu:Ag:CO $\simeq$ 1:1:1000), showing the formation of *only* $Cu(CO)_3$ (*a*) and $Ag(CO)_3$ (*b*). [Note that the absence of CO stretching modes assignable to mixed metal species confirms the mononuclear formulation of $Cu(CO)_3$ and $Ag(CO)_3$.]

10. HIGHER CLUSTERS

The techniques described above are, of course, not limited to synthesizing mono- and dinuclears but may be applied to the creation and identification of metal clusters or metal-cluster compounds containing more than two metal atoms.

Interest in species of this sort is considerable and spans many areas of chemistry. For example, theoretical chemists such as Blyholder[48] and Baetzold[31] have performed CNDO and extended Hückel calculations on nickel and silver clusters of various sizes as models for bulk metal and metal surfaces and in order to gain insight into vapor-phase nucleation processes and dispersed metal catalysts. Politzer[49] and Blyholder[50] have also performed molecular orbital calculations on small metal particles to which one or more CO molecules are bonded, in this case as a model for chemisorbed CO on either larger metal particles or bulk surfaces.

Among inorganic chemists, cluster compounds have also been of interest. Carbonyl compounds such as $Co_4(CO)_{12}$ and $Rh_6(CO)_{16}$ have been reported,[51] while more recently Cotton and his coworkers[52] have shown that the carbonyl groups around the cluster compound, $Rh_4(CO)_{12}$, are fluxional, mimicking thereby the mobility of many absorbed molecules on metal surfaces. It appears, therefore, that the gap between bona fide coordination compounds and molecules chemisorbed on bulk metal can be bridged, sealing up in the process the chasm separating homogeneous and heterogeneous catalysis.

Although much may be accomplished using conventional techniques, the synthesis of certain cluster compounds (as well as small, unsupported metal clusters) can only be achieved using matrix isolation. For example, it is generally assumed that CO chemisorbs on large metal particles in such a manner that only one admolecule resides on each surface metal atom or, when bridging, only one CO between any two surface metal atoms. In making cluster compounds that serve as models for the chemisorbed molecule-metal system, one wishes, therefore, to restrict the number of COs to one per metal atom or, what is more easily accomplished, to limit the number of COs to one per cluster compound, that is, to synthesize molecules of the form, M_xCO, where x is an integer. All conventionally generated cluster compounds have more than one CO bonded to each metal atom. Recently, compounds of the form, M_xCO, (where $x > 2$ and M = Cu and Ni) have been successfully synthesized[53] using matrix techniques similar to those described in earlier sections of this chapter. In characterizing these complexes one must expand the statistical and kinetic treatment of Section 9.6 to include higher clusters. This was done by means of two computer programs, the first of which calculated the statistical formation of metal clusters by randomly placing metal atoms in fcc lattice (such as that of solid argon) and counting the number of metal dimers, trimers, and so forth, so formed. The process is repeated for different metal/argon ratios. In the second program, a set of simultaneous differential equations similar to those in Section 9.6 was solved numerically for the concentrations of the various product species of the form, $M_x(CO)_y$, which may result from the cocondensation reaction of M with a CO/Ar mixture. This too is repeated for various metal/argon concentrations. The results of these calculations suggest that for metal/argon ratios less than a few percent and for CO in excess of metal, the equation

$$\frac{[M_x(CO)_y]}{[MCO]} = [M]_0^{x-1}$$

is approximately obeyed. In the above equation the square brackets indicate concentration and $[M]_0$ is the metal/argon ratio. One can, therefore, determine the size of the metal cluster bonded to a CO giving rise to a particular absorption by noting the manner in which the intensity of that band varies with metal concentration.

Using this technique, compounds of the form, Ni_2CO and Ni_3CO, have been identified.[53] In fact, two ir absorptions, at 1973 cm^{-1} and 1938 cm^{-1}, were assigned to Ni_2CO. (The number of COs was determined in a $^{13}CO/^{12}CO$ mixed-isotope experiment.) The presence of two types of Ni_2CO may be explained in several ways. For example, the higher frequency mode may belong to a linear molecule, while the lower frequency band arises from a bridge-bonded CO. If so, one must explain the surprising drop in frequency on going from NiCO (which absorbs at 1999 cm^{-1}) to Ni_2CO (terminal). More importantly, such a low frequency value for terminally bonded CO taken together with those attributed[53] to CO terminally bonded to Ni_3, which come at 1970 and 1963 cm^{-1}, causes some problems in the interpretation of infrared spectra for CO adsorbed on silica-supported nickel particles.[54] These show several bands above 2000 cm^{-1} that have been attributed to linear bonded CO. In view of the matrix results, it is possible that the commonly held view that only one CO molecule bonds to a surface atom is wrong or that silica-supported nickel is not really zero-valent metal: either or both of these conditions tend to increase the CO stretching frequency.

Metal clusters themselves are worthy of a study even in the absence of "adsorbate" molecules. Although no systematic study of metal clusters in excess of two metal atoms has yet been reported, these particles may be synthesized by the matrix methods described in earlier parts of this chapter and observed via uv-visible spectroscopy. In this manner, one may settle some of the discrepant theoretical results reported by Blyholder[48] and Baetzold.[31] The latter reports, for instance, that his quantum-mechanical calculations predict that linear clusters are, in general, more stable than two- or three-dimensional ones, while the former finds three-dimensional metal clusters to be the most stable. Baetzold also finds that as the number of atoms in the cluster increases, the highest occupied molecular orbital and the cluster stability alternate in a zig-zag fashion. Blyholder finds no such behavior, but rather a monotonic trend in these quantities with cluster size. Clearly, only experimental data will decide conclusively between the one or the other view.

11. CONCLUSION

The information presented in this chapter demonstrates that the diffusion properties of metal atoms may be probed and their aggregation processes, at least to dimeric molecules, controlled by utilizing matrix-isolation techniques. The chemistry of the transition metal diatomic molecules, their spectroscopy, and the characterization of the reaction products should prove very fruitful indeed in providing a large number of new molecules and in helping to understand the chemical nature of diatomic molecules. Moreover, the possibility exists for tailor-making new polynuclear molecules with features of interest to those interested in compounds containing metal-metal bonds, the reactivity and selectivity of transition metal-based catalysts, or the relationship of the binuclear complex, M_2L_n to L, chemisorbed on bulk metal, M. Finally, a great deal of knowledge concerning the energetics of metal formation might be obtained by carefully studying metal atom aggregation processes using matrix isolation infrared, laser Raman, uv-visible, and esr spectroscopy.

REFERENCES

1. (a) A. Kant and B. Strauss, *J. Chem. Phys.*, **41**, 3806 (1964); ibid., **45**, 3161 (1966); (b) A. Kant, *J. Chem. Phys.*, **41**, 1872 (1964); (c) A. Kant, S. S. Lin and B. Strauss, *J. Chem. Phys.*, **49**, 1983 (1968); (d) S. S. Lin and A. Kant, *J. Phys. Chem.*, **73**, 2450 (1969).
2. P. Schissel, *J. Chem. Phys.*, **26**, 1276 (1957).
3. M. Ackerman, F. E. Stafford, and J. Drowart, *J. Chem. Phys.*, **33**, 1784 (1960).
4. D. N. Travis and R. F. Barrow, *Proc. Chem. Soc., London*, 64 (1962).
5. B. Siegel, *Quart. Rev. Chem. Soc.*, **19**, 77 (1965).
6. D. M. Mann and H. P. Broida, *J. Chem. Phys.*, **55**, 84 (1971).
7. F. Schoch and E. Kay, *J. Chem. Phys.*, **59**, 718 (1973).
8. L. Brewer, B. A. King, J. L. Wang, B. Meyer, and G. F. Moore, *J. Chem. Phys.*, **49**, 5209 (1968).
9. L. Brewer and J. L. F. Wang, *J. Mol. Spectrosc.*, **40**, 95 (1971).
10. L. Brewer and C. Chang, *J. Chem. Phys.*, **56**, 1728 (1972).
11. L. Brewer and B. King, *J. Chem. Phys.*, **53**, 3981 (1970).
12. L. Andrews and G. Pimentel, *J. Chem. Phys.*, **47**, 2905 (1967).
13. O. Schnepp, *J. Phys. Chem. Solids*, **17**, 188 (1961).
14. W. F. Cooper, G. A. Clarke, and C. R. Hare, *J. Phys. Chem.*, **76**, 2268 (1972), and references therein.
15. B. Meyer, *Low Temperature Spectroscopy*, Elsevier, New York, 1971; N. B. Hannay, *Solid State Chemistry*, Prentice-Hall, New Jersey, 1967.
16. J. M. Brom, W. D. Hewett, and W. Weltner, *J. Chem. Phys.*, **62**, 3122 (1975).
17. J. F. L. Wang, Ph. D. thesis, University of California, Berkeley, 1969.
18. A. Bos and A. T. Hare, *J, Chem. Soc.*, *Faraday II*, **70**, 451 (1974).
19. E. P. Kündig, M. Moskovits, and G. A. Ozin, *Angew. Chem.* (*Int. Edit.*), **14**, 314 (1975).
20. H. Huber, A. Ford, W. Klotzbücher, M. Moskovits, and G. A. Ozin, *J. Chem. Phys.*, 1976, in press.
21. E. P. Kündig, M. Moskovits, and G. A. Ozin, *Nature*, **254**, 503 (1975).
22. W. Klotzbücher and G. A. Ozin, unpublished work.
23. E. P. Kündig, M. Moskovits, and G. A. Ozin, *J. Amer. Chem. Soc.*, **97**, 2097 (1975).
24. D. McIntosh and G. A. Ozin, unpublished work.
25. C. E. Moore, *Atomic Energy Levels*, Nat. Bur. Stand. Circular 467.
26. L. Andrews and G. C. Pimentel, *J. Chem. Phys.*, **47**, 2905 (1967).
27. L. Hanlan and G. A. Ozin, unpublished work.
28. D. W. Green and D. M. Gruen, *J. Chem. Phys.*, **60**, 1797 (1974); ibid., **57**, 4462 (1972).
29. R. Hoffmann, *J. Chem. Phys.*, **39**, 1397 (1963).
30. W. F. Cooper, G. A. Clark, and C. R. Hare, *J. Phys. Chem.*, **76**, 2268 (1972).
31. R. C. Baetzold, *J. Chem. Phys.*, **55**, 4363 (1971).

32. R. C. Baetzold, *J. Catal.*, **29**, 129 (1973); R. C. Baetzold and R. E. Mack, *J. Chem. Phys.*, **62**, 1513 (1975).
33. P. Joyes and M. Leleyter, *At. Mol. Phys.*, **6**, 150 (1973).
34. C. R. Hare, T. P. Sleight, W. F. Cooper, and G. A. Clarke, *Inorg. Chem.*, **7**, 669 (1968).
35. A. Kant and S. S. Lin, *J. Chem. Phys.*, **51**, 1644 (1969).
36. E. Clementi and D. C. Raimondi, *J. Chem. Phys.*, **38**, 2086 (1963).
37. L. C. Cusachs, *J. Chem. Phys.*, **43**, 5157 (1965).
38. H. Huber, E. P. Kündig, G. A. Ozin, and A. J. Poë, *J. Amer. Chem. Soc.*, **97**, 308 (1975).
39. L. Hanlan, H. Huber, E. P. Kündig, B. McGarvey and G. A. Ozin, *J. Amer. Chem. Soc.*, **97**, 7054 (1975).
40. L. Hanlan and G. A. Ozin, *J. Amer. Chem. Soc.*, **96**, 6324 (1974).
41. O. Crichton, M. Poliakoff, A. J. Rest, and J. J. Turner, *J. Chem. Soc.*, *Dalton*, 1321 (1973).
42. E. P. Kündig and G. A. Ozin, *J. Amer. Chem. Soc.*, **96**, 5585 (1974).
43. M. A. Graham, M. Poliakoff, and J. J. Turner, *J. Chem. Soc.*, *A*, 2939 (1971).
44. P. A. Breeze and J. J. Turner, *J. Organometal. Chem.* **7**, C44 (1972).
45. E. P. Kündig and G. A. Ozin, unpublished results.
46. L. Andrews, *J. Chem. Phys.*, **54**, 4935 (1971); ibid., **53**, 1327 (1972).
47. D. McIntosh and G. A. Ozin, *J. Amer. Chem. Soc.*, in press.
48. G. Blyholder, *Surf. Sci.*, **42**, 249 (1974).
49. P. Politzer and S. D. Kasten, *J. Phys. Chem.*, (in press).
50. G. Blyholder and M. C. Allen, *J. Amer. Chem. Soc.*, **91**, 3158 (1969).
51. F. A. Cotton and G. Wilkinson, *Advanced Inorganic Chemistry*, 3rd Edit., p. 689, Wiley-Interscience, New York, 1972.
52. F. A. Cotton, L. Kruczynski, B. L. Shapiro, and L. F. Johnson; *J. Amer. Chem. Soc.*, **94**, 6191 (1972).
53. M. Moskovits and J. Hulse, *Surf. Sci.*, *57*, 125 (1976).
54. R. P. Eischens, S. A. Francis, and W. A. Pliskin, *J. Phys. Chem.*, **60**, 194 (1956).
55. R. K. Nesbet, *Phys. Rev.*, **135**, *A*460, (1964).

Spectroscopic Identification and Characterization of Matrix Isolated Atoms

10

Dieter M. Gruen

1. INTRODUCTION

Electronic spectroscopies (uv-visible absorption, luminescence, esr, and Mössbauer) are widely used in matrix-isolation studies. One can discern four major areas of interest. The first is that of fundamental spectroscopy

Work performed under the auspices of the USERDA.

where the matrix-isolation technique has contributed importantly to an understanding of the electronic structure, particularly of high-temperature molecules, free radicals, and molecular ions. The second area deals with the perturbation of spectral features by the matrix surroundings. It is becoming clear that matrix-isolated species, atoms for example, are sensitive probes of their environment and that data derived from matrix studies can furnish detailed information on the interaction of trapped species with the matrix cage. Electronic spectroscopies are of prime importance in a third area of matrix studies that is of relatively more recent origin, that of chemistry in matrices. Because the products of a reaction that has been carried out in a matrix generally are not and very frequently could not be recovered or subjected to conventional kinds of analyses, spectroscopic methods are virtually the only means at hand for their identification and characterization. Electronic spectroscopies are important adjuncts to vibrational spectroscopies, and in the case of atomic reactants or products, are the only techniques available for accomplishing the task of identification. Finally, the recent application of matrix-isolation spectroscopy to quantitative sputtering studies has made the spectroscopic identification of sputtered products into a sensitive new tool for studying the sputtering process.[1]

The contributions that electronic spectroscopies can make to chemistry in matrices will become apparent as a result of the discussions in the present chapter. However, the fundamental aspects of research in this field are also discussed. The application of spectroscopic results to chemical problems relies in the first instance on basic data concerning the electronic structures of atoms and molecules in matrices. The results of the efforts to reach a fundamental understanding of the effects of matrix perturbations on the spectra of matrix-isolated atoms are treated, but not in detail. These effects have a profound influence on the appearance of spectral features and have to be understood in order to interpret the matrix data in terms of well-known atomic transitions.[2]

2. ELECTRONIC SPECTROSCOPY OF ATOMS IN MATRICES

The study of the properties of atoms isolated in matrices is of great fundamental interest because of the wealth of information that is obtained concerning guest-host interactions. Fortunately for matrix chemistry, this motivation has resulted in the accumulation of uv-visible, esr, and Mössbauer data on 42 atoms isolated, for the most part, in noble gas matrices. Table 1 lists the atoms that have been studied, the nature of the investigations, as well as suitable literature references.

Atoms are of particular importance to carrying out chemical reactions at cryogenic temperatures in matrices for several reasons. Atoms can undergo reactions requiring little or no activation energy with a very wide variety of

Table 1. Matrix-Isolated Atoms

Group	Element	Type of Studies[a]	Ref.
IA	H	A, E	57, 58, 121, 124–126
	D	A, E	56, 123–125
	Li	A, E	59–61, 64, 130, 131
	Na	A, E	9, 61–64, 130, 131
	K	A, E	61–64, 130, 131
	Rb	A, L, E	63, 64, 73, 76, 130, 131
	Cs	A, E	63, 64, 130, 131
IIA	Mg	A	16, 65, 66
	Cu	A	13, 66, 67, 69
	Ba	A	66, 68
IIIA	Al	A, E	71
	Ga	A, E	71
	In	A	72
	Tl	A	73
IVA	Sn	A	21
	Pb	A	13, 66
VA	N	L, E	7, 55, 86, 87, 104–114, 123, 124, 127, 128
	P	E	147, 148
	As	E	148
VIA	O	L	7, 55, 86, 87, 104–114, 159
VIIIA	Ar	A	77
	Kr	A	77, 80
	Xe	A	77, 79–82
IB	Cu	A, E	2, 33, 83, 84
	Ag	A, E	33, 83, 85
	Au	A, E	1, 33, 83
IIB	Zn	A	95–97
	Cd	A	61, 66, 84, 95–97
	Hg	A	9, 61, 75, 81, 82, 89–94
IIIB	Sc	A	99
	Y	A	99
IVB	Ti	A, L	10
VB	Nb	A, L	17
	Ta	A	20
VIB	Cr	A	21, 37
	Mo	A	12
VIIB	Mn	A	16, 21, 100
VIIIB	Fe	A, M	19, 21, 34, 84, 101, 102, 155, 156
	Co	A	21
	Ni	A	21, 104
	Pd	A	21
	Pt	A	2

[a] A, absorption spectroscopy; L, luminescence spectroscopy; E, electron spin resonance; M, Mössbauer spectroscopy.

molecular species. Examples are the recently discovered reactions of metal atoms with CO, O_2, and N_2 to form metal carbonyl, dioxygen, and dinitrogen complexes. It was found in a study of the reaction of Pt atoms with N_2 molecules[3] that the diffusion of Pt atoms in the matrix could be monitored by following the disappearance of the uv-visible atomic absorption spectrum. Simultaneously, the appearance of spectral features associated with the formation of PtN_2 or $Pt(N_2)_2$ complexes was observed. Clearly, uv-visible spectroscopy provides another means, in addition to vibrational spectroscopy, for monitoring the course of an important class of reactions. It should be pointed out that uv-visible spectroscopy is a very sensitive technique for measuring small concentrations of atoms because many of the "resonance" transitions have oscillator strengths near unity.

Photolysis experiments are often impeded by the cage effect, which prevents the escape of the reaction products and may thus lead to the regeneration of the parent molecule. The cage effect is an extremely important factor in determining which photochemical process occurs. Since the generation of atoms, free radicals, or molecular ions in matrices are frequently accomplished via *in situ* photolysis, production of a fragment small enough to diffuse away from the photolysis site is crucially important to the occurrence of subsequent reaction. Electrons and atoms are, in general, able to diffuse through a noble-gas matrix under conditions where diatomic and polyatomic species are immobile. Photochemical experiments resulting in the formation of atoms, therefore, can be monitored in a sensitive way by following the intensity of the appropriate spectrum. Interesting examples are the production of the diatomic CuO and MnO molecules via the *in situ* photolysis of noble gas matrices containing codeposited Cu atoms and O_2 molecules or Mn atoms and O_2 molecules. Prior to photolysis, only the atomic absorption bands and esr spectra of Cu and Mn atoms were observed. Formation of O atoms by photolysis, their subsequent diffusion in the matrices, and reaction to form the diatomic metal oxides resulted in disappearance of the metal atomic spectra and the appearance of the characteristic CuO and MnO molecular absorption bands.[4]

In the burgeoning field of reactions of metal atoms with organic molecules to yield organometallic compounds, matrix-isolation studies will become increasingly important in establishing reaction intermediates and identifying reaction products. Here too, knowledge of the absorption, fluorescence, esr, and Mössbauer spectra of matrix-isolated metal atoms will be an important aid in establishing reactant and product stoichiometries as well as other reaction parameters. As a case in point, monitoring the concentration of Cr atoms reacting with benzene in an argon matrix to form dibenzene chromium may be mentioned.[5]

The history of work on matrix-isolated atoms can be divided into four periods. The first began in the 1920s with the work of Vegard, whose studies included electron or ion excitation of low-temperature matrices of atmospheric gases.[6,7] Vegard and later other workers were attempting to elucidate atmospheric and steller phenomena. The second phase occupied the 1950s and is exemplified by the work at the NBS (National Bureau of Standards), where matrix-isolated products of electrical discharges through gases or gas mixtures were studied.[8] The third stage began in 1959 and consisted of extensive work on metals that could be easily vaporized, such as Na and Hg.[9] Finally, work was begun during the last five years on studies of refractory atoms such as Ti, a class that includes most of the transition metals.[10] The refractory metal atoms not only present special experimental difficulties but also yield complex spectra, whose interpretation is a topic of considerable interest in its own right. A review of optical studies of matrix-isolated atoms appeared in 1971, which covered approximately 20 different atoms.[11] The growth of the field, however, is such that twice that number of atoms could be included for discussion in the present chapter, which begins with a discussion of AMCOR, a scheme for correlating atomic transition intensities with the intensities of atomic absorption bands in low-temperature matrices.[2] A detailed description of the spectra of Mo atoms in matrices is given to illustrate the application of AMCOR to the analysis of the optical spectra of matrix-isolated atoms.[12] Matrix-perturbation effects are discussed in a separate section devoted to the spectra of matrix-isolated Au atoms.[1] No attempt is made to discuss matrix effects elsewhere in depth, although they are mentioned when appropriate. The chapter continues with a review of the optical spectroscopy, both absorption and luminescence, of atoms in matrices. This is followed by sections on esr and Mössbauer spectra.

One might expect the spectra of matrix-isolated atoms to mirror, to a degree, the spectra of the analogous gaseous atoms. Spectral changes due to imbedding an atom in a matrix are usually ascribed to "matrix perturbations," which can affect band positions, widths, and intensities.

Bands arising from atomic transitions usually shift in energy by only relatively small amounts, of the order of a few percent, on going from the gaseous to the matrix environment. However, the energy shifts can be of the order of 10–20%, particularly when the excited atomic state involves an electron in an *S* orbital.[13,71] The shift varies from matrix to matrix, often in a predictable manner.

Perhaps the most obvious differences are seen in the band widths. Matrix effects can often greatly increase these values, and widths of 200 cm^{-1} or more have been reported for individual transitions. Band splittings caused by crystal field,[15] multiple site effects, or spin-orbit coupling interactions[1,12] are also often noted. All of these effects, in spite of considerable effort, are as yet not completely understood.

A comparison between the atomic and matrix intensities of transition metal atoms was first made by Schnepp and applied to Mn.[16] Mn displays two strong resonance bands arising from the $a^6S \rightarrow z^6P^0$, Y^6P° transitions, each transition being split into two or three components in the matrices. Schnepp estimated the ratio of the intensities of these two bands in the matrix and compared the value he found, 4.0 ± 0.3, to the ratio of the intensities of the gaseous atomic lines, 3.7.

A generalized intensity correlation method, herein after referred to as AMCOR, has been elaborated[2] and has been applied with considerable success to a number of atoms including Ti,[10] Nb,[17,18] Mo,[12] Pt,[2] Cu,[2] Fe,[19] and Ta.[20] Because of its quite general applicability and usefulness, the AMCOR method is discussed more fully below.

2.1 AMCOR—The Correlation between Gaseous Atomic Spectra and the Absorption Spectra of Atoms Isolated in Noble-Gas Matrices

Until relatively recently, studies of atomic absorption spectra in matrices have been confined primarily to atoms with symmetric (S state) ground states. Gruen and Carstens[10] as well as Mann and Broida[21] have extended the experiments to include transition metal atoms isolated in noble gas matrices. This work has shown that even very complex spectra, such as the spectrum of Ti atoms trapped in Ar and Xe matrices, consisting of over 20 overlapping, partially resolved bands, can be correlated with the spectra of the gaseous atoms. It was shown that the intensities and energies of the various bands seen in the spectrum of the matrix-isolated Ti atoms follow the general pattern of intensities and energies of bands seen in the arc spectrum of neutral Ti. Since the matrix-isolated atoms are at low temperatures, only transitions from the ground state manifold are observed in the absorption spectrum.

A series of charts and tables have been prepared[2] listing the most intense transitions of 64 elements for which intensity and energy data are available. Line graphs depict the positions and intensities of all transitions involving the ground state of each atom. Tables that include the wavelength, wavenumber, probability, intensity, electronic configuration, and state designations, including J values, of the states involved in each transition are also presented for these elements. The intensities and energies of the transitions were taken from the extensive tables of Corliss and Bozman[22] and were in turn correlated with the state designations of Moore and other authors.[23] Data are included for all transitions listed in the tables of Corliss and Bozman in the 10,000–50,000 cm^{-1} regions and often to energies somewhat above this latter value.

The intensity data of Corliss and Bozman[22] were used for this compilation because they represent the most extensive tabulation of this sort. It is now recognized that the *gf* values of Corliss and Bozman, which are based on arc emission spectra using copper electrodes to which known amounts of the element under study had been added, are by and large not the most reliable values available. Better numbers for the first twenty elements are to be found in two volumes of numerical tables of critically evaluated transition probabilities,[24] and these tables should be consulted for the most up-to-date intensity information. Also useful are the series of bibliographies on atomic transition probabilities issued by the NBS.[25]

The intensity data in Ref. 2 are listed in terms of the probability, $g_u A$ and the intensity, $g_u f_{ul}$, of the transition, where g_u is the degeneracy of the upper state from which emission occurs, f_{ul} is the oscillator strength for the emission line, and A is the Einstein probability. In Ref. 10 it was shown that

$$f_{lu} = \frac{g_u f_{ul}}{gl} \tag{1}$$

where f_{lu} is the oscillator strength for the analogous transition in absorption and gl is the degeneracy of the ground state. Since in matrix-isolation experiments the transitions are all from the ground state, the intensity of the various bands are related to the intensity of the gaseous emission lines by the factor, gl. Thus, the absorption spectrum can be expected to correlate directly with a plot of $g_u f_{ul}$ values whenever the matrix perturbations are relatively small, as is the case in noble gas matrices.[10,21]

To illustrate the application of AMCOR to the interpretation of the spectra of atoms isolated in noble-gas matrices, the Mo spectrum is discussed at this point.[12,26]

The transitions with measurable gf values involving the ground state of atomic Mo are listed in Table 2. The table also includes columns labeled wavelength, λ in Å units, wavenumber (cm^{-1}), probability (gA), and intensity (gf) of the transition. The electronic configuration and state designations—including S, the spin degeneracy; L, the angular momentum quantum number; and J, the total momentum quantum number—are also tabulated. The *gf* values are shown graphically at the bottom of Figure 1. The heights of the lines are proportional to the measured intensity of each atomic transition ($gf > 0.1$) listed in Table 2. The spectra of Mo atoms isolated in Xe, Kr, and Ar matrices are displayed successively above the *gf* lines.

The well-defined low-energy triplet observed in each of the matrices can be assigned to the three components of the $a^7S \rightarrow z^7P$ atomic transition. It also appears reasonable on the basis of the atomic frequencies and oscillator strengths (Table 2) to assign the high-energy bands that show a well-developed triplet structure in a Xe matrix, to the $a^7S \rightarrow y^7P$ and $a^7S \rightarrow z^7D$ atomic transitions. Because of the complexity of the high-energy band, the remainder of the Mo-atom discussion focusses on the three peaks assigned to the

Table 2. Oscillator Strengths of Free Neutral Mo Atoms

	Wavelength	cm^{-1}	*gA*	*gf*	4*d*	5*s*	5*p*	*S*	*L*	*J*
0	—	0.0	—	—	5	1	0	7	0	3
1	2944.21	33,955	0.16000	0.02100	4	1	1	5	1	3
2	3002.21	33,299	0.19000	0.02500	4	1	1	5	1	2
3	3112.12	32,123	0.98000	0.14000	4	1	1	7	2	4
4	3132.59	31,913	9.80000	1.40000	4	1	1	7	1	4
5	3158.16	31,655	3.80000	0.57000	4	1	1	7	2	3
6	3170.35	31,533	5.40000	0.82000	4	1	1	7	1	3
7	3193.97	31,300	4.40000	0.68000	4	1	1	7	1	2
8	3208.83	31,155	1.70000	0.26000	4	1	1	7	2	2
9	3456.39	28,924	0.31000	0.05600	5	0	1	5	1	3
10	3466.83	28,836	0.10000	0.01800	5	0	1	5	1	2
11	3798.25	26,321	4.40000	0.94000	5	0	1	7	1	4
12	3864.11	25,872	3.40000	0.77000	5	0	1	7	1	3
13	3902.96	25,614	2.10000	0.47000	5	0	1	7	1	2

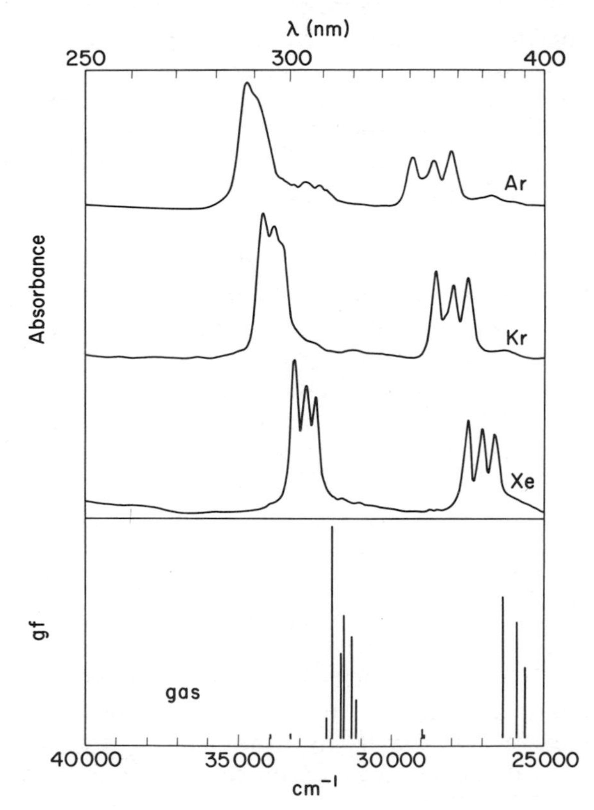

Figure 1. Absorption spectra of Mo atoms isolated in Ar, Kr, and Xe matrices compared to the gas phase transitions involving the a^7S ground state.

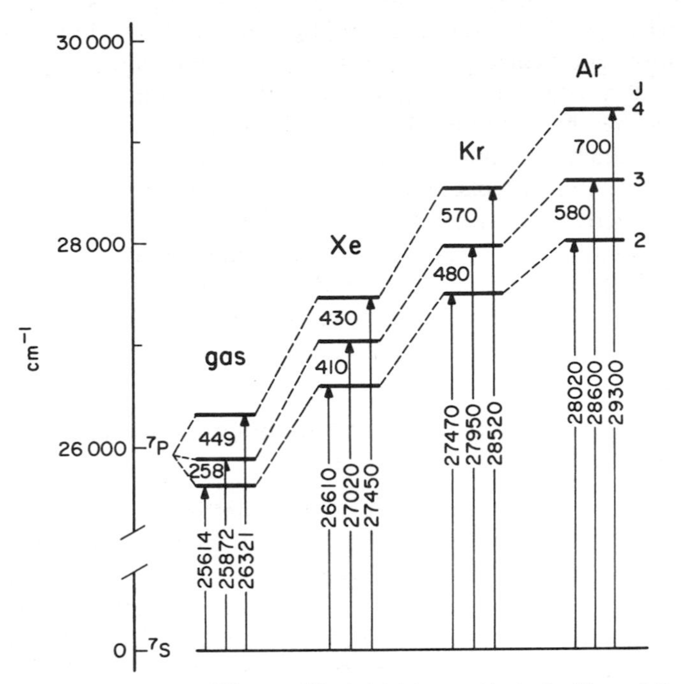

Figure 2. Comparison of the $z^7P^0_{4,3,2} \leftarrow a^7S$ triplet frequencies in Ar, Kr, and Xe matrices with the gas-phase values for Mo atoms.

$a^7S \rightarrow z^7P$ transition. Figure 2 compares the z^7P_4, z^7P_3, and z^7P_2 energy levels relative to the a^7S level observed in each of the matrices and in the atomic spectrum.

For pure L—S coupling the sublevel splitting of a regular ^7P should be such that the $^7P_4$–$^7P_3$ splitting is 1.33 times that of the $^7P_3$–$^7P_2$. In the gas phase this ratio is observed to be 1.74 for Mo so that spin-orbit coupling is only a partial explanation of the origin of the sublevel width. The total sublevel width, $^7P_4$–$^7P_2$, increases in going from gaseous Mo to Mo isolated in Xe, Kr, and Ar matrices. This means that there is an appreciable effect of the matrix on the z^7P spin-orbit interaction or the extent of configuration interaction or both. The ratio of sublevel spacings does not change from the gas phase value of 1.74 in a regular manner—1.1 for Xe and 1.2 for both Kr and Ar. The observation of a regularly increasing total width, $^7P_4$–$^7P_2$, is consistent with the observations of the increasing $^2P_{3/2}$–$^2P_{1/2}$ splitting in Au which will be discussed in the next section. However, it is clear from the $(^7P_4–^7P_3)/(^7P_3–^7P_2)$ matrix frequency ratios compared to the gaseous atom ratio that, at least in the case of Mo, the change cannot be ascribed simply to changes in the spin-orbit interaction.

It is evident from the discussion of the Mo spectrum that, in favorable cases, a quite good correspondence is observed between the atomic and the matrix spectra. The effect of matrix perturbations on the spin-orbit-coupling interaction has been pointed out. Other spectral changes due to matrix interactions including the shifting and broadening of the spectral lines, crystal-field splitting effects, and multiple-site occupation are well illustrated in the Au spectrum. Matrix perturbations of atomic-transition probabilities have also been studied in the case of Au atoms. Because of their general relevance to the interpretation and understanding of the spectra of matrix-isolated atoms, all of the various matrix-perturbation effects are discussed in connection with the Au spectrum and are referred to only in passing in the subsequent portions of the chapter.

2.2 Matrix Perturbation Effects on the Spectra of Atoms: Au Atoms in Ar, Kr, and Xe Matrices

The effects on the spectra of atoms due to matrix perturbations have frequently been discussed in terms of considerations originally developed for pressure broadening and shifting of spectral lines.[9,28] The treatment of spectral perturbations due to pressure of a foreign gas is complicated by the more or less random distribution of the interacting particles in space and the Boltzmann distribution of their kinetic energies.[29] By contrast, in a matrix the intermolecular distance is fixed. For the case of an impurity atom at a substitutional site the cluster geometry is well defined, and the motions of the particles about their equilibrium positions are given by the lattice dynamics of a rare-gas solid at a low temperature. Current interpretations of matrix-perturbation effects based on simple models using Lennard–Jones (6–12) and (6–8–12) potential functions to describe interactions between trapped atoms and their rare-gas cages have been used to evaluate van der Waals interaction constants, for excited electronic states of impurity centers.[30] These interaction constants evaluated from empirical energy shift data are in qualitative agreement with the electronic dielectric constant theory of Heinrichs.[31] It appears, therefore, that a satisfactory first-order theory exists capable of rationalizing at least the gross effects of energy-level shifts due to matrix perturbations. Studies of the absorption spectra of atoms in matrices have, however, revealed splittings of degenerate energy levels[15] and temperature effects on linewidths,[14,32] which require refinements of present theoretical interpretations for a more complete understanding. Electron spin resonance[33] and Mössbauer[34] experiments are also providing detailed information on the interaction of electronic ground states of atoms with their noble gas surroundings. Finally, in the case of Au atoms,[1] data on oscillator strengths and a detailed examination of the energy shifts of multiplet components of spectral lines have been shown to provide additional information bearing on the nature of atomic interactions in solids.

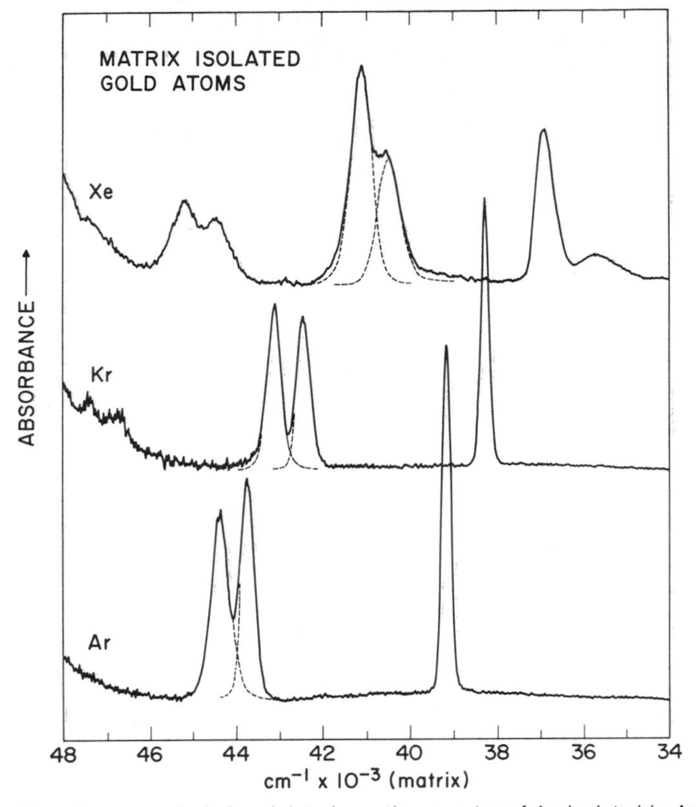

Figure 3. Baseline corrected ultraviolet absorption spectra of Au isolated in Ar, Kr, and Xe matrices.

The baseline-corrected ultraviolet absorption spectra of gold isolated in Ar, Kr, and Xe matrices are shown in Figure 3. These spectra are qualitatively similar to those obtained by photographic methods,[35,36] but show improved resolution through the use of a double-beam spectrophotometer. In Ar and Kr, only the gold atom transitions, $^2S_{1/2} \rightarrow {}^2P_{1/2}$ and $^2S_{1/2} \rightarrow {}^2P_{3/2}$, are well resolved. In Xe, a number of additional features are present. The fact that the $^2S_{1/2} \rightarrow {}^2P_{3/2}$ transition is observed to be split into two components in each matrix indicates that the matrix environment does not possess the cubic field symmetry anticipated from the esr results.[33] Moreover, the effect of the matrix interaction on the relative absorbances of the two components in this transition does not appear to be equal. In calculating the matrix oscillator strength, $f_{lu}^{(m)}$, for the $^2S_{1/2} \rightarrow {}^2P_{3/2}$ transition, the integrated area was obtained for the combined components without regard for their relative contribution to the total peak intensity. In the last column of Table 3, the $f_{lu}^{(m)}$ values in the various matrices are listed. These should be compared to

Table 3. Absorptions of Au in Noble Gas Matrices Compared with AMCOR Data

Matrix	Transition	Gas λ (nm)	Gas $\bar{\nu}$ (cm^{-1})	Matrix λ (nm)	Matrix $\bar{\nu}$ (cm^{-1})	Half-width (cm^{-1})	Matrix gas $\Delta\bar{\nu}$ (cm^{-1})	$f_{lu}^{(m)}$
Ar	$^2P_{1/2} \leftarrow {}^2S_{1/2}$	267.59	37,359	255.19	39,175	220	+1816	0.132
	$^2P_{3/2} \leftarrow {}^2S_{1/2}$	242.80	41,174	228.43	43,763	410	+2589	0.293
				225.28	44,375	480	+3210	
Kr	$^2P_{1/2} \leftarrow {}^2S_{1/2}$	267.59	37,359	261.10	38,288	200	+ 929	0.090
	$^2P_{3/2} \leftarrow {}^2S_{1/2}$	242.80	41,174	235.36	42,475	390	+1301	0.200
				231.88	43,113	380	+1939	
	$^4F_{7/2} \leftarrow {}^2S_{1/2}$?	219.53	45,537	—	ca. 46,750	—	+1213	—
	$^4F_{5/2} \leftarrow {}^2S_{1/2}$?	216.50	46,174	—	ca. 47,450	—	+1276	—
Xe	?	—	—	326.44	30,625	1000	—	—
	?	—	—	280.22	35,675	—	—	—
	$^2P_{1/2} \leftarrow {}^2S_{1/2}$	267.59	37,359	271.11	36,875	470	− 484	0.061
	$^2P_{3/2} \leftarrow {}^2S_{1/2}$	242.80	41,174	246.68	40,525	580	−1183	0.145
				243.29	41,090	620	− 84	
	$^4F_{7/2} \leftarrow {}^2S_{1/2}$?	219.53	45,537	224.90	44,450	730	−1087	—
	$^4F_{5/2} \leftarrow {}^2S_{1/2}$?	216.50	46,174	221.46	45,140	680	−1034	—

Table 4. Oscillator-Strength Values Reported in the Literature for the Au Atomic Transitions $^2P_{1/2}$–$^2S_{1/2}$, $^2P_{3/2}$–$^2S_{1/2}$

	Method of Measurement			
Transition	Hook[a]	Atomic Beam[b]	Emission[c]	Absorption[d]
$6^2P_{1/2}$–$6^2S_{1/2}$	0.19	0.125	0.06	0.076
$6^2P_{3/2}$–$6^2S_{1/2}$	0.41	—	0.08	0.18

[a] Reference 45.
[b] Reference 133.
[c] Reference 22.
[d] Reference 157.

reported oscillator strength values for gaseous Au atom transitions listed in Table 4.

Table 3 also lists the observed frequencies for the matrix-isolated Au spectra, together with the corresponding AMCOR data. In addition, the relative shifts in transition energy on going from the gas phase to the matrix are given.

Matrix Perturbation of Energy Levels

The ground state of the neutral gold atom is [Xe]$5d^{10}6s$, $^2S_{1/2}$. The two "resonance" transitions in the neutral gaseous atom at 37,358.9 and 41,174.3 cm^{-1} are to the $^2P_{1/2}$ and $^2P_{3/2}$ states, respectively, of the $5d^{10}6p$ configuration. These are the only transitions listed in the intensity tables of Corliss and Bozman[2,22] involving the atomic ground state in the 0–50,000 cm^{-1} energy interval. The major spectral feature for Au atoms in Ar, Kr, and Xe matrices (Figure 3) are, therefore, most reasonably assigned to ^2S ground state to $^2P_{1/2}$ and $^2P_{3/2}$ excited-state transitions. Interactions of the Au atoms with the surrounding matrix atoms manifest themselves in line broadening, as well as energy-level shifts and splittings. The half-widths of the lines are almost the same in Ar and Kr but are considerably broader in Xe. The shifts in the energy-levels are shown graphically in Figure 4. Both the $^2P_{1/2}$ and $^2P_{3/2}$ transitions are shifted to lower energies in Xe but to higher energies in Kr and Ar matrices. Energy-level shifts of electronic transitions of atoms in matrices have been frequently observed and have been widely discussed in the literature, usually on the basis of an intermolecular potential of the Lennard–Jones (6–12) type.[9] It is to be noted from an examination of Table 3 and of Figures 3 and 4 that the $^2P_{3/2}$ states are shifted to higher energies relative to the shifts experienced by the $^2P_{1/2}$ states. Specifically, it can be seen that the $^2P_{1/2}$–$^2P_{3/2}$ separation increases continuously in going from

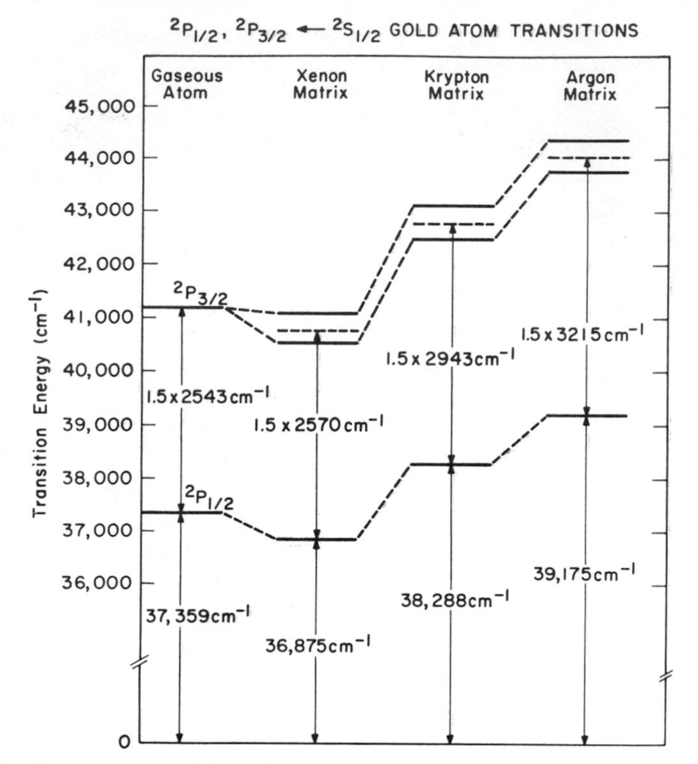

Figure 4. Resonance transitions of Au atoms in the gas and in Xe, Kr, and Ar matrices.

the gaseous Au atom to Au atoms isolated in Xe, Kr, and Ar matrices. Presumably, somewhat larger values of the ε^* and σ^* Lennard–Jones parameters are to be associated with the excited $^2P_{3/2}$ than the $^2P_{1/2}$ state because of the larger spatial extension of the wavefunction of the $^2P_{3/2}$ compared to the $^2P_{1/2}$ states. Since the $^2P_{1/2}$–$^2P_{3/2}$ separation is determined largely by spin-orbit forces, the variations in energy between the two states in the different matrices apparently reflects changes in the spin-orbit coupling parameter, ζ.

The intimate connection between the spin-orbit coupling interaction and the effective nuclear charge is discussed more fully in the next section. Suffice it to say here, that the sizable changes in the case of Au have also been observed in Cr[37] and Mo spectra.[12] Such changes may provide an additional way of investigating interactions, particularly of p electrons with the matrix cage. It is, therefore, of interest to obtain reliable values of ζ in the different matrices. Because of the splitting of the $^2P_{3/2}$ state evident from an examination of the spectra (Figure 3), one cannot simply take the $^2P_{1/2}$–$^2P_{3/2}$ separation as equal to $\frac{3}{2}\zeta$. Rather, one must first inquire into the origin of the splitting and determine its relationship to the spin-orbit interaction. Kasai and McLeod[33] have concluded, on the basis of esr measurements, that Au atoms are located at substitutional sites in Xe, Kr, and Ar matrices. They interpreted the sharp and isotropic esr signals as indicating little deviation of the symmetry of the trapping site from an octahedral type. The two findings are contradictory because the van der Waals diameter of Au (4.6 Å) is larger than the nearest neighbor spacings in Ar, Kr, and Xe lattices, which are 3.83, 4.05, and 4.41 Å, respectively. A possible rationalization of these results, which would also be consistent with the optical data, is to postulate a distorted tetradecahedral substitutional site. The distortion is large enough to cause a "crystal"-field splitting[15] of $\sim 600\ \mathrm{cm}^{-1}$ in the excited $^2P_{3/2}$ state, $41{,}000\ \mathrm{cm}^{-1}$ above ground, but not sufficiently large to affect the isotropic nature of the esr signals obtained on ^2S ground state atoms. Anisotropic contributions to the hyperfine tensor can be estimated from $\langle r^{-3}\rangle_{nl}$ parameters calculated from Hartree–Fock wavefunctions. Reasonable values of 0.1–0.01 for p-type admixing coefficients give approximately 1% p character to the ground state. The experimental line width of the esr signal would then hardly be exceeded, and a distortion of the substitutional site would go undetected.

The following calculations are based on a model of a tetradecahedral substitutional site axially distorted by elongation of the distances between the Au atom and the three matrix atoms above and below the plane relative to the distances to the six matrix atoms in the plane.

One simultaneously diagonalizes the spin-orbit and crystal field interactions by finding the eigenvalues of the matrix[38]

$$\begin{array}{cccc}
M_L, M_S & 1, \pm\frac{1}{2} & -1, \mp\frac{1}{2} & 0, \pm\frac{1}{2} \\
 & \dfrac{A_2}{2} + \dfrac{\zeta_2}{2} - E & 0 & 0 \\
 & & \dfrac{A_2}{2} - \dfrac{\zeta_2}{2} - E & \dfrac{2^{1/2}}{2}\zeta_2 \\
 & & & -A_2 - E
\end{array}$$

The three roots of the secular equation are given by the expressions

$$E_1 = E(^2\pi_{3/2}) = \tfrac{1}{2}(\zeta + A_2) \tag{2}$$

$$E_2 = E(^2\Sigma^+) = -\tfrac{1}{4}\{\zeta + A_2 - 3(\zeta^2 + A_2^2 - \tfrac{2}{3}\zeta A_2)^{1/2}\} \tag{3}$$

$$E_3 = E(^2\pi) = -\tfrac{1}{4}\{\zeta + A_2 + 3(\zeta^2 + A_2^2 - \tfrac{2}{3}\zeta A_2)^{1/2}\} \tag{4}$$

In these expressions, A_2 is the single parameter characterizing the magnitude of the axial electric field.

If one defines the energy differences

$$\Delta E_1 = E_2 - E_3 \tag{5}$$

$$\Delta E_2 = E_1 - E_3 \tag{6}$$

it can easily be shown that

$$\zeta_2 = \tfrac{1}{3}(4\Delta E_2 - 2\Delta E_1) - A_2 \tag{7}$$

$$A_2 = \tfrac{1}{3}\{2\Delta E_2 - \Delta E_1 - [\Delta E_1^2 - 2\Delta E_2(\Delta E_1 - \Delta E_2)]\} \tag{8}$$

The values of ζ_2 and A_2 entered in Table 5 were calculated from the experimental data using equations (7) and (8). Another calculation of ζ and A

Table 5. Spin-Orbit Coupling Constants and Crystal-Field Parameters Determined for Au in Various Noble Gas Matrices

		Matrix Gas		
	Gaseous Atom	Ar	Kr	Xe
$E(^2\pi_{1/2})$	37359[d]	39175	38288	36875
$E(^2\Sigma^+)$	41174	43763	42475	40525
$E(^2\pi_{3/2})$	41174	44375	43113	41090
ΔE_1[a]	3815	4588	4187	3650
ΔE_2[b]	3815	5200	4825	4215
ζ_1	2543	3263	3004	2622
$A_1 = \Delta E_3$[c]	—	612	638	565
ζ_2	2543	3215	2943	2570
A_2	—	660	697	618

[a] $\Delta E_1 = E(^2\Sigma^+) - E(^2\pi_{1/2})$.
[b] $\Delta E_2 = E(^2\pi_{3/2}) - E(^2\pi_{1/2})$.
[c] $\Delta E_3 = E(^2\pi_{3/2}) - E(^2\Sigma^+)$.
[d] All values given above are in units of cm^{-1}.

can be carried out in the M_J system of quantization. This approximate calculation gives the expressions

$$\zeta_1 = \frac{2}{3}\left(\Delta E_1 + \frac{A_1}{2}\right) \tag{9}$$

$$A_1 = \Delta E_3 = E_1 - E_2 \tag{10}$$

The values, ζ_1 and A_1, entered in Table 5 were calculated from the optical data using equations (9) and (10). It can be seen that the values of ζ_2 obtained by simultaneous diagonalization of spin-orbit and crystal-field interactions are ~2% lower than the values obtained from the more approximate calculations, while the A_2 values are ~10% higher than the A_1 values.

The A_2 crystal-field-splitting parameters have total variability of 12% going from 660 cm^{-1} in Ar to 697 cm^{-1} in Kr to 618 cm^{-1} in Xe. The induced multipole-induced multipole interactions responsible for the splitting of the $^2P_{3/2}$ state may be functions, $f\left(\frac{\alpha^x}{r^y}\right)$, such that the increase in internuclear distance, r, is approximately compensated for by the increasing polarizability, α, of the heavier matrix atoms.

The spin-orbit coupling constant changes by 23%. From 2543 cm^{-1} in the gaseous Au atom, ζ increases to 2570 cm^{-1} in Xe, to 2943 cm^{-1} in Kr, and to 3215 cm^{-1} in Ar.

Spin-Orbit Coupling

The calculation of the ζ parameter given in the previous section is strictly true only in the limit of zero configuration interaction (CI). There are two lines of evidence that indicate that configuration interaction may be important in Au. A strong two-electron transition is observed in the emission spectra of Au(I)[22,23] between the ^2P and ^2D states ($gf = 0.27$) whose enhanced intensity could be due to CI of the type:

$$^2P = a(5d^{10}6_p) + b(5d^9 6s6p) \tag{11}$$

$$^2D = c(5d^9 6s^2) + d(5d^9 6p^2) \tag{12}$$

Relativistic Hartree–Fock calculations without CI give too low an energy separation for the $^2P_{1/2} - {}^2P_{3/2}$ states and oscillator strengths for the $^2S \rightarrow {}^2P$ transitions that are too high by a factor of two compared with experimental values.[39] It is likely, although the appropriate calculations have still to be performed, that the energy and intensity discrepancies between theory and experiment will be minimized when CI is properly included in the calculations.

In the absence of quantitative data on this mixing of the $5d^{10}6p$ and $5d^9 6s6p$ configurations, no estimates can be made of the contribution to the $^2P_{1/2}$–$^2P_{3/2}$ splitting energy due to spin-spin and spin–other-orbit interactions.[40] Furthermore, the Z dependence of the spin-orbit coupling interaction is a matter of considerable complexity and is still a subject of theoretical study.[41] In the following discussion, the $6p$ electron is assumed to be moving in a hydrogenic potential field.

It can be shown[42] that ζ is proportional to Z_{eff}^4 for hydrogenic wavefunctions. The experimentally determined $^2P_{1/2}$–$^2P_{3/2}$ separations for the isoelectronic series, Au(I), Hg(II), Tl(III), Pb(IV), and Bi(V) are 3815, 9123,

14,813, 21,061, and 27,920 cm^{-1}, respectively.[23] With ζ scaling as Z_{eff}^4, these energies can be calculated to an accuracy of $\pm 10\%$ using the expression

$$\Delta E(^2P_{1/2}-{}^2P_{3/2})\ \text{cm}^{-1} = \frac{3}{2}\zeta_{\text{Au(I)}}\left(\frac{Z_{\text{eff}}[\text{IonX}]}{Z_{\text{eff}}[\text{Au(I)}]}\right)^4$$
$$= \frac{3815.4}{(4.8)^4}(Z_{\text{eff}}[\text{IonX}])^4$$

with the following values of Z_{eff}; Au(I) = 4.8, Hg(II) = 5.8, Tl(III) = 6.65, Pb(IV) = 7.35, and Bi(V) = 7.9. Using Z_{eff} = 4.8 for the neutral gaseous Au atom, one can calculate Z_{eff} for Au atoms imbedded in noble gas matrices from the experimentally determined ζ values (Table 4). In this way, one obtains $Z_{eff}[M]$ for Au atoms in the various matrices, M: $Z_{eff}[Xe] = 4.83$, $Z_{eff}[Kr] = 4.99$, and $Z_{eff}[Ar] = 5.11$. The 26% change in ζ going from gaseous Au atoms to Au atoms isolated in an Ar matrix, for example, can be rationalized in terms of a 6% increase in Z_{eff}.

The increase in Z_{eff} is presumably due to the increase in electronic charge density of the $6p$ electron near the nucleus of the Au atoms due to compression of the $6p$ wavefunction (primarily the wavefunction of the $^2P_{3/2}$ state) by the noble gas cage. The degree of $6p$ wavefunction compression appears to be related to internuclear spacing, to the polarizabilities of the noble gas atoms, and to site symmetry considerations.

The Au atoms are located in substitutional sites in Ar, Kr, and Xe matrices. The esr spectrum of Au atoms[33] in Ne matrices shows quite a different situation, however, in that Au atoms occupy at least two different sites in contradistinction to single-site occupation in Ar, Kr, and Xe. The hyperfine coupling constants determined by Kasai and McLeod[33] display parallel behavior to the spin-orbit coupling constants in Ar, Kr, and Xe matrices. The reversal in trend for the hyperfine coupling constant in Ne, a value that is almost the same as that in Kr, suggests that similar behavior will be found for the spin-orbit coupling constant if the packing is such as to compensate for the low polarizability of N3.

The results obtained on Au atoms in noble gas matrices is reminiscent of studies by Robin and Robin[43] on the effect of foreign gas pressure on the energies of the resonance lines of alkali atoms. These workers found that below 300 atm of Ar or N_2, the $^2P_{1/2}-{}^2P_{3/2}$ separation of the Rb resonance lines, for example, remains almost constant. However, as the pressure of the perturbing gas is raised above 300 atm, the $^2P_{1/2}-{}^2P_{3/2}$ separation increases rapidly. (See also the discussion of Rb atoms isolated in Kr matrices.[76]) Takeo and Ch'en have derived an expression for the spin-orbit interaction that specifically takes into account changes in the atomic field in which the valence electron moves due to pressure of a foreign gas.[29,44] At high pressures, the term involving the influence of pressure on the effective nuclear charge dominates the expression. The model of Takeo and Ch'en invokes distortion of the atomic core, with a resulting smaller screening effect as the mechanism for an increase in ζ.

The model proposed to account for the increase in ζ due to interaction with 12 nearest neighbor noble gas atoms in solid noble gas matrices postulates compression of the Au $6p$ wavefunction due to overlap with the 3, 4, and $5p$ wavefunctions of Ar, Kr, and Xe. The degree of compression would appear to be proportional to the repulsive term in the interatomic potential between the Au atom and the noble gas atoms.

Oscillator Strengths of Au Atoms in Matrices

The experimental and calculational procedures employed in arriving at the $f_{\mathrm{lu}}^{(\mathrm{m})}$ values for the ${}^2P_{1/2} \leftarrow {}^2S_{1/2}$ and ${}^2P_{3/2} \leftarrow {}^2S_{1/2}$ transitions listed in Table 3 have been described in detail in the literature.[1] As already stated, the ${}^2P_{3/2} \leftarrow {}^2S_{1/2}$ intensity was taken as the sum over the ${}^2\pi_{3/2}$ and ${}^2\Sigma^+$ components, whose relative intensities (see Figure 3) remain almost constant at 3:2 in all three matrices. Although $f_{\mathrm{lu}}^{(\mathrm{m})}$ values decreases by a factor of two in going from Ar to Xe, the intensity ratios for the two resonance transitions range narrowly between 2.2 and 2.4. In order to afford a comparison of $f_{\mathrm{lu}}^{(\mathrm{m})}$ values obtained in noble gas matrices with f_{lu} of gaseous Au atoms, a compilation of literature data on the latter is given in Table 4. The numbers vary by factors of 3–5 as between emission[22] and "hook" measurements.[45] Furthermore, the gaseous-atom intensity ratios for the two resonance transitions vary in the range, 1.5–2.3. The experimental approaches to oscillator strength determinations of gaseous atoms have been critically assessed.[46] The "hook" method, which depends on the interferometric measurement of the anomalous dispersion of spectral lines, was developed by Roschdestwensky.[47] When combined with absorption measurements this method has led to the determination of absolute values of f_{lu} by Ostrovskii, Penkin, and Shabanova.[48] Using these methods, Penkin and Slavenas[45] obtained the values, $f_{\mathrm{lu}} = 0.19$ for the transition to ${}^3P_{1/2}$ and $f_{\mathrm{lu}} = 0.41$ for the transition to ${}^2P_{3/2}$. The f_{lu} ratio for the two transitions in matrices are quite close to the gas-phase ratio. In an Ar matrix, the absolute values are lower by about 30% compared to the gas-phase values as determined by the "hook" method. In the other matrices, the values are considerably lower than the "hook" values, but still within the range of variability of the gas phase values listed in Table 4.

Quantum-mechanical calculations of the oscillator strengths of the Au(I) resonance doublets have been carried out by Slavenas,[49] giving f_{lu} values of 0.30 and 0.65. A more recent relativistic Hartree–Fock calculation[39] also gives f_{lu} values that are a factor of about two higher than the "hook" numbers. As mentioned earlier, the major discrepancy between theory and experiment may be resolved when CI is properly included in the calculations.[39] It is of interest to note that the matrix results may provide support for this conjecture. Examination of Figure 3 and Table 3 shows that absorption bands in the 45,000–47,000 cm^{-1} region, which are most reasonably assigned to states of the $5d^9 6s6p$ configuration, become increasingly intense in the series, Ar → Kr → Xe. Mixing of this configuration with the $5d^{10} 6p$ configuration, resulting in intensity borrowing, could explain the decreasing intensity of the resonance transitions in the heavier noble gas matrices.

Measurements of oscillator strengths on matrix-isolated Ti atoms[10] and on matrix-isolated Cr, Fe, Ni, Cu, Pd, and Sn atoms[21] have given values 10–100 times lower than f_{lu} values based on emission measurements for the respective gaseous atoms,[22] while $f_{lu}^{(m)}$ values for the Au resonance lines obtained in the present series of measurements are fairly close to the gaseous-atom oscillator strength. By analogy with results of oscillator strength determinations on metal atoms perturbed by high pressures of a foreign gas, f_{lu} values can be expected to increase, decrease, or remain relatively unaffected. Judging from the results of the few relevant experiments that have been performed, the effect on the oscillator strengths seems to depend on the particular electronic transitions under investigation and for a given transition, to vary with the perturbing gas. The oscillator strength of the $^3P_1 \leftarrow {}^1S_0$ transition in Hg has been found to decrease slightly at pressures up to 50 atm of Ar, CO_2, H_2, and H_2O.[50] At higher pressures, the decrease is as much as a factor of four in He up to 500 atm[51] and a factor of five in Ne up to 1400 atm.[52] On the other hand, the decrease in Ar up to 50 atm is reversed at higher Ar pressures, and the oscillator strength is larger than the unperturbed value by a factor of three at 500 atm. That large changes in oscillator strength due to perturbing gases at high pressures are not restricted to forbidden transitions such as the $^3P_1 \leftarrow {}^1S_0$ transition in Hg is illustrated by the work of Ch'en,[53] who showed that f_{lu} values of the Rb resonance doublet decrease by factors of two to eight in He and Ar at pressures up to 100 atm.

Oscillator strengths calculations for the LiHe and NaHe $X^2\Sigma^+ \leftarrow A^2$ molecular transitions show decreasing f_{lu} values with decreasing Li—He or Na—He distance along the molecular axis.[54] In connection with these calculations, it is of interest to note that Kraus, Maldanado, and Wahl[54] find the potential curves for the $^2\pi$ states of LiHe and NaHe to be attractive in nature.

3. OPTICAL SPECTROSCOPY OF MATRIX-ISOLATED ATOMS

In this section, the absorption spectra, primarily in the 5000–50,000 cm^{-1} region, of matrix-isolated atoms is reviewed group by group, and where possible the spectra of two or more electronically similar atoms are compared.

3.1 Group IA: The Alkali Metals and Hydrogen

Matrix-isolated H atoms were apparently first observed[55] by bombarding solid H_2 with positive ions and measuring the resulting luminescence spectrum.

The H absorption spectrum in solid Ar has the $1s \rightarrow 2p$ transition[56,57] at 10.6 eV (85,200 cm^{-1}). It is shifted to higher energies from the atomic value (10.20 eV) by almost 3000 cm^{-1} and has a half-width of about 2000 cm^{-1}. The matrix shift was computed to first order using a tight-binding treatment,[58] resulting in calculated values of 10.6 and 12.3 eV for the $1s \rightarrow 2p$ and $1s \rightarrow 2s$ transitions, respectively. Unfortunately, the higher energy transition is obscured in the matrix by the rare gas absorptions, but the agreement of the theoretical model with experiment for the $1s \rightarrow 2p$ transition is good.

The spectrum of D atoms in solid Ar[56] shows the expected isotope shift, with a band maximum for D atoms about 0.015 eV above that of H atoms. The band width for D in Ar is about 15% less than that of H in Ar, possibly due to changes in the localized vibrational modes involving the impurity atoms. It should be noted that the isotope shift is only slightly larger than the quoted experimental error (0.01 eV).

The atomic spectra of the alkali metals are reasonably simple. AMCOR predicts two absorption band systems in the uv-visible spectra of matrix-isolated Na—Cs corresponding to the $ns \rightarrow (n + 1)p$ and $ns \rightarrow (n + 2)p$ transitions. In the case of Li, a third absorption band arising from the $ns \rightarrow (n + 3)p$ transition is expected. Each atomic ^2P excited state is split by spin-orbit coupling into its $J = \frac{1}{2}, \frac{3}{2}$ components. The magnitude of the splitting increases from a few wavenumbers in the case of Li to several hundred wavenumbers for atomic Cs.

Because of the expected simplicity of their spectra and because of the relative ease with which atomic beams of the alkali metals can be produced, one would expect them to be ideal candidates for studying their matrix spectra.

Indeed, the alkali metals have been studied more widely than any other group. Thus, Na along with Hg appear to have been the first metal atoms to be doped into matrices.[9] Li has been studied in Ar,[59,60] Kr,[59,60] and Xe[59–61] matrices. Na and K have received similar attention in all three matrices.[61–63] The absorption spectrum of Rb in Ar[63,64] and of Cs in Ar, Kr, and Xe[63] have been reported. Data have also been obtained on the Rb^+ "shake up" spectrum after the β decay of ^{85}Kr in solid Kr[73,74] and on optical emission spectra of Rb atoms isolated in solid Kr.[76]

As was discussed above, two (or three for Li) $s \rightarrow p$ transitions are predicted by AMCOR,[2] with the low-energy system expected to have an intensity about an order of magnitude higher than the high-energy transitions. Only the low-energy systems have been observed in matrices up to now. No bands have so far been reported for the higher energy transitions for any of the alkali metals. Such bands should be observable even though of low intensity.[62,64]

In all cases except Li the $s \rightarrow p$ transition appears as two triplets, one (the "red" triplet) centered in energy near the atomic value and a second (the "blue" triplet) at higher energies.

There are a number of discrepancies both in reported band positions and, in some cases, the number of observed bands. Band positions tend to agree to within about 50 cm^{-1} with errors due mainly to the difficulty of determining the maxima of the broad bands. Experimental conditions such as the deposition temperature also affect band position. For example, an energy shift of about 2 $cm^{-1}/°K$ for Na and K spectra has been reported.[62]

The "blue" triplet is not always reported, perhaps due to its relatively low spectral intensity. A large decrease in intensity of the "blue" triplet observed on going from a sapphire to a glass substrate, illustrates an unusual sensitivity of this band to deposition temperature.[63] The site that causes the "red" triplet is unstable against matrix annealing.[62,134] The "blue" triplet has in fact a quartet or an even more complicated structure.[64,134] The epr spectrum, related to the "blue" absorption band, shows a large positive shift of the hyperfine coupling constant:[64] this supports the supposition that the site that causes the "blue" band and is the more stable one is an interstitial site.[135] The observed multiplet structure of these absorption bands may be explained by a removal of the orbital degeneracy of the excited alkali atom state.[9,62,63] This can be caused by either a static distortion of the cubic lattice symmetry by the alkali atom as theoretically analyzed by Brith and Schnepp,[15] or by a dynamic interaction of the excited alkali atom state with vibrational modes (Jahn–Teller effect).[9] An interpretation of the multiplet structure by interactions with non-nearest-neighbor alkali atoms[59,61] has been experimentally disproved by Brewer and King.[36] Since the excited alkali atom state can only split into two or three levels, the observed quartet structure of the "blue" absorption band is evidence for at least two different configurations of the corresponding trapping site.

However, in recently published papers,[60] where the absorption as well as the luminescence spectra of alkali atoms in rare gas solids are reported, it is claimed that the multiplet structure in the absorption bands is only caused by a Jahn–Teller effect. The authors come to this conclusion due to the observation of a single line, that is, no multiplet structure in emission. The interpretation of the observed multiplet structure in the spectra of alkali atoms in rare-gas solids, therefore, is still under discussion, and the whole topic must be reconsidered in view of the rather significant changes in the spin-orbit coupling parameter of Au atoms isolated in noble gas matrices that have recently been observed.[1]

In recent work, Micklitz and Luchner[76] interpreted the emission spectrum of Rb atoms isolated in Kr matrices in terms of an interstitial site with C_{4v} symmetry having a small orthorhombic distortion.

The $5p \leftarrow 5s$ resonance transition in Rb atoms isolated in solid krypton was observed in emission by studying the recombination process in a ^{85}Kr crystal: $^{85}Kr \xrightarrow{\beta^-} Rb^+$; $Rb^+ + e^- \rightarrow Rb^* \rightarrow Rb + h\nu$. The observed emission lines were much narrower than the usually observed absorption lines.

Micklitz and Luchner[76] assumed that the spin-orbit coupling parameter was unchanged from the free Rb atom value. They chose the form of their crystalline potential on the basis of the observed triplet structure that showed a large singlet-doublet splitting relative to the splitting within the doublet. Such a crystal-field symmetry could be produced in different ways: for example, either by a vacancy neighboring the Rb atom on a octahedral interstitial site (as discussed by Brith and Schnepp[15]), or by a "dumbbell" configuration of the interstitial site that appears to be the interstitial configuration in fcc metals[136,137] together with a slight orthorhombic distortion of these lattice sites. The analysis of the triplet gives $a = 2430 \pm 40$ cm^{-1} and $b = 270 \pm 40$ cm^{-1}. The crystalline field parameter "a" has the order of magnitude calculated by Brith and Schnepp[15] for a vacancy neighboring an alkali atom. However, it should be pointed out, particularly in view of the results of Robin and Robin[43] mentioned earlier, that the assumption of an unchanged spin-orbit coupling parameter is probably unwarranted.

Th results for Li differ from the other alkali metals in that only bands corresponding to the "red" triplet are observed with reports as to the number of components varying from four or five[59,60] to seven.[61] Whether or not the extra features are due to impurity bands is still an open question.

Li appears to undergo diffusional dimerization in matrices with Li_2 bands observed near 20,000 cm^{-1}. Andrews and Pimentel estimated the Li/Li_2 ratio ranges from 10^{-4} to 1 depending on dilution.[59] Only at extremely low Li concentrations was Li_2 unobservable, thus illustrating the high mobility of the small Li atom and the difficulties associated with its isolation in noble gas matrices.

3.2 Group IIA: Mg, Ca, Ba

In this group, Mg,[16,65,66] Ca,[13,66,67] and, less extensively, Ba[66,68] have been isolated and studied spectroscopically in rare-gas matrices. Much of the work[13,65–67] has been done at hydrogen temperatures (16–20°K). Both for Mg and Ca, diffusion during the deposition process plays an important role, causing difficulties in isolation and spectral interpretation. Thus, for deposits made at 20°K, Mg and Ca atoms cannot be isolated in Ar as monomers, in Kr both atoms and dimers are found in the spectra, while in Xe the predominant species are isolated atoms. It has been noted that the spectrum of Ca in Ar bears no resemblance to that in Kr and Xe, presumably because of the formation of dimers and higher polymers (Fl). On the other hand, the absorption spectra of isolated Mg atoms appear to have been obtained without the difficulties mentioned above by operating at liquid helium temperatures.[16]

For Mg and Ca only one absorption band is expected to occur with appreciable intensity arising from the $^1S_0 \rightarrow {}^1P_1$ ($ns^2 \rightarrow nsnp$) transition. In most studies, this $s \rightarrow p$ transition is reported to be split into three components similar to the results for the alkali metals. The matrix shifts of the absorption bands decreases along the series, Ar to Kr to Xe, becoming negative for the latter compared to the AMCOR values. The negative shift for Mg in Xe is unusually high, -1010 cm^{-1}.[16] The shifts become smaller in going from Mg to Ca to Ba.

The temperature dependence of the band maximum for Ca and Mg has been reported for several matrices.[66] As the temperature of the deposit is raised, the center of gravity of the $s \rightarrow p$ transition shifts to higher energies and the splitting between components increases. Francis and Webber also noted that the temperature dependence can be of the same order of magnitude as the band splitting,[67] and changes in Ca band widths have been noted for various deposition temperatures and concentrations.[69]

As with the alkali metals, there are considerable discrepancies in the reported data. Although the differences between the Mg data could be due to variations in deposition temperatures, the discrepancies in the three sets of Ca data cannot be rationalized in this way.

In addition to the major triplet, two other bands were observed, in the Mg spectrum in Ar and Kr, but not in Xe spectra.[16] These extra bands were attributed to unstable trapping sites. Additional bands were also noted in the spectrum of Ca in Kr and Xe,[67] which were also tentatively assigned to Ca occupying nonequivalent matrix sites.

Also reported were a progression of lines believed to arise from Ca_2.[67] The bands occur only when high concentrations of metal are used and occur in regions expected for such a molecule. Dimer formation has also been noted for Mg_2 in Ar and Kr matrices.[67] From temperature studies it was concluded that dimer formation occurs on deposition as a result of surface diffusion.[67] Dimer formation apparently does not take place on isochronal annealing of Kr to 45°K and of Xe to 65°K.

An extra band has been reported in the spectrum of Ca, which was assigned to the $^1S \rightarrow {}^1D$ atomic transition,[67] although this transition has a low oscillator strength in the atomic spectrum. This assignment may be questioned particularly in view of a number of other observed bands for which no assignments are given.[67]

The spectrum of Ca atoms has also been studied in a number of hydrocarbons matrices at various temperatures.[70]

3.3 Groups IIIA to VIIA: Al, Ga, In, Tl, Sn, Pb

For the Group IIIA metals, matrix spectra have been reported for Al,[71] Ga,[71] In,[72] and Tl.[75] Although studies of C_2,[74] C_3,[74,75] and C_2^-[138] species have been published, the spectrum of isolated C atoms has apparently not yet been reported. Among the remaining atoms in Group IVA, only Sn[21] and Pb[13,66] have been studied in the matrix. Finally in Groups VA, VIA, and VIIA, only N and O have been examined, much of the work being directed towards studies of their luminescence spectra discussed in a later section.

The spectra of Group IIIA elements differ from those discussed previously in that $P \rightarrow S$ transitions are observed in absorption, and one would not expect to observe splittings in the excited state.

The uv-visible as well as the esr spectra of Al and Ga atoms isolated in Ne, Ar, Kr, and Xe matrices have been studied by Ammeter and Schlosnagle.[71] The lowest energy band in each matrix is assigned to the $^2P \rightarrow {}^2S$ transition corresponding to one-electron jumps out of the singly occupied np shells into the empty $(n + 1)s$ shells. The higher energy bands with more complex structure are assigned to $^2P \rightarrow {}^2D$ transitions. The spectra of Al and Ga atoms bear a pronounced resemblance to each other. The two lowest energy "resonance" transitions in the free Al and Ga atoms are quite close to one another. Furthermore, as pointed out by Ammeter and Schlosnagle, the Hartree–Fock valence orbital radii for these two atoms are closely similar and, thus, they would be expected to occupy similar sites in the various matrices. The $^2P \rightarrow {}^2S$ transitions consisting of a single well-shaped band with a half-width of $\sim$600 cm^{-1} for both metals in argon and neon, but exhibiting superpositions of bands in the behavier matrices, shows a matrix and temperature behavior very similar to the closely related $^3P_0 \rightarrow {}^3P_1$ transition of matrix-isolated Pb atoms, studied by Brewer and Chang.[13] The matrix shifts for both Al and Ga in all matrices were calculated for the $^2P \rightarrow {}^2S$ transitions and compared to the shifts for Pb obtained by Brewer and Chang (Table 6). The matrix shifts are towards the blue and increase strongly from the heavy to the light rare gas matrices. Also in agreement with the observations by Brewer and Chang, the $^2P \rightarrow {}^2S$ bands for Al and Ga shift reversibly to the red on warming in all matrices; for example, temperature increases of $\sim$20°K resulted in shifts of -110 ± 20 cm^{-1} for Ga in argon and -90 ± 20 cm^{-1} for Ga in krypton, while no significant shifts in the short-wavelength part of the spectra were noted. When the

Table 6. Matrix Shifts for $(n + 1)s \leftarrow np$ Transitions

	Al[a]	Ga[a]	Pb[b]
Transition	$4s(^2S) \leftarrow 3p(^2P_{1/2})$	$5s(^2S) \leftarrow 4p(^2P_{1/2})$	$6p7s(^3P_1) \leftarrow 6p^2(^3P_0)$
in Xe	+1,640[c]	(+520, +2090)	+2,480
in Kr	+1,820	+1,860	+2,970
in Ar	+4,150	+4,360	+5,550
in Ne	+5,900	+5,740	—

[a] Reference 71, calculated from strongest band observed, at 4.2°K.
[b] Reference 13, T = 20°K.
[c] All values given above are in units of cm^{-1}.

temperature was increased above about one-third of the melting point of the rare-gas matrices, irreversible loss of the uv spectra resulted, while new absorption bands appeared over the whole visible range, attributable to dimer and cluster formation.

The results in Table 6 show that the matrix shifts for Al and Ga at 4.2°K are very similar, but are significantly smaller than the corresponding shifts for Pb at 20°K. The close similarity of the *p*-shell diameters suggests that these three atoms occupy similar sites, and the observed matrix shifts reflect the expected strongly repulsive matrix interaction of the excited states having the common characteristic feature of a single electron in an outer *s* orbit that penetrates the valence shells of the rare gas cage substantially. Furthermore, the observed trends (largest repulsion in the lightest rare gases, red-shift upon lattice expansion by warming) are in agreement with what would be expected from atoms occupying substitutional sites. This point will be discussed in a later section on esr results. In any case, the matrix shifts, and temperature dependences of these bands are larger and more pronounced than atoms with $ns \rightarrow np$ resonance transitions characteristic of the alkali, alkaline earth, and Group IB metals, for example.

The reported vacuum-uv spectrum of In in Ar, Kr, and Xe is dominated by two groups of bands.[72] The low-energy system appears as a triplet in Ar, but is unresolved in Kr or Xe, and has been assigned to the $^2P^0_{1/2} \rightarrow {}^2P_{1/2,3/2}$ transition. All of these bands lie below the energies of the corresponding transitions of the gaseous atoms. The lack of the high-energy bands in Xe is not understood.

The uv region of the spectrum is more perplexing. Indium in the gas phase[2] displays six resonance lines, eight with gf values ranging from 0.015 to 0.99 and two with lower intensity: $^2P^0_{1/2} \rightarrow {}^4P$ ($gf = 0.0055$) and $^2P^0_{1/2} \rightarrow {}^4P_{3/2}$ ($gf = 0.0091$). Duley[72] assigns one band occurring at about the same energy in all matrices to the former transition. Another band appearing only in Ar is assigned to the latter. Other bands with higher *gf* values and, therefore, higher expected intensities are not observed. It seems clear that more experimental work is needed on this element.

A preliminary study of Tl atoms in Kr and Ar has been made,[75] but the matrix spectrum could not be adequately interpreted in terms of a correlation with the atomic spectrum.

Two low-lying transitions in Sn have been assigned to $^3P_0 \rightarrow {}^3P^0_{0,1}$ ($5s^25p^2 \rightarrow 5s^25p6s$),[21] while higher energy bands have been assigned to $^3P_0$ ($5s^25p^2$) $\rightarrow$ $^1P^0_1$ ($5s^25p6s$), $^3D^0_1$ ($5s^25p5d$) transitions. Relative *gf* values were estimated and found to be qualitatively correct.

In the case of Pb atoms, a single absorption band has been observed[13] that decreases in energy in the series, Ar to Kr to Xe. The band is broad and shifts to the red reversibly on warming. The band, assigned to the $^3P_0 \rightarrow {}^3P_1$ ($6p^2 \rightarrow 6p7s$) transition, has no resolvable structure, in contrast to several of the other matrix spectra already discussed. This particular transition has *gf* values that are a factor of three to six larger than those of other four lines listed in the AMCOR tables.[2] Detailed studies of the Pb spectrum in the uv or near-vacuum uv would probably reveal these higher energy transitions.

The atomic absorption band of Pb was not observed in SF_6 matrices or mixed noble-gas–SF_6 matrices.[13] The formation of a charge-transfer complex or a chemical reaction could explain the absence of the transition, but no evidence for the occurrence of either phenomenon has as yet been adduced.[13]

Much of the work reported on Pb[13] has dealt with the formation of Pb_2 that was observed in all matrices. The results indicated that diffusion of Pb occurred only during the initial stages of the matrix formation, and little or no dimerization occurred as a result of isochronal annealing experiments.

Two attempts have been made to obtain the absorption spectra of N and O atoms. Harvey and Brown[86] studied spectra of condensed N_2 obtained from a discharge in the gas phase. Bass and Broida[87] studied the discharge products of N_2 and O_2 in the near-ir, visible, and uv regions, finding only weak absorption features due to N_2. Neither study has yielded data on the atomic absorption spectra of N or O in matrices. In any event, the absorption spectra of C, N, O, and F atoms in low-temperature matrices would be expected to lie in the vacuum uv.

3.4 Group VIIIA: The Noble Gases

The atomic absorption spectra of Ar, Kr, and Xe lie in the vacuum-uv region at energies higher than above 66,000, 80,000, and 96,000 cm^{-1}, respectively. Spectra of both pure rare gas matrices[77,78] and matrix-isolated noble gas impurities in noble gas matrices[79–82,159,160] have been studied.

Generally, only the two lowest absorption bands are observed because of instrumental limitations or solvent absorption. Baldini,[78] however, has reported the spectra of solid Ar, Kr, and Xe to 900 Å (113,000 cm^{-1}) and noted a number of additional bands at higher energies. Schnepp and Dressler[77] reported four bands for solid Xe that were correlated with transitions to four bound excited states of Xe_2. Only two bands are expected in this region on the basis of the atomic model, and, in fact, Baldini found only two.

For the limited studies reported, the bands follow the usual trend, decreasing in energy in the series, Ar to Xe. In Xe matrices the energies are very close to the gas-phase values. This conclusion holds even for homogeneous matrices, that is, transitions for Xe atoms in a Xe matrix or for Kr atoms in a Kr matrix.

A pronounced temperature effect has also been found in the spectra of solid rare gases.[78] The absorption bands shifted to lower energies by amounts ranging from 200 to 500 cm^{-1} upon annealing the matrices at 40–50°K. These results may account for the disagreement in band energies quoted by various authors.[77,78]

3.5 Group IB: Cu, Ag, Au

Spectra of matrix-isolated Cu, Ag, and Au are similar to those previously discussed in that the spectra are dominated by intense bands arising from $s \rightarrow p$ transition: $^2S \rightarrow {}^2P[nd^{10}(n+1)s \rightarrow nd^{10}(n+1)p]$. The spectra for matrix-isolated Au atoms have already been discussed in an earlier section of this chapter. The results differ for Ag and Au in that the spin-orbit splitting of the atomic $^2P_{1/2}$ and $^2P_{3/2}$ states becomes larger than the half-width of the absorption bands in the spectra of the matrix-isolated atoms. The splitting for Cu, Ag, and Au is 249, 921, and 3815 cm^{-1}, respectively, in the gaseous atoms.[2] Three bands are observed in the case of Cu[2,83,84] and Au,[1,83] corresponding to the $^2S \rightarrow {}^2P_{1/2}, {}^2P_{3/2}$ transitions, the latter being split by crystal-field effects in a manner similar to the Group IA elements. For matrix-isolated Au, however, the splitting between the $^2P_{1/2}$ state and the average of the split $^2P_{3/2}$ state is 3800 cm^{-1}, whereas for Cu it is only 410 cm^{-1}.

The spectrum of Ag is complicated by an additional band.[83–85] Four bands were observed[85] in the spectrum of Ag isolated in Ar, Kr, and Xe, rather than the expected three. The extra band may be due to a dimer transition of Ag_2.[85] In subsequent work, the spectrum of Ag was remeasured and found to consist of four bands in Ar or Xe, but only three, as expected, in Kr.[83] The extra band in Xe (the third highest in energy) was tentatively assigned to the forbidden $^2S \rightarrow {}^2D_{3/2}$ transition. Unfortunately, the matrix shift, $-4000\ cm^{-1}$, does not support this assignment when compared to the shifts of the remaining bands, 100–500 cm^{-1}. No reasonable assignment could be found for the extra low-energy band in Ar. Thus, the spectrum of Ag has not yet been completely explained.

In a series of experiments the spectra of Kr matrices containing both Ag and Au atoms or Cu and Au atoms, were studied.[83] The resulting spectra were shown to be identical to the spectra of the metals deposited separately under the same conditions and at comparable concentrations. On the basis of these results, it was concluded that solute-solute interactions are not responsible for the observed splitting patterns.

Matrix-isolated Cu atoms display a number of transitions arising from $3d^{10}4s^1 \rightarrow 3d^9 4s^1 4p^1$ at energies somewhat higher than the $^2S \rightarrow {}^2P$ transition but still below 50,000 cm^{-1}. In both Xe[84] and Ar[2] the major band system displays a reasonably well-resolved triplet structure. Additional structure above and below the major triplet has been observed in Cu deposited in Ar near 15°K.[2] Other lines at energies lower than the triplet have also been seen in Ar, Kr, and Xe matrices deposited at 4.2°K, but were absent or very weak when deposited at 20°K.[21] In the latter study the bands, which annealed with increasing temperature, were assigned to Cu atoms in unstable sites, possibly in small amounts of hcp phase stabilized by the metal impurity.[27]

Four bands occurring regularly in the 35, 320–40, 270 cm^{-1} region when using 2% Cu in Xe have been tentatively assigned to Cu_2.[83] Two of these lines have also been observed at presumably lower concentrations.[84] Although Shirk and Bass[84] did not estimate their atomic concentrations, one would expect low yields from their sputtering apparatus used to prepare the Cu matrices. On the other hand, the sputtering process is known to result in the ejection Cu_2 molecules. Furthermore, recent work[158] at high Cu concentrations again has resulted in the observation of weak features in the 280–230 nm region, which have been assigned to Cu_2 molecules.

Carstens and Gruen[2] estimated the absolute absorption intensity of the $^2S \rightarrow {}^2P$ transition of Cu in an Ar matrix and found it to agree within a factor of two with that predicted from the gaseous emission spectrum of Cu. The intensities of the remaining bands ($d \rightarrow p$) were a factor of about five times more intense than expected from the gf values of the respective atomic transitions. These results suggest stronger matrix perturbation of the $d \rightarrow p$ compared to the $s \rightarrow p$ oscillator strengths.

The first reported low-temperature spectra of isolated Ag atoms were obtained in solid ethanol and ice at 77°K.[132] Frozen solutions containing silver salts were bombarded with X-rays to produce Ag atoms. The resulting absorption spectrum was independent of the anion used and consisted of a broad band centered near 25,400 cm^{-1}. Electron paramagnetic resonance measurements indicated the presence of Ag atoms. Weaker bands were attributed to Ag_2^+. Above 130°K, the broad band split into two narrower bands centered at 27,000 and 23,000 cm^{-1}, which persisted to 150–160°K before disappearing.

Brewer, Chang, and King[88] have used SF_6 and n-C_7F_{16} matrices for studies of Cu, Ag, and Au atoms. The Au bands in SF_6 were narrower than in rare-gas matrices, but the Cu and Ag bands were broader in SF_6. Isochronal annealing studies could be carried out in SF_6 to 77°K, but atomic transitions were not observed when the codeposition was carried out at 77°K. Nor could atomic transitions be observed in n-C_6F_{16} matrices.

3.6 Group IIB: Zn, Cd, Hg

From the AMCOR data all three of the Group IIB metals are expected to display intense bands in the near-uv region (40,000–50,000 cm^{-1}), but the bands for Hg are of a fundamentally different character. Both Zn and Cd atoms have a resonance transition in this region arising from the $^1S_0 \rightarrow {}^1P_1(ns^2 \rightarrow nsnp)$ promotion. This transition for Hg, however, occurs in the vacuum ultraviolet: instead, the 2537 Å line of Hg arises from the spin-forbidden $^1S_0 \rightarrow {}^3P_1^0(6s^2 \rightarrow 6s6p)$ transition that has gained intensity due to configuration interaction. The magnitude of this phenomenon in Hg is illustrated by the gf values of the $^1S_0 \rightarrow 3P_1$ transitions, which are 0.0018, 0.0014, and 0.34 for the gaseous Zn, Cd, and Hg atoms, respectively. The spin-forbidden transition of Hg is two orders of magnitude more intense than in the case of Zn or Cd. Probably because of its ease of vaporization, Hg was one of the first metals to be studied using matrix-isolation techniques,[9] and has been studied more extensively than any other metal atom.[61,75,81,82,89–94] The spectra of Zn[95–97] and Cd[61,66,84,95–97] have also received considerable attention.

Both Zn and Cd generally display the $^1S_0 \rightarrow {}^1P_1^0$ transition as a partially resolved triplet, although at high metal concentrations the bands broaden considerably.[97] The matrix shifts for both Zn and Cd follow the usual trend, with Ar showing the largest blue shift, Kr a smaller blue shift, and Xe a slight red shift.

In addition to the triplet, a number of subsidiary features have been reported.[97] First, a doublet band at energies higher than the triplet is noted in most spectra. Correlating these results with those on Hg atoms, the doublet was tentatively assigned to a $^3P_0^0 \rightarrow {}^3P_1(5s5p \rightarrow 5p^2)$ transition. Population of the $^3P_0^0$ state is assumed to occur by continuous uv excitation of higher excited singlet states followed by intersystem crossing to the triplet manifold. This assignment is discussed more fully below in connection with results on Hg atoms. A number of additional weaker, diffuse bands observed in all the spectra of Zn and Cd (except for Cd in Xe) were correlated with the known spectra of gaseous Zn_2 and Cd_2, but alternative assignments of these extra features to Zn or Cd atoms in unstable sites were also proposed.[97] Implied in the assignment of the extra spectral features to Zn_2 and Cd_2 dimers are matrix shifts of 2000–7000 cm^{-1}, which are unusually large for molecular species.

Other authors have reported only the triplet bands, probably because higher M/R ratios were employed and the measured spectra were less intense. However, several of the extra bands for Cd in Xe have been noted in other work.[84]

The $^1S_0 \rightarrow {}^3P_1^0$ band of Hg was shown to be a triplet in Ar, Kr, and Xe at 4.2°K with splittings of about 120 cm^{-1}.[9] These results were later contradicted when only a single band 370 cm^{-1} to lower energies was found at 20°K.[98] In still later work,[91] the transition was said to be a doublet. By varying several experimental parameters such as the Hg atom concentration, it, however, had been possible[90] to duplicate the earlier spectra.[9,98] At low Hg concentrations, the band is partially resolved. With increasing metal concentration, the spectrum red-shifts and merges into one broad band. These studies further showed that the band width and position is highly dependent on a number of experimental parameters, including temperature, the nature of the matrix, as well as small admixtures of impurities such as H_2O, N_2, O_2, and so forth.

Isolation of Hg atoms in matrices of pure Ne, N_2, and O_2 was found to give atomic transitions only in the former two.[90] In O_2 no spectra were obtained, probably because of oxide formation. The vacuum-uv transition, $^1S_0 \rightarrow {}^1P_1^0$, has also been studied.[94,98] The band appears with triplet structure in Ar, Kr, and Xe for M/R = 2000,[89] but for M/R = 200 it is broadened and red-shifted.[91]

A number of additional bands in the Hg spectrum[91] were attributed to transitions from excited triplet states, similar to those mentioned earlier for Zn and Cd.[97] Duley's analysis of the Hg bands implies that the $^3P_0^0$ and $^3P_2^0$ states are separated by 9010 and 8070 cm^{-1} in Ar and Kr, respectively. These numbers must be compared to the gas-phase value, 6397 cm^{-1}. Butchard et al.[94] have reported luminescence bands for Hg in NH_3, N_2, H_2O, and Xe, which were attributed to an eximer band of Hg_2. Comparison with gaseous Hg_2 indicates the bands in the matrices fall in the expected region. The emission appeared as two broad bands that merged into one after annealing an NH_3 matrix to 65°K.

3.7 Group IIIB: Sc, Y

In work on the optical and esr spectra of ScO and YO, matrices were produced containing Sc and Y atoms as a result of decomposition of the oxides.[99] No attempt was made to analyze the spectra, but mention was made that La atoms had also been observed.

The spectrum of Sc in Ne consists of two broad featureless bands. In Ar at least six bands are seen, but upon annealing all but two disappear. In Ne matrices, Y displays a more complex spectrum consisting of at least six overlapping, partially resolved bands. AMCOR predicts these two atoms to have many more bands than were reported.

In view of the number of expected but unreported bands, it would be very much worthwhile to restudy the spectra of Sc and Y atoms in matrices containing higher concentrations of the metals.

3.8 Group IVB: Ti

In this group, only Ti has been studied.[10] Relatively complex spectra are observed in both Ar and Xe matrices. The luminescence spectrum emitted when Ti in an Ar matrix is excited with uv light is discussed in more detail in the next section. In both Ar and Xe matrices, the absorption spectrum consists of a large number of overlapping bands, but there is a general correspondence between the matrix and the atomic spectrum. Half-widths range from 300 to 1000 cm^{-1} in Ar, with somewhat smaller values in Xe.

The matrix shifts in Ar vary from 600 to 1300 cm^{-1}, increasing with increasing transition energy. Shifts in Xe vary irregularly, decreasing from 600 to -200 cm^{-1} as the transition energy increases.

An estimate of the absolute band intensities of Ti atoms was found to be a factor of about 10 less in the matrices compared to the atomic values. The estimate, however, has rather large error limits associated with it.

3.9 Group VB: Nb, Ta

Both Nb[17] and Ta[20] atoms have been studied in noble gas matrices, the former in Ar, Kr, and Xe. The reported spectra of Nb and Ta metal atoms generally follow thc AMCOR predictions, but Ta gives a much more complex spectrum than Nb. Nonetheless, matrix bands due to Ta atoms were assigned on the basis of an AMCOR scheme.[20]

Tentative assignments for Nb have been made by Green and Gruen.[17] The matrix shifts of Nb follow the usual trends, decreasing from Ar (2100–2600 cm^{-1}) to Kr (1600–1900 cm^{-1}) to Xe (100–900 cm^{-1}). Interestingly, the matrix shifts of Ta in Ar appear to be somewhat less pronounced than those of Nb and vary from 900 to 1500 cm^{-1} with an average value of $\sim$1600 cm^{-1}.[20]

A number of bands were observed in the visible region at relatively high Nb concentrations that cannot be assigned on the basis of the AMCOR scheme. These low-intensity bands are not due to reaction with impurities such as O_2, H_2, or N_2, which had been added intentionally. It is possible that the bands are due to very weak atomic transitions, but no attempt at interpretation was made.

Nb also gives evidence for dimer formation. On warming some matrices containing Nb atoms, bands were observed in the visible region, which are thought to arise from Nb_2.[17] The characterization of metal dimers and clusters containing more than two metal atoms promises to be an interesting area for fundamental matrix studies. Much work remains to be done since most observations on metal dimer formation up to now have been qualitative in nature. When dimerization is observed, as in the case of Cu_2 and Pb_2, it seems to occur only at the stage of initial matrix formation and not on warming of the matrix. The formation of Nb_2 by bulk diffusion, therefore, is somewhat unusual.

3.10 Group VIB: Cr, Mo

From AMCOR considerations one expects two major triplet systems, due to $a^7S \rightarrow z^7P$ and $a^7S \rightarrow y^7P$ transitions, to be observed in the spectrum of matrix-isolated Cr atoms. In gaseous Cr, the first, arising from $3d^54s \rightarrow 3d^54p$ promotions, occurs in the range 23,300–23,500 cm^{-1}; the second due to $3d^54s \rightarrow 3d^44s4p$ transitions lies in the uv between 27,700 and 27,900 cm^{-1}. The *gf* values for the second triplet are about five times larger than for the first. Both triplets are spin allowed; however, a number of spin-forbidden transitions also have listed *gf* values that are lower by factors of 10^2–10^3 compared to the spin-allowed transitions.

Mann and Broida[21] observed the second higher energy triplet of Cr atoms isolated in Ar, but not the first. The three observed components in Ar had half-widths of $\sim$500 cm^{-1} and were blue-shifted 1900–2700 cm^{-1} from the gas-phase values. The shift increased with the band energy.

Two weaker bands were seen in the spectrum that were tentatively assigned to spin-forbidden transitions, one of which is not observed in the atomic spectrum. These two bands are a factor of approximately 15 more intense than expected. The matrix shifts implied by the assignments of these two bands, $\sim$4000 cm^{-1}, as pointed out by the authors, seem large and may indicate the bands are caused by the presence of impurities. In recent work,[37] the spectra of matrix-isolated Cr atoms have been restudied. In addition to the $^7S \rightarrow y^7P$ transition identified for Cr atoms in Ar by Mann and Broida,[21] the more recent work has also located the $^7S \rightarrow z^7P$ triplet in Ar, Kr, and Xe matrices. The spectra are shown in Figure 5. It is of interest to note that the two absorption bands appear with an intensity ratio of approximately

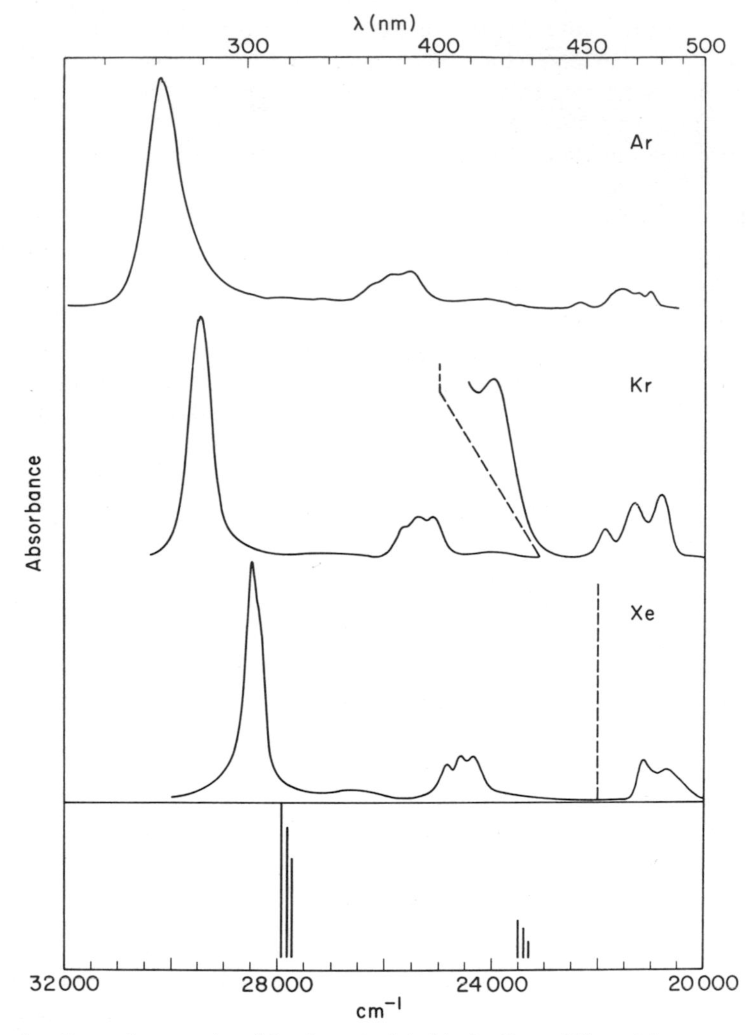

Figure 5. Absorption spectra of Cr atoms isolated in Ar, Kr, and Xe matrices compared to the gas phase transitions involving the ^7S ground state.

four, in agreement with the intensity ratio for the free gaseous Cr atom $^2S \rightarrow z^7P/^2S \rightarrow y^7P$ transitions. From a careful examination of the locations of the band maxima for the $^2S \rightarrow y^7P$ transition, one can show that the transition is blue-shifted progressively on going from the free atom to one isolated in Xe, Kr, and Ar matrices. Furthermore, the separation between the various J components also increases in this series in a manner similar to that already noted for matrix-isolated Mo atoms.[12] In Cr, this effect is particularly pronounced. The overall splitting of the ^7P state has increased more than a factor of three on going from the free gaseous atom to Cr isolated in Ar matrices. Calculations are being performed to examine whether combined crystal-field and spin-orbit interactions can account for the magnitude of the splitting and for the observed band shapes. The spectra of matrix-isolated Mo atoms has already been discussed in an earlier section of the chapter.

3.11 Group VIIB: Mn

Mn was one of the first metals to be studied using matrix-isolation techniques. Schnepp[16] reported the spectra of Mn atoms in Ar, Kr, and Xe matrices. Further work in Ar was done by Lee and Gutmacher[100] and later by Mann and Broida.[21]

For the more intense bands of the Mn spectrum, the agreement between the data of the various authors is quite good, particularly in view of the variety of different experimental techniques used to produce the matrices and measure the spectra. The average differences between reported band maxima is of the order of 50 cm^{-1}. For the weaker bands, the differences are larger, and, in fact, all authors reported some bands not observed by the others.

Considerations based on AMCOR would lead one to expect absorption bands in matrices near 24,800, 35,700, and 50,000 cm^{-1} arising from $^6S \rightarrow {}^6P$ transitions. The three components of each of these three triplets are split by 20, 80, and 200 cm^{-1}, respectively, due to spin-orbit coupling interactions. A number of weaker lines should also be observed, but their gf values are lower by one to three orders of magnitude. Only the two lowest energy triplets have been studied in matrices, and no attempts have been made to observe the highest energy triplet that undoubtedly occurs in the vacuum-uv region in most matrices.

The two triplets of Mn are observed as three fairly broad bands in the matrix. Since the splitting, 400–600 cm^{-1}, is higher than in the atomic spectrum, it must arise from matrix-perturbation effects. Lee and Gutmacher[100] have, however, reported a sharpening of the bands on annealing an argon matrix.

The average width and matrix shift in Ar are, respectively, 300 and 490 cm^{-1} for the lower triplet ($a^6S_{5/2} \rightarrow z^6P^0_{3/2,5/2,7/2}$) and 200 and 300 cm^{-1} for the upper ($a^6S_{5/2} \rightarrow y^6P^0_{3/2,5/2,7/2}$).[21]

The low-energy triplet occurs with highest energy in Kr and lowest in Ar with Xe intermediate. In this respect, it does not follow the usual pattern.

The high-energy triplet, however, displays the usual order, the band energy increasing in going from Xe to Kr to Ar.

3.12 Group VIIIB: Fe, Co, Ni, Pd, Pt

Fe

AMCOR predicts a complex spectrum for Fe involving 28 lines having measurable *gf* values corresponding to $d^6s^2 \rightarrow d^6sp$, d^7p, and d^5s^2p electron promotions.

A number of workers have studied the spectrum of Fe in noble gas matrices.[19,21,84,101,102] Mann and Broida[21] have reported the spectrum of Fe in Ar, Kr, and Xe at 4°K. These workers were unable to isolate Fe in Ar at 20°K, undoubtedly because of higher diffusion rates. However, successful isolation has been achieved in Ar at 14°K.[19] Micklitz and Barrett performed a thorough temperature study of the Fe spectrum in Kr.[102]

The spectrum of Fe in Ar superimposed on plotted *gf* values is given in Ref. 19. By shifting the *gf* values 1500 cm^{-1} to higher energies, a reasonable correspondence with AMCOR data was obtained. Mann and Broida[21] measured shifts of 300–1800 cm^{-1} in Ar, 200–1100 cm^{-1} in Kr, and −700–400 cm^{-1} in Xe. These values follow the usual trend. In Ar and Kr matrices, within a given electronic configuration the matrix shift appears to increase with increasing band energy. This interesting correlation seems to hold for Fe atoms and to a lesser extent for Co and Ni atoms.

The Fe spectrum displays less structure in Kr and Xe than in Ar matrices.[21,102] In a carefully annealed matrix of Kr, a one-to-one correspondence between the atomic lines and the matrix bands was found.[102] In Ar, however, annealing did not result in a simplification of structural features.[19] These observations probably indicate that there are several stable sites in Ar matrices.

A narrow band with half-widths of 80–90 cm^{-1} in Ar[19] and 42 ± 1 cm^{-1} in Kr[102] was observed and assigned to the $3d^64s^2\ {}^6D_4 \rightarrow 3d^7(a^4F)^4P\ {}^5D_4^0$ transition. The effect of temperature[102] in red-shifting this band 22 cm^{-1} and increasing its half-width on going from 4 to 40°K was interpreted on the basis of a theory of Lax. This theory[103] of electron-phonon interaction rationalizes the observed band-width. Additional broadening due to unresolved crystal-field splitting is negligible. The Debye temperature of solid Kr as probed by Fe atoms yields $\theta_D = 60 \pm 5$°K from the temperature dependence of the linewidth and $\theta_D = 53 \pm 6$°K from the line asymmetry.[102]

The relative intensities of the various band systems of matrix-isolated Fe atoms follow the atomic *gf* values to within a factor of two for most of the bands.

Co

Co also displays a fairly complex spectrum of partially resolved bands. The spectrum has been measured in Ar,[21] but only tentative assignments have been made for most of the bands. Matrix shifts vary from 400 to 1400 cm^{-1}.

Ni, Pd, Pt

Ni displays three groups of bands,[104] each consisting of several partially resolved bands. In view of the complexity of the spectrum, definite assignments are difficult, but an average shift of about 1300 cm^{-1} is evident.

Mann and Broida[21] also studied the spectrum of Ni in Ar at 4°K and made band assignments. The resulting matrix shifts varied from 441 to 2100 cm^{-1} with an average of 1540 cm^{-1}.

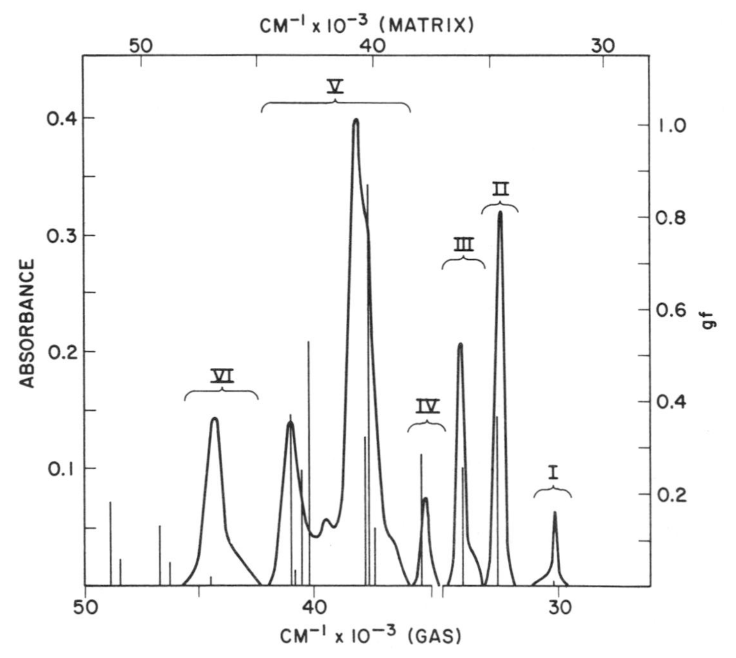

Figure 6. Absorption spectrum of Pt atoms in an Ar matrix. (The energy scale for the *gf* values has been up-shifted 2000 cm^{-1} below 35,000 cm^{-1} and 2500 cm^{-1} above 35,000 cm^{-1}.)

The spectrum of Pt in Ar has been previously reported[2] and is shown in Figure 6. Assignments for the four lowest energy bands can be made in a straightforward way on the basis of AMCOR data. They are listed in Table 7. For the higher energy bands, assignments become more difficult because of the increased spectral complexity, and none have been attempted.

Table 7. Analysis of the Low-Energy Spectrum of Pt in Ar

Assignment	E_{gas} (cm^{-1})	E_{mat} (cm^{-1})	ΔE (cm^{-1})
$^3D_3 \rightarrow z^5D_4^0$	30057	32100	1940
$[1]_2^0$	32620	34550	1930
$[3]_3^0$	34122	36250	2130
$[4]_3^0$	35322	37800	2480

As can be seen, the matrix shifts of Pt transitions in Ar vary from 2000 to 2500 cm^{-1} and tend to increase with the transition energy. Several bands are predicted to lie at energies higher than 50,000 cm^{-1}.

Because of the distinct separation of the four lower energy bands of Pt, this metal atom could well serve as a model system for further studies of matrix effects. At least two of these bands show structure in the Ar matrix. Detailed studies at various temperatures and in several rare-gas matrices, therefore, would be of interest.

AMCOR predicts the Pd spectrum to occur at energies higher than 36,000 cm^{-1}. Thus, if Pd behaves like Ni or Pt, its absorption spectrum in Ar matrices should lie at still higher energies. However, Pd atoms in solid Ar have been reported to have a spectrum in the 29,000–34,000 cm^{-1} region.[21] Red-shifts of the order of -7000 cm^{-1} would be indicated on the basis of such an analysis. Not only are the bands shifted to lower energies contrary to expectations, but the magnitude of the shifts are larger than for any other metals studied up to now.

It is possible that bands arising from metal impurities were observed, and a reexamination of the Pd spectrum is indicated.

3.13 Lanthanides and Actinides

Except for the preliminary report of the spectrum of La,[99] spectra of matrix-isolated atoms of none of the elements in these two groups have been reported. The spectra are expected to be quite complicated because of the large number of lines listed in the AMCOR tables for Nd and Eu, for example.[2]

3.14 Summary

To date, the atomic absorption spectra in low-temperature matrices of somewhat less than half of the elements have been recorded. Work on many of those studied has been very limited, with atoms of the elements in Groups IA, IB, IIA, and IIB having been investigated most extensively.

Areas of research that will undoubtedly prove fruitful include additional work on atoms of the transition metals, as well as totally new studies on atoms of the actinides and the lanthanide elements. It is to be hoped that work on these elements will be reported in the near future. Other elements for which very little data are available include those in Groups IIIA to VIIIA.

4. LUMINESCENCE SPECTRA OF MATRIX-ISOLATED ATOMS

Luminescence spectra of atoms isolated in low-temperature matrices appear to have been the first spectroscopic phenomena studied in such systems. In the classic work of Vegard[7,55,104,106] and of McLennan[107–109] initiated in the 1920s, the general procedure was to bombard solid N_2, O_2, H_2, or Ar with positive ions or electrons and to record the resulting emission spectra photographically. Under these conditions, molecular dissociation occurred, and luminescence spectra from excited N or O atoms in their respective N_2 or O_2 host matrices were obtained.

Although the experimental data gathered in this early work was substantiated by more recent results, some of the conclusions arrived at by the earlier investigators have been found to be in error, largely due to the belated recognition of the importance of small amounts of impurities in determining spectral characteristics. Indeed, as is discussed later, even recent experiments have been plagued by similar problems.

Because of the impurity effects, it is difficult to discuss and evaluate the work on individual atoms separately. Therefore, the literature on the emission spectra of solidified gases is discussed together in more or less chronological order.

Most of the early, and much of the more recent, work has been done on solid-nitrogen matrices. The studies began with Vegard, who, attempting to explain the 5577 Å line of the Aurora Borealis, hypothesized that the line was due to emission from N_2 crystals bombarded at low temperatures by charged particles in the upper atmosphere. In attempts to reproduce this spectral line, Vegard[7] performed a series of experiments varying a number of experimental parameters and exciting N_2 or N_2–inert-gas mixtures with electrons or positive ions. Vegard believed that under suitable conditions the position and contour of the auroral line could be experimentally reproduced, thus proving the hypothesis. McLennan and coworkers,[103,107,108] using similar experimental techniques, were unable to duplicate all of Vegard's results and later proved that this particular line was due to O atom impurities in the nitrogen samples.[106,109] Nonetheless, Vegard introduced an important new method in the field of low-temperature spectroscopy that has come to be known as matrix-isolation spectroscopy.

The luminescence spectra of N atoms trapped in N_2 matrices have been restudied using a variety of experimental methods. The N atoms were generally produced using an electrodeless discharge[110,111] and were deposited together with undissociated N_2 on a cold surface. During deposition the surface of the solid deposit displayed three major, intense, luminescence systems attributed to N atoms: (a) The α system consisting of five sharp lines (half-width of ~ 1 cm^{-1}) with lifetimes of 10–20 sec; (b) The β system consisting of broad diffuse bands with half-widths of about 100 cm^{-1} whose upper states decay with very short lifetimes; (c) The A system consisting of sharp, intense lines in the blue. On the basis of a crystal-field model,[112] the α lines were assigned to the $(2p^3)^2D_{5/2,3/2} \rightarrow (2p^3)^4S_{3/2}$ transition of N atoms. This theory was later refined,[113] achieving an excellent agreement with experimental results. The diffuse β system was originally, and incorrectly, assigned to another transition of N atoms. Although the assignment was only tentative, the emitter was believed to be atomic in nature.

Upon warming a N_2 matrix containing N atoms, after decay of the phosphorescence an intense emission is observed near 35°K. This luminescence, which had been noted by Vegard,[105] was assigned to a transition of N_2, presumably formed in an excited electronic state by diffusion and recombination of two N atoms.

The agreement between the NBS work[110,111] and the earlier work of Vegard and of McLennan is quite good in that all authors reported the three major luminescence systems. Points of difference, however, have remained. Thus, McLennan et al.[108] reported up to eight lines in the α system, indicating that N atoms were formed in two or more sites.[112]

A series of careful experiments also have cast doubt on some of the earlier interpretations[114] by showing that the emission was considerably more extensive and complex than had originally been believed. The β system was reassigned[114,115] to the $^1D \rightarrow {}^1S$ transition of O atoms, and the assignment of the α system was confirmed. It was also found that the α and β systems were each only the strongest members of two series of three or more band systems.[115] The members of each series were separated from one another by energies that agreed quantitatively with the vibrational frequency of N_2. The α systems were assigned, therefore, to an N—N_2 complex and the β systems to an O—N_2 complex. Also found were other emission bands involving two N—N_2 complexes in other excited electronic states.

The importance of impurities in modifying the luminescence spectra was clearly brought out in this work.[114] For example, several bands assignable to NO_2 and NH were observed in the emission spectrum. Additions of small amounts of various gases greatly changed the relative intensities of the various bands. Also studied were intensity effects caused by the use of different rare gas matrices and varying concentrations of N_2.[114]

The thermoluminescence spectra were found to be considerably more complex than had previously been believed.[115] Many bands observed in the luminescence spectrum immediately after deposition were observed also on warming the matrix. The A system was the only band system not seen in the thermoluminescence spectrum. The assignment of the A bands to N_2 transitions was placed in doubt, but no alternate assignment has been proposed.[114]

Spectra produced by electron bombardment of solid N_2 as well as other matrices[115,117] revealed no luminescences that had not been reported previously.

Tinti and Robinson[118] bombarded N_2 in rare gas matrices with X-rays and measured the emission spectra and decay lifetimes. In addition to the molecular emission of N_2, atomic emission from the $^2D \rightarrow {}^4S$ system of N were observed. The radiative lifetime of this transition was roughly 10, 14, and 340 sec in Kr, Ar, and Ne matrices, respectively. This can be compared to 40 sec in N_2 matrices and 12 hr for the free atom. The decay curves did not obey first-order kinetics in the noble gas matrices, and different mechanisms appear to be involved in the relaxation processes occurring in the various environments.

The emission of matrix-isolated O_2 excited by electron bombardment has also been studied.[119] In addition to molecular luminescence, bands were observed that could be correlated with the $^1S \rightarrow {}^1D$ system of atomic oxygen in N_2, Ne, Ar, Kr, and Xe matrices. The bands moved to lower energies and become more diffuse with increasing atomic number of the matrix gas.

More recently,[159] luminescence from solid krypton, which has been grown from a melt, has been studied during X-irradiation. Peaks near 3000 Å (4.1 eV) have been observed that are not enhanced upon selective impurity dopings, grow with X-irradiation, and are only present at temperatures below 25°K; these peaks are thought to be caused by intrinsic color centers. This luminescence, which may be due to the excited states of the Kr_2 molecule, and the emission near 5700 Å (2.2 eV), which is attributed to an atomic oxygen impurity, can be described in terms of the theory of zero-phonon transitions which has been applied to color centers in the alkali halides.[161]

Luminescence studies on matrix-isolated metal atoms have been much more limited compared to the work on atmospheric gases. Carstens and Gruen[10] have described the luminescence of Ti atoms excited with uv light. The emission appears as an intense red luminescence in an Ar matrix, but is not observed in Xe. The spectrum in an Ar matrix displays two band systems. The first, consisting of at least ten relatively narrow lines (half-widths of 30–140 cm^{-1}), was assigned to the $z^5G^0 \rightarrow a^3F$ transition on the basis of agreement in number and position with predictions assuming $j - j$ selection rules. The second, low-energy, band system consists of seven or more broader bands (half-widths 40–300 cm^{-1}). No assignment was given, although it was believed that the bands arise from transitions between two or more excited states of Ti atoms.

The emission spectrum of Nb in an Ar matrix has also been reported,[17] but no interpretation has been given.

Finally, Au and Cu are said to luminescence[11] in the visible region of the spectrum when excited with uv light. Work on these metal atoms has apparently not yet been published.

5. ELECTRON SPIN RESONANCE AND MÖSSBAUER SPECTRA OF MATRIX-ISOLATED ATOMS

Electron spin resonance and Mössbauer spectroscopy are important adjuncts to optical absorption and emission spectroscopy in that such measurements can, in principle, yield very detailed information on the ground states of the atoms under study. Up to now little attempt has been made to correlate results obtained using these various spectroscopies. Only rarely have both esr and absorption spectral measurements been made on matrix-isolated atoms in the same study.[64,71] Hopefully, this situation will be remedied in the future, since considerably more information on matrix interactions could be gained by performing several kinds of measurements on the same system.

The history of esr studies of matrix-isolated atoms is similar to that of absorption spectroscopy. Electron spin resonance studies began with research on N or H atoms isolated in their respective molecular matrices in 1956.[125] Later studies of these atoms in rare gas matrices were undertaken, and work was begun on the esr spectra of the alkali metals. Finally, more recently, esr studies have been reported on Cu, Ag, Au, Al, and Ga atoms.

The first esr spectra of H atoms isolated in solid matrices at low temperatures was reported by Livingston, Zeldes, and Taylor,[122] who irradiated H_2SO_4, $HClO_4$, and H_3PO_4 at 77°K with a Co^{60} source. Wall, Brown, and Florin[123] used a similar technique to produce H atoms in solid H_2 or CH_4 matrices and also to produce N atoms in N_2, Ar, and Xe matrices. A very useful and versatile way to produce the atoms is the use of a silent electrical discharge. This has been employed by Jen, Foner, Cochran, and Bowers for studies of H in H_2,[120,124,125] CH_4 and N_2,[120] Ar,[121,124,126] and Kr and Xe.[121] The same workers have studied N atoms in N_2, H_2, and CH_4.[124] Finally, Cole and coworkers have produced N atoms in N_2[127,128] and in NH_3[129] using this method. D atoms produced in a silent discharge have also been studied in D_2 matrices.[123–125]

The esr spectra of H or N atoms in matrices are similar to those of the free atoms. For H atoms the shift in the g factor is very small, and the change in the hyperfine coupling constant is less than 1% from the free-atom value.[121,124,125] Similar results hold for N atoms, although the change in the hyperfine coupling constant approaches 30%.[124] The change in the hyperfine coupling constant increases for both H and N in the series H_2, Ar (for H) or N_2 (for N), CH_4.[120,124] The lines are broadened in the matrix compared to their widths for the free atoms. This is believed to be a result of anisotropic matrix sites,[124] although several other hypotheses have been advanced.[123]

Some esr experiments indicate that H in H_2 or D in D_2 occupy only one site,[125] which is believed to be substitutional.[121] However, *in situ* production of H atoms by photolysis of HI, gives esr spectra indicating multiple site occupation, some of which may be interstitial.[121] It is believed that H atoms produced by this latter technique possess sufficient energy to occupy these less stable sites.[121] The lines attributed to different sites, at least in Ar,[126] disappear during isochronal annealing experiments.

It was not possible to isolate H produced in a discharge in Ne matrices, although H atoms could be formed in the matrix by photolysis. This is probably due to high surface-diffusion rates during Ne deposition. Additional hyperfine interactions increase the complexity of the spectrum of H atoms in Xe.

The esr spectrum of N atoms generally appears as a triplet; however, weaker satellite bands are often observed in N_2 matrices.[123,127,128] These are apparently caused by crystal-field interactions.[128]

Similar results, small shifts in g and the hyperfine coupling constant from the free-atom values, are found for the alkali metals. Jen et al.[130] studied the esr spectra of Li, Na, K, Rb, and Cs in Ar, Kr, and Xe matrices. Other authors[64,131] have made less extensive studies of some of the same systems.

The esr spectra of the alkali metals indicate the presence of two or more occupied sites in the matrix. Usually, the results can be interpreted on the basis of two stable sites, although in Ar matrices all alkali metals except Li occupy three or more sites. In one case, Na in Ar,[130] various lines disappeared at discrete temperatures on warming the matrix. Exposure of a matrix to 300°K black body radiation[131] resulted in spectral changes. Both of these observations support multiple site occupation.

Results for Rb in Ar were interpreted on the basis of a model involving three occupied sites, each consisting of a substitutional site in which one or more nearest neighbors were missing.[64]

The esr results support conclusions based on an analysis of optical absorption spectra. For example, the absorption spectra of alkali metal atoms also give evidence for multiple-site occupation in noble gas matrices. Furthermore, multiple-site occupation appears to be more probable in Ar than in Kr or Xe matrices.

The behavior of the Group IB metals differs from that of the alkali metal atoms. Although Ag atoms apparently occupy only one stable site when deposited at 20°K, multiple trapping sites have been postulated to explain the absorption spectra of Cu.[21] Kasai and McLeod,[33] however, except for Cu and Au in Ne matrices, found no evidence for multiple-site occupation in the esr spectra of Cu, Ag, and Au in Ne, Ar, Kr, and Xe. For Cu and Au in Ne evidence of two occupied sites is deduced from the esr spectra.

Electron spin resonance spectra of these atoms have been observed by Zhitnikov et al.,[139–142] who succeeded in isolating the atomic species in various molecular matrices at liquid-nitrogen temperature. The spectra observed by these authors, using paraffin as a matrix, are very close to those expected on the basis of the hyperfine coupling constants known from the atomic-beam experiments.[143,144]

Hyperfine coupling constants of Cu, Ag, and Au atoms were also determined in noble gas matrices[33] and examined for the matrix effect. Adrian[145] and Jen et al.[130] have shown that the effect of a rare gas matrix upon the hyperfine coupling constant of a trapped atom may be considered in three parts; (a) the attractive, van der Waals interaction between a trapped atom and the surrounding rare gas atoms, (b) the repulsive interaction between them, and (c) the "overlap" effect. Taking A_0 to be the coupling constant of a free atom, and ΔA to be the deviation of the observed coupling constant from A_0, the combined effect of the three interactions is such that, if the van der Waals diameter of the trapped atom is near or larger than the nearest neighbor spacing of the host lattice, $\Delta A/A_0$ is positive, and the larger it is, the smaller the nearest neighbor spacing. The van der Waals diameters of Cu, Ag, and Au atoms are 4.4, 4.7, and 4.6 Å, respectively. The nearest neighbor spacings in Ne, Ar, Kr, and Xe lattices are 3.20, 3.83, 4.05, and 4.41 Å, respectively. In compliance with the theory, except the case of Au in Xe, the observed $\Delta A/A_0$ are all positive and are progressively larger in the order of Xe, Kr, and Ar.

Interestingly, the trend in the hyperfine coupling constant in Au atoms, for example, going from Xe to Kr to Ar, is the same as observed for the spin-orbit coupling parameter.[1] With Xe matrices a superhyperfine structure with magnetic Xe nuclei was partially resolved, and by means of computer simulation, it was shown that these atoms are substitutionally incorporated within the Xe lattice. Also, evidence for atom-vacancy pairing was observed for Cu in Ne matrix.[33]

Aluminum and gallium atoms have been trapped in Ne, Ar, Kr, and Xe matrices and studied by optical and esr spectroscopy at 4.2°K and slightly higher temperatures.[71] The results indicate that both metal atoms occupy axially distorted substitutional sites in all rare gas lattices. This elongated tetradecahedral MeX_{12} coordination is particularly stable for rare gas complexes of Group III metal atoms exhibiting a single unpaired electron in their outermost *p* shell. From the esr data large splittings of the aluminum and gallium p shells have been derived increasing from $\sim 1600\ cm^{-1}$ in neon to ~3200 cm^{-1} in xenon for both atoms. The corresponding Jahn–Teller stabilization energies E_{JT} (increasing from ~1.5 kcal/mole for $MeNe_{12}$ to ~3.0 kcal for $MeXe_{12}$) can be explained by the "σ-π" effect. The van der Waals interatomic correlation energy is maximized, and the repulsive exchange energy is minimized by attraction of the equatorial ligand atoms to the metal center and repulsion of the remaining ligands from the σ antibonding axial positions. The esr spectra exhibit axial symmetry, show effects of preferential orientation, and demonstrate almost complete quenching of the free-atom angular momentum in each case. The basic features of the g values and the metal hyperfine tensor (and of their strong dependence on the matrix and on temperature) can be understood within a simple crystal-field model, but there are significant deviations. The introduction of orbital angular momentum and spin-orbit reduction factors resulting from orthogonalization of the metal p orbitals to the valence shells of the surrounding rare gas atoms removed a large part of the discrepancies, but quantitative agreement with experiment could be obtained only when the dynamic Jahn–Teller effect was taken into account. In order to establish the geometries of the rare gas cages surrounding the trapped metal atoms, numerical calculations of orbital and spin-orbit reduction factors were performed for various sites in the rare gas lattices. For the determination of the vibronic quenching parameters a slight extension of Ham's second-order theory of an orbital triplet interacting with an e_g vibrational mode was required. The results of Ammeter and Schlosnagle[71] taken together with the semiempirical calculations of Baylis[146] show that atoms with single occupied p shells form the strongest van der Waals complexes with rare gas atoms among all atoms in the periodic table.

Electron spin resonance measurements of P atoms in Ar[147] and Kr[148] and of As atoms in Kr[148] matrices have yielded g factors and hyperfine coupling constants. In the case of P atoms, where a comparison with the free-atom value[149] can be made, the hyperfine coupling constants increase in the series, free atom → Ar matrix → Kr matrix.

In 1961 Jaccarino and Wertheim[150] proposed a Mössbauer experiment with rare-gas–matrix-isolated ^{57}Fe atoms to solve the problem of iron-isomer-shift calibration. According to the results of optical spectroscopy, the Mössbauer isomer shift of a matrix-isolated ^{57}Fe atom would be approximately that of the free ^{57}Fe atom with the atomic configuration, $3d^6 4s^2$. Several investigators have used the Mössbauer-effect technique to study the hyperfine interactions of ^{83}Kr in solid krypton;[151–153] however, these are not rare-gas–matrix-isolation experiments, since an impurity atom is not being studied. The first successful rare-gas–matrix-isolation Mössbauer experiment was carried out with iodine molecules (I_2) imbedded in a solid argon matrix at 22°K.[154] The results obtained in this experiment indicated that the isolated molecule indeed can be considered as an almost free entity where matrix interactions are negligible. The first results of a Mössbauer experiment with ^{57}Fe in an argon matrix at 4.2°K were reported by Barrett and his collaborators.[34,155,156]

The Mössbauer absorption spectra of ^{57}Fe were measured in argon, krypton, and xenon matrices with iron atom concentrations from 0.3 to 3% and matrix temperatures between 1.45 and 20.5°K. All of the spectra show an absorption line with an isomer shift of $\delta = -0.75 \pm 0.3$ mm/sec with respect to an iron foil at 300°K. This isomer shift is independent of rare gas matrix, iron concentration, and matrix temperature. The line is ascribed to an isolated ^{57}Fe atom (monomer) with an atomic configuration of $3d^6 4s^2$. The measured isomer shift gives a new calibration point in the isomer-shift versus electron-density plot for ^{57}Fe. The observed $1/T$ temperature dependence of the monomer linewidth shows that the direct phonon process is dominant in the spin-lattice relaxation mechanism. Spin-lattice relaxation times of the order of 2.5×10^{-10} sec were obtained by assuming a hyperfine field of 1.1×10^6 Oe at the ^{57}Fe nucleus due to an iron atom with unquenched orbital momentum. From the temperature dependence of the Mössbauer f factor, the Mössbauer temperatures, θ_M, in the Debye model were calculated and compared with the values expected from specific-heat measurements. The Mossbauer spectra showed in addition to the monomer absorption line, a pair of narrow lines ($\Gamma = 0.22$–0.3 mm/sec) that intensified with increasing iron concentrations. These lines were interpreted as the result of the quadrupole splitting of the $I = \frac{3}{2}$ excited state of ^{57}Fe in the axial field produced by an iron nearest neighbor (dimer). The measured quadrupole splitting is $\Delta E = 4.05 \pm 0.4$ mm/sec, and the isomer shift for the dimer was $\delta = -0.14 \pm 0.02$ mm/sec, corresponding to an effective atomic configuration of $3d^6 4s^x$ with $x = 1.47 \pm 0.04$. For ^{57}Fe in krypton and xenon, the measured dimer/monomer ratio was that expected from probability considerations, but in argon it was a factor of approximately three higher than expected.

REFERENCES

1. D. M. Gruen, S. L. Gaudioso, R. L. McBeth, and J. L. Lerner, *J. Chem. Phys.*, **60**, 89 (1974).
2. D. H. W. Carstens, W. Brashear, D. R. Eslinger, and D. M. Gruen, *Appl. Spectrosc.*, **26**, 184 (1972).
3. D. W. Green, J. Thomas, and D. M. Gruen, *J. Chem. Phys.*, **58**, 5433 (1973).
4. K. R. Thompson, W. C. Easley, and L. B. Knight, *J. Phys. Chem.*, **77**, 49 (1973).
5. J. Boyd, J. Lavoie, and D. M. Gruen, *J. Chem. Phys.*, **60**, 4088 (1974).
6. L. Vegard, *Nature*, **113**, 716 (1924).

7. L. Vegard, H. Kamerlingh Onnes, and W.-H. Keesom, *C.R. Acad. Sci., Paris*, **180**, 1084 (1925).
8. A. M. Bass and H. P. Broida, *Formation and Trapping of Free Radicals*, Academic, New York, 1960.
9. M. McCarty and G. W. Robinson, *Mol. Phys.*, **2**, 415 (1959).
10. D. M. Gruen, and D. H. W. Carstens, *J. Chem. Phys.*, **54**, 5206 (1971).
11. B. Meyer, *Low Temperature Spectroscopy*, Elsevier, New York, 1971.
12. D. W. Green, and D. M. Gruen, *J. Chem. Phys.*, **60**, 1797 (1974).
13. L. Brewer and C. Chang, *J. Chem. Phys.*, **56**, 1728 (1972).
14. R. W. Zwanzig, *Mol. Phys.*, **3**, 305 (1960).
15. M. Brith and O. Schnepp, *J. Chem. Phys.*, **39**, 2714 (1963).
16. O. Schnepp, *J. Phys. Chem. Solids*, **17**, 188 (1961).
17. D. W. Green and D. M. Gruen, *J. Chem. Phys.*, **57**, 4462 (1972).
18. D. W. Green, D. M. Gruen, F. Schreiner, and J. L. Lerner, *Appl. Spectrosc.*, **28**, 34 (1974).
19. D. H. W. Carstens, J. F. Kozlowski, and D. M. Gruen, *High Temp. Sci.*, **4**, 301 (1972).
20. W. R. M. Graham and W. Weltner, *J. Chem. Phys.*, **56**, 4400 (1972).
21. D. M. Mann and H. P. Broida, *J. Chem. Phys.*, **55**, 84 (1971).
22. C. H. Corliss and W. L. Bozman, *Nat. Bur. Stand. Monograph*, **53**, (1962).
23. C. E. Moore, *Nat. Bur. Stand.* Circular No. **467**, Vol. I, 1949; ibid., Vol. II, 1952; ibid., Vol. III, 1958.
24. W. L. Wiese, M. W. Smith, and B. M. Glennon, *Atomic Transition Probabilities*, Vol. I, NSRDS-NBS-4, 1966.
25. W. L. Wiese et al., *Bibliography on Atomic Transition Probabilities*, NBS Special Publication 320, 1970; ibid., Suppl. 1, 1971; ibid., Suppl. 2, 1973.
26. F. Schoch and E. Kay, *J. Chem. Phys.*, **59**, 718 (1973).
27. I. Lefkowitz, K. Kramer, M. Shields, and G. Pollack, *J. Appl. Phys.*, **38**, 4867 (1967).
28. J. -Y. Rconin, *J. Quant. Spectrosc. Radiat. Transfer*, **11**, 1151 (1971).
29. S. Y. Ch'en and M. Takeo, *Rev. Mod. Phys.*, **29**, 20 (1957).
30. J. -Y. Roncin, *Chem. Phys. Lett.*, **3**, 408 (1969).
31. J. Heinrichs, *Phys. Rev., B*, **2**, 518 (1970).
32. H. Micklitz and P. H. Barrett, *Phys. Rev., B*, **4**, 3845 (1971).
33. P. H. Kasai and D. McLeod, *J. Chem. Phys.*, **55**, 1566 (1971).
34. T. K. McNab, H. Micklitz, and P. H. Barrett, *Phys. Rev., B*, **4**, 3787 (1971).
35. A. Morelle, Ph. D. thesis, University of Washington, 1970.
36. L. Brewer and B. King, *J. Chem. Phys.*, **53**, 3981 (1970).
37. D. M. Gruen, J. Lavoie, and J. Boyd, unpublished work.
38. J. S. Griffith, *The Theory of Transition Metal Ions*, Cambridge U. P., Cambridge, England, 1964.
39. Yong-Ki Kim, private communication.
40. M. Blume and R. E. Watson, *Proc. Roy. Soc., Ser. A*, **270**, 127 (1962); ibid., **271**, 565 (1963).
41. J. L. Dehmer, *Phys. Rev., A*, **7**, 4 (1973).
42. L. I. Schiff, *Quantum Mechanics*, McGraw-Hill, New York, p. 280, 1949.

43. J. Robin and S. Robin, *Compt. Rend.*, **233**, 1019 (1951).
44. M. Takeo and S. Y. Ch'en, *Phys. Rev.*, **93**, 420 (1954).
45. N. P. Penkin and I. -Yu. Yu. Slavenas, *Opt. Spectrosc.*, **15**, 3 (1963).
46. W. L. Wiese, Proc. *Xth International Conference on Spectroscopy*, p. 37, Spartan, Washington, D.C., 1963.
47. D. Roschdestwensky, *Ann. Phys.*, **39**, 307 (1912).
48. Iu. I. Ostrovskii, N. P. Penkin, and L. N. Shabanova, *Soviet Phys. Dokl.*, **3**, 538 (1958).
49. J. Slavenas, *Liet. TSR Aukst. Mokyklos* **7**, 619 (1967); C. A. **59**, 101160g (1968); Yu. Slavenas, *Liet. Fiz Rinkinys*, **7**, 619 (1967); C. A. **59**, 7085h (1963); N. P. Penkin, I. Yu, and Yu. Slavenas, *Optika i Spektroskopia*, **15**, 9 (1963).
50. C. Fuchtbauer, G. Joos, and O. Dinkelacker, *Ann. Phys.*, **71**, 204 (1923).
51. A. Michels and H. DeKluiver, *Physica*, **22**, 919 (1956).
52. A. Michels, H. DeKluiver, and B. Castle, *Physica*, **23**, 1131 (1957).
53. S. Y. Ch'en, *Phys. Rev.*, **58**, 1051 (1940).
54. M. Krauss, P. Maldonado, and A. C. Wahl, *J. Chem. Phys.*, **54**, 4944 (1971).
55. L. Vegard and W. H. Keesom, *Verslag Akad. Wetensch. Amsterdam*, **36**, 364 (1927).
56. G. Baldini, *Phys. Rev.*, **136**, A248 (1964).
57. J. -Y. Roncin, N. Damany, and B. Vodar, *C. R. Acad. Sci.*, *Paris*, **260**, 96 (1965).
58. T. H. Keil and A. Gold, *Phys. Rev.*, *A*, **136**, 252 (1964).
59. L. Andrews and G. C. Pimentel, *J. Chem. Phys.*, **47**, 2905 (1967).
60. A. A. Belyaeva, Y. B. Predtechnenskii, and L. D. Shcherba, *Opt. Spectrosc.* (*USSR, Engl. Trans.*), **24**, 233 (1968); ibid., **34**, 21 (1973); ibid., *Izv. Akad. Nauk SSSR, Ser. Fiz.*, **33**, 895 (1969).
61. R. B. Merrithew, G. V. Marusak, and C. E. Blount, *J. Mol. Spectrosc.*, **29**, 54 (1969).
62. B. Meyer, *J. Chem. Phys.*, **43**, 2986 (1965).
63. W. Weyhnmann and F. M. Pipkin, *Phys. Rev.*, **137**, A490 (1965).
64. S. L. Kupperman and F. M. Pipkin, *Phys. Rev.*, **166**, 207 (1968).
65. L. Brewer and J. L. Wang, *J. Mol. Spectrosc.*, **40**, 95 (1971).
66. J. L. -F. Wang, Ph.D. thesis, University of California, Berkeley, 1969.
67. J. E. Francis, Jr., and S. E. Webber, *J. Chem. Phys.*, **56**, 5879 (1972).
68. R. L. Barger and H. P. Broida, Nat. Bur. Stand. Rep. 8200, 1963.
69. W. W. Duley, *Nature*, **218**, 153 (1968).
70. W. W. Duley and W. R. M. Graham, *Nature*, **224**, 785 (1969).
71. J. H. Ammeter and D. C. Schlosnagle, *J. Chem. Phys.*, **59**, 4784 (1973).
72. W. W. Duley and W. R. S. Garton, *Proc. Phys. Soc.*, **92**, 830 (1967).
73. H. Micklitz and K. Luchner, *Z. Naturforsch.*, *A*, **22**, 1650 (1967); ibid., **227**, 301 (1969).
74. H. Micklitz, *Z. Phys.*, **215**, 302 (1968).
75. G. D. Brabson, Ph.D. thesis, University of California, Berkeley, 1965.
76. H. Micklitz and K. Luchner, *Z. Phys.*, **270**, 79 (1974).
77. O. Schnepp and K. Dressler, *J. Chem. Phys.*, **33**, 49 (1960).
78. G. Baldini, *Phys. Rev.*, **128**, 1562 (1962).
79. J. -Y. Roncin, V. Chandrasekharan, N. Damany, and M. B. Vodar, *J. Chem. Phys.*, **60**, 1212 (1963).

80. J. -Y. Roncin, V. Chandrasekharan, and N. Damany, *C. R. Acad. Sci., Paris*, **258**, 2513 (1964).

81. J. -Y. Roncin, N. Damany, and J. Romand, *J. Mol. Spectrosc.*, **22**, 154 (1967).

82. K. Dressler, *J. Opt. Soc. Amer.*, **50**, 501 (1960).

83. L. Brewer and B. King, *J. Chem. Phys.*, **53**, 3981 (1970).

84. J. S. Shirk and A. M. Bass, *J. Chem. Phys.*, **49**, 5156 (1968).

85. L. Brewer, B. A. King, J. L. Wang, B. Meyer, and G. F. Moore, *J. Chem. Phys.*, **49**, 5209 (1968).

86. K. R. Harvey and H. W. Brown, *J. Chim. Phys.*, **56**, 745 (1959).

87. A. M. Bass and H. P. Broida, *J. Mol. Spectrosc.*, **2**, 42 (1958).

88. L. Brewer, C. Chang, and B. King, *Inorg. Chem.*, **9**, 814 (1970).

89. W. W. Duley, *Phys. Lett.*, **19**, 361 (1965).

90. L. Brewer, B. Meyer, and G. D. Brabson, *J. Chem. Phys.*, **43**, 3973 (1965).

91. W. W. Duley, *Proc. Phys. Soc.*, **90**, 263 (1967).

92. W. W. Duley, *Proc. Phys. Soc.*, **88**, 1049 (1966).

93. M. McCarty, Jr., *J. Chem. Phys.*, **52**, 4973 (1970).

94. J. A. Butchard, R. F. C. Claridge, and L. F. Phillips, *Chem. Phys. Lett.*, **8**, 139 (1971).

95. W. W. Duley, *Nature*, **210**, 624 (1966).

96. M. Guy, J. -Y. Roncin, and N. Damany, *C. R. Acad. Sci., Paris, Ser. B*, **263**, 546 (1966).

97. W. W. Duley, *Proc. Phys. Soc.*, **91**, 976 (1967).

98. J. -Y. Roncin and N. Damany-Astoin, *C. R. Acad. Sci., Paris*, **253**, 835 (1961).

99. W. Weltner, Jr., D. McLeod, Jr., and P. H. Kasai, *J. Chem. Phys.*, **46**, 3172 (1966).

100. E. L. Lee and R. G. Gutmacher, *J. Phys. Chem. Solids*, **23**, 1823 (1962).

101. W. Keune and E. Luescher, *Solid State Commun.*, **8**, 811 (1970).

102. H. Micklitz and P. H. Barrett, *Phys. Rev.*, *B*, **4**, 3845 (1971).

103. M. Lax, *J. Chem. Phys.*, **20**, 1752 (1952).

104. D. H. W. Carstens, J. Kozlowski, and D. M. Gruen, unpublished results.

105. L. Vegard, *Nature*, **114**, 357 (1924).

106. L. Vegard and G. Kvifte, *Nature*, **162**, 967 (1948).

107. J. C. McLennan and G. M. Shrum, *Proc. Roy. Soc., Ser. A*, **106**, 138 (1924).

108. J. C. McLennan, H. J. C. Ireton, and K. Thompson, *Nature*, **118**, 408 (1926).

109. J. C. McLennan, H. J. C. Ireton, and E. W. Samson, *Proc. Roy. Soc., Ser. A*, **120**, 303 (1928).

110. A. M. Bass and H. P. Broida, *Phys. Rev.*, **101**, 1740 (1956).

111. H. P. Broida and J. R. Pellman, *Phys. Rev.*, **95**, 845 (1954).

112. C. M. Herzfeld and H. P. Broida, *Phys. Rev.*, **101**, 606 (1956).

113. C. M. Herzfeld, *Phys. Rev.*, **107**, 1239 (1957).

114. M. Peyron and H. P. Broida, *J. Chem. Phys.*, **30**, 139 (1959).

115. M. Peyron, E. M. Horl, H. W. Brown, and H. P. Broida, *J. Chem. Phys.*, **30**, 1304 (1959).

116. L. J. Schoen and R. E. Hebbert, *J. Mol. Spectrosc.*, **3**, 417 (1959).

117. E. M. Horl, *J. Mol. Spectrosc.*, **3**, 425 (1959).

118. D. S. Tinti and G. W. Robinson, *J. Chem. Phys.*, **49**, 3229 (1968).

119. L. J. Schoen and H. P. Broida, *J. Chem. Phys.*, **32**, 1184 (1960).

120. S. N. Foner, C. K. Jen, E. L. Cochran, and V. A. Bowers, *J. Chem. Phys.*, **28**, 351 (1958).

121. S. N. Foner, E. L. Cochran, V. A. Bowers, and C. K. Jen, *J. Chem. Phys.*, **32**, 963 (1960).

122. R. Livingston, H. Zeldes, and E. H. Taylor, *Discuss. Faraday Soc.*, **19**, 166 (1955).

123. L. A. Wall, D. E. Brown, and R. E. Florin, *J. Phys. Chem.*, **63**, 1762 (1959).

124. C. K. Jen, S. N. Foner, E. L. Cochran, and V. A. Bowers, *Phys. Rev.*, **112**, 1169 (1958).

125. C. K. Jen, S. N. Foner, E. L. Cochran, and V. A. Bowers, *Phys. Rev.*, **104**, 846 (1956).

126. E. L. Cochran, V. A. Bowers, S. N. Foner, and C. K. Jen, *Phys. Rev. Letts.*, **2**, 43 (1959).

127. T. Cole, J. T. Harding, J. R. Pellam, and D. M. Yost, *J. Chem. Phys.*, **27**, 593 (1957).

128. T. Cole, and H. M. McConnell, *J. Chem. Phys.*, **29**, 451 (1958).

129. T. Cole, and J. T. Harding, *J. Chem. Phys.*, **28**, 993 (1958).

130. C. K. Jen, V. A. Bowers, E. L. Cochran, and S. N. Foner, *Phys. Rev.*, **126**, 1749 (1962).

131. J. P. Goldsborough and T. R. Koehler, *Bull. Amer. Phys. Soc.*, **7**, 449 (1962).

132. R. A. Zhitnikov and N. I. Melnikov, *Opt. Spectrosc.* (*USSR, Engl. Trans.*), **24**, 53 (1968).

133. G. M. Lawrence, J. K. Link, and R. B. King, *Astrophys. J.*, **141**, 293 (1965).

134. H. J. Coufal, U. Nagel, M. Burger, and F. Lüscher, *Phys. Lett.*, *A*, **47**, 327 (1974).

135. M. A. Belogovskii and K. B. Tolpygo, *Sov. Phys., Solid State*, **14**, 2571 (1973).

136. A. Scholz and C. Lehmann, *Phys. Rev.*, *B*, **6**, 813 (1972).

137. P. Erhart and W. Schilling, *Phys. Rev.*, *B*, **8**, 2604 (1973).

138. D. E. Milligan and M. E. Jacox, *J. Chem. Phys.*, **51**, 1952 (1969).

139. R. A. Zhitnikov, N. V. Kolesnikov, and V. I. Kosyakov, *Zh. Eksp. Teor. Fiz.*, **43**, 1186 (1962) [*Sov. Phys., JETP*, **16**, 839 (1962)].

140. R. A. Zhitnikov, N. V. Kolenikov, and V. I. Kosyakov, *Zh. Eksp. Teor. Fiz.*, **44**, (1963) [*Sov. Phys. JETP*, **17**, 815 (1963)].

141. R. A. Zhitnikov and N. V. Kolesnikov, *Zh. Eksp. Teor. Fiz.*, **46**, 89 (1964) [*Sov. Phys. JETP*, **19**, 65 (1964)].

142. R. A. Zhitnikov and N. V. Kolesnikov, *Fiz. Tverd. Tela*, **6**, 3307 (1964) [*Sov. Phys., Solid State*, **6**, 2645 (1965)].

143. Y. Ting and H. Lew, *Phys. Rev.*, **105**, 581 (1957).

144 G. Wessel and H. Lew, *Phys. Rev.*, **92**, 641 (1953).

145. F. J. Adrian, *J. Chem. Phys.*, **32**, 972 (1960).

146. W. E. Baylis, *J. Chem. Phys.*, **51**, 2665 (1969).

147. F. J. Adrian, E. L. Cochran, and V. A. Bowers, *Advan. Ser.*, **36**, 50 (1962).

148. R. L. Morehouse, J. J. Christiansen, and W. Gordy, *J. Chem. Phys.*, **45**, 1747 (1966).

149. H. Dehmelt, *Phys. Rev.*, **99**, 527 (1955).

150. V. Jaccarino and G. K. Wertheim, in *Proceedings of the Second International Conference on the Mössbauer Effect, Saclay, France, 1961*, D. M. J. Compton and A. H. Schoen, Eds., p. 260, Wiley, New York, 1962.

151. Y. Hazony, P. Hillmann, M. Pasternak, and S. Ruby, *Phys. Lett.*, **2**, 337 (1962).

152. M. Pasternak, A. Simopoulous, S. Bukshpan, and T. Sonnino, *Phys. Lett.*, **22**, 52 (1966).

153. G. Gilbert and C. E. Violet, *Phys. Lett.*, *A*, **28**, 285 (1968).

154. S. Bukshpan, C. Goldstein, and T. Sonnino, *J. Chem. Phys.*, **49**, 5477 (1968).

155. P. H. Barrett and T. K. McNab, *Phys. Rev. Lett.*, **25**, 1601 (1970).

156. T. K. McNab and P. H. Barrett, *Mössbauer Effect Methodology*, Vol. 7, p. 59, Plenum, New York, 1971.

157. N. L. Moise, *Astrophys. J.*, **144**, 774 (1966).

158. H. Huber, E. P. Kündig, M. Moskovits and G. A. Ozin, *J. Amer. Chem. Soc.*, **97**, 2097 (1975).

159. W. F. Lewis and K. J. Teegarden, *Phys. Rev., B*, **8**, 3024 (1973).

160. O. Cheshnovsky, B. Raz, and J. Jortner, *J. Chem. Phys.*, **57**, 4628 (1972).

161. D. B. Fitchen, Chapter V, in *Physics of Color Centers*, W. B. Fowler, Ed., Academic, New York, 1968.

Photochemistry in Low-Temperature Matrices

11

Jeremy K. Burdett and James J. Turner

1. INTRODUCTION

Most of the chapters in this volume are concerned with the preparation of interesting molecular entities by cocondensation at low temperature of two or more independent species. Such preparations may be *macro* (e.g., the chapter by Timms in this volume) in the sense that the products can be handled by conventional vacuum-line techniques and the properties examined by conventional laboratory methods. The preparation may also be *micro* in the sense that minute quantities of matrix-isolated species are produced that can only be examined *in situ* by appropriate spectroscopic techniques. The important point about such methods is that species that are highly reactive in the gas phase (e.g., Li atoms) are rapidly quenched with reactant substrate. Reaction probably occurs at a relatively high temperature on the surface; the very rapid quenching helps to stabilize otherwise unstable compounds, since the low temperature makes unsurmountable even a small activation energy for decomposition. In matrix-isolation experiments, of course, reaction is also physically prevented by the surrounding isolating matrix. Thermodynamic arguments can also be employed, but care is required, since describing individual events in such terms runs counter to the statistical nature of thermodynamics.

Rapid quenching of unstable species generated in the gas phase is also effective. For example, condensation at 77°K of the gases in an electric discharge involving Br_2 and O_2 or F_2 and O_2 leads to formation of $(BrO_2)_x$[26] and O_2F_2,[111] respectively. Condensation of the gaseous products of a Xe/Cl_2 discharge produces $XeCl_2$ isolated in a matrix, identified by both infrared[87] and Raman spectroscopy.[10] Careful design of spray-on nozzle geometry has allowed the isolation[53] of BH_3 in argon by heating a gas-phase mixture of BH_3CO and argon.

In each of these methods the reaction producing the unstable species is rapidly quenched. It is, however, possible to generate such species *in situ* at low temperature by supplying the necessary energy by photolysis. For example, photolysis of a low-temperature liquid solution of OF_2 and O_2 produces a large concentration of the O_2F radical.[116] Similarly, photolysis of many species in matrices leads to the stabilization of some very interesting intermediates, radicals, and molecular fragments. In this chapter, we describe various photochemical experiments in matrices.

2. PHOTOCHEMICAL SYNTHESIS

There are three prime and obvious requirements to be met before photolytic changes can be observed:

a. The parent molecule(s) must have an absorption(s) in the accessible vacuum-uv or uv-visible region.

b. The energy of such an absorption band must be at least equal to the appropriate bond dissociation energy or the energy associated with the desired chemical change.

c. The product(s) of photolysis must have infrared or uv-visible absorptions of comparable extinction coefficient to the parent for these methods to be used in detection.

A further requirement is that in any mechanism that relies on movement through the matrix of an atom or molecule, there must be enough excess energy to overcome the physical restriction of the matrix. This point is discussed in much greater detail later.

Production of matrix species by photochemical techniques has both advantages and disadvantages, as we hope will be clear from the later sections of this chapter. However, it is appropriate to consider briefly the kind of reactions which can be photochemically promoted in matrices. The techniques can be divided somewhat arbitrarily into four categories: fission of parent compound, isomerization, bimolecular reactions both with and without 'tunneling', and electron transfer.

2.1 Fission of Parent Compound

Some examples illustrate this method:

(1) $$\text{O (=O)} \xrightarrow{h\nu} CO_2 + \square \xrightarrow{h\nu} CH{\equiv}CH \quad \text{(Refs. 28, 61)}$$

(2) $$OF_2 \xrightarrow{h\nu} OF + F \xrightarrow{\Delta} OF_2 \quad \text{(Ref. 6)}$$

(3) $$Cr(CO)_6 \xrightarrow{h\nu} Cr(CO)_5 + CO \quad \text{(Ref. 43)}$$

In (1) the primary products (CO_2 and □) do not react with each other at the low temperature, since a nonzero activation energy exists for the reverse reaction. Warming to 35°K is sufficient to allow recombination. In (2) the F escapes from the surrounding matrix cage and only recombines with OF when the matrix is annealed. In (3) it is probable that the photoejected CO is stereochemically unfavorably arranged with respect to $Cr(CO)_5$. If species cannot escape from the cage and the energetics are favorable for the reverse reaction we observe no change, for example,

$$CrO_2Cl_2 \underset{}{\overset{h\nu}{\rightleftharpoons}} [CrO_2Cl + Cl?] \quad \text{(Ref. 23)}$$

However, use can be made of this property, for example,

$$CHCl_3 \xrightarrow{h\nu} [CCl_3 + H?] \rightarrow CCl_3 + H \quad \text{(Refs. 51, 52)}$$

where diffusion away from the photolysis site of the small H atom produced is a much more likely event than diffusion of the large chlorine atom. CCl_3 is produced much more efficiently than $CHCl_2$.

It is clear from the above that the cage effect plays a very significant role in matrix photochemistry, and we devote a whole section to this shortly.

2.2 Isomerization

This is a fairly obvious photochemically induced process illustrated by some work of Goldfarb et al.[31]

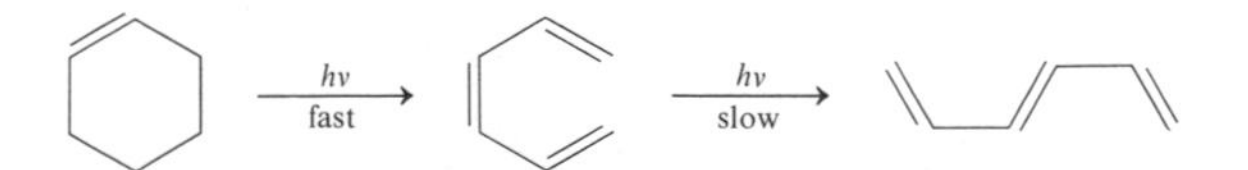

However, it is important to note that this reaction scheme is different from the corresponding photochemistry in the vapor phase. Thus, to extrapolate matrix observations to room temperature systems can sometimes be very misleading.

2.3 Bimolecular Reactions

This can be achieved by turning the matrix technique on its head and photolyzing a parent molecule in a *reactive* matrix, for example,

$$HI + \underset{\text{(matrix)}}{CO} \xrightarrow{h\nu} HCO + I \qquad \text{(Ref. 84)}$$

(HCO was the first radical to be obtained by the matrix technique.)

$$Co(CO)_3NO + \underset{\text{(matrix)}}{CO} \rightarrow Co(CO)_4 + NO \qquad \text{(Ref. 30)}$$

$$Ni(CO)_4 + \underset{\text{(matrix)}}{N_2} \rightarrow Ni(CO)_3N_2 + CO \qquad \text{(Ref. 101)}$$

There is usually no need to use a pure matrix of the reactant. For example, 10% of NO in an argon matrix allows

$$Os(CO)_5 + \underset{\text{(matrix)}}{NO/Ar} \rightarrow Os(CO)_2(NO)_2 + 3CO + Ar \qquad \text{(Ref. 30)}$$

If the 'dopant' is in low concentration, then it may be necessary for the photolytically generated species to force its way through the matrix to reach the dopant, for example,

$$F_2/Ar + O_2/Ar \xrightarrow{h\nu} 2F/Ar + O_2/Ar \qquad \text{(Ref. 116)}$$
$$\downarrow$$
$$F/Ar + FO_2/Ar$$
$$\downarrow$$
$$Ar + F_2O_2/Ar$$

2.4 Electron Transfer

This technique is common enough in solution, even without any photochemical assistance. For example, many aromatic anions can be studied by esr on simply adding the parent hydrocarbon to a solution of sodium in liquid ammonia. Kasai[56] was the first to observe charged species in matrices, for example,

$$B_2H_6 + Na/Ar \xrightarrow{h\nu} Na^+ + B_2H_6^- \qquad \text{(Ref. 56)}$$

Vacuum-ultraviolet irradiation is another source of electrons.

$$C_2H_2 \xrightarrow[\text{uv}]{\text{vac}} C_2^- + ? \qquad \text{(Ref. 11)}$$

$$Cr(CO)_6 \left.\begin{array}{l} + Na/Ar \\ + \text{vac uv} \end{array}\right\} \begin{array}{c} \xrightarrow{h\nu} \\ \rightarrow \end{array} Cr(CO)_5^- \qquad \text{(Refs. 13, 14, 17)}$$

Some of the subtleties of the above techniques are considered shortly. However, this is an appropriate place to describe the experimental methods.

3. EXPERIMENTAL TECHNIQUE

The matrix is prepared by depositing the reactant(s) highly dilute in (say) argon on to the low-temperature window. If the reactant is volatile ($>10^{-3}$ mm pressure at room temperature), then a gas mixture may readily be made up. It can then be deposited slowly onto the cold window, or rapidly using the pulsed technique originally described by Rochkind.[105] Recently Perutz and Turner[93] have shown that the isolation of appropriate parent compounds by the pulse technique is at least as good as that by slow spray-on. Moreover, it is usually easier to obtain clear matrices by pulsing, whereas slow spray-on often produces scattering matrices. Since the pulsed technique is very much quicker than slow spray-on, it is generally to be preferred. In addition, there is substantially less danger of impurities being incorporated in the matrix by the faster method. For example, at 10^{-6} torr the time taken to condense a monolayer on the cold window is of the order of 2 sec.[39] The effect of impurities is strikingly demonstrated by studies on CO in argon; recently, it has been shown that the absorption traditionally assigned to CO monomer,[32] because of its increasing intensity with dilution and hence spray-on time, is in fact CO/H_2O, the H_2O being so important because of the very long time for spray-on.[35]

However, if the parent is not very volatile (e.g., $Fe_2(CO)_9$),[96] then it is necessary to deposit this separately, sometimes from an oven, while allowing the matrix gas to enter from a different nozzle. In Lewis's original work[63] and the extension by Norman and Porter,[88] the matrix was a hydrocarbon or similar glass prepared by cooling a solution of some parent in this liquid 'matrix' down to 77°K. Norman and Porter using this method were able to photochemically generate a whole series of free radicals trapped in the glass. More recently, glasses have been very successfully used for organometallic systems by Sheline[110] and Braterman.[9] Since the glass is transparent in the uv-visible, the spectra of species in this range are easily observed, but only in unusually favorable cases does the infrared spectrum of the matrix not dominate that of the reactive species. For example, Chapman,[27] working at 77°K in organic glasses has detected ketenes by their very strong ir absorptions. Transition metal carbonyls are another favorable case.

The most common photolysis source is the mercury arc, suitably filtered by either appropriate solutions or intereference filters. It is also possible with very photosensitive systems to photolyze with monochromatic light produced by a mercury arc plus a monochromator. Line sources include the Cd lamp and discharges in I_2 or Hg. For photolysis in the vacuum ultraviolet a gas discharge is used (e.g., H_2, Ar, Kr), and light enters the cryostat via a LiF window transparent to about 110 nm. There is an extensive discussion of lamps and filters by Calvert and Pitts.[25]

All the usual spectroscopic techniques except nmr and X-ray difffraction can be used for isolated systems. Infrared, uv-visible and esr have been the most popular, but more recently Raman spectroscopy, Mössbauer spectroscopy, and magnetic circular dichroism have been applied to matrix isolation.

Before considering in more detail some of the subtleties of photochemistry in matrices, it is appropriate to briefly compare this method of sample preparation with that of cocondensation synthesis.

4. PHOTOCHEMICAL AND COCONDENSATION MATRIX SYNTHESES COMPARED

In 1966 Andrews and Pimentel[5] condensed Li atoms with methyl iodide in argon and assigned the resulting infrared spectrum to the CH_3 radical plus LiI. The radical has also been produced by the vacuum-uv photolysis of CH_4 in argon.[71] The ir frequencies of the methyl radicals produced by the two routes were different, and it was subsequently shown[113] by using a variety of alkali metal M and CH_3X species that in the cocondensation experiment the MX molecule stays sufficiently close to the CH_3 to form a system best described as a methyl alkali halide radical. The vibrational frequencies of this perturbed methyl radical are significantly different from those of the unperturbed variety.[22] An example of a misinterpretation from matrix photochemistry comes from metal carbonyl chemistry. Rest and Turner[103] photolyzed $Ni(CO)_4$ in various noble gas matrices and assigned the two CO-stretching infrared bands produced to a pyramidal $Ni(CO)_3$ molecule. DeKock[33] condensed Ni atoms with CO at 4°K, and on warming to ~20°K obtained infrared evidence for $Ni(CO)_x$ ($x = 1$–4); $Ni(CO)_3$ had only one band. Subsequently, it has been shown[104] that $Ni(CO)_3$ generated photochemically in a matrix has only one band, at the same frequency as DeKock's. The extra band in the earlier work was due to the presence of dimers arising from the use of too high a concentration of the parent $Ni(CO)_4$ in argon (1:160). With high dilution (1:2000) the dimer band is unobservable on photolysis.

The two methods can, of course, be complementary. Photolysis of $Fe(CO)_5$ in argon produces stepwise loss of CO groups so that the atomic spectrum of Fe atoms can be detected;[98] the spectrum agrees closely with atom spectra obtained by condensing Fe atoms with excess argon.[65] Similarly, the intermediate $Cr(CO)_x$ species produced on photolysis of $Cr(CO)_6$ can be identified with fragments from Cr atom/CO synthesis.[42]

It is worthwhile to summarize some of the major differences between the photolytic and cocondensation methods of approach.

a. As recently elegantly demonstrated by Ozin and Moskovits,[91] the synthesis of species by cocondensation of transition metal atoms with substrates (e.g., CO, O_2, N_2) can be misleading because of the tendency of the metal atoms to polymerize, and hence produce polynuclear molecules rather than the mononuclear species expected. This is a serious problem because it is often difficult to determine the number of metal atoms in the molecule from infrared data (see Chapter 9). Generation of fragments by photolysis overcomes this difficulty because the dilution of parent in matrix can be made so large that dimerization of fragments is unimportant.

b. Ultraviolet-visible spectra of photolytically generated species are usually easy to obtain, and these spectra are often of particular relevance to room-temperature solution photochemistry. Like the scattering matrices often produced by slow spray-on, the uv-visible spectrum of cocondensed species may be dominated by the atom spectrum.

c. The great disadvantage of photochemical generation is that it is necessary to start with some stable compound. For example, although $Ni(N_2)_x$ species are easily prepared by cocondensation of Ni atoms with N_2,[20,21,48] there is no way by which they can be produced photochemically. Surprisingly, however, there is evidence[42] that $Cr(N_2)_x$ can be produced on photolysis of $Cr(CO)_6$ in a N_2 matrix. Because of perturbation by metal atoms, IR bands of species made by cocondensation are often much broader than those obtained by photochemical methods.

d. The range of photochemically produced species is limited by the cage effect, although this can occasionally be turned to advantage.

e. It is important to ask whether photochemically generated species have the same structure as those generated by atom synthesis, that is, is the matrix acting as a rigid clamp in the photochemical experiments. As yet, there is no evidence for any distinction, and there is strong evidence (see later) that there is sufficient localized matrix perturbation to allow the photolytic products to adopt their lowest energy structure. One possible exception was $Co(CO)_4$; Crichton et al.[30] produced this compound by photolysis of $Co(CO)_3NO$ in a CO matrix, and on the basis of isotopic analysis assigned the infrared bands to a C_{3v} structure; Ozin et al.[89,90] detected $Co(CO)_4$ by condensation of Co atoms with CO and on the basis of Raman polarization data initially deduced the structure to be D_{2d}. It should be noted, however, that both authors get identical infrared spectra for the nonisotopically substituted molecule. More recently Ozin et al.[119] have obtained the matrix esr spectra of $Co(^{12}C^{16}O)_4$ and $Co(^{13}C^{16}O)_4$ using the atom cocondensation technique, and there is now general agreement that the molecule has the C_{3v} structure. Molecular-orbital calculations[16] suggest that the C_{3v} and D_{2d} structures are very close in energy.

f. Surprisingly, if transition metal carbonyls in high concentration in Ar or N_2 are photolyzed with white light for a long period, the spectrum[42,97] shows bands not assignable to mononuclear or even binuclear carbonyls. These bands occur close to where absorptions are observed when CO or N_2 is adsorbed on a metal surface, and thus represent the interaction of the diatomic molecule with a cluster of metal atoms. For example, Figure 1 shows the result of prolonged photolysis of $Mo(CO)_6$ in argon.[42] The ramifications of this type of experiment in the fields of surface chemistry and catalysis are discussed further in Chapter 9.

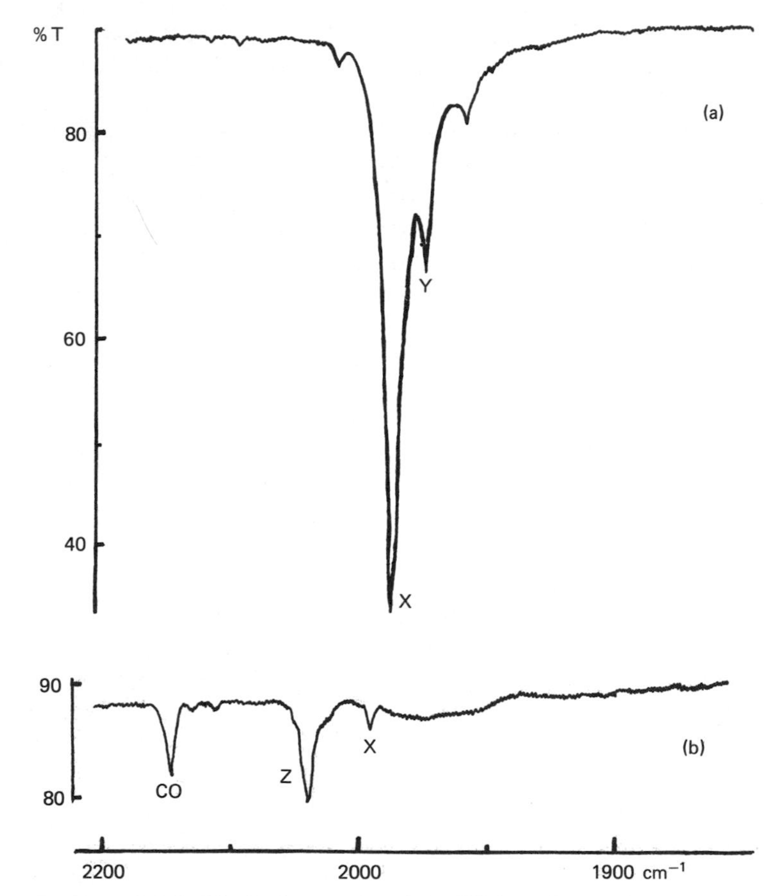

Figure 1. Extended photolysis of concentrated $Mo(CO)_6$/Ar mixtures (1:150). (*a*) Initial spectra of slow sprayed on sample showing band due to dimer, X = $Mo(CO)_6$, Y = $[Mo(CO)_6]_2$. (*b*) After $\frac{1}{2}$ hr photolysis with an unfiltered medium pressure mercury lamp. Z = CO chemisorbed on a cluster of Mo atoms. (From Ref. 42.)

5. THE CAGE AND MATRIX PHOTOCHEMISTRY

We have briefly considered the cage above. This concept, useful in solution photochemistry, is a vital factor in understanding matrix photochemistry. It forms a framework upon which much of the rationalization of the photochemical observations may be hung.

The rate and products of photolysis of molecules in solution or isolated in matrices are often strikingly different from the corresponding processes occurring in the gas phase. Many metal carbonyls, for example, after uv irradiation in the gaseous phase give metal and carbon monoxide, intermediate carbonyls not being detected. In solution the stepwise loss of CO may be followed. For example, with Group VI hexacarbonyls dissolved in a suitable solvent containing a small amount of a ligand, L,

$$M(CO)_6 \xrightarrow[L]{h\nu} M(CO)_5L \xrightarrow[L]{h\nu} M(CO)_4L_2$$

In the matrix the intermediates, $M(CO)_x$, produced after stepwise carbonyl loss may often be isolated.[43,100] Clearly then, the host material, whether matrix or solvent, is exerting a determining effect on the reaction, probably acting in this example as a 'third body' to remove the excess energy, Δ, imparted to the fragments of the primary photolysis step, to prevent further bond fission.

$$A \xrightarrow{h\nu} B + C + \Delta$$

However, diffusion away of these energetic fragments, B and C, of the photochemical process is inhibited by collisions with the surrounding molecular sheath. In solution photochemistry this is known as the Franck–Rabinowitch cage effect.[37] One possible reaction of B and C is simple recombination to form the parent molecule,

$$B + C \rightarrow A$$

and will readily occur if the activation energy for the process is relatively low and the two fragments are prevented from moving apart. On the other hand, the more energetic the fragment, the higher the probability of it escaping from the immediate vicinity of its partner. The cage effect should thus decrease in importance as Δ increases. Once a fragment has moved a significant distance away from the photolysis site (i.e., out of the cage), its chance of return is negligible. Assuming that the same electronic transition is being irradiated, the amount of fragment produced should then increase with decreasing excitation wavelength. This is indeed found to be the case; for example, the quantum yield for X^- loss on irradiation of $[Co^{III}(NH_3)_5X]^{+2}$ in aqueous solution increases as λ decreases.[3] The important difference between the cage of solution photochemistry and the matrix cage is that in the latter the host material is considerably more rigid. As will be seen, it is difficult to eject species larger than first-row atoms from the cage of most matrices; this limits the scope of the sort of photochemistry that may be examined using the matrix technique.

The dependence of the rate of photolysis of a matrix-isolated species upon excitation energy, analogous to the solution example above, is neatly illustrated by the studies[6] of Arkell et al. on OF_2 mentioned previously. This molecule has a broad continuum with ill-defined maxima at 421, 358, and 294 nm.[116] In the gaseous phase, photolysis anywhere within this band envelope leads to fission of an OF bond with production of OF and F radicals. However, in the matrix OF is not observed unless the excitation wavelength is less than 365 nm (3.4 eV). With increasing photon energy the yield of OF increases. Since the OF bond dissociation energy is about 2.2 eV, the OF and F fragments under these conditions must have a combined residual energy of about 1.2 eV. By conservation of momentum this means that the F atom needs a threshold energy of about 0.8 eV to 'escape' from the matrix cage. If escape cannot occur, then the energy of activation is sufficiently low (~ 0 in these radical reactions) to allow rapid recombination to give OF_2. The ejected fluorine atom probably does not travel very far away from the photolysis site. In the photolysis[47,15] of NF_2 and PF_2, infrared absorption bands due to NF_3 and PF_3, respectively, are observed due to reaction of diffused fluorine atoms and unphotolyzed AF_2. In the phosphorus case the amount of PF_3 produced with $\lambda > 250$ nm could be accounted for by reaction of F atoms with a PF_2 fragment in an adjacent site in the matrix. This indicates that the F atom traveled no further than just outside the cage of matrix atoms containing the photolyzed PF_2 radical. Photolysis of a fluoride is a general route to characterization of lower fluorides.

$$AF_x \xrightarrow{h\nu} AF_{x-1} + F$$

Photolytic processes involving the ejection of first-row atoms other than F have not received very much attention, but photolysis[85] of C_3O_2 is a useful source of carbon atoms and leads to CCO. HBO can be observed after photolysis[64] of matrix-isolated $H_2B_2O_3$ by ejection of B and O atoms. In the latter process, a hydrogen atom is also produced. H atoms are readily detected in matrices using esr, and along with CH_3 are often observed on photolysis of samples containing hydrocarbon (grease) impurities. Because of their small size, diffusion away from the photolysis site through interstices in the matrix cage is expected to be easier than for fluorine. This is evidenced by the large number of matrix-characterized species prepared by *in situ* photolysis of a parent hydrogen-containing molecules. Over one half of the photolysis processes observed to date in matrices are of two types

a. $$AH_x \rightarrow AH_{x-1}, AH_{x-2} \cdots A$$

(Not always has the naked A atom been identified). Examples include

$$MH_4 \rightarrow MH_3, MH_2, MH \qquad (M = C, Si) \qquad (Refs. 71, 72)$$

$$MH_2 \rightarrow MH \qquad (M = O, S) \qquad (Refs. 1, 2)$$

$$PH_3 \rightarrow PH_2, P \qquad (Ref. 86)$$

b.

$$A_nB_mH_x \rightarrow A_nB_mH_{x-1}$$

where typical examples are

$$H_xMHal_{4-x} \rightarrow MHal_{4-x}$$

($x = 1, 2$: Hal = Cl, F for M = Si,[73–76] Hal = Cl for M = Ge;[45] $x = 1, 2, 3$, Hal = Cl for M = C[51,52]).

$$CH_2ClF \rightarrow CClF \qquad (Ref. 108)$$

$$GeH_3Br \rightarrow GeHBr, GeH_2Br \qquad (Ref. 49)$$

$$C_2H_2 \rightarrow C_2 \qquad (Ref. 38)$$

and include the production of charged species (to be discussed at more length below)

$$(HCl)_2 \rightarrow HCl_2^- \qquad (Ref. 77)$$

While it has been noted that the hydrogen atom needs little energy to escape from the matrix cage, quite often high-energy vacuum-uv radiation needs to be employed in photolysis to eject a hydrogen atom. This is simply due to the fact that it is only in this region of the spectrum that some of these hydrogen-containing species absorb. For example, methane is transparent above 150 nm and has an absorption maximum at about 95 nm. Therefore, 122 nm hydrogen radiation can be used[71] to produce CH_3 from CH_4. H_2S and HI, by contrast, readily lose a hydrogen atom at wavelengths in the ultraviolet, and are good sources of this atom.[1]

High irradiation energies are also necessary to eject larger atoms from the matrix cage. Thus, vacuum-uv photolysis[52] of $CHCl_3$ in Ar at 14°K gives, among other things, $CHCl_2$ arising through photoejection of a Cl atom. The infrared spectrum of the CN radical may similarly be observed[36,78,79] after vacuum-uv photolysis of XCN species in Ar at 14°K, where X = H, F, Cl, Br, but only in the cases of X = H, F is there any evidence for formation of X_2CN species formed by diffusion of the X atom. In addition, the yield of CN is much lower for the Cl and Br cases than for H and F. If ICN is used as precursor, the amount of CN produced is very small, and the more sensitive tool of esr needs to be used to detect the product. Very low yields of CN are also seen on photolysis of $(CN)_2$/Ar samples. Thus Cl, Br, I and CN may escape from the influence of the CN fragment only with difficulty even with high-energy irradiation. Even then, diffusion away from the photolysis site to react with close XCN molecules does not occur. At higher temperatures and using a hydrocarbon glass the CS radical is readily produced by photolysis of CS_2, but in solid nitrogen at 20°K where the interstices are much smaller, no evidence for such photolytic production is observed.

In fact, the very sensitive esr technique has been extensively used to detect paramagnetic species produced photolytically in matrices. In many of these experiments no change is detectable in the infrared spectrum of the parent molecule. Use of even higher energies (γ irradiation) has led to the observation of, for example, the esr spectra of CF_3 (from C_2F_6),[107] PCl_2 (from PCl_3),[60] and, interestingly, the species $CH_3 \cdots X^-$ (from CH_3X, X = Cl, Br).[109] In this last example the four-line spectrum of the CH_3 radical is split (Figure 2) by the nuclear spin of the Br^- ion close to it. An analysis of the spectrum indicates that the spin density on the Br^- ion is approximately 0.1. The two species are thus close together in the same matrix cage, the Br atom being unable to escape. Similar complex esr spectra have been observed[117] in the uv photolysis of PCl_3 in Ar at 20°K, where a similar but neutral species $PCl_2 \cdots Cl$ is formulated. Whether the perturbed or unperturbed PCl_2 radical is observed depends to a significant extent on the matrix material itself. In Kr and Xe matrices where the interstices between the cage atoms are relatively large, the esr spectrum of the unperturbed PCl_2 species is seen, indicating escape of the Cl atom. This is the same spectrum observed during spray-on photolysis, irrespective of matrix material. In argon where Cl escape is restricted by the smaller mesh, the perturbed radical is seen, and in an N_2 matrix with the smallest cage interstices, no esr signal is observed even after prolonged photolysis, indicating complete recombination. This series of observations, while in keeping with the order of the mesh size in the various matrices, is not in keeping with a model of photoejection based on lattice energies as the determining factor. The idea of the matrix cage in terms of the size of the interstices compared with the size of the ejected atom is, thus, a useful model to work with. The failure to isolate cyclobutadiene from the *in situ* photolysis of $CbFe(CO)_3$, for example,[9] is thus readily understood, although this might be a two-photon process.

There are several ways of avoiding the cage problem and the very short diffusion length of most photoejected species in these rigid media. As hinted above, a spray-on photolysis is sometimes the answer, but often this may not produce what is required (cf. carbonyl photolysis earlier). The cage effect, however, is only important if the energy of activation for the reverse process is small, that is, if the photogenerated fragments have a significant chance to recombine on the rebound after a collision with the cage walls. With a low barrier to recombination, PCl_3 was regenerated in the PCl_3/N_2

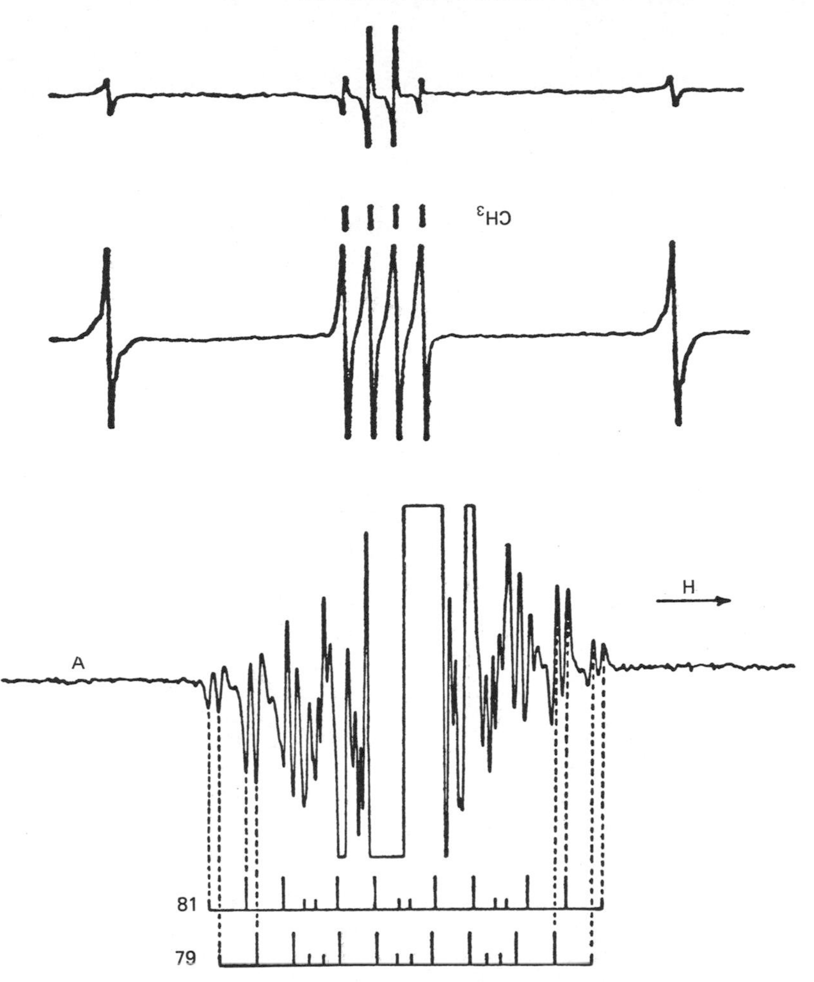

Figure 2. (*a*) Electron spin resonance spectrum of the CH_3 radical in an argon matrix at 4°K. (*b*) Electron spin resonance spectrum of the matrix-isolated CH_3 radical strongly perturbed by a neighboring Br^- ion.

photolysis example above. By contrast, photolysis of an azide, XN_3, should lead to isolation of the XN species, since an inert N_2 unit is produced in the same cage.

$$XN_3 \rightarrow XN + N\equiv N$$

XN radicals (X = H, F, Cl, Br) have been made[80] in this way, as well as CNN,[81] NCN[82] (from N_3CN), and HNC[83] (from CH_3N_3). Similar arguments may be used to rationalize the formation[12] of CH_2 from diazirine,

$$\begin{array}{c}H\quad N\\ \diagdown\;\diagup\!\|\\ C\;\;\|\\ \diagup\;\diagdown\!\|\\ H\quad N\end{array}$$

, and CF_2[70] from CF_2N_2, where N_2 is also produced. Surprisingly, the CH_2 produced by photolysis of diazomethane (CH_2—N_2) rapidly reacts with N_2 (as evidenced by $^{15}N_2$ doping experiments), and the carbene may not be isolated. The barrier to recombination need not involve a reorganization of the bonding electrons as in these N_2 examples. For example, two PF_2 radicals are formed within the same matrix cage when P_2F_4 is photolyzed[15] ($\lambda > 250$ nm) in Ar at 4°K. (There is no chance at these wavelengths that the PF_2 unit will be ejected). On warming the matrix to 20°K, P_2F_4 is regenerated. Whether the barrier to recombination is overcome by a softening of the matrix at 20°K relative to 4°K, or whether it is due to the two PF_2 groups being rotated unfavorably with respect to one another is not clear.

One class of photolysis reactions that is almost unaffected by the presence of the matrix cage is that of isomerization. For example,

$$PF_2 - PF_2 \rightarrow PF_3 - PF \qquad \text{(Ref. 15)}$$

$$CH_3N_3 \rightarrow [CH_3N] \rightarrow CH_2 = NH^{69} \qquad \text{(Ref. 69)}$$

$$Cl - A - Cl \rightarrow A - Cl - Cl \quad (A = P, O) \quad \text{(Refs. 117, 106)}$$

$$XCN \rightarrow XNC \quad (X = H, F, Cl, Br) \qquad \text{(Refs. 78, 79)}$$

All of these processes require uv or visible irradiation, but Pimentel and coworkers have discovered[46] a fascinating infrared photolysis, the only genuine photolysis at these wavelengths observed to date.

$$cis\text{-HONO} \underset{\text{near-ir}}{\overset{\text{uv}}{\rightleftharpoons}} trans\text{-HONO}$$

The frequency of the near-ir shows an isotopic dependence; 3200–3650 cm^{-1} (for HONO) and 3500–4100 cm^{-1} (for DONO). A detailed analysis of the energy-transfer processes ocurring here has been made.[41] As mentioned previously, the cage effect can be of great importance when considering bimolecular reactions. If one species can move relatively freely in the matrix, then the cage effect is less important. Thus if F_2/Ar (1:150) and HN_3(1:200) mixtures are cocondensed at 14°K and photolyzed, the infrared and ultraviolet spectra of the HNF radical may be observed[50] via the processes on site (1)

$$F_2 \rightarrow F + F$$

on site (2)

$$HN_3 \rightarrow HN + N_2$$
$$F + HN \rightarrow HNF$$

also,

$$F + HNF \rightarrow HNF_2$$

If, for bimolecular reactions, no species moves freely in the matrix then the reacting entities must be adjacent, either by using one reactant as the matrix or having it present at high concentration in an inert matrix.

6. CHARGED SPECIES IN MATRICES

Some of our previous examples have included the photolytic production of charged species. Kasai[56–58] first observed that if electron-donating and electron-accepting species were isolated in the same matrix, then electron transfer between them may be promoted by photoejection of an electron. Return of the electron from the anion so formed to the cation is prevented by the local potential trap in the form of the ionization potential of the anion (electron affinity of the neutral species), and migration of the ions towards each other is not allowed by the rigid nature of the matrix itself. Thus, for example, if HI and a small concentration of metal atoms (M = Na, Cd, Mn, Cr) are cocondensed with Ar at 4°K, on irradiation with $\lambda > 500$ nm the esr spectra of hydrogen atoms and the ions, Cd^+, Mn^+, and Cr^+, are seen.[56] In the case of Na no esr signal is expected from the unipositive ion, but here the esr signal due to Na itself disappears on irradiation. The results are interpreted along the following lines. Since HI does not photolyze with $\lambda > 500$ nm when M atoms are not present, fission of the HI bond cannot be a primary photochemical step. A probable process is dissociative electron capture by HI, namely,

$$HI + M \xrightarrow{\lambda > 500\text{ nm}} M^+ + [HI]^- \rightarrow M^+ + H + I^-$$

Similarly, the following processes have been observed:

$$B_2H_6 \xrightarrow[\lambda > 500\text{ nm}]{Na} B_2H_6^- \qquad \text{(Refs. 55, 56)}$$

$$Na + PyX \xrightarrow{\lambda > 500\text{ nm}} Na^+ + [PyX]^- \rightarrow Py\cdot + X^- + Na^+$$
$$(Py = \text{pyridyl})$$

In the last experiment the Na:PyX:Ar ratio was 1:10:1000. If attempts are made to produce the Py radicals with straight-forward photolysis ($\lambda <$ 400 nm) from PyX, ring rupture is all that is observed. Several important points arise from Kasai's studies. Firstly, in the pyridyl case, the Py· and X^- species must lie quite close to each other within the matrix cage, there not being enough excess energy to move the halogen atom very far. This is not incompatible with the previous discussion, since only one of the partners is a radical species and the pair was formed by dissociation of $[PyX]^-$ itself. That is, there is probably a barrier to the reverse process, and the Py/X^-

pair is probably more stable than the PyX^- ion anyway. Secondly, the photon energy used ($\lambda > 500$ nm) is less than the energy needed to ionize Na atoms. In fact, at these wavelengths there is only sufficient energy to excite the $3s \rightarrow 3p$ transition. This implies that the electron-transfer process arises through a cooperative effect between alkali metal and substrate via

$$M + AB \xrightarrow{h\nu} M^* + AB \rightarrow M^+ + [AB]^-$$
$$AB = B_2H_6, PyX, HI$$

where the M and AB species must be close enough together for such an interaction to occur, but end up far enough apart so that the influence of the nuclear spin of M^+ is not seen in the esr spectrum of Py· or $B_2H_6^-$.

An alternative route to the production of charged species in matrices is via vacuum-ultraviolet photolysis. At these energies there is often sufficient energy either to photoionize a matrix-isolated species to produce a cation and electron, or to photolyze trace impurities to give a source of electrons that may then lead to anion formation. Thus, when $HCCl_3$ is isolated in an argon matrix at 14°K and is irradiated with $\lambda = 122$ nm, 10.2 eV (H lamp), CCl_3 is produced.[52] This may be identified by its infrared spectrum. On receipt of another photon CCl_3 is ionized, and the species CCl_3^+ is observed. The ionization potential of CCl_3 is about 8.7 eV, and so there is sufficient energy for this process. When higher energy radiation, $\lambda = 107$ nm, 11.6 eV (Ar lamp), is used the $HCCl_2^+$ ion is identified, although this energy is slightly below the threshold for direct production of $HCCl_2^+$ from $HCCl_3$ (probably between 11.7 and 12.4 eV). More difficult to understand is the production of ionic species with $\lambda = 147$ nm, 8.4 eV radiation. However, if it is noted that the ejected counterion will not travel too far away from the cation, especially if the anion is the large Cl^- species, then the threshold photon energy for production of $HCCl_2^+$ is reduced by the Coulombic interaction between the charged particles. If the ions were separated by 4 Å, then the threshold energy for production of $HCCl_2^+ \cdots Cl^-$ is reduced by about 3.6 eV to 8.1–8.8 eV, that is, within the observed range of photolysis energies. The species, $HCCl_2^-$, is also observed after the vacuum-uv photolysis of chloroform. In this particular case, it is probable that dissociative electron attachment occurs to a significant extent.

$$HCCl_3 + e \rightarrow HCCl_2^- + Cl$$

Here the electron is provided by photoionization of, for example, CCl_3 and impurities in the matrix. If an alkali metal is also trapped in the matrix, then this ion may also be observed using longer wavelength photolysis, since here the metal acts as a source of electrons.

The NO_2^- ion has been produced[67,68] by vacuum-uv photolysis of matrix-isolated NO_2. In this case, no infrared data was obtained indicative of a counterion, and thus trace impurity photolysis must be regarded as the dominant electron source. The ion may also be made by cocondensing the alkali metals, Cs, Rb, K, or Na, with an Ar/NO_2 mixture at 4°K. In this case, no photolysis is necessary for production of anion, and the electron transfer process occurs spontaneously during spray-on. The infrared spectrum of the ion is independent of the metal used, illustrative of negligible mutual perturbation of the ions in the matrix. At high metal/NO_2 ratios, however, new sets of bands may be observed due to such interactions. The NO_2^- ion may be readily photoionized by irradiation with an unfiltered mercury lamp if no alkali metal atoms are present. If metal atoms are present in the matrix, then a constant source of electrons is available.

Another interesting ion is the HX_2^- ($X = Cl, Br$) species, which can be produced[77] under a variety of conditions. These ions may be produced in matrices using $\lambda = 122$ nm, 10.2 eV, even though the ionization potentials of the HBr and HCl are 11.6 and 12.7 eV, respectively. However, the yield of anion decreases rapidly as the Ar/HX ratio becomes larger than 100:1. Juxtaposition of two HX molecules is thus necessary for anion formation. Photoionization of trace impurities [or $(HX)_2$ itself] may act as an electron source and

$$(HX)_2 + e \rightarrow X\text{—}H\text{—}X^- + H$$

or there is the possibility of direct excitation of HX itself

$$HX + h\nu \rightarrow H^+ \cdots X^- + HX \rightarrow H^+ + XHX^-$$

A highly ionic state (the V state) has been experimentally observed at excitation energies accessible by the lamp energy. This mechanism is similar to that postulated to rationalize Kasai's production of ionic species with Na and $\lambda > 500$ nm.

$$M + h\nu \rightarrow M^* + AB \rightarrow M^+ + AB^-$$

This scheme also requires the two HX species to be close together. HX_2^- may also be produced if alkali atoms are included in the matrix as sources of electrons. In this case,

$$M^* + HX \rightarrow M^+ + HX^-$$
$$HX^- \rightarrow H + X^-$$
$$X^- + HX \rightarrow XHX^-$$

This scheme does not require the HX molecules to be as close together initially. Infrared bands due to the perturbation $M^+ \cdots XHX^-$ that arise on spray-on may also be observed in the spectrum. On irradiation with a mercury lamp they decrease in intensity. The possibility arises that sufficient energy may be imparted to the complex to eject the M^+ ion away from the vicinity of the XHX^- species; $\lambda = 340$ nm only is needed to overcome the electrostatic energy of two ions 4 Å apart.

Charged transition metal carbonyls may also be produced either by vacuum-uv photolysis or by using a matrix doped with Na atoms. The ions, $M(CO)_5^-$ (M = Cr, Mo, W), have been identified after cocondensing Na atoms with $M(CO)_6$ followed by visible photolysis.[13,14] (A small amount of anion is formed during spray-on.) Similarly, the ions, $Ni(CO)_3^-$ and $Fe(CO)_4^-$, may be identified by this route. Vacuum-uv photolysis using Ar and H radiation also leads[17] to the observation of the infrared spectra of these anions as well as $V(CO)_6^-$. Here, although CO-stretching absorptions are observed at high frequency in the infrared, where cationic species are expected to occur, no definite assignments have been made. For $Ni(CO)_3^-$ the counterion is tentatively assigned[17] as CO^+. The ionization potential of CO is very large (~14 eV), and calculation shows that if production of CO^+ does occur the two ions should be quite close together in the matrix. The ionization potentials of the $M(CO)_6$ species are around 8.5 eV, and thus photoionization with the H lamp is to be expected. Photoionization of the anionic species formed, however, can occur with uv or mercury lamp photolysis. Since no CO^+ is seen in the experiments involving the eighteen electron carbonyls of Cr, Mo, W, or Fe, the anion is probably produced by dissociative electron capture. (No binary carbonyl with more than eighteen electrons has been observed.)

$$M(CO)_6 + e \rightarrow M(CO)_5^- + CO$$

$$Fe(CO)_5 + e \rightarrow Fe(CO)_4^- + CO$$

Vacuum-uv photolysis of matrix-isolated $V(CO)_6$ gives $V(CO)_6^-$ and not $V(CO)_5^-$. Since $V(CO)_6^-$ is isoelectronic with the stable $M(CO)_6$ species, this parent anionic species does not lose CO. A secondary means of production under certain conditions is via electron capture by $M(CO)_5$ and $Fe(CO)_4$ present, which has been produced by uv photolysis via the uv component of the vacuum-uv lamp. As noted above, these carbonyls are incredibly photosensitive. The ejected CO almost certainly ends up in the same matrix cage as the anion, but is prevented from returning to the ion by virtue of the fact that this would create a 19 electron species. In support of this idea, these anions (17 electrons) may be readily produced in pure CO matrices, whereas the corresponding neutral species (16 electrons) can only be obtained in low yields relative to their 18 electron parents.

Presumably some sort of barrier to recombination exists between NO^- and $\pi CpNi^+$ produced[29] photolytically with low-energy uv radiation from CpNiNO. Possibly there is enough energy in the photolytic process for the two ions to end up some way away from each other.

7. PHOTOCHEMISTRY OF TRANSITION METAL CARBONYLS

There have been many room-temperature photochemical studies involving transition metal carbonyls. Early work by Strohmeier[112] showed that in the reaction

$$Cr(CO)_6 + L \xrightarrow[\text{hydrocarbon solvent}]{h\nu} Cr(CO)_5L + CO$$

(L, for example, being a phosphine) the quantum yield is very close to unity and independent of L. These results suggest that, at least at low concentration, the primary photochemical species is $Cr(CO)_5$. Massey and Orgel[66] and later Sheline and coworkers[110] examined the photochemical behavior of $Cr(CO)_6$ in a polymethylmethacrylate room-temperature glass and in a hydrocarbon glass at 77°K. On the basis of infrared spectra in the CO-stretching region, it was concluded that $Cr(CO)_5$ has a C_{4v} structure. Work was originally undertaken in these laboratories using inert gases at 4–20°K as matrix in order to learn more about such intermediates. Because of the extreme photosensitivity of these carbonyls, not only was it relatively easy to generate lower carbonyl fragments, but also profitable to use the carbonyls as probes for examining the basic processes occurring in matrix photochemistry.

The fragmentation results can be summarized

$$M(CO)_6/\text{argon} \rightarrow \underset{C_{4v}}{M(CO)_5} + \underset{C_{2v}}{M(CO)_4} + \underset{C_{3v}}{M(CO)_3} + ?$$

M = (Cr, Mo, W) (Refs. 42, 43, 92, 94, 120, 122)

$$Fe(CO)_5 \rightarrow \underset{C_{2v}}{Fe(CO)_4} + \underset{C_{3v}}{Fe(CO)_3} + Fe(CO)_2 \cdots + Fe$$ (Refs. 95, 98–100)

$$Ni(CO)_4 \rightarrow \underset{D_{3h}}{Ni(CO)_3} + CO$$ (Refs. 103, 104)

The structures of the fragments at first sight is surprising, but a theoretical model has been developed by one of us[18] that indicates that symmetrical structures are not always to be expected for these fragments. It is also perhaps surprising that the photoejected CO (which from its infrared spectrum is clearly present in the matrix) does not recombine instantly with the fragments; it is even more surprising when it is recalled that photolysis of C_2N_2 gives very little CN radical, presumably because of the cage effect. The CO molecule is presumably of similar size to CN. We return to this question below.

One of the fascinating observations in the field of matrix carbonyl photochemistry is the reversibility of the forward-photolysis process by either annealing techniques or irradiation in an absorption band of the fragment.

Specifically,

$$Cr(CO)_6/\text{argon} \underset{h\nu'(\text{visible})}{\overset{h\nu(uv)}{\rightleftarrows}} Cr(CO)_5 + CO \qquad \text{(Ref. 43)}$$

$$\overset{\Delta}{\nleftarrow}$$

$$Ni(CO)_4/\text{argon} \underset{h\nu'}{\overset{h\nu}{\rightleftarrows}} Ni(CO)_3 + CO \qquad \text{(Ref. 103)}$$

$$\longleftarrow$$

(The Δ refers to annealing the matrix to $\sim 35°K$ and recooling to $20°K$.) Such reversal behavior has also been observed in several other carbonyl systems, for example, $HMn(CO)_5$.[102] In the first case, forward and backward photolysis proceeds easily; in the $Ni(CO)_4$ case it is very difficult to generate any $Ni(CO)_3$, and no wavelength can be found that easily removes it. By analogy with Callear's gas phase work,[24] one possible explanation is that $Ni(CO)_4$ and $Ni(CO)_3$ have similar absorption maxima so that it is not possible to irradiate $Ni(CO)_4$ without at the same time irradiating $Ni(CO)_3$ and hence promoting the back reaction. The observations concerning thermal and photolytic reversal to the parent molecule are reproduced by a model we describe in the next section.

Since

$$Cr(CO)_6 \xrightarrow{h\nu} Cr(CO)_5 + CO \xrightarrow{h\nu} Cr(CO)_6$$

it is not surprising if

$$\underset{(1)}{Cr(CO)_6 \xrightarrow{h\nu} Cr(CO)_5 + CO} \underset{(2)}{\xrightarrow[L]{h\nu} Cr(CO)_5L}$$

where stages (1) and (2) may occur in one stage or two separately observable stages. This process has been extensively used to synthesize interesting molecules. Some examples are

$$Ni(CO)_4 \xrightarrow[N_2]{h\nu} Ni(CO)_3N_2 \qquad \text{(Ref. 101)}$$

(with frequencies close to those observed by Ozin[62] on condensation of $Ni/CO/N_2$)

$$Cr(CO)_6 \xrightarrow[N_2\text{ matrix}]{h\nu} Cr(CO)_5N_2 + Cr(CO)_4N_2 \qquad \text{(Ref. 42)}$$

$$Fe(CO)_5 \xrightarrow[N_2]{h\nu} Fe(CO)_4N_2 \qquad \text{(Ref. 99)}$$

$$Os(CO)_5 \xrightarrow[NO/Ar]{h\nu} Os(CO)_2(NO)_2 \qquad \text{(Ref. 99)}$$

$$Co(CO)_3NO \xrightarrow[CO]{h\nu} Co(CO)_4 + NO \qquad \text{(Ref. 30)}$$

(These may not be one-step processes, of course.) More subtle is the generation of species with new ligands that are a considerable surprise. For example, photolysis of $Fe(CO)_5$ in CH_4 or Xe leads to the isolation[99] of both 'naked' $Fe(CO)_4$ and $Fe(CO)_4$ attached to CH_4 or Xe, according to the following scheme

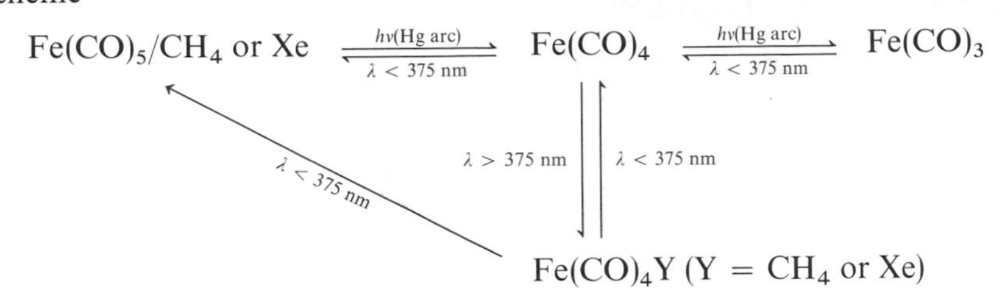

(The source of $\lambda < 375$ nm is radiation from unfiltered Nernst glower of infrared spectrometer.) The structures of $Fe(CO)_4$ and $Fe(CO)_4Y$ have been shown by[99] ^{13}CO isotopic studies to be C_{2v} with bond angles 120°, 145° for $Fe(CO)_4$, and 125°, 174° for $Fe(CO)_4CH_4$. The behavior is readily explained if $Fe(CO)_5$, $Fe(CO)_4$, and $Fe(CO)_4Y$ absorb at different wavelengths.

Photolysis of $Cr(CO)_6$ in a wide variety of matrices produces $Cr(CO)_5$ with a C_{4v} structure. However, although the infrared spectral shifts $Cr(CO)_6$, $Cr(CO)_5$ are very similar in all matrices, the visible absorption band of $Cr(CO)_5$ is astonishingly sensitive to the matrix,[121] varying from 624 nm in Ne to 490 nm in Xe. Distinctly different infrared and visible spectra for the $Cr(CO)_5[Ne]$ and $Cr(CO)_5[Xe]$ species are observed in mixed Xe/Ne matrices. The probable explanation of this is that the $Cr(CO)_5$ fragment is very weakly attached to Xe (and even less so to Ne), the major part of the spectral shift being due to a change in axial/radial bond angle in the $Cr(CO)_5$ fragment due to the presence of a matrix molecule in the sixth coordination site of the $M(CO)_5$. The significance of this observation in the present context is that it is thus not possible to obtain truly 'naked' $Cr(CO)_5$, but only a perturbed species. Since $Cr(CO)_5$ in CH_4 absorbs at 489 nm but in CF_4 absorbs at 580 nm, Kelly[59] suggested that the strength of interaction is weaker in the second case than in the first. Thus, in hydrocarbon room-temperature solutions the photochemically generated $Cr(CO)_5$ forms $Cr(CO)_5[C_6H_{12}]$, which is more stable than $Cr(CO)_5[C_7F_{14}]$. In fact, $Cr(CO)_5[C_7F_{14}]$ reacts with CO 10^2 times faster than does $Cr(CO)_5[C_6H_{12}]$ to reform $Cr(CO)_6$. In mixed matrices (Ne/Xe in particular) where the visible bands due to $Cr(CO)_5[Ne]$ and $Cr(CO)_5[Xe]$ can be well resolved, irradiation in the $Cr(CO)_5[Ne]$ band at 620 nm removes this species by reversal to $Cr(CO)_6$.[92,19] Also, direct conversion of $Cr(CO)_5[Xe]$ to $Cr(CO)_5[Ne]$ and vice versa occurs on visible irradiation without going through the $Cr(CO)_6$ stage. Such experiments are of great value in interpreting the basic chemical physics occurring in low-temperature matrices.

8. CHEMICAL PHYSICS OF THE PHOTOLYSIS PROCESS

There are one or two further observations from carbonyl photochemistry that are relevant to this section.

1. Figure 3 reproduces schematically the splitting occurring in the t_{1u} CO-stretching mode of matrix-isolated $Cr(CO)_6$ and its behavior on photolysis.[43] Identical splitting and photolytic behavior also occurs[92] with the low-frequency a_1 vibration of the $Cr(^{12}CO)_5(^{13}CO)$ molecule, ruling out a distortion effect and thus indicating the occurrence of two trapping sites for $Cr(CO)_6$ on deposition. Photolysis removes $Cr(CO)_6$ trapped in both sites, but on reversal, localized events have allowed the reformed $Cr(CO)_6$ itself to determine the immediate matrix structure and occupy one type of site only. Not surprisingly, annealing the matrix to 35°K and recooling allows the whole argon matrix to reimpose the two different sites.

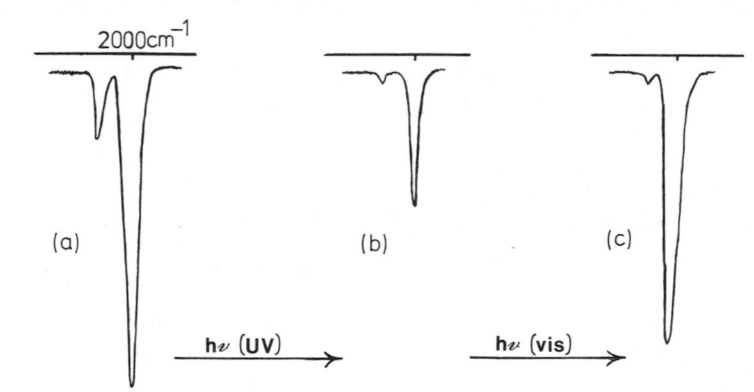

Figure 3. The "doublet" behavior on photolysis of matrix-isolated Group VI hexacarbonyls. (*a*) The infrared spectrum around 2000 cm^{-1} of an $Ar/M(CO)_6$ matrix. (*b*) After photolysis with monochromatic uv radiation showing decrease in intensity of both components of the doublet. (*c*) After reversal, using monochromatic visible radiation when only the lower frequency component of the doublet is regenerated.

2. It has recently been demonstrated[19] by polarized photochemistry and by polarized ir and uv-visible spectroscopy that among several processes occurring in the matrix there is one of particular relevance. This photo-orientation process on visible photolysis is schematically illustrated in Figure 4. Thus it appears that the C_{4v} fragment is actually turned over in the matrix. Several years ago Albrecht described photolysis experiments[4] with polarized light that generated oriented molecules by the preferential photolysis of a randomly oriental sample; however, there was no evidence for photoorienta-

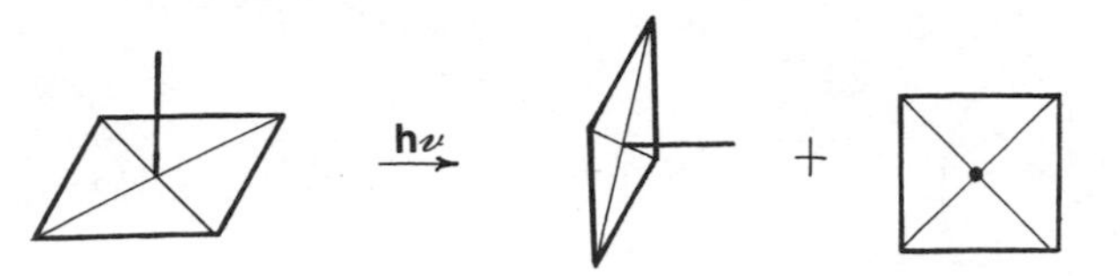

Figure 4. Photoorientation processes detected on visible photolysis of square pyramidal $M(CO)_5$ species (M = Cr, Mo, W).

tion, (i.e., physically changing the direction that the principal axes of the molecule point in space). Thus in our experiment the photoorientation is more likely to proceed through a trigonal bipyramid intermediate, rather than a true physical rotation (cf. the Berry mechanism for ground state equatorial-axial ligand exchange in five coordinate trigonal bipyramidal systems, e.g., PF_5, $Fe(CO)_5$). Such an explanation fits the *cis-trans* isomerization work of Braterman and Black[8] on C_{4v} $Mo(CO)_4P(cy)_3$ and Poliakoff[97] on $Cr(CO)_4CS$, where equatorial-axial ligand exchange has been demonstrated. The implication is that the matrix is not so rigid as to prevent such movements, and hence that species (at least with reference to the ligand geometry) generated photochemically find themselves in an immediate environment that is sufficiently flexible to allow the fragment to adopt its lowest energy geometrical configuration.

3. Brief photolysis[100] of, for example, $Fe(CO)_5$ in an argon matrix produces $Fe(CO)_4$ and CO: this is very easily reversed simply by exposing the matrix to light from the ir spectrometer's Nernst glower. However, prolonged photolysis increases the proportion of $Fe(CO)_4$ and CO, which have slightly different spectra from the initially observed fragment, and it is not possible to reverse these species to $Fe(CO)_5$. It has been postulated that in the first case the CO remains in the immediate environment of the $Fe(CO)_4$ and gives rise to small frequency shifts in both the CO vibrations of the $Fe(CO)_4$ molecule and the vibration of the 'free' CO molecule. In the second case, prolonged photolysis causes sufficient local matrix rearrangement to allow the CO to escape 'beyond the reach' of the $Fe(CO)_4$. Similar behavior occurs[92] on extended photolysis of $Cr(CO)_6$. We are now in a position to propose a photochemical summary of these events (Figure 5).

Visible irradiation of $M(CO)_5$ or uv irradiation of the parent hexacarbonyl followed by loss of CO leads to an electronically excited $M(CO)_5$ unit. Molecular orbital considerations[16,18] predict a trigonal bipyramidal geometry for this species. The electronic ground state of these d^6 $M(CO)_5$ molecules is predicted[16,18] and observed[43,92] to be square pyramidal. Decay to the electronic ground state may take place to give one of five orientations dependent upon the direction in which the unique axial CO ligand points

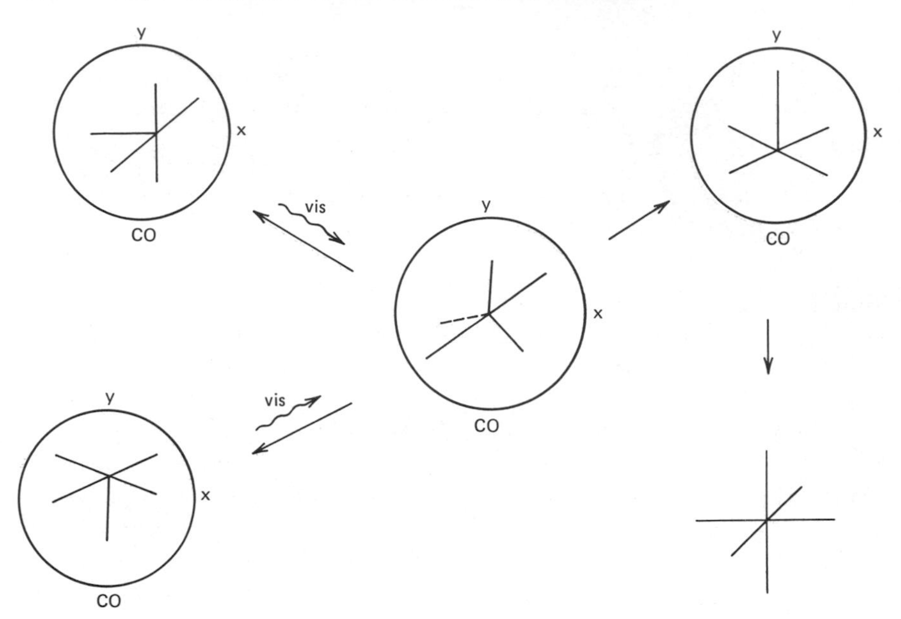

Figure 5. Photochemical pathways of $Cr(CO)_5$ through a trigonal bipyramidal excited electronic state.

in space. In all but one of these the sixth coordinate site does not contain CO, and in all five geometrical environments the CO is located such that thermal recombination into this site is very unfavorable. Photochemical recombination is readily understood since visible radiation produces the trigonal bipyramidal geometry, which has a one in six chance of decaying to a square pyramidal arrangement with a CO close to the sixth site. The photoorientation and $M(CO)_5[X] \rightarrow M(CO)_5[Y]$ processes can also be readily understood on the model that would also be applicable to the isoelectronic $HMn(CO)_4$ species.[102] A similar approach may be used to rationalize the behavior of $Fe(CO)_5$ on photolysis. The difficulty with which $Ni(CO)_4$ (and indeed all tetrahedral transition metal CO/NO compounds) photolyzes in the matrix is explicable along similar lines. The photolytically produced $Ni(CO)_3$ will end up with the photoejected CO just above the plane of the fragment in an ideal position for recombination, either during the time the photon energy is being dissipated from the site or later during an annealing process. Another factor influencing the facile thermal recombination is that in early experiments it was virtually impossible to obtain a matrix-isolated sample of $Ni(CO)_4$ without CO contamination. Such CO doping increases the number of CO molecules close to a photolytically produced $Ni(CO)_3$ fragment.

The presence of a CO molecule adjacent to the $M(CO)_x$ fragment, but physically prevented from recombining with it due to attached CO groups lying between the ejected CO and the metal as in Figure 5, thus rationalizes the splittings noted in the infrared spectrum in observation 3, above. On extended photolysis the proportion of ejected CO molecules lying further away from the fragment (and thus photolytically irreversible) increases. On each uv photolysis there is a small chance that the ejected CO group will lie out of reach of the $M(CO)_x$ fragment (cf. the small amount of CN produced on photolysis of $(CN)_2$). Due to the very large extinction coefficients and high quantum yields for the ejection process on extended photolysis this photolytically irreversal process is amplified.

After considering the behavior on photolysis of the molecule itself within the solid-matrix cage, it is pertinent to enquire how much disruption of the surrounding matrix molecules is caused by the absorption of a photon by the molecule and the dissipation of the excess energy of the photolytic process. This is an extremely complex question, and at present we have little detailed knowledge of the energy-transfer processes occurring close to the 'epicenter' of the energy release.

The absorbed photon energy is eventually removed from the matrix by the cryogenic source, and this 'thermal decay' has been elegantly demonstrated by Abouaf-Marguin and coworkers,[35] using laser induced fluorescence in the infrared part of the spectrum. Let us use a classical continuum model to decide how fast the energy is removed from the photolysis site itself and absorbed by the matrix material. Assume that Q joules of energy are instantaneously released at a point in the matrix. Classical concepts would suggest that the area around this 'hot spot' would become molten and allow diffusion of the species located initially at the epicenter. To obtain a relationship between the temperature (T), the time (t) after the release, and the distance (r) from the photolytic site, we require solution of the diffusion equation

$$\frac{\partial T}{\partial t} - D^2 T = \frac{Q}{\sigma}\,\delta(t)\,\delta(r)$$

D $(=\lambda/\sigma)$ is the thermal diffusivity of the matrix material, λ its thermal conductivity, and σ its specific heat. The solution may readily be shown to be

$$T(r, t) = \frac{Q}{\sigma}\left(\frac{1}{4\pi Dt}\right)^{3/2} \exp\left(-\frac{r^2}{4Dt}\right)$$

and is shown graphically in Figure 6. In Figure 6*b* we see a thermal spike decreasing in intensity diffusing away from the energy epicenter. It has been suggested that diffusion may readily occur in matrices if the matrix temperature exceeds a third of the melting temperature. Using this equation we

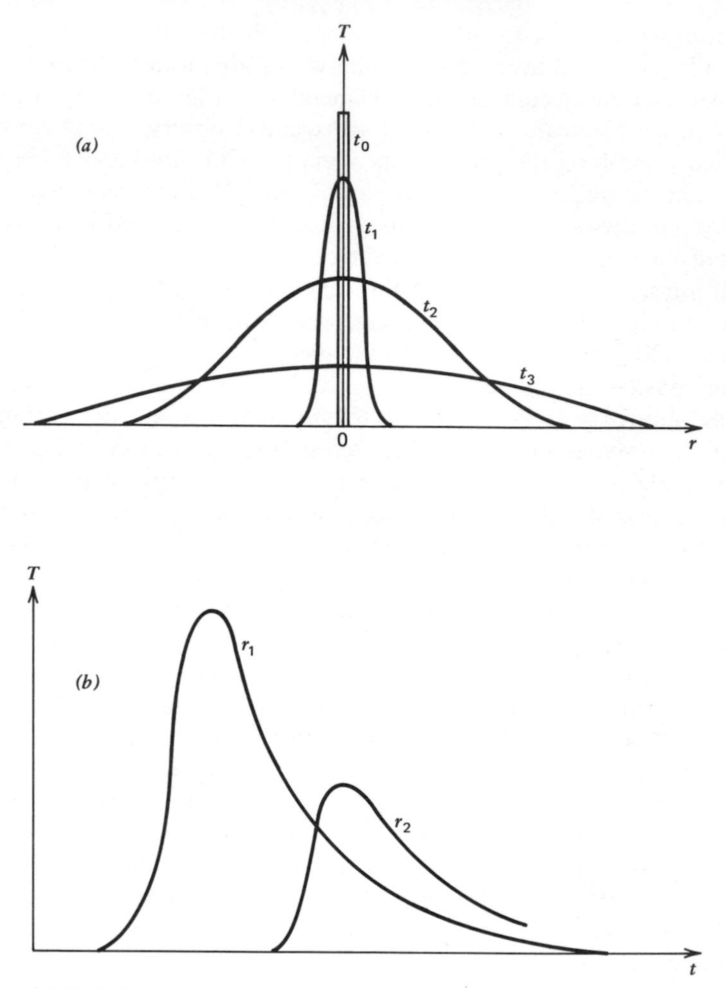

Figure 6. (*a*) Variation of matrix temperature as a function of distance from photolysis site at various times after release of energy ($t_0 - t_3$). (*b*) Variation of matrix temperature as a function of time at two different distances from the photolysis site ($r_2 > r_1$).

may readily calculate the time for which $T > T_{\text{melting}}/3$ for any value of r. Using typical data for an argon matrix at 4°K; for $r = 2.5\text{Å}$ $\Delta t = 1.1 \times 10^{-12}$ sec, $r = 0$ $\Delta t = 9.7 \times 10^{-12}$ sec, and similar results are found for other matrix materials. Δt corresponds to 10–100 periods of vibration for and A—B bond ($\nu \simeq 500$ cm^{-1}, $\tau = 6 \times 10^{-14}$ sec) and is long enough for the A—B bond to break and the B fragment to disappear into the "local soup" and diffuse away from the photolysis site. Similar arguments using the molten matrix concept to rationalize induced dimerization have been used by DeMore and Davidson.[34] Obviously the larger the size of Q, the longer the matrix remains molten, and the greater the chance of A and B being physically separated when the matrix locally "freezes." Alternatively, if no photodissociation occurs, then the larger Q, the further the absorber molecule itself is able to diffuse in the matrix before it is retrapped.

The disadvantage of such a classical and readily visualized approach to the problem is the following. The concept of temperature is a statistical one and, therefore, relies for its validity on the treatment only of systems that contain a large number of particles. The number of matrix molecules in the immediate vicinity of the photolysis site is very small by statistical standards, and so the approach is open to a considerable amount of criticism. What is certain, however, is that there is enough energy produced in the photolytic act to give the matrix molecules juxtaposed to the absorber enough energy that a significant rearrangement of the environment surrounding the trapped species occurs, as indicated by the 'doublet' behavior of $M(CO)_6$ species above. Recent views concerning the local disturbance in a solid medium after release of a large amount of energy (compared to the binding forces between matrix molecules) tend to suggest that in general only a relatively small number of local molecules become very energetic and only under very favorable conditions may an energy-rich species move more than a few atomic diameters away from the epicenter of the process. Computer simulation of the process has been performed[40,114,115] for such processes occurring in alkali halide molecules and simple metallic structures. For the case of NaCl, only the small Na^+ ion was able to escape from the epicenter through the lattice after high energy irradiation; the larger Cl^- ion always remained trapped close to its original lattice position. These results support the ideas surveyed above concerning the size criterion for escape from the matrix cage. Experiments involving "hot atom" chemistry (chemical effects of neutron capture) have also indicated[54] that the hot spot discussed classically above need not contain more than a few molecules, and that the effect of point high-energy release in such systems is a much more localized phenomenon than earlier ideas concerning the local soup.

Some other evidence concerning the local rearrangement process comes from studies on trapped phthalocyanins.[7] With reference to Figure 7 the absorptions due to the transitions, X → A and X → B, may be recorded. The transition energy, X → B, was found to be dependent upon whether the sample had been annealed. If the sample was pumped with radiation, $\lambda = 360$ nm, then the emission, B → X, was found to occur at the same frequency as the X → B absorption of the annealed sample. One conclusion, therefore, is that the nonradiative transition, A → B, locally annealed the matrix. However, this experiment did not eliminate the possibility that the

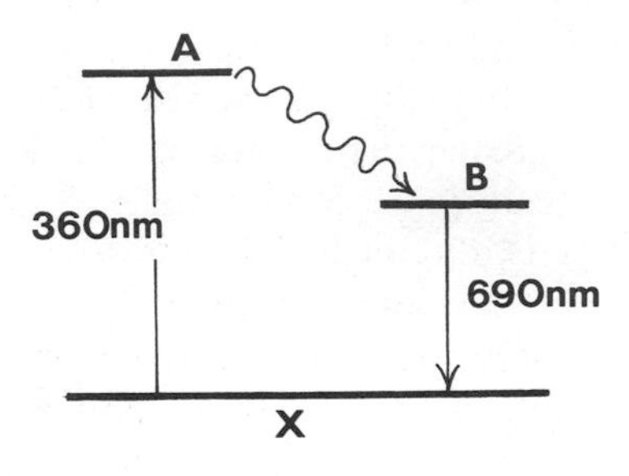

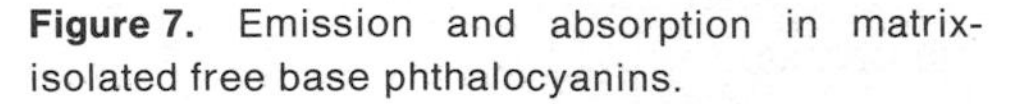

Figure 7. Emission and absorption in matrix-isolated free base phthalocyanins.

Franck–Condon factors are probably different for absorption and emission as would be the case if the potential curves of X and B were shifted with respect to one another.

Weltner, McLeod, and Kasai have reported[118] an intriguing bleaching process associated with trapped ScO, YO, and LaO molecules during white-light irradiation. If these monoxide molecules are excited electronically by the incident radiation, then after a period of a few hours their esr, infrared, and visible spectra disappear. Apart from a weak visible band, no new spectral features appear. This behavior has been interpreted by assuming that "hot spots" are created in the matrix at the sites of the MO molecules by the absorption of light. The trapped species may then slowly move through the locally softened matrix and polymerize. Unfortunately, we have no information concerning the concentration of the trapped molecules themselves to estimate how extensive such radiation-induced dimerization must be and thus whether the observed data can be explained on the basis of a local perturbation.

REFERENCES

1. N. Acquista and L. J. Schoen, *J. Chem. Phys.*, **53**, 1290 (1970).
2. N. Acquista, L. J. Schoen, and D. R. Lide, *J. Chem. Phys.*, **48**, 1534 (1968).
3. A. W. Adamson, *Discuss. Faraday Soc.*, **29**, 163 (1960).
4. A. C. Albrecht, *J. Mol. Specrosc.*, **6**, 84 (1961).
5. L. Andrews and G. C. Pimentel, *J. Chem. Phys.*, **47**, 3637 (1967).
6. A. Arkell, R. R. Reinhard, and L. P. Larson, *J. Amer. Chem. Soc.*, **87**, 1016 (1965),
7. L. Bajema, M. Gouterman, and B. Meyer, *J. Mol. Spectrosc.*, **27**, 225 (1968).
8. J. D. Black and P. S. Braterman, *J. Organometal. Chem.*, **63**, C19 (1973).
9. J. D. Black, M. J. Boylan, P. S. Braterman, and W. J. Wallace, *J. Organometal, Chem.*, **63**, C21 (1973).
10. D. Boal and G. A. Ozin, *Specrosc. Lett.*, **4**, 43 (1971).
11. V. Bondybey and J. W. Nibler, *J. Chem. Phys.*, **56**, 4719 (1972).
12. R. A. Bernheim, H. W. Bernard, P. S. Wang, L. S. Wood, and P. S. Skell, *J. Chem. Phys.*, **53**, 1280 (1970); ibid., **54**, 3223 (1971).

13 P. A. Breeze, Ph.D. thesis, University of Cambridge, 1974.
14. P. A. Breeze and J. J. Turner, *J. Organometal. Chem.*, **44**, C7 (1972).
15. J. K. Burdett, L. Hodges, V. Dunning, and J. H. Current, *J. Phys. Chem.*, **74**, 4053 (1970).
16. J. K. Burdett, *J. Chem. Soc., Faraday II*, **70**, 1599 (1974).
17. J. K. Burdett, *Chem. Commun.*, 763 (1973).
18. J. K. Burdett, *Inorg. Chem.*, **14**, 375 (1975); and *J. C. S. Faraday II*, **70**, 1599 (1974).
19. J. K. Burdett, R. N. Perutz, M. Poliakoff and J. J. Turner; *Chem. Commun.*, 157 (1975).
20. J. K. Burdett, M. A. Graham, and J. J. Turner, *J. Chem. Soc., A*, 1620 (1972).
21. J. K. Burdett and J. J. Turner, *Chem. Commun.*, 885 (1971).
22. J. K. Burdett, *J. Mol. Spectrosc.*, **36**, 365 (1970).
23. W. G. Burton and J. J. Turner, unpublished data.
24. A. B. Callear, *Proc. Roy. Soc., Ser. A*, **265**, 71, 88 (1961).
25. J. G. Calvert and J. N. Pitts, *Photochemistry*, Wiley, New York, 1966.
26. C. C. Campbell and J. J. Turner, unpublished data.
27. O. L. Chapman, C. L. McIntosh, and L. L. Barber, *Chem. Commun.*, 1162 (1971).
28. O. L. Chapman, C. L. McIntosh, and J. Pacansky, *J. Amer. Chem. Soc.*, **95**, 614 (1973).
29. O. Crichton and A. J. Rest, *Chem. Commun.*, 407 (1973).
30. O. Crichton, M. Poliakoff, A. J. Rest, and J. J. Turner, *J. Chem. Soc., Dalton*, 1321 (1973).
31. P. Datta, T. D. Goldfarb, and R. S. Boikess, *J. Amer. Chem. Soc.*, **93**, 5189 (1971).
32. J. B. Davies and H. E. Hallam, *J. Chem. Soc., Faraday II*, **68**, 509 (1972).
33. R. L. DeKock, *Inorg. Chem.*, **10**, 1205 (1971).
34. W. B. DeMore and N. Davidson, *J. Amer. Chem. Soc.*, **81**, 5689 (1959).
35. H. Dubost and L. Abouaf-Marguin, *Chem. Phys. Lett.*, **17**, 269 (1972).
36. W. C. Easly and W. Weltner, *J. Chem. Phys.*, **52**, 197 (1970).
37. J. Franck and E. Rabinowitch, *Trans. Faraday Soc.*, **30**, 120 (1934).
38. R. P. Frosch, *J. Chem. Phys.*, **54**, 2660 (1971).
39. R. P. H. Gasser, *Quart. Rev.*, **25**, 223 (1971).
40. J. B. Gibson, A. N. Goland, M. Milgram, and G. H. Vineyard, *Phys. Rev.*, **120**, 1221 (1960).
41. M. F. Goodman, H. H. Seliger, and J. M. Minkowski, *Photochem. Photobiol.*, **12**, 355 (1970).
42. M. A. Graham, Ph.D. thesis, University of Cambridge, 1970.
43. M. A. Graham, M. Poliakoff, and J. J. Turner, *J. Chem. Soc., A*, 2939 (1971).
44. M. A. Graham, R. N. Perutz, M. Poliakoff, and J. J. Turner, *J. Organometal. Chem.*, **34**, C34 (1972).
45. W. A. Guillory and C. E. Smith, *J. Chem. Phys.*, **53**, 1661 (1970).
46. R. T. Hall and G. C. Pimentel, *J. Chem. Phys.*, **38**, 1889 (1963).
47. M. D. Harmony and R. J. Myers, *J. Chem. Phys.*, **37**, 636 (1962).
48. H. Huber, E. P. Kündig, M. Moskovits, and G. A. Ozin, *J. Amer. Chem. Soc.*, **95**, 332 (1973).
49. R. J. Isabel and W. A. Guillory, *J. Chem. Phys.*, **57**, 1116 (1972).

50. M. E. Jacox and D. E. Milligan, *J. Chem. Phys.*, **46**, 184 (1967).
51. M. E. Jacox and D. E. Milligan, *J. Chem. Phys.*, **53**, 2688 (1970).
52. M. E. Jacox and D. E. Milligan, *J. Chem. Phys.*, **54**, 3935 (1971).
53. A. Kaldor and R. F. Porter, *J. Amer. Chem. Soc.*, **93**, 2140 (1971).
54. W. Kanellakopulos-Drossopulos and D. R. Wiles, *Can. J. Chem.*, **52**, 894 (1974).
55. P. H. Kasai and D. McLeod, *J. Amer. Chem. Soc.*, **92**, 6085 (1970).
56. P. H. Kasai, *Phys. Rev. Lett.*, **21**, 67 (1968).
57. P. H. Kasai and D. McLeod, *J. Chem. Phys.*, **51**, 1250 (1969).
58. P. H. Kasai, E. B. Whipple and W. Weltner, *J. Chem. Phys.*, **44**, 2581 (1966).
59. J. M. Kelly, unpublished ata.
60. G. F. Kokoszka and F. E. Brinckman, *J. Amer. Chem. Soc.*, **92**, 1199 (1970).
61. A. Krantz, C. Y. Lin, and M. D. Newton, *J. Amer. Chem. Soc.*, **95**, 2744 (1973).
62. E. P. Kündig, M. Moskovits, and G. A. Ozin, *Can. J. Chem.*, **51**, 2737 (1973).
63. G. N. Lewis and D. Lipkin, *J. Amer. Chem. Soc.*, **64**, 2801 (1942).
64. E. R. Lory and R. F. Porter, *J. Amer. Chem. Soc.*, **93**, 6301 (1971).
65. D. M. Mann and H. P. Broida, *J. Chem. Phys.*, **55**, 84 (1971).
66. A. G. Massey and L. E. Orgel, *Nature*, **191**, 1386 (1961).
67. D. E. Milligan, M. E. Jacox, and W. A. Guillory, *J. Chem. Phys.*, **52**, 3864 (1970).
68. D. E. Milligan and M. E. Jacox, *J. Chem. Phys.*, **55**, 3404 (1971).
69. D. E. Milligan and M. E. Jacox, *J. Chem. Phys.*, **35**, 1491 (1961).
70. D. E. Milligan, D. E. Mann, M. E. Jacox, and R. A. Mitsch, *J. Chem. Phys.*, **41**, 1199 (1964).
71. D. E. Milligan and M. E. Jacox, *J. Chem. Phys.*, **47**, 5146 (1967).
72. D. E. Milligan and M. E. Jacox, *J. Chem. Phys.*, **52**, 2594 (1970).
73. D. E. Milligan and M. E. Jacox, *J. Chem. Phys.*, **49**, 4269 (1968).
74. D. E. Milligan and M. E. Jacox, *J. Chem. Phys.*, **49**, 5330 (1968).
75. D. E. Milligan and M. E. Jacox, *J. Chem. Phys.*, **49**, 1938 (1968).
76. D. E. Milligan and M. E. Jacox, *J. Chem. Phys.*, **49**, 3130 (1968).
77. D. E. Milligan and M. E. Jacox, *J. Chem. Phys.*, **53**, 2034 (1970).
78. D. E. Milligan and M. E. Jacox, *J. Chem. Phys.*, **47**, 278 (1967).
79. D. E. Milligan and M. E. Jacox, *J. Chem. Phys.*, **47**, 5157 (1967).
80. D. E. Milligan and M. E. Jacox, *J. Chem. Phys.*, **40**, 2461 (1964).
81. D. E. Milligan and M. E. Jacox, *J. Chem. Phys.*, **44**, 2850 (1966).
82. D. E. Milligan, M. E. Jacox, and A. M. Bass, *J. Chem. Phys.*, **43**, 3149 (1965).
83. D. E. Milligan and M. E. Jacox, *J. Chem. Phys.*, **39**, 712 (1963).
84. D. E. Milligan and M. E. Jacox, *J. Chem. Phys.*, **51**, 277 (1969).
85. N. G. Moll and W. E. Thompson, *J. Chem. Phys.*, **44**, 2684 (1966).
86. R. L. Morehouse, J. J. Christiansen, and W. Gordy, *J. Chem. Phys.*, **45**, 1747 (1966).
87. L. Y. Nelson and G. C. Pimentel, *Inorg. Chem.*, **6**, 1758 (1967).
88. I. Norman and G. Porter, *Nature*, **174**, 508 (1954).
89. G. A. Ozin and A. Vander Voet, *Account. Chem. Res.*, **6**, 313 (1973).
90. G. A. Ozin, in *Vibrational Spectroscopy of Trapped Species*, H. E. Hallam, Ed., Chapter 9, p. 373 Wiley, London, 1973.

91. G. A. Ozin and M. Moskovits, this volume, Chapter 1.
92. R. N. Perutz, Ph.D. thesis, University of Cambridge, 1974.
93. R. N. Perutz and J. J. Turner, *J. Chem. Soc., Faraday Trans. II*, **69**, 452 (1973).
94. R. N. Perutz and J. J. Turner, *Inorg. Chem.*, **14**, 262 (1975).
95. M. Poliakoff, *J. Chem. Soc., Dalton*, 210 (1974).
96. M. Poliakoff and J. J. Turner, *J. Chem. Soc., A*, 2403 (1971).
97. M. Poliakoff, to be published.
98. M. Poliakoff and J. J. Turner, *J. Chem. Soc., Faraday Trans. II*, **70**, 93 (1974).
99. M. Poliakoff and J. J. Turner, *J. Chem. Soc., Dalton*, 1351 (1973); 2276 (1974).
100. M. Poliakoff and J. J. Turner, *J. Chem. Soc., Dalton*, 1351 (1973).
101. A. J. Rest, *J. Organometal. Chem.*, **40**, C76 (1972).
102. A. J. Rest and J. J. Turner, *Chem. Commun.*, 375 (1969).
103. A. J. Rest and J. J. Turner, *Chem. Commun.*, 1026 (1969).
104. A. J. Rest, private communicatıon.
105. M. M. Rochkind, *Spectrochim. Acta*, **27**A, 547 (1971).
106. M. M. Rochkind and G. C. Pimentel, *J. Chem. Phys.*, **46**, 4481 (1967).
107. L. J. Schoen and D. E. Mann, *J. Chem. Phys.*, **41**, 1514 (1964).
108. C. E. Smith, D. E. Milligan and M. E. Jacox, *J. Chem. Phys.*, **54**, 2780 (1971).
109. E. D. Sprague and F. Williams, *J. Chem. Phys.*, **54**, 5425 (1971).
110. I. W. Stolz, G. R. Dobson and R. K. Sheline, *J. Amer. Chem. Soc.*, **84**, 3589 (1962); ibid., **85**, 1013 (1963).
111. A. G. Streng, *Chem. Rev.*, **63**, 607 (1963).
112. W. Strohmeier and K. Gerlach, *Chem. Ber.*, **94**, 398 (1961).
113. L. Y. Tan and G. C. Pimentel, *J. Chem. Phys.*, **48**, 5202 (1968).
114. I. McC. Torrens, and L. T. Chadderton, *Phys. Rev.*, **159**, 671 (1967).
115. I. McC. Torrens, L. T. Chadderton and D. V. Morgan, *J. Appl. Phys.*, **37**, 2395 (1966).
116. J. J. Turner, in *Comprehensive Inorgsmic Chemistry* V. C. Bailor, H. J. Emeleus, R. S. Nyholm, A. F. Trotman-Dickenson, Eds., p. 685, Pergamon, Oxford, 1973.
117. M. S. Wei, Ph.D. thesis, University of Michigan, Ann Arbor (1970).
118. W. Weltner, D. MacLeod, and P. H. Kasai, *J. Chem. Phys.*, **46**, 3172 (1967).
119. H. Huber, L. Hanlan, B. McGarvey, E. P. Kündig, and G. A. Ozin, *J. Amer. Chem. Soc.*, **97**, 7054 (1975).
120. R. N. Perutz and J. J. Turner *J. Amer. Chem. Soc.*, **97**, 4800 (1975).
121. R. N. Perutz and J. J. Turner *J. Amer. Chem. Soc.*, **97**, 4791 (1975).
122. There has been some controversy about the structure of $Cr(CO)_5$ (E. P. Kündig and G. A. Ozin., *J. Amer. Chem. Soc.*, **96**, 3820 (1974)), but the square pyramidal (C_{4v}) structure now seems well established. (J. K. Burdett et al. *J. Amer. Chem. Soc.*, **97**, 4805 (1975)).

Index

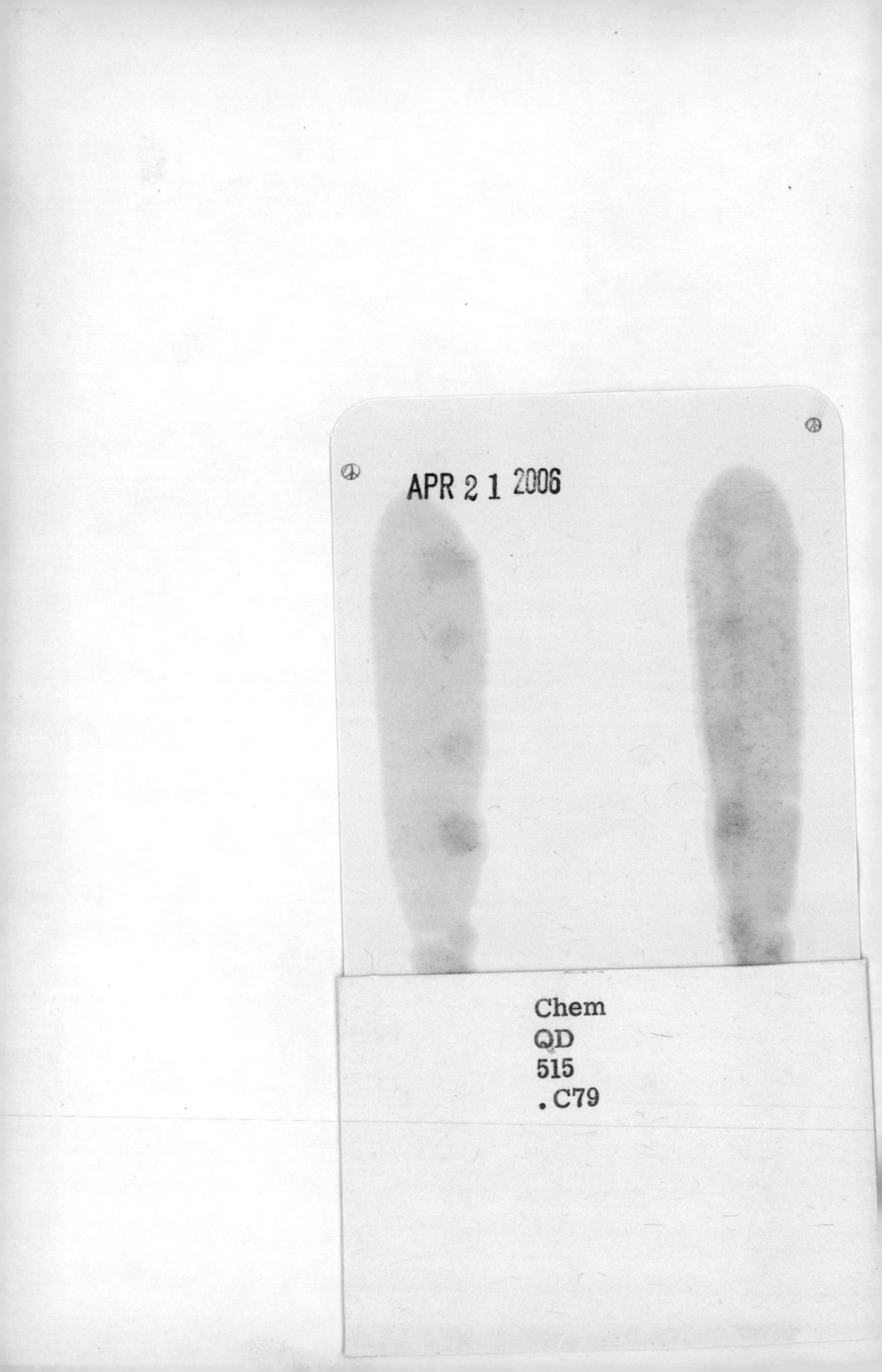